363.7394 SCI   6/08
Desbonnet, Alan.
Science of ecosystem-based
management : Narragansett
Bay in the 21st century
32148001446052

MIDDLETOWN PUBLIC LIBRARY

W9-BOA-688

Science for Ecosystem-based Management

Springer Series on Environmental Management

Volumes published since 1992

Food Web Management: A Case Study of Lake Mendota (1992)
J.F. Kitchell (ed.)

Restoration and Recovery of an Industrial Region: Progress in Restoring the Smelter-Damaged
Landscape near Sudbury, Canada (1995)
J.M. Gunn (ed.)

Limnological and Engineering Analysis of a Polluted Urban Lake: Prelude to Environmental
Management of Onondaga Lake, New York (1996)
S.W. Effler (ed.)

Assessment and Management of Plant Invasions (1997)
J.O. Luken and J.W. Thieret (eds.)

Marine Debris: Sources, Impacts, and Solutions (1997)
J.M. Coe and D.B. Rogers (eds.)

Environmental Problem Solving: Psychosocial Barriers to Adaptive Change (1999)
A. Miller

Rural Planning from an Environmental Systems Perspective (1999)
F.B. Golley and J. Bellot (eds.)

Wildlife Study Design (2001)
M.L. Morrison, W.M. Block, M.D. Strickland, and W.L. Kendall

Selenium Assessment in Aquatic Ecosystems: A Guide for Hazard Evaluation and Water Quality
Criteria (2002)
A.D. Lemly

Quantifying Environmental Impact Assessments Using Fuzzy Logic (2005)
R.B. Shepard

Changing Land Use Patterns in the Coastal Zone: Managing Environmental Quality in Rapidly
Developing Regions (2006)
G.S. Kleppel, M.R. DeVoe, and M.V. Rawson (eds.)

The Longleaf Pine Ecosystem: Ecology, Silviculture, and Restoration (2006)
S. Jose, E.J. Jokela, and D.L. Miller (eds.)

Linking Restoration and Ecological Succession (2007)
L. Walker, J. Walker, and R. Hobbs (eds.)

Ecology, Planning, and Management of Urban Forests: International Perspective (2008)
M. Carreiro, Y. Song, and J. Wu (eds.)

Science for Ecosystem-based Management: Narragansett Bay in the 21st Century (2008)
Alan Desbonnet and Barry A. Costa-Pierce (eds.)

363.7394
SCI

3 2148 00144 6052

Alan Desbonnet · Barry A. Costa-Pierce
Editors

Science for Ecosystem-based Management

Narragansett Bay in the 21st Century

Middletown Public Library
700 West Main Rd
Middletown, RI 02842

Springer

6/08

Editors

Alan Desbonnet
Rhode Island Sea Grant College
Program, University of Rhode Island,
Narragansett, RI 02882
aland@gso.uri.edu

Barry A. Costa-Pierce
Rhode Island Sea Grant College
Program, University of Rhode Island,
Narragansett, RI 02882
bcp@gso.uri.edu

ISBN: 978-0-387-35298-5   e-ISBN: 978-0-387-35299-2

Library of Congress Control Number: 2007932596

© 2008 Springer Science+Business Media, LLC
All rights reserved. This work may not be translated or copied in whole or in part without the written
permission of the publisher (Springer Science+Business Media, LLC, 233 Spring Street, New York,
NY 10013, USA), except for brief excerpts in connection with reviews or scholarly analysis. Use in
connection with any form of information storage and retrieval, electronic adaptation, computer
software, or by similar or dissimilar methodology now known or hereafter developed is forbidden.
The use in this publication of trade names, trademarks, service marks, and similar terms, even if they
are not identified as such, is not to be taken as an expression of opinion as to whether or not they are
subject to proprietary rights.

Cover Illustration: Cover Photo of the Sloop "Providence" by Janice Raynor, JAR Images, York,
ME(USA).

Printed on acid-free paper.

9 8 7 6 5 4 3 2 1

springer.com

Preface

Narragansett Bay was the site of the origin of the Industrial Revolution in the United States, and as such has been provided a life of royalty, but at great environmental cost. The bay early on was dressed daily in various colorful hues and tints being discharged from the burgeoning textile plants gracing its shores and tributaries. As the rainbow hues faded with loss of that industry to southern states, the bay was showered anew with precious metals discharged to its waters by a rapidly expanding jewelry industry thriving in Providence. As the population of adorers of the bay increased along its shores, they fed the bay with a diet rich in nutrients from their wastewater discharges, to the point that it can be considered as fat, obese, overfed, or in the scientific lingo—"eutrophic."

For the most part, we consider eutrophic to be a bad, undesirable thing, but from an ecological perspective being "well fed" can be a very good thing. Every species is driven by the basic need to eat—lions gorge to the point of immobility over a kill, and were prey not so difficult to come by on the Serengeti, we would see fat lions. In some human cultures, obesity is considered the epitome of success as any person who can be so well fed is obviously ingenious at attaining the resources needed for a good life.

And so it is in nature. From a grand ecological vantage high atop the bluffs of Newport, the Narragansett Bay ecosystem appears a peaceful setting. It is much to our distaste, however, to find that our beloved bay has some "health issues," mainly by way of regions of low dissolved oxygen in bottom waters and a preponderance of opportunistic and nuisance species—weeds if you will—in the upper reaches of the bay. These health issues have been considered to be a result of too many nutrients, and so Narragansett Bay is being "put on a diet."

This book is the offspring of the 3rd Rhode Island Sea Grant Annual Science Symposium titled "State of Science Knowledge of Nutrients in Narragansett Bay" convened on Block Island (RI) during November of 2004. Over 50 scientists and resource managers locked themselves away at the Hotel Manisses for two long days of intense and spirited debate about the status of nutrient enrichment in Narragansett Bay, and what changes might be expected from the implementation of a nitrogen reduction program. Detailed scientific presentations were given, followed by heated arguments and passionate debates, which led to agreement on some issues, but raised infinitely more questions regarding the current ecological functioning of the bay relative to nutrient availability and climate change.

This review volume, the first since Hale (1980) over 25 years ago, strives to capture the knowledge base that blossomed at the symposium, and dives into the waters of a 21st century Narragansett Bay to establish a new baseline so that we can better understand the recent events that have brought the bay to its current condition, and what we might expect as Narragansett Bay "slims down" as a result of its forced diet of nutrient reduction. It is unclear what the bay will be like 10, 20 or 50 years into the future, but this book points the way toward some new trajectories for change. When we dive into the bay 50 years hence, will the waters be clear? Will our future bay ever be as it was before that first set of cloth sails appeared on the horizon several centuries ago? Will the species we desire—eelgrass, winter flounder, quahogs, and oysters—proliferate as the bay adjusts to a lower nutrient input? Or will we see some very different, very unexpected bay?

When it comes to the process of ecological change that the bay will undergo as the nutrients are taken away, our predictive power is fairly poor. While the scientific chapters in this book cannot accurately predict the outcomes of reducing the flow of nutrients to Narragansett Bay, they can help to better understand what the bay ecosystem is like at the onset of the new millennium, before starting its reduced nutrient diet. Predictions are made in this book by scientists based on best available information and models constructed on existing trends seen in the bay ecosystem. Predictions are, at best, well informed guesses, much like what blind men describing an elephant might provide. But the blind men, as told by Nixon *et al.* in Chapter 5, had it relatively easy. In our parable for Narragansett Bay, the description being given by the blind men takes place while the elephant changes from tapir-like organism, to wooly mammoth, and to the modern day pachyderm. Climate change is shifting the ecology of the bay rapidly, but at the same time local anthropogenic changes are occurring. These rapid changes are making it very challenging to form a solid description of bay ecology.

This volume makes a significant contribution to our understanding of the Narragansett Bay ecosystem, and will improve our descriptive abilities and scope of our vision. With this newfound wisdom, perhaps we can move forward in forging a new ecosystem-based approach to the management of Narragansett Bay. We think an "ecofunctional approach" may have merit as stated in Chapter 19. This approach will acknowledge and embrace change, and take full account of the opportunities and challenges it presents to understanding this unique and wonderful shallow water, coastal plain estuarine ecosystem that is the heart and soul of Rhode Island.

April 2007   Alan Desbonnet

Reference

Hale, S.O. 1980. Narragansett Bay: A friend's perspective. Rhode Island Sea Grant #42. Narragansett, RI. 42 pp.

Acknowledgments

No book, especially one delving so deeply into complex, multi-disciplinary environmental science can make it to print without an immense amount of time and effort on behalf of readers and reviewers. This volume has only been possible because of the intense effort and hard work of a very long list of reviewers who, because of their dedication to the advancement of marine science, willingly devoted themselves to assist with the development of this book. Below we list the grand cast of experts who have provided guidance and assistance in holding the chapters herein to a high standard of scientific rigor. Each provided insights, corrections, suggestions, and criticisms that have helped improve the science contained within these pages. To each reviewer we give our deepest thanks and utmost gratitude. While the editors share any praise for the science presented with all reviewers, we take sole responsibility for all mistakes and oversights.

Daniel B. Albert—University of North Carolina Chapel Hill
Marc J. Alperin—University of North Carolina Chapel Hill
Shimon Anisfeld—Yale University
Kenneth Black—Scottish Institute of Marine Science (United Kingdom)
Donald F. Boesch—University of Maryland Chesapeake Biological Laboratory
James D. Bowen—University of North Carolina Charlotte
Matthew Bracken—University of California, Davis Bodega Marine Laboratory
Suzanne Bricker—National Oceanic and Atmospheric Administration
Kenneth Brooks—Aquatic Environmental Sciences
Marius Brouwer—University of Southern Mississippi
Tom Brosnan—National Oceanic and Atmospheric Administration
David M. Burdick—University of New Hampshire
Christopher Buzzelli—National Oceanic and Atmospheric Administration
Francis Chan—Oregon State University
Robert Chant—Rutgers University
Feng Chen—University of Maryland Biotechnology Institution
James Churchill—Woods Hole Oceanographic Institution

Jeffrey C. Cornwell—University of Maryland Horn Point Laboratory
Kevin J. Craig—Duke University
Byron Crump—University of Maryland Horn Point Laboratory
Hans G. Dam—University of Connecticut Avery Point
Oliver B. Fringer—Stanford University
Ann Giblin—Marine Biological Laboratory
Patricia M. Glibert—University of Maryland Horn Point Laboratory
Christopher Gobler—Stony Brook University
Allen Gontz—University of Massachusetts Boston
Peter Groffman—Institute for Ecosystem Studies
James Hagy—US Environmental Protection Agency
Carlton Hunt—Battelle Scientific, Inc.
Russell Isaac—Massachusetts Department of Environmental Protection
Samantha Joye—University of Georgia
Michael Kemp—University of Maryland Horn Point Laboratory
Paul F. Kemp—Stony Brook University
David Kimmel—University of Maryland Horn Point Laboratory
Patricia Kremer—University of Connecticut Avery Point
Kevin Kroeger—US Geological Survey
Fabien Laurier—University of Maryland Chesapeake Biological
    Laboratory
Gary Lovett—Institute for Ecosystem Studies
Daniel MacDonald—University of Massachusetts Dartmouth
Laurence P. Madin—Woods Hole Oceanographic Institution
Roberta L. Marinelli—University of Maryland Chesapeake Biological
    Laboratory
James E. Perry, III—Virginia Institute of Marine Science
Chris Peter—University of New Hampshire
Len Pietrafesa—North Carolina State University
Antonietta Quigg—Texas A&M University
Stewart A. Rounds—US Geological Survey
David Rudnick—South Florida Water Management District
Jennifer Ruesink—University of Washington
Andrew Seen—University of Tasmania (Australia)
Rochelle D. Seitz—Virginia Institute of Marine Science
Kipp Shearman—Oregon State University
Jan Smith—Massachusetts Department of Environmental Protection
Peter E. Smith—US Geological Survey
Stephen V. Smith—Centro de Investigacion Cientifica (Mexico)
Juliane Struve—Environmental Scientist (United Kingdom)
Gordon T. Taylor—Stony Brook University
Chris Turner—US Coast Guard Research and Development Center
Johan Varekamp—Wesleyan University
George Waldbusser—University of Maryland Chesapeake Biological
    Laboratory

Michael M. Whitney—University of Connecticut
Robert Wilson—Stony Brook University
G. Lynne Wingard—US Geological Survey
Eric K. Wommack—University of Delaware

We also gratefully acknowledge the contribution of Ms. Jen Riley, who spent many hours copy editing and doing grammatical correction on the manuscripts contained in this volume. Her journalistic and editorial skills helped in getting chapters more readable as well as more grammatically correct. We also thank Ms. Sara Schroeder for her diligent efforts in reformatting a number of figures and graphs in the "eleventh hour" of our efforts to move this book through to publication. We are also immensely grateful to Eivy Monroy at the URI Coastal Resources Center for generating the two-page "overview" map of the bay, which plays a key role in tying together locations across a suite of chapters. This publication is sponsored in part by Rhode Island Sea Grant, under NOAA Grant No. NA040AR4170062. The views expressed herein are those of the authors and do not necessarily reflect the views of NOAA or any of its sub-agencies.

Contents

Contributors xv

1. **Geologic and Contemporary Landscapes of the Narragansett Bay
    Ecosystem**. 1
    Jon C. Boothroyd and Peter V. August

2. **Narragansett Bay Amidst a Globally Changing Climate** 35
    Michael E. Q. Pilson

3. **Estimating Atmospheric Nitrogen Deposition in the Northeastern
    United States: Relevance to Narragansett Bay** 47
    Robert W. Howarth

4. **Groundwater Nitrogen Transport and Input along the Narragansett
    Bay Coastal Margin** 67
    Barbara L. Nowicki and Arthur J. Gold

5. **Nitrogen and Phosphorus Inputs to Narragansett Bay: Past, Present,
    and Future** 101
    Scott W. Nixon, Betty A. Buckley, Stephen L. Granger,
    Lora A. Harris, Autumn J. Oczkowski, Robinson W. Fulweiler,
    and Luke W. Cole

6. **Nitrogen Inputs to Narragansett Bay: An Historical Perspective** 177
    Steven P. Hamburg, Donald Pryor and Matthew A.
    Vadeboncoeur

7. **Anthropogenic Eutrophication of Narragansett Bay: Evidence from
    Dated Sediment Cores** 211
    John W. King, J. Bradford Hubeny, Carol L. Gibson,
    Elizabeth Laliberte, Kathryn H. Ford, Mark Cantwell,
    Rick McKinney, and Peter Appleby

8. **Circulation and Transport Dynamics in Narragansett Bay** 233
    Malcolm L. Spaulding and Craig Swanson

9. **Critical Issues for Circulation Modeling of Narragansett Bay and
    Mount Hope Bay** 281
    Changsheng Chen, Liuzhi Zhao, Geoff Cowles,
    and Brian Rothschild

10. **The Dynamics of Water Exchange Between Narragansett Bay and
    Rhode Island Sound** 301
    Christopher Kincaid, Deanna Bergondo, and Kurt Rosenberger

11. **Summer Bottom Water Dissolved Oxygen in Upper Narragansett Bay** 325
    Emily Saarman, Warren L. Prell, David W. Murray,
    and Christopher F. Deacutis

12. **Evidence of Ecological Impacts from Excess Nutrients in Upper
    Narragansett Bay** 349
    Christopher F. Deacutis

13. **An Ecosystem-based Perspective of Mount Hope Bay** 383
    Christian Krahforst and Marc Carullo

14. **Natural Viral Communities in the Narragansett Bay Ecosystem** 419
    Marcia F. Marston

15. **Nutrient and Plankton Dynamics in Narragansett Bay** 431
    Theodore J. Smayda and David G. Borkman

16. **Narragansett Bay Ctenophore-Zooplankton-Phytoplankton
    Dynamics in a Changing Climate** 485
    Barbara K. Sullivan, Dian J. Gifford, John H. Costello,
    and Jason R. Graff

17. **Coastal Salt Marsh Community Change in Narragansett Bay in
    Response to Cultural Eutrophication** 499
    Cathleen Wigand

18. **Impacts of Nutrients on Narragansett Bay Productivity: A Gradient
    Approach** 523
    Candace A. Oviatt

19. **An "Ecofunctional" Approach to Ecosystem-based Management for
    Narragansett Bay** 545
    Barry A. Costa-Pierce and Alan Desbonnet

Index 563

Contributors

Peter Appleby
Department of Applied
Mathematics Theoretical Physics,
University of Liverpool,
Liverpool, UK

Peter V. August
Coastal Institute and Department of
Natural Resources Science,
University of Rhode Island, Bay
Campus, 124A Coastal Institute
Building, Narragansett, RI 02882

Deanna Bergondo
Graduate School of Oceanography,
University of Rhode Island,
Narragansett, RI 02882

Jon C. Boothroyd
Department of Geology,
University of Rhode Island, 314
Woodward Hall, Kingston,
RI 02882

David G. Borkman
Graduate School of Oceanography,
University of Rhode Island,
Kingston, RI 02881

Betty A. Buckley
Graduate School of Oceanography,
University of Rhode Island,
Narragansett, RI 02882

Mark Cantwell
Environmental Protection Agency,
Atlantic Ecology Division,
Narragansett, RI 02882

Marc Carullo
Massachusetts Office of Coastal Zone
Management, 251 Causeway Street,
Suite 800, Boston, MA 02114

Changsheng Chen
The School for Marine Science and
Technology, University of
Massachusetts at Dartmouth,
706 South Rodney French Blvd.,
New Bedford, MA 02744

Luke W. Cole
Department of Environmental
Sciences, University of Virginia,
Charlottesville, VA 22904

John H. Costello
Biology Department, Providence
College, Providence, RI 02918

Geoffrey Cowels
The School for Marine Science and
Technology, University of
Massachusetts at Dartmouth, 706
South Rodney French Blvd.,
New Bedford, MA 02744

Christopher F. Deacutis
Narragansett Bay Estuary
Program, University of
Rhode Island, Graduate School
of Oceanography, Box 27 Coastal
Institute Building, Narragansett,
RI 02882

Kathryn H. Ford
Massachusetts Division of
Marine Fisheries, Pocasset,
MA 02559

Robinson W. Fulweiler
Graduate School of Oceanography,
University of Rhode Island,
Narragansett, RI 02882

Carol L. Gibson
Graduate School of Oceanography,
University of Rhode Island,
Narragansett, RI 02882

Dian J. Gifford
University of Rhode Island,
Graduate School of Oceanography,
11 Aquarium Road, Narragansett,
RI 02882

Arthur J. Gold
Department of Natural Resources
Science, 110 Coastal Institute,
University of Rhode Island,
Kingston, RI 02881

Jason R. Graff
University of Rhode Island,
Graduate School of Oceanography,
11 Aquarium Road, Narragansett,
RI 02882

Stephen L. Granger
Graduate School of Oceanography,
University of Rhode Island,
Narragansett, RI 02882

Steven P. Hamburg
Center for Environmental Studies,
Brown University, Providence,
RI 02912-1943

Lora A. Harris
Ecosystems Center, Marine
Biological Laboratory,
Woods Hole, MA 02543

Robert W. Howarth
Department of Ecology and
Evolutionary Biology,
Cornell University, Ithaca,
NY 14853

J. Bradford Hubeny
Department of Geological Sciences,
Salem State College,
Salem, MA 01970

Christopher Kincaid
Graduate School of Oceanography,
University of Rhode Island,
Narragansett, RI 02882

John W. King
Graduate School of Oceanography,
University of Rhode Island,
Narragansett, RI 02882

Christian Krahforst
Massachusetts Bays National
Estuary Program, 251 Causeway
Street, Suite 800, Boston,
MA 02114

Elizabeth Laliberte
Graduate School of Oceanography,
University of Rhode Island,
Narragansett, RI 02882

Marcia F. Marston
Department of Biology, Roger
Williams University, Bristol,
RI 02809

Rick McKinney
Environmental Protection Agency,
Atlantic Ecology Division,
Narragansett, RI 02882

David W. Murray
Department of Geological Sciences,
Brown University, Providence
RI 02912-1846

Scott W. Nixon
Graduate School of Oceanography,
University of Rhode Island,
Narragansett, RI 02882

Barbara L. Nowicki
University of Rhode Island
School of Education, Chafee Hall,
Room 242, Kingston, RI 02881

Autumn J. Oczkowski
Graduate School of Oceanography,
University of Rhode Island,
Narragansett, RI 02882

Candace A. Oviatt
University of Rhode Island, Graduate
School of Oceanography, 11 Aquarium
Road, Narragansett, RI 02882

Michael E. Q. Pilson
University of Rhode Island,
Graduate School of Oceanography,
Narragansett, RI 02882

Warren L. Prell
Department of Geological Sciences,
Brown University, Providence
RI 02912-1846

Donald Pryor
Center for Environmental Studies,
Brown University, Providence,
RI 02912-1943

Kurt Rosenberger
USGS Pacific Science Center, 400
Natural Bridges Drive, Santa Cruz,
CA 95060

Brian Rothschild
The School for Marine Science and
Technology, University of
Massachusetts at Dartmouth,
706 South Rodney French Blvd.,
New Bedford, MA 02744

Emily Saarman
Department of Geological
Sciences, Brown University,
Providence
RI 02912-1846

Theodore J. Smayda
Graduate School of
Oceanography, University of
Rhode Island, Kingston,
RI 02881

Malcolm L. Spaulding
University of Rhode Island,
Department of Ocean Engineering,
Box 40, 25 Sheets Building,
Narragansett, RI 02882

Barbara K. Sullivan
University of Rhode Island, Graduate
School of Oceanography,
11 Aquarium Road, Narragansett,
RI 02882

Craig Swanson
Applied Science Associates, Inc.,
70 Dean Knauss Drive, Narragansett,
RI 02882

Matthew A. Vadeboncoeur
Center for Environmental Studies,
Brown University, Providence,
RI 02912-1943

Cathleen Wigand
US Environmental Protection
Agency, Office of Research and
Development, National Health and
Environmental Effects Research
Laboratory, Atlantic Ecology
Division, 27 Tarzwell Drive,
Narragansett, RI 02882

Liuzhi Zhao
The School for Marine Science
and Technology, University
of Massachusetts at Dartmouth,
706 South Rodney French
Blvd., New Bedford,
MA 02744

Chapter 1
Geologic and Contemporary Landscapes of the Narragansett Bay Ecosystem

Jon C. Boothroyd and Peter V. August

1.1 Geologic Setting

1.1.1 Bedrock Geology

Figure 1.1 shows a general overview of Narragansett Bay, its major bays and inlets, islands, and other features, both naturally placed and human created. Figure 1.1 provides map of Narragansett Bay and its' features, and should be returned to often to put into perspective detail that is brought out in this, and other chapters of this volume.

The character of Narragansett Bay sediment is determined in large part by the rock types in surrounding watersheds, particularly the Blackstone and Pawtuxet Rivers. Figure 1.2 shows a simplified bedrock map of Rhode Island (RI; Hermes *et al.,* 1994). Narragansett Bay lies within the Avalon Zone of southeastern New England, a body of rock that was accreted to eastern North America some time during the Permian Period (290–250 million years ago [mya]). The lighter patterned rocks in Fig. 1.2 are granites, weathering of which creates an abundance of quartz-rich sand-sized sediment.

Present-day Narragansett Bay lies within metamorphosed sedimentary rock from late Devonian (370 mya) to Carboniferous coal-bearing rocks (320–300 mya). These metamorphic rocks vary from metaquartzites to phyllites to mica-rich schists, and are less resistant to physical and chemical weathering than the granites and granite gneisses to the east and west, thus the basin is topographically lower than the surrounding uplands.

The geologic history of the Narragansett Bay area, including adjacent watersheds, is less clear from the end of emplacement of Avalon Zone rocks and the accompanying Alleghenian orogeny (290–250 mya) to the latest Pleistocene glaciations (140,000–20,000 years before present [yBP]). Rhode Island was

Jon C. Boothroyd
Department of Geosciences, University of Rhode Island, 314 Woodward Hall, Kingston, RI 02882
jon_boothroyd@uri.edu

A. Desbonnet, B. A. Costa-Pierce (eds.), *Science for Ecosystem-based Management.*
© Springer 2008

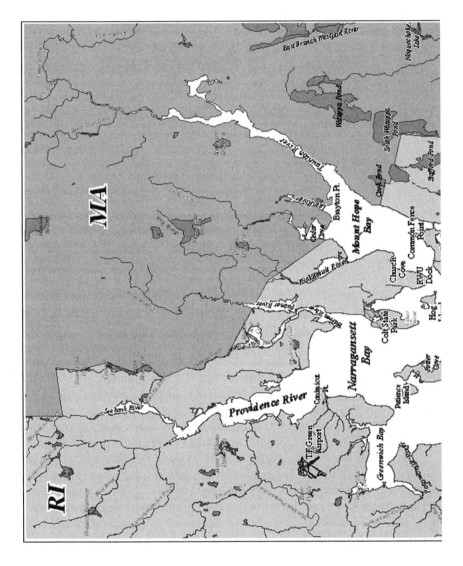

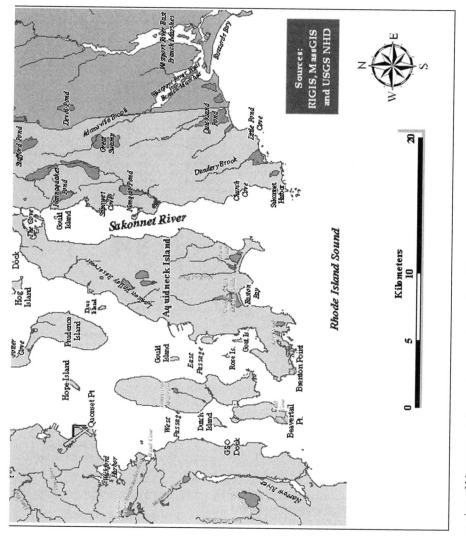

Fig. 1.1 A bird's eye view of Narragansett Bay and surrounding regions. Map by Eivy Monroy at URI Coastal Resources Center.

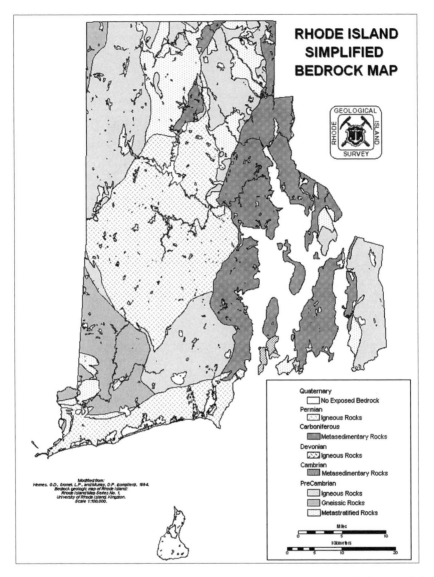

Fig. 1.2 A simplified bedrock map of Rhode Island. Narragansett Bay is underlain mostly by metamorphosed sedimentary rocks of Carboniferous age that are less resistant to weathering than the granites and gneisses to the east and west.

probably covered by Coastal Plain sediment of Cretaceous (145–65 mya) to Tertiary (65–1.8 mya) origin, but all in-place evidence have been eroded. Interpretation of the general geomorphology of New England (Denny, 1982) indicates that there was a tectonic tilt of the landscape toward the southeast during part of the Tertiary Period, resulting in flow of major river systems from northwest to southeast. These river systems were subsequently eroded down through Coastal Plain sediment to the older, basement rocks below, in many cases cutting across the folds and faults of the older rocks. When major river systems encountered less resistant rocks such as those of the Narragansett Basin, they reoriented to the trend of those rocks. This explains the deep bedrock valleys of the West and East Passages and the Sakonnet River. These deep, preglacial valleys were largely formed by fluvial erosion of the Blackstone and Taunton Rivers.

Deep chemical weathering during the Tertiary Period undoubtedly resulted in a deep, weathered sediment and soil cover, much like the saprolite of the southeastern United States. Subsequent glaciations have eroded most of this cover down to bedrock, and provided abundant material for glacial deposits.

1.1.2 Quaternary Geology

The Quaternary Period (1.8 mya to present) is characterized by multiple episodes of glaciation of the northern hemisphere during the Pleistocene Epoch (1.8 mya to 11,000 yBP). Rhode Island was undoubtedly covered by glacial ice many times, but subsequent ice advances erased the evidence of all but perhaps the last glacial stage—the Wisconsinan ~70,000–11,000 yBP—on mainland RI.

Most of mainland RI independent of Narragansett Bay is characterized by low, rolling hills separated by moderate-width, flat-floored valleys. The hills are mantled by till deposited during the late Wisconsinan Stage that culminated about 25,000 yBP; the valleys are filled with glacial river and glacial lake stratified sediment, deposited in discrete bodies, called morphosequences, beneath, adjacent to, and in front of, a retreating Laurentide continental ice sheet.

Rhode Island can be divided into four general glacial provinces (Fig. 1.3): (1) thick stratified deposits (south and central), including the western and northern parts of Narragansett Bay, (2) granitic/gneissic gravelly till upland (northwest), (3) compact till upland (east), and (4) Block Island, which is a complex of till and stratified material. The thick stratified deposits adjacent to Narragansett Bay are of particular importance because of numerous high-yield wells for municipal water supply and turf irrigation, and because of hazardous materials buried in old landfills or disposed of directly on or into stratified material.

1.1.3 Late Quaternary Deglacial History of Southern Rhode Island

The Laurentide ice sheet reached its maximum extent approximately 25,000 yBP when ice extended about three miles south of Block Island (Schafer and Hartshorn, 1965; Stone and Borns, 1986). World-wide sea level dropped

QUATERNARY DEPOSITS of RHODE ISLAND

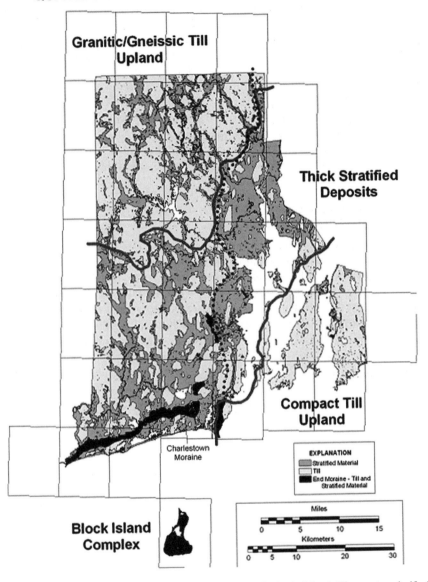

Granitic/Gneissic Till Upland

Thick Stratified Deposits

Compact Till Upland

Charlestown
Moraine

EXPLANATION
Stratified Material
Till
End Moraine - Till and
Stratified Material

Block Island Complex

Miles

| 0 | 5 | 10 | 15 |

Kilometers

| 0 | 5 | 10 | 20 | 30 |

Fig. 1.3 Quaternary (glacial and postglacial) deposits of Rhode Island. The western half of Narragansett Bay is bordered by, and underlain by, thick stratified deposits of gravel, sand, silt, and some clay. The eastern half of the bay is bordered by, and underlain in part by, a dense glacial till derived from the metamorphosed shales and sandstones of the Narragansett Basin. The western edge of the Narragansett Basin is shown by the dotted line.

over 120 m (Fairbanks, 1989) because seawater was locked up in the world's ice sheets. The oceanic shoreline was near the edge of the continental shelf, some 48 km south of Block Island. The Earth's continental lithosphere (crust and uppermost mantle) was depressed beneath the load of glacial ice displacing the asthenosphere (the layer beneath the lithosphere), which deformed plastically by flowing to the edges of the ice sheets. A rough estimate is 1 m of lithospheric depression for every 3 m of ice thickness. When ice melted and the Laurentide ice margin retreated northward, water was added back to the world ocean and a eustatic rise in sea level took place. Removal of the continental ice mass also caused an isostatic rebound of the lithosphere.

A composite family of relative sea-level curves based on the work of Oldale and O'Hara (1980) (southeastern New England shelf), Fairbanks (1989), Bard *et al.* (1990) (Barbados isostatic), Uchupi *et al.* (1996) (Boston), and Stone *et al.* (2005) (central Long Island Sound) is shown in Fig. 1.4. Note that the ages are expressed as radiocarbon years and not calendar years. For a comparison of the two, go to *http://radiocarbon.Ideo.columbia.edu/research/radiocarbon.htm*. Eustatic sea-level was rising by 18,500 14^C yBP ($\sim$22,000 yBP) as the Laurentide ice margin retreated from Block Island across what is now Block Island Sound as glacial melt water was added back to the ocean. A large glacial lake, or lakes, formed in RI, Block Island, and Long Island Sounds, is still far from the rising sea (Needell *et al.,* 1983; Needell and Lewis, 1984), as the ice retreated to an ice margin defined by the Charlestown (Fig. 1.5) and Buzzards Bay end moraines. Balco *et al.* (2002) and Balco and Schaefer (2006), using Be^{10} dating techniques, indicate the deposition of the moraines during 19,400–18,900 yBP ($\sim$16–15,500 14^C yBP).

Laurentide ice retreated from the Charlestown–Buzzards Bay morainal position to the Ledyard position by 19,000–18,600 yBP (Balco and Schaefer, 2006). A distinct lobe of Laurentide ice, the Narragansett Bay Lobe (Stone and Borns, 1986), retreated up Narragansett Bay with a lake, or lakes, forming behind the Charlestown–Point Judith–Buzzards Bay connecting morainal segment that crops out on the seafloor as accumulations of boulders (Fig. 1.5). The Narragansett Bay lobe may have retreated more slowly than the mainland ice (western RI lobe) because the ice was thicker in the topographically lower Narragansett Basin. Accumulations of boulders mark the position of end moraine segments in the West Passage of Narragansett Bay at the approximate latitude of Bonnet Shores. Later moraines, marking stillstands of Laurentide ice may be at Fox Island, Prudence Island, and just south of Greenwich Bay. Laurentide ice had probably retreated north of Narragansett Bay by 17,500 cal yBP ($\sim$14,500 14^C yBP).

1.1.4 Deglacial Deposits in Narragansett Bay and the Western Watersheds

The retreat of Laurentide ice through mainland Rhode Island and up Narragansett Bay was marked by the deposition of morphosequences at the margin of

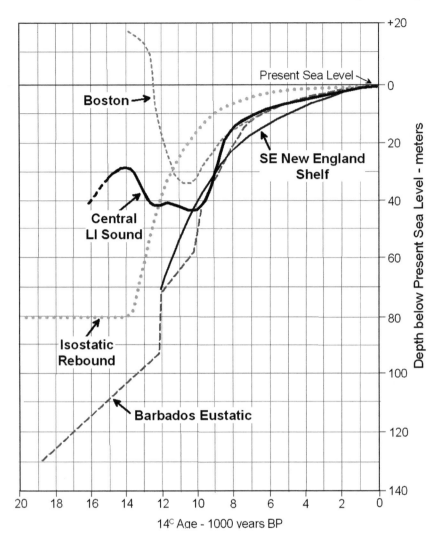

Fig. 1.4 A composite group of relative sea-level curves from the work of Fairbanks (1989) and Bard et al. (1990) (Barbados eustatic), Oldale and O' Hara (1980) (SE New England shelf), Uchupi et al. (1996) (Boston), and Stone et al. (2005) (Central Long Island Sound plus isostatic rebound).

active ice, particularly on the west side of the bay and adjacent watersheds. Morphosequences are bodies of stratified material that may range from coarse-grained gravel near the glacier margin to sand-sized or finer material down gradient from the margin. Landforms display slopes collapsed toward the former ice margin at the proximal (near ice) end of the deposit, collapse features generated as the glacier ice melts, and a preserved river channel and floodplain surface sloping away from the glacier margin (Koteff and Pessl, 1981; Stone and

END MORAINES OF SOUTHEASTERN NEW ENGLAND

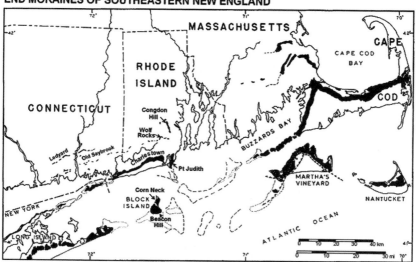

Fig. 1.5 End moraines of southern New England. The Point Judith and Congdon Hill moraines mark the western boundary of the glacier lobe that occupied Narragansett Bay.

Stone, 2005). Morphosequences are identified and mapped using 1:24,000 scale topographic maps. The Quaternary (surficial) quadrangle maps germane to the discussion of deglacial deposits in Narragansett Bay are Narragansett Pier (Schafer, 1961b), Wickford (Schafer, 1961a), East Greenwich (Smith 1955a, Boothroyd and McCandless, 2003), Providence (Smith, 1955b, Boothroyd and McCandless, 2003), East Providence (Boothroyd, 2002), and Bristol (Smith, 1955c). The key to understanding glacial geology lies not only in the study and mapping of landforms (morphosequences) but also in the understanding of sediment texture, how the sediment was deposited, and in what habitat the sediment was deposited (Gustavson and Boothroyd, 1987).

1.1.5 Alluvial Fans and Fan Deltas

The most prominent glacial features of western Narragansett Bay and adjacent watersheds are the large alluvial fans and fan deltas (alluvial fans that end in standing bodies of water) that drained into a glacial lake, or lakes, in what is now Narragansett Bay. Melt water flowed from englacial and subglacial tunnels into a developing Glacial Lake Narragansett to first form lacustrine fans (sediment bodies formed beneath the lake surface), and later, large deltas as in Fig. 1.6, an example depicting deposition in the Providence River area. The first delta in the developing sequence formed across what is present-day Narragansett Beach. A succession of deltas was deposited as the

PROVIDENCE RI AREA ~ 17,000 YEARS BP

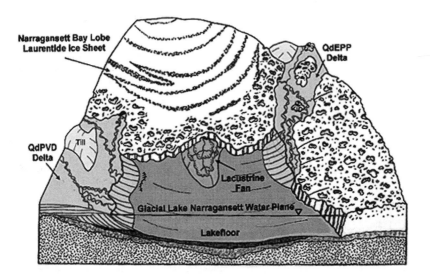

Fig. 1.6 An interpretation of the landscape in the Providence, RI area approximately 17,000 years ago when a glacier lobe was situated at the head of the present day Providence River and covered what is today downtown Providence. Melt water from the ice was depositing sediment into Glacial Lake Narragansett which existed before the bay was flooded by marine water.

ice retreated northward into an enlarging glacial lake. They filled much of what is the West Passage and account for the present shallow water depths as opposed to the East Passage. Notable deltas are at Quonset Point, Potowomut, and Warwick north of Greenwich Bay, and Edgewood at Cranston/Providence (Fig. 1.3). The delta-plain surfaces dip below present sea level; a good example is the Warwick Plains delta at Greenwich Bay (Boothroyd and McCandless, 2003). T.F. Green airport sits on the delta-plain surface and the Pawcatuck River flows around the northern margin of the delta. There is a now submerged delta plain filling the West Passage from Fox Island to Prudence Island.

   Boring logs from water wells and borings for bridge abutments indicates a deep, proglacial valley extending from Providence south at least to Greenwich Bay beneath the westernmost area of the large deltas, and close to the boundary of the Narragansett Basin (dotted line in Fig. 1.3). This valley was probably the course of an earlier Blackstone River that is now forced by later glacial deposition to flow around the delta deposits and into Narragansett Bay at the Seekonk River. Subglacial flow, beneath melting glacier ice, probably flowed down the Blackstone valley in the Narragansett Basin to deposit some of the large deltas in eastern Narragansett Bay. The ancestral course of the Taunton River was probably through Mount Hope Bay and down the present-day East Passage. Glacial Taunton River sediment was deposited as large deltas now filling Mount Hope Bay and across the northern end of Aquidneck Island. The lower East Passage received almost no sediment from melting glaciers which accounts for its present deep depth.

The West Passage delta systems also received sediment from the granitic-till upland as the western RI lobe retreated northward. Flow down braided rivers such as the glacial Annaquatucket, Hunt, Hardig, Pawtuxet, Pocasset, and Woonasquatucket, deposited large alluvial fans where they debouched from the upland and into the Narragansett Basin (boundary is dotted line in Fig. 1.3). Many other streams that are just small brooks today carried melt water and sediment for a short time from the upland onto the developing alluvial fans and delta plains. Postglacial rebound has now reversed the drainage in some of these smaller valleys.

1.1.6 Isostatic Adjustment

As mentioned previously, the weight of the Laurentide ice sheet depressed the lithosphere (crust and upper mantle). Depression was greatest under the center of ice mass, near Hudson Bay, and became less in all directions toward the ice margin. Measurements of shoreline elevations in former glacial lakes in the Connecticut River valley indicated an isostatic uplift profile of 0.889 m km^{-1} in a compass azimuth direction of $339.5°$ (a line toward Hudson Bay), while glacial marine deltas in northeastern Massachusetts and southern New Hampshire indicated an uplift profile of 0.852 m km^{-1} along a $331.5°$ azimuth (Koteff et al., 1993). We have used, and generalized the latter calculation for postglacial isostatic adjustment in RI, to 0.85 m km^{-1} and a $332°$ azimuth because of the easterly position of the state. Measurements imply that postglacial isostatic uplift, or rebound upward, in New England was delayed for several thousand years after the initiation of deglaciation, until after 16,800 cal yBP (14,000 ^{14}C yBP) (Fig. 1.4). Postglacial rebound isolines (lines of equal rebound) trend ENE-SSW ($62°$–$242°$) across present-day Narragansett Bay at an angle to the general north–south trend of the bay. The Providence area was isostatically depressed 20 m lower than the mouth of the bay at Narragansett. This means that the now uplifted glacial delta plain in the Edgewood section of Cranston was depressed 20 m *relative* to the delta plain at the mouth of Narrow River.

Isostatic rebound may have been the reason for the draining of Glacial Lake Narragansett, which would have allowed it to persist after the glacial lakes in Block Island and RI Sounds had already drained. Tilting of the water plane of the Lake allowed water to escape over a temporary dam to the south. Glacial Lake Taunton, in the upper Taunton River watershed, also drained at this time. It seems likely to have drained through the location of the present-day Sakonnet

River because there are no large incisions in the delta at Mount Hope, and the lower East Passage still retains a deep depression with shallower depths in the Castle Hill area.

1.1.7 Postglacial Drainage Systems

Isostatic rebound plus the cessation of melt water drainage deranged many of the rivers supplied by meteoric water. The Blackstone River was forced to flow around the large glacial deltas in the Providence area and found a new path into Narragansett Bay via the Seekonk River. The Woonasquatucket, Pocasset, and Pawtuxet Rivers flow around the now-uplifted northern edges of delta plains that marked former ice margins. These rivers combined to form an incised drainage in the Providence River that was directed to the east and down the East Passage.

The newly exposed lakefloor of glacial lake Narragansett is comprised of relatively flat gravelly delta plains (Greenwich Bay), more steeply dipping (up to 20°) sandy delta slopes, and an undulating glacial lakefloor mantled by varved silt and clay over end moraines (till), lacustrine fans (sand and gravel), and bedrock. Lowering of the ground-water table contributed to spring sapping and headward erosion into the delta slopes around the bay. The present bathymetry of Greenwich Bay illustrates the result of this process. Spring sapping as an important process incising and redistributing sediment from the deltas has been documented for presently emergent deltas in Maine (Ashley et al., 1991). Spring sapping as a major process differs from the interpretation of McMaster (1984), who attributes most of the postglacial incision to an organized tributary system supplied by meteoric water.

1.1.8 Late Glacial to Postglacial Sea-Level Rise

Eustatic, or world-ocean melt water addition, sea-level curves based on Barbados data by Fairbanks (1989) and Bard et al. (1990) show that world-wide sea level was rising by 22,000 calendar years BP indicating that ice melt, and hence Laurentide deglaciation, had begun by that time and probably sooner.

Stone et al. (2005) present a relative sea-level curve for central Long Island Sound that corresponds to a similar distance from the terminal glacier position as was central Block Island Sound (Fig. 1.4). They argue that central Long Island Sound was flooded by marine water by 15,500 ^{14}C BP, before isostatic rebound began. If this interpretation is correct, then the Block spill-way, now at a depth of −30 m, had to remain isostatically depressed below its present depth. Block Island would have been an island separated from Long Island by an estuarine entrance channel into Block Island Sound. Marine water did not penetrate into Narragansett Bay at this time because the exposed glacial lake floor south of the bay was at too shallow a depth (less than 30 m below present sea level) for the northward advancing sea to transgress into the lower East Passage.

Marine water probably penetrated the lower East Passage by 9,000 ^{14}C BP (10,200 cal yBP) (McMaster, 1984) when advancing sea level topped the sill (about −30 m below present sea level) and flowed into what may have been a postglacial lake. McMaster (1984) used the sea-level curve developed by Oldale and O'Hara (1980) for the southeastern New England shelf to determine reference shorelines at 5-m depth increments below present sea level that marked the transgression of marine water into Narragansett Bay. These reference shorelines are important because they describe times when certain parts of Narragansett Bay were flooded with marine water and hence how much estuarine sediment may have been deposited on top of the glacial sediment. This is important because the type, thickness, and depth of sediment substrate determine, in part, the present biological habitats of the bay.

At the −30-m shoreline, McMaster depicts an estuary in the East Passage extending to Gould Island. He shows small postglacial lakes in the West Passage near Dutch Island, northwest of Gould Island, and in upper Mount Hope Bay. By 8,350 ^{14}C BP (~9,300 cal yBP), sea-level had risen 5 m to lengthen the East Passage estuary to the confluence of the early Holocene Blackstone and Taunton Rivers near Hog Island (McMaster, 1984). McMaster assumed that the course of the Taunton River was southeast to the East Passage and not down the Sakonnet River.

McMaster (1984) depicts a −20-m below present sea-level stage at ~7,500 ^{14}C BP (~8,300 cal yBP), in which estuarine conditions extended northward in the East Passage to the vicinity of Ohio Ledge; a marine bay formed near Gould Island, flooding the former freshwater lake, and marine water penetrated the West Passage to the morainal high near Bonnet Point. A −15-m stage is shown for ~6,250 ^{14}C BP (~7,100 cal yBP) that has marine transgression around Conanicut Island southward to flood the freshwater lake near Bonnet Point, and estuarine conditions north to Conimicut Point and into Mount Hope Bay, transforming the former postglacial lake to estuarine conditions (McMaster, 1984). Timing of the −20- and −15-m stages agrees quite well with later work by Peck and McMaster (1991) who report dates on *Crassostrea virginica* (7,140 ^{14}C BP, 7,962 cal yBP) at −14.6 m below present mean sea level (msl), and on fresh-water peat (6,200 ^{14}C BP, 7,108 cal yBP) at −12.8 m below present msl. These dates plot close to the sea-level rise curve of Oldale and O'Hara (1980).

McMaster notes that most of the upper West Passage, including Greenwich Bay, as still emergent at this time. The emergent surfaces are the glacial delta plains discussed earlier and mentioned by Peck and McMaster (1991). By ~4,750 ^{14}C BP (~5,500 cal yBP), sea level had risen to −10 m below present, flooding most of the West Passage and the Sakonnet River.

Most sea-level rise curves indicate a slowing of the rate of rise from about 5,000 cal yBP to the present (Fig. 1.4, Peck and McMaster, 1991). This was caused by a decrease in the rate of release of glacial melt water to the world ocean, probably the result of a general global cooling after a relatively warm period from 8,000 to 5,000 cal yBP (Ruddiman, 2001). The rate of sea-level rise slowed to approximately 2–3 cm 100 yr^{-1} for the last 5,500 years (Peck and McMaster, 1991) and slowly inundated the glacial delta plains of western Narragansett Bay and elsewhere within Mount Hope Bay and the Sakonnet River. The slowing rate of sea-level rise allowed fringing salt marshes to become established along the lower energy shorelines of the open bay and within now flooding protected coves. Basal dates on most salt-marsh peat cluster around 2,500 ^{14}C BP (~2,600 cal yBP) (Donnelly and Bertness, 2001).

1.2 Late Holocene (Present) Geologic Framework

The present geologic framework of Narragansett Bay is heavily dependent on the configuration of glacial processes, landforms, and sediment type as discussed earlier. Postglacial (Late Pleistocene to Holocene) sediment began accumulating as soon as Glacial Lake Narragansett drained and a rudimentary fresh-water drainage system became established on the newly emergent floor of Narragansett Bay. Holocene sediment accumulation accelerated as marine water entered the bay and submerged the former glacial lacustrine environments. Most workers, as summarized by Peck and McMaster (1991), give a Holocene sedimentation rate ranging from 0.65–1.3 mm yr^{-1} (6.5–13 cm 100 yr^{-1}) for fine-grained sediment in the deeper channels and basins. This sedimentation rate would allow the accumulation of 5–10 m of sediment in the deeper channels and 3–5 m in the shallower low-energy basins.

The bay shoreline contributed all sediment sizes to subtidal environments as the high-energy areas of the shoreline eroded and receded under the impact of storm events. The eroded silt and minor clay was deposited in the deeper, low-energy channels and basins along with organic silt-sized sediment formed from decaying plant material. The sand- and gravel-sized sediments were deposited adjacent to the shoreline as depositional platforms and erosional terraces, and in coves as spits and flood-tidal deltas.

It is common for geologists to classify areas, such as Narragansett Bay, by depositional environment. A depositional environment is defined as a locale where geologic processes shaped geologic materials (the sediment) into morphologic forms (e.g., barrier spit). A sedimentary facies is sedimentary geologic material with certain identifiable characteristics such as particle size, composition, color, fabric, stratification, and biologic content. Depositional environments are comprised of sedimentary facies. Because many nongeologists are unfamiliar with the concept of depositional environments, the term "benthic geologic habitat" has been substituted for subtidal depositional environments. Intertidal to supratidal geologic habitats are self-explanatory. It will become apparent in the following discussion that some of the URLs will lead to maps that confuse the two concepts—sedimentary facies and geologic habitat.

1.2.1 Particle Size of Narragansett Bay Sediment

Landmark work by McMaster (1960) resulted in a comprehensive set of sediment grab samples from Narragansett Bay. His work was reproduced in digital format by the US Geological Survey (Poppe *et al.*, 2003; *http://pubs.usgs.gov/of/ 2003/of03-001*). Although McMaster sampled on a grid pattern, the sampling density managed to include most of the geologic habitats in the open bay but not the coves or the Providence River north of Pawtuxet Neck. McMaster (1960) presented a map that illustrated a generalized distribution of sediment types or particle sizes. The map has been widely reproduced, for instance, by the Narragansett Bay Project (French *et al.*, 1992), and can be downloaded at the following web sites *http://www.edc.uri.edu/fish/image_maps/sediment.jpg* or *http://www.narrbay.org/static.htm*.

Sediment particle sizes are indicated using the ternary plot method of Shepard (1954), which uses sand, silt, and clay as end members on the diagram. Fig.1.7 illustrates particle sizes of surface sediment in Greenwich Bay obtained by grab sampling (Boothroyd and Oakley, 2005). Data from McMaster (1960) are also indicated. The range of sediment sizes—sand to silty sand to sandy silt

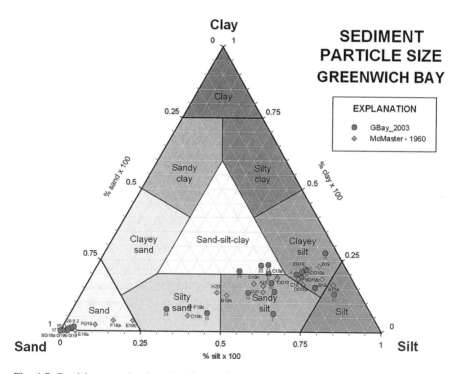

Fig. 1.7 Particle or grain size of sediment from surface grab samples from locations in Greenwich Bay. Grain sizes range from sand to clayey silt. Locations of the samples are given in McMaster (1960) and Boothroyd and Oakley (2005).

to clayey silt—reflects an overall grouping for Narragansett Bay as a whole. In other words, Greenwich Bay is a representative subset of Narragansett Bay sediment distribution in geologic habitats.

1.2.2 Benthic Geologic Habitats

Geologic habitats combine information from geological processes (e.g., tidal currents and wave motion), morphologic form (basin, tidal flat), particle size and biota (infauna and epifauna) and lastly, alteration by humans. Table 1.1 illustrates an array of geologic habitats found in Greenwich Bay (Boothroyd and Oakley, 2005).

1.2.2.1 Estuarine Bay Margin

The estuarine bay margin consists of habitats that are sandy and gravelly with geologic features such as tidal bedforms and wave-formed bars reflecting the high energy of these locales. We define platforms as depositional; that is, coarse material is transported from elsewhere and deposited to form the platform. An example is the area bayward of Buttonwoods. Terraces are erosional in that material is being removed to form the terrace and transported to another location. An example is the area on the southwest margin of Warwick Neck.

1.2.2.2 Estuarine Bay Floor

The estuarine bay floor includes deeper areas: sand sheets, gravel pavements, basins, and channels. Sand sheets are the expression of depositional platforms left behind by sea-level rise but still shallow enough to be subjected to wave processes. The eastern part of Greenwich Bay, exclusive of the deeper channel is an example. Gravel pavements are terraces also left behind by sea-level rise and subject to modification only by hurricane events, if at all. An example is the pavement that extends south of the till headland of Warwick Neck. Basins are generally low-energy areas that have accumulated Holocene silt and some clay, including organic-rich sediment. A good example is the western part of Greenwich Bay. Channels formed during the initial incision after glacial lake drainage and before sea-level rise, are now areas of fine-grained sediment accumulation. An example is the axial channel in eastern Greenwich Bay.

1.2.2.3 Estuarine Cove

Estuarine coves are usually drowned stream valleys or ice-block basins existing as appendages to the smaller bays or to Narragansett Bay proper. Cove

Table 1.1 Benthic geologic habitats in Narragansett Bay.

Estuarine Bay	Estuarine Cove	Geologic features	Other features
Subtidal Bayfloor	*Subtidal*	Tidal bedforms	Quahog harvesting
Bayfloor boulder	Cove floor organic	Wave-formed bars	(rake) trails
gravel pavement	Silt	Isolated boulders	Sunken Boats
Bayfloor sand sheet	Dredged channel	(within other	
Bayfloor basin silt	Inlet channel	habits)	
(fine)	Ebb channel		
Bayfloor basin silt	Channel		
(coarse)	(undifferentiated)		
Bay channel silt	Sand flat		
(coarse)	Mud flat		
Bay channel silt	Vegetated flat		
(patches)	(undifferentiated)		
Bay channel organic	Dredged flat		
silt (undiff)			
Bay channel organic	*Intertidal/Subtidal*		
silt (patches)	Flood-tidal delta		
	sand flats		
Subtidal Bay Margin	Channel-margin bars		
Depositional	Mud flat		
platform sand			
Sheet	*Intertidal*		
Depositional	Fringing flat		
platform	(undifferentiated)		
(vegetated)	Vegetated flat		
	(undifferentiated)		
Intertidal/Subtidal Bay			
Margin			
Distributary delta			
Distributary delta			
platform			
Erosional gravel			
terrace			
Erosional gravel			
terrace (vegetated)			
Fringing tidal flat			
(undifferentiated)			

habitats are a mix of habitats ranging from sandy tidal deltas and sandy tidal flats, intertidal and subtidal, to mudflats, intertidal and subtidal, to axial channels, probably the path of the original stream. Many of the flats are vegetated.

1.2.2.4 Altered Habitats

Humans have altered benthic geologic habitats by dredging axial channels for navigation by commercial shipping and recreational boating; they have dredged fringing tidal flats and estuarine basins for ports and marinas; they have disturbed the bottom in mooring fields; and they have altered the habitat of basins and sand sheets by fishing activity.

1.2.2.5 Shoreline Geologic Habitats

Just as for the benthic habitats, shoreline geologic habitats (depositional environments) combine information from geological processes (e.g., tidal currents and wave motion), morphologic form (beach, tidal flat), particle size and biota (infauna and epifauna) and lastly, alteration by humans. Most of the shoreline in Narragansett Bay subject to wave energy is fronted by a beach; however, narrow and coarse grained. However, if the habitat directly landward of the beach is subject to constant change by geologic processes, then that shoreline habitat takes precedence in the following classification scheme.

Table 1.2 is a geological classification of shoreline types in Narragansett Bay exclusive of low-energy coves where marshes and tidal flats dominate. However, included in the study were the shoreline from Point Judith to Cormorant Point and the Newport–Middletown shoreline. Table 1.2 was compiled in 1978 from hundreds of low-level, low-oblique aerial photographs taken from the State of Rhode Island helicopter. Photos were obtained on four separate flights between May and August, 1978. Shoreline distance was measured on 1:24,000 scale topographic quadrangles using a planimeter. Changes since 1978 would likely be an increase in shoreline protection structures along stratified material bluffs and beaches which would lessen the percentage of the latter types.

A comprehensive classification that includes coves is in the environmental sensitivity index (ESI), a scheme to assess the impact of oil spills on shoreline habitat, done for NOAA and RI Department of Environmental Management by Research Planning Institute *http://www.researchplanning.com/index.html*. Maps can be found at *http://www.edc.uri.edu/riesi/*. Illustrated descriptions of RI shoreline types given in the introduction at the web site are particularly

Table 1.2 Geological classification of shoreline habitats in Narragansett Bay.

Shoreline type	Percent of total shoreline
Beach and barrier spit	27
Glacial stratified material—bluff	10
Till bluff	23
Meta-sedimentary bedrock	8.5
Igneous and other metamorphic bedrock	5.5
Discontinuous bedrock	1.5
Shoreline protection structure	24.5

well done. The shoreline habitat ranking for sensitivity to oil spills is given in Table 1.3.

1.2.2.6 Beach and Barrier Spit

Beaches and barrier spits make up 27% of the total noncove shoreline. A beach (or berm) is a sand or gravel constructional feature subject to wave processes extending from mean lower low water (mllw) landward to a foredune, bluff, bedrock cliff, or shoreline protection structure. Some beaches in long-term depositional areas may become very wide (up to 50 m). If the beach persists through major storm events without major erosion impacting the landward bluff, then the shoreline is classified as a beach habitat. An example is the west side of Common Fence Point in Portsmouth, where the beach is developed on dredged material place there in the 1960s (Fig. 1.8a).

A barrier spit is a constructional feature (usually sand) comprised a beach, foredune zone, and back-barrier flat, enclosing an aquatic habitat such as a small estuarine embayment, coastal lagoon or salt marsh, and connected at one or both ends to a headland bluff. Barrier spits formed in Narragansett Bay in response to waves over fetch distances of up to 15 km that eroded headland bluffs and transported sediment alongshore to enclose an aquatic habitat. Most

Table 1.3 Shoreline habitat rankings for the ESI index. Sensitivity increases from top to bottom, with habitats at table bottom the most sensitive.

Ranking	Shoreline habitat
1a	Exposed rocky shores
1b	Exposed man-made structures
2a	Exposed wave-cut platforms in bedrock
3a	Fine-to-medium grained sand beaches
3b	Scarps and steep slopes in sand
4	Coarse-grained sand beaches
5	Mixed sand and gravel beaches
6a	Gravel beaches
6b	Riprap
7	Exposed tidal flats
8a	Sheltered rocky shores
8b	Sheltered, solid man-made structures
8c	Sheltered riprap
9a	Sheltered tidal flats
9b	Sheltered vegetated low banks
10a	Salt and brackish water marshes
10b	Freshwater marshes
10c	Scrub-shrub wetlands

Research Planning Institute; *http://www.researchplanning. com/services/envir/esi.html*

Fig. 1.8 An array of aerial images illustrating shoreline types in Narragansett Bay. (a) Wide beach developed on dredged material on the west side of Common Fence Pt., Portsmouth; (b) Barrier spit (Barrington Beach), Barrington; (c) Bluff comprised of glacial stratified material near Nyatt Pt., Barrington; (d) Bluff comprised of glacial till at Warwick Pt., Warwick Neck, Warwick; (e) Cliff comprised of metamorphosed sedimentary rock at Beavertail, Conanicut Island, Jamestown; (f) Cliff comprised of igneous bedrock (granite) at Cormorant Pt., Newport; (g) An array of shoreline protection structures on a glacial stratified-material bluff at Bullock Neck, East Providence. The structures from left to right are: a riprap revetment under construction, a steel sheet-pile seawall, three reinforced concrete seawalls, and another riprap revetment.

undergo frontal erosion and the overwash of sediment onto the back barrier and into marshes and small lagoons. A good example is Barrington Beach adjacent to the RI Country Club in Barrington (Fig. 1.8b).

1.2.2.7 Glacial Stratified Material—Bluff

Bluffs of glacial stratified material are a type of headland and make up 10% of Narragansett Bay shoreline. The stratified bluffs are comprised of easily erodible gravel, sand, silt, and minor clay. The bluffs supplied coarser material to adjacent beaches and barrier spits and finer sediment to bay basins and axial channels. All bluffs (stratified and till—see below) are fronted by a beach but direct wave impact during storms is able to erode the bluffs. The easily erodible nature of the stratified sediment makes them prime candidates for shoreline protection structures where allowed. An example is the eastern Nyatt Point area of Barrington (Fig. 1.8c).

1.2.2.8 Till Bluff

Till bluffs are a type of headland that comprises 23% of the bay shoreline. Till bluffs in the western part of the bay are composed of sandy till and are more easily erodible than the till bluffs of the eastern bay including Conanicut and Aquidneck Islands and the eastern shoreline of the Sakonnet River. The till bluffs of the eastern areas contain more silt and clay that forms a dense, compact till. All till bluffs are fronted by a narrow beach, often composed of gravel eroded from the adjacent bluff. Spring sapping contributes to the erosion of the higher bluffs. Bluffs range in height from just a few meters to the 15 + m (Warwick Point on Warwick Neck) shown in Fig. 1.8d.

1.2.2.9 Meta-Sedimentary Bedrock

Metamorphosed sedimentary rocks of the Narragansett Bay Group, plus older rocks at Beavertail, make up 8.5% of the bay shoreline. Bedrock is usually exposed along the shoreline as 3–15 m high cliffs. Small, gravelly, pocket beaches are sometimes present. Rock types are schist and phyllite. The bedrock shorelines are backed by bluffs of either glacial stratified material or till that are protected from wave erosion by all but the largest storms (hurricanes). A great example is the Beavertail shoreline illustrated in Fig. 1.8e.

1.2.2.10 Igneous and Other Metamorphic Bedrock

Igneous and metamorphosed igneous (gneiss) bedrock comprises 5.5% of the bay shoreline. Small, gravelly pocket beaches may be present. These bedrock shorelines are the most resistant of all types in the bay. A good example is Cormorant Point shown in Fig. 1.8f.

1.2.2.11 Discontinuous Bedrock

Discontinuous bedrock shorelines consist of scattered outcrops just seaward of a bluff that serve to buffer wave action on the bluff. These outcrops function as a kind of shoreline protection structure. An example is shown at the tip of Common Fence Point, Portsmouth in Fig. 1.8a.

1.2.2.12 Shoreline Protection Structures

Where an above shoreline type has been modified by the construction of a shoreline protection structure that is viable and working, i.e., the structure either traps sediment or offers protection from direct wave action on a bluff or foredune, the shoreline is reclassified to reflect the shoreline protection structure. These shorelines are sometimes called "hardened shorelines." Great care was taken in the original study in 1978 to ensure that the structure actually was viable. If not, the shoreline was classified as to geologic habitat even though a structure may have been present. Shoreline protection structures (working) comprised 24.5% of the bay shoreline in 1978. Ongoing studies of shoreline change in the bay using 2003 orthophotography suggest that 30% may more closely represent actual length.

Shoreline protection structures may be revetments, bulkheads, seawalls, groins, breakwaters, or jetties. See Section 300.7 of the RI Coastal Resources Management Plan (RI CRMC, 1995 as amended) for a fuller discussion of structures. Other structures, such as piers, are not strictly protection structures but often have a protection element such as a seawall incorporated in the facility. Structure installation ideas have evolved through time; pre-1954 seawalls were usually concrete; newer walls are sheet pile or have been super ceded by riprap revetments. Many older structures were destroyed in the hurricanes of 1938 and 1954; the remnants of many that still exist no longer offer shoreline protection or provide their original function. Figure 1.8g, a stratified material bluff on Bullock Neck in East Providence protected by an array of structures is a good example of changes in structural materials through time.

1.3 Watershed Environment

The statistical summaries provided here were obtained from the most current publicly available data for Narragansett Bay and its watershed. All data were retrieved from their source web sites in the Fall of 2005. Whenever possible, single datasets that spanned the states of RI and Massachusetts (MA) were used over state-specific data that are not comparable with neighboring states. All geospatial analyses were conducted using ArcGIS version 9.1 software (Environmental Systems Research Institute, 2005, Redlands, CA, USA) using coverage and grid data formats whenever possible. All data were projected into RI

State Plane feet coordinates in NAD83. Statistical summaries were obtained using SPSS version 13.0 software (SPSS Inc., 2004). Areas are expressed in square kilometers and linear measurements are expressed in kilometers.

1.3.1 Sources of Data and Processing Methodology

1.3.1.1 Narragansett Bay Watershed

The "SENEHUC" (Southeastern New England Hydrologic Unit Code) dataset was obtained from the RIGIS web site (August *et al.*, 1995; *http://www.edc.uri.edu/rigis*). The SENEHUC dataset consists of level 12 HUC units (Seaber *et al.*, 1987) developed by the US Department of Agriculture Natural Resources Conservation Service in Warwick, RI. The Narragansett Bay watershed was defined by HUC8 codes 0109003 and 0109004. These were extracted from the full SENEHUC dataset and dissolved to HUC10 detail. This resulted in 11 watershed regions in RI and MA (Fig. 1.9). We used the southern boundary in the SENEHUC dataset to close Narragansett Bay. This was a line running from Sakonnet Point to Sachuest Point on Aquidneck Island, along the southern shore of Aquidneck Island, from Brenton Point on Aquidneck Island to Beavertail Point on Conanicut Island, to the western shore of Narragansett Bay just north of where the Narrow River empties into Narragansett Bay. We used the watershed names in the attribute "WATERSHED" in the SENEHUC dataset to identify drainage sub-basins.

1.3.1.2 Coastline

The land–sea boundary in the watershed dataset was used to measure the length of coastline for Narragansett Bay. These data were used for coastline analyses because the dataset spanned RI and MA coastal regions of Narragansett Bay. The positional accuracy of the data was evaluated against 2003–2004 large scale (0.6 m pixel size) true color orthophotography and found to be very good. In a number of cases, small islands in Narragansett Bay and some of the shallow embayments were missing. Missing islands were obtained from the RITOWN5K from the RIGIS web site. RITOWN5K was created from 1:5000 scale 1997 panchromatic orthophotography (0.6 m pixel size).

1.3.1.3 Land Cover

The National Land Cover Dataset (NLCD) was used for assessment of land cover within the Narragansett Bay watershed (Hollister *et al.*, 2004). NLCD data for the watershed were obtained from the US Geological Survey Seamless Data Distribution System (*http://seamless.usgs.gov/*). NLCD data are developed from Landsat Thematic Mapper satellite imagery obtained in 1992 and are

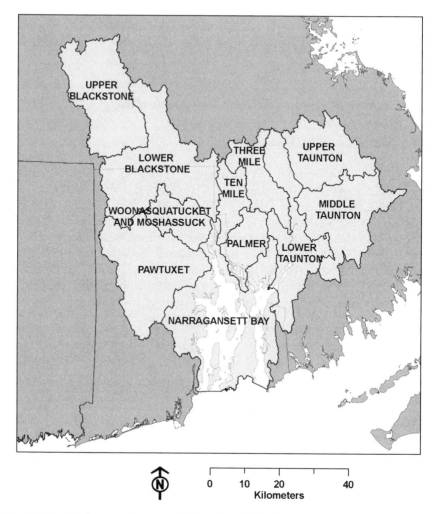

Fig. 1.9 The Narragansett Bay watershed and its sub-basins.

represented in raster format with 30 m pixel sizes. The Narragansett Bay watershed data contain 17 different land cover classes. NLCD was chosen for the land use assessment because it is the best available land cover database that comes from a single source and spans the entire watershed.

1.3.1.4 Roads

Road data for the entire watershed were obtained using the USGS Seamless Data Distribution system and consist of the Bureau of Transportation Statistics (BTS) roads dataset from The National Map database (Kelmelis *et al.*, 2003). This is a nationwide roads database for use at a nominal map scale of 1:100,000.

BTS roads were visually inspected against the 2003–2004 digital orthophoto-graphy for RI and were found to be of excellent quality. Ferry routes were deleted from the dataset prior to the calculation of road statistics.

1.3.1.5 Human Population Density

Year 2000 census data from the US Census Bureau were obtained for RI from the RIGIS web site and MA from the MassGIS web site (*http://www.mass.gov/mgis/*). Block-level resolution Summary File 1 (SF1) data were used for the analyses (US Census Bureau, 2005). Prior to merging the census data with the watershed boundaries, population density was calculated for each census block polygon. After the merger with watershed boundaries, a new estimate of popu-lation was calculated by multiplying density by census block polygon area. This procedure adjusts population estimates in polygons that were reduced in size by watershed boundaries. For example, if only 30% of a census block fell within a watershed, only 30% of the original population estimate for that census block was retained for subsequent analyses.

1.3.1.6 Bathymetry

Bathymetric data for Narragansett Bay were obtained from the Narragansett Bay data portal web site (*http://www.narrbay.org*). This dataset is a raster grid (15.2 m pixel size) of depths developed by NOAA Coastal Services Center (CSC) using a composite of the bathymetric information gathered through the National Ocean Service hydrographic surveys in RI waters from 1934 to 1996. Depths are recorded relative to mean low water or mean lower low water. NOAA CSC's processing steps to develop this dataset are well documented in the metadata available with the data. The bathymetric data only cover the RI portion of Narragansett Bay. All data falling outside the southern extent of the Narragansett Bay boundary (defined in the watersheds dataset) were excluded from analysis; thus, the bathymetric summary statistics refer to RI waters in Narragansett Bay north of the southern boundaries of Aquidneck and Conanicut Islands.

1.3.2 Bay and Watershed Statistics

The size of the Narragansett Bay watershed is given in Table 1.4. The bay is comprised of 658.5 km of shoreline of which 544.9 km (82.8% total shoreline) occurs in RI and 113.6 km (17.2% total shoreline) occurs in MA. The median depth of Narragansett Bay is 6.4 m (Table 1.5, Fig. 1.10). One quarter of the bay is less than 3.7 m deep and 90% of the bay is less than 16.5 m deep.

The Ten Mile and Woonasquatucket/Moshassuck River sub-basins were the most heavily developed of all the sub-basins in the Narragansett Bay watershed (Table 1.6). The Lower Blackstone sub-basin was the least developed. The

Table 1.4 Aerial statistics for the Narragansett Bay watershed.

Region	Area (km^2)	Percent of total watershed
Including estuarine waters		
Total Watershed	4,766.2	100%
Watershed in RI	2,077.6	43.6%
Watershed in MA	2,688.6	56.4%
Excluding estuarine waters		
Total Watershed	4,384.0	100%
Watershed in RI	1,720.4	39.2%
Watershed in MA	2,663.6	60.8%

Table 1.5 Bathymetric profile of the Rhode Island portion of Narragansett Bay.

Mean (±SD) depth	8.1 ± 7.2 m
Median depth	6.4 m
Percentile categories	
25	3.7 m
75	10.4 m
90	16.5 m
95	22.9 m
99	36.3 m

Note: Statistics based upon 491,455 pixels of bathymetric data.

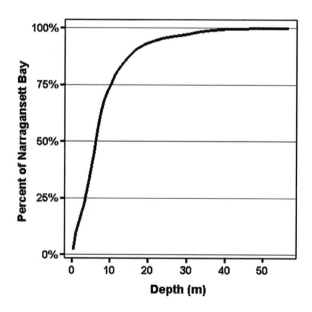

Fig. 1.10 Cumulative distribution function of depths in the Rhode Island portion of Narragansett Bay.

Table 1.6 Land use, human population density, and road density in the Narragansett Bay watershed and sub-basins. Areas are expressed in km^2.

Land use/population class, road metric	Sub-basin											
	Narr Bay	% Total	Pawtux	% Total	Upper Black	% Total	Lower Black	% Total	Woon and Mosh	% Total	Ten Mile	% Total
Developed												
Low Res	147.7	15.6	62.5	10.4	86.1	16.2	67.6	9.7	39.9	20.7	37.7	26.2
High Res	15.6	1.6	9.2	1.5	18.0	3.4	5.8	0.8	13.3	6.9	4.9	3.4
Commercial	45.4	4.8	20.1	3.4	30.4	5.7	15.1	2.2	13.0	6.8	9.1	6.4
Grass urban	63.6	6.7	12.1	2.0	15.0	2.8	10.4	1.5	6.9	3.6	6.0	4.2
Quarries	0.9	0.1	4.8	0.8	1.5	0.3	2.1	0.3	0.2	0.1	0.9	0.6
Transition	0.4	0.0	0.5	0.1	1.6	0.3	1.0	0.1	0.2	0.1	0.1	0.1
Total developed	273.5	28.9	109.2	18.2	152.7	28.7	101.9	14.6	73.6	38.2	58.8	40.9
Forest												
Decid	127.4	13.4	277.6	46.3	217.6	41.0	363.9	52.2	74.0	38.4	44.3	30.8
Conif	11.0	1.2	24.0	4.0	9.1	1.7	17.2	2.5	1.9	1.0	2.2	1.5
Mixed	78.3	8.3	99.9	16.7	54.1	10.2	107.4	15.4	18.3	9.5	20.0	13.9
Shrub	0.4	0.0	0.0	0.0	1.0	0.2	0.8	0.1	0.3	0.2	0.0	0.0
Total Forest	217.1	22.9	401.6	67.0	281.9	53.1	489.3	70.2	94.5	49.1	66.4	46.2
Agricultural												0.0
Orchard	0.2	0.0	0.6	0.1	0.5	0.1	0.5	0.1	0.2	0.1	0.2	0.1
Pasture	20.3	2.1	10.4	1.7	13.8	2.6	17.0	2.4	3.5	1.8	2.9	2.0
Row Crop	2.7	0.3	15.7	2.6	18.2	3.4	27.7	4.0	5.0	2.6	3.4	2.4
Total Ag	23.3	2.5	26.7	4.4	32.5	6.1	45.3	6.5	8.7	4.5	6.4	4.5

Table 1.6 (continued)

Land use/ population class, road metric	Sub-basin											
	Narr Bay	% Total	Pawtux	% Total	Upper Black	% Total	Lower Black	% Total	Woon and Mosh	% Total	Ten Mile	% Total
Water												
Open Water	370.4	39.1	27.8	4.6	25.0	4.7	16.3	2.3	3.9	2.0	2.9	2.0
Wetland												
Wetland Woody	47.4	5.0	28.1	4.7	28.7	5.4	33.1	4.7	9.4	4.9	4.4	3.1
Wetland Herb	13.3	1.4	6.1	1.0	9.9	1.9	11.0	1.6	2.4	1.3	4.6	3.2
Total Wetland	60.7	6.4	34.1	5.7	38.6	7.3	44.1	6.3	11.9	6.2	9.0	6.3
Other												
Bare Rock	2.8	0.3	0.2	0.0	0.7	0.1	0.3	0.0	0.0	0.0	0.1	0.0
Total Area	947.9	100.0	599.6	100.0	531.3	100.0	697.2	100.0	192.6	100.0	143.6	100.0
Km Roads	3380.3		2032.2		2169.9		2045.8		1228.7		770.8	
Km Road/Sq Km Land	5.85		3.55		4.29		3.00		6.51		5.48	
Persons RI	328,379	87.8	201,405	100	273,790	100	129,343	69.0	202,825	100	34,371	33.6
Persons MA	45,532	12.2					58,042	31.0			67,981	66.4
Total Pop	373,911		201,405		273,790		187,385		202,825		102,352	
Pop Density	648		352		541		275		1,075		728	

Table 1.6 (continued)

Land use/ population class, road metric	Sub-basin											
	Three Mile	% Total	Palm	% Total	Upper Taunt	% Total	Mid Taunt	% Total	Lower Taunt	% Total	Total	% Bay Basin
Developed												
Low Res	35.0	15.8	20.2	11.1	89.9	25.2	38.4	8.3	64.9	15.1	689.8	14.5
High Res	0.4	0.2	0.9	0.5	2.7	0.7	0.3	0.1	5.7	1.3	76.9	1.6
Commercial	10.9	4.9	6.9	3.8	13.7	3.8	11.7	2.5	14.7	3.4	191.1	4.0
Grass urban	6.7	3.0	12.3	6.8	21.6	6.1	25.8	5.5	18.2	4.2	198.5	4.2
Quarries	0.4	0.2	0.1	0.0	0.1	0.0	5.0	1.1	1.3	0.3	17.3	0.4
Transition	0.2	0.1	0.3	0.1	0.2	0.1	0.5	0.1	0.5	0.1	5.5	0.1
Total developed	53.5	24.2	40.6	22.4	128.2	36.0	81.7	17.6	105.4	24.6	1,179.1	24.7
Forest												
Decid	63.0	28.5	57.5	31.7	96.0	27.0	92.3	19.8	114.3	26.7	1,528.0	32.1
Conif	8.7	4.0	3.7	2.1	10.8	3.0	33.4	7.2	14.9	3.5	137.0	2.9
Mixed	63.5	28.8	43.5	24.0	72.7	20.4	166.3	35.7	112.5	26.2	836.4	17.6
Shrub	0.2	0.1	0.1	0.1	0.2	0.0	0.2	0.0	0.2	0.0	3.4	0.1
Total Forest	135.5	61.4	104.8	57.8	179.7	50.5	292.2	62.8	241.9	56.4	2,504.9	52.6
Agricultural		0.0		0.0		0.0		0.0		0.0		
Orchard	0.1	0.1	0.3	0.1	0.2	0.1	0.4	0.1	0.2	0.0	3.5	0.1
Pasture	4.3	2.0	9.8	5.4	2.1	0.6	13.5	2.9	7.9	1.8	105.5	2.2
Row Crop	4.9	2.2	5.0	2.7	1.4	0.4	7.7	1.7	6.7	1.6	98.4	2.1
Total Ag	9.3	4.2	15.0	8.3	3.8	1.1	21.6	4.7	14.8	3.4	207.3	4.4

Table 1.6 (continued)

Land use/ population class, road metric	Sub-basin											% Bay Basin
	Three Mile	% Total	Palm	% Total	Upper Taunt	% Total	Mid Taunt	% Total	Lower Taunt	% Total	Total	
Water												
Open Water	3.7	1.7	5.6	3.1	10.3	2.9	29.2	6.3	32.2	7.5	527.5	11.1
Wetland												
Wetland Woody	11.0	5.0	9.9	5.5	22.2	6.2	20.3	4.4	20.9	4.9	235.3	4.9
Wetland Herb	7.6	3.5	5.2	2.9	11.7	3.3	20.1	4.3	13.6	3.2	105.6	2.2
Total Wetland	18.7	8.5	15.1	8.3	33.9	9.5	40.3	8.7	34.5	8.0	340.9	7.2
Other												
Bare Rock	0.1	0.1	0.1	0.1	0.2	0.0	0.1	0.0	0.2	0.0	4.8	0.1
Total Area	220.8	100.0	181.3	100.0	356.1	100.0	465.2	100.0	428.9	100.0	4,764.5	100.0
Km Roads	689.6		516.9		1,457.2		1,050.2		1,382.4		16,724.2	
Km Road/Sq Km Land	3.18		2.94		4.21		2.41		3.48		3.95	
Persons RI			19,769	51.4					3,596		919,688	49.2
Persons MA	61,357	100	18,656	48.6	194,872	100	73,360	100	157,445	2.2	951,035	50.8
Total Pop	61,357		38,425		194,872		73,360		161,041		1,870,723	
Pop Density	283		219		564		168		406		442	

Note: Road and population densities were calculated using the total area of each sub-basin excluding open water (fresh and estuarine).

Abbreviations: Res – Residential, Decid – Deciduous, Conif – Coniferous, Pop – Population. Narr Bay = Narragansett Bay; Pawtux = Pawtuxet River; Upper Black = Upper Blackstone River; Lower Black = Lower Blackstone River; Woon and Mosh = Woonasquatucket and Moshashuck Rivers; Palm = Palmer River; Upper, Mid and Lower Taunt = Upper, Mid and Lower Taunton River, respectively.

Woonasquatucket/Moshassuck River sub-basin has the highest density of roads (6.51 km km^{-2} land area). The overall road density in the watershed was 3.95 km km^{-2} land area. The Woonasquatucket/Moshassuck River sub-basin was the most densely populated (1,075 people km^{-2} land area) and the Middle Taunton sub-basin was the least densely populated (168 people km^{-2} land area, Table 1.6). The overall population density of the whole watershed was 442 people km^{-2} land area. The total human population of the watershed is 1,870,723 persons and is distributed equally between MA and RI.

References

Ashley, G.M., Boothroyd, J.C., and Borns, H.W., Jr. 1991. Sedimentology of Late Pleistocene (Laurentide) deglacial-phase deposits, eastern Maine: an example of a temperate marine, grounded ice-sheet margin. *In* Anderson, J.B., and Ashley, G.M. (eds.) Glacial marine sedimentation – paleoclimatic significance. *Geological Society of America Special Paper* pp. 141–193, Boulder, CO.

August, P.V., McCann, A., and LaBash, C. 1995. Geographic information systems in Rhode Island. *University of Rhode Island Cooperative Extension Fact Sheet* 95–1:1–12.

Balco, G., and Schaefer, J. 2006. Cosmogenic-nuclide and varve chronologies for the deglaciation of southern New England. *Quaternary Geochronology* 1:15–28.

Balco, G., Stone, J.O.H., Porter, S.C., and Caffee, M. 2002. Cosmogenic-nuclide ages for New England coastal moraines, Martha's Vineyard and Cape Cod, Massachusetts, USA. *Quaternary Science Reviews* 21:2127–2135.

Bard, E., Hamelin, B., Fairbanks, R.G., and Zindler, A. 1990. Comparison of ^{14}C and Th ages obtained by mass spectrometry in corals from Barbados. Implications for the sea level during the last glacial cycle and for the production of ^{14}C by cosmic rays during the last 30,000 years. *Nature* 345:405–410.

Boothroyd, J.C., 2002. Quaternary geology of the Providence-East Providence Quadrangle, RI: Open File Map 2002–01, Rhode Island Geological Survey STATEMAP Program, Glacial morphosequence Map (scale: 1:24,000, 7 ½ x 15 min), Report, 15pp, Kingston, RI.

Boothroyd, J.C., and McCandless, S.J. 2003. Quaternary geology of the East Greenwich and parts of the Bristol and Crompton Quadrangles, RI: Open File Map 2003-01, Rhode Island Geological Survey STATEMAP Program, Glacial morphosequence Map (scale: 1:24,000, 7 ½ min), Kingston, RI.

Boothroyd, J.C., and Oakley, B.A. 2005. Benthic geologic habitats of Greenwich Bay, RI. In Special Area Management Plan for Greenwich Bay and Watershed, Rhode Island Coastal Resources Management Council, Wakefield, RI. 1:10,000 scale maps and side-scan sonar images posters.

Denny, C.S. 1982. Geomorphology of New England. US Geological Survey Professional Paper 1208.

Donnelly, J.P., and Bertness, M.D. 2001. Rapid shoreward encroachment of salt marsh cordgrass in response to accelerated sea-level rise. *Proceedings of the National Academy of Sciences* 98:14218–14223.

Environmental Systems Research Institute (ESRI). 2005. ArcGIS Release 9.1., Redlands, CA.

Fairbanks, R.G. 1989. A 17,000 year glacio-eustatic sea level curve: influence of glacial melting rates on the Younger Dryas event and deep-ocean circulation. *Nature* 342:637–642.

French, D., Rines, H., Boothroyd, J., Galagan, C., Harlin, M., Keller, A., Klein-McPhee, G., Pratt, S., Gould, M., Villalard-Bohnsack, M., Gould, L., Steere, L., and Porter, S. 1992. Atlas and habitat inventory/resource mapping for Narragansett Bay and associated coastlines, Rhode Island and Massachusetts: Final Report for the Narragansett Bay Project, Providence, RI.

Gustavson, T.C., and Boothroyd, J.C. 1987. A depositional model for outwash, sediment sources, and hydrologic characteristics, Malaspina Glacier, Alaska: A modern analog of the southeastern margin of the Laurentide Ice Sheet. *Geological Society of America Bulletin*, 99, pp. 187–200.

Hermes, O.D., Gromet, L.P., and Murray, D.P. 1994. Bedrock geology map of Rhode Island. Rhode Island Map Series No. 1, University of Rhode Island, Kingston, RI.

Hollister, J., August, P., Copeland, J., and Gonzales, L. 2004. Assessing the accuracy of the National Land Cover dataset at multiple spatial extents. *Photogrammetric Engineering and Remote Sensing* 70:405–414.

Kelmelis, J. A., DeMulder, M. L., Ogrosky, C. E., VanDriel, N. J., and Ryan, B. J. 2003. The National Map – From geography to mapping and back again. *Photogrammetric Engineering and Remote Sensing* 69:1109–1118.

Koteff, C., and Pessl, F. Jr. 1981. Systematic ice retreat in New England. U.S. Geological Survey Professional Paper 1179, Washington, DC.

Koteff, C., Robinson, G.R., and Goldsmith, R. 1993. Delayed postglacial uplift and synglacial sea levels in coastal central New England. *Quaternary Research* 49:46–54.

McMaster, R.L. 1960. Sediments of the Narragansett Bay system and Rhode Island. *Journal of Sedimentary Petrology* 30(2):249–274.

McMaster, R.L. 1984. Holocene stratigraphy and depositional history of the Narragansett Bay System, Rhode Island, USA. *Sedimentology* 31:777–792.

Needell, S.W., and Lewis, R.S. 1984. Geology and shallow structure of Block Island Sound, Rhode Island and New York. US Geological Survey Miscellaneous Field Studies Map MF-1621, 4 sheets.

Needell, S.W., O'Hara, C.J., and Knebel, H.J. 1983. Maps showing geology and shallow structure of western Rhode Island Sound. US Geological Survey Miscellaneous Field Studies Map MF-1537, 4 sheets.

Oldale, R.N., and O'Hara, C.J. 1980. New radiocarbon dates from the inner continental shelf off southern Massachusetts and a local sea-level-rise curve for the past 12,000 years. *Geology*, 8(2):102–106.

Peck, J.A., and McMaster, R.M. 1991. Stratigraphy and geologic history of Quaternary sediments in lower West Passage, Narragansett Bay, Rhode Island. *In* Gayes, P.T., Lewis, R.S., and Bokuniewicz, H.J. (eds.) Quaternary geology of Long Island Sound and adjacent coastal areas. *Journal of Coastal Research* Special Issue No. 11, pp. 25–37.

Poppe, L.J., Paskevich, V.F., Williams, S.J., Hastings, M.E., Kelley, J.T., Belknap, D.F., Ward, L.G., FitzGerald, D.M., and Larsen, P.F. 2003. Surficial sediment data from the Gulf of Maine, Georges Bank, and vicinity: A GIS compilation. US Geological Survey Open-File Report 03-001.

Rhode Island Coastal Resources Management Council (RICRMC). 1995. Rhode Island Coastal Resources Management Plan, as amended, Providence, RI.

Ruddiman, W.F. 2001. Earth's Climate: past and future. New York: W.H. Freeman & Sons.

Schafer, J.P. 1961a. Surficial geology of the Wickford quadrangle, Rhode Island. U.S. Geological Survey Quadrangle Map GQ-136, Washington, DC.

Schafer, J.P. 1961b. Surficial geology of the Narragansett Pier quadrangle, Rhode Island. U.S. Geological Survey Quadrangle Map GQ-140, Washington, DC.

Schafer, J.P., and Hartshorn, J.H. 1965. The Quaternary of New England. *In* Wright, H.E., Jr., and Frey, D.G. (eds.) The Quaternary of the United States, A review volume for the VII Congress of the International Association for Quaternary Research, Princeton, NJ: *Princeton University Press*, pp. 113–128.

Seaber, P.R., Kapinos, F., and Knapp, G. 1987. Hydrologic Unit Maps. US Geological Survey Water-Supply Paper 2294, 63 pp.

Shepard, F.P. 1954. Nomenclature based on sand-silt-clay ratios. *Journal of Sedimentary Petrology* 24:151–158.

Smith, J.H. 1955a. Surficial geology of the East Greenwich quadrangle, Rhode Island. U.S. Geological Survey Quadrangle Map GQ-62 (scale 1:31,680), Washington, DC.

Smith, J.H. 1955b. Surficial geology of the Bristol quadrangle and vicinity, Rhode Island. U.S. Geological Survey Quadrangle Map GQ-70 (scale 1:31,680), Washington, DC.

Smith, J.H. 1956c. Surficial geology of the Providence quadrangle, Rhode Island. U.S. Geological Survey Quadrangle Map GQ-84 (scale 1:31,680), Washington, DC.

SPSS Inc. 2004. SPSS for Windows, Release 13.0 Chicago, IL.

Stone, B.D., and Borns, H.W., Jr 1986. Pleistocene glacial and interglacial stratigraphy of New England, Long Island, and adjacent Georges Bank and Gulf of Maine. *In* Sibrava, V., Bowen, D.Q., and Richmond, G.M. (eds.) Quaternary glaciations in the Northern Hemisphere, pp. 39–52, Oxford, UK: Pergamon Press.

Stone, B.D., and Stone, J.R. 2005. Sedimentary facies and morphosequences of glacial melt water deposits. *In* Stone, J.R., Schafer, J.P., London, E.H., DiGiacomo-Cohen, M.L., Lweis, R.S., and Thompson, W.B. Quaternary geologic map of Connecticut and Long Island Sound basin. US Geological Survey Scientific Investigations Map (2784, scale 1:125,000, 2 sheets), Washington, DC.

Stone, J.R., Schafer, J.P., London, E.H., DiGiacomo-Cohen, M.L., Lewis, R.S., and Thompson, W.B. (2005). Quaternary geologic map of Connecticut and Long Island Sound basin: U.S. Geological Survey Scientific Investigations Map 2784 (scale 1:125,000, 2 sheets), 72 p. report, Washington, DC.

Uchupi, E., Giese, G.S., Aubrey, D.G., and Kim, D.J. 1996. The late quaternary construction of Cape Cod, Massachusetts: A reconsideration of the W. M. Davis model. Geological Society of America Special Paper 309, Boulder, CO.

US Census Bureau. 2005. Census 2000 Summary File 1 Technical Documentation. US Census Bureau. Washington, D.C. 635 pp.

Chapter 2
Narragansett Bay Amidst a Globally Changing Climate

Michael E. Q. Pilson

2.1 Introduction

Various features of the weather over Narragansett Bay and the surrounding watershed are important when considering the ecology and management of this ecosystem. Evidence for changes in climate, in both short- and long-term weather patterns, continues to accumulate, and there is firm expectation of significant changes in the future. This chapter contains brief summaries of some historical aspects of local weather affecting Narragansett Bay, which are drawn from materials presented in Pilson (1989), and updated here with additional materials.

Many of the original sources of data contained extremely detailed records. For example, there were hourly records of temperature at weather stations many decades ago, and daily records of water flow at six or seven gauging stations. In order to keep this report to a manageable size, most of the data were averaged by month, or monthly totals were extracted from original records. When appropriate, the varying lengths of months were taken into account in making the averages. One disadvantage in using monthly totals or monthly averages is that the response time of the bay to various perturbations will most often be on shorter time scales. Those who would investigate such phenomena in detail may have to acquire the original records in their complete form, using this chapter as an entry point. The monthly averages are, however, convenient in most cases for the examination of bay processes at the seasonal scales explored here.

The records of the US Weather Bureau from Providence, Rhode Island, began on October 22, 1904 with the establishment of an office in the University Hall at Brown University on Prospect Street (US Weather Bureau; NOAA, various years). On January 1, 1909, the station was moved to the Banigan Building at 10 Weybosset Street in Providence. Since 1913, however, the measurements were affected by the presence of a higher building to the west, and so

Michael E. Q. Pilson
University of Rhode Island, Graduate School of Oceanography, Narragansett, RI 02882
pilson@gso.uri.edu

A. Desbonnet, B. A. Costa-Pierce (eds.), *Science for Ecosystem-based Management.*   35
© Springer 2008

the instruments were moved to the Turks Head Building at 11 Weybosset Street, where they could be fixed 74–86 ft higher (wind gauge at 251 ft above ground level, thermometers at 215 ft). On June 10, 1940, the recording station was moved again, this time to the Post Office Annex Building, though the wind gauge was moved to the new site the following June. Instrument exposure at this site was quite poor; hence, on November 10, 1941, the official Providence observation program was transferred to the State Airport in Warwick, where there had been a subsidiary weather station since June 16, 1932. The official temperature recording station for Providence was transferred back to the Post Office Annex Building on January 1, 1942, and the precipitation station in March of the same year, and both remained there until May 20, 1953, when they were moved again to the T.F. Green State Airport in Warwick. There have also been a number of probably minor moves within the grounds of the airport (NOAA, 1998). All these moves should be kept in mind when considering the possibility of long-term trends of temperature and precipitation.

2.2 Temperature

Earlier data for temperature and precipitation from other sources, in some cases going back to 1832, are tabulated in Pilson (1989). These data are not reproduced here because, where overlap with the Weather Bureau data exists, there are systematic differences that have not been reconciled.

The Weather Bureau data from 1905 to 2005 (Fig. 2.1) suggest the very likely existence of a long-term trend in the annually averaged temperatures at this

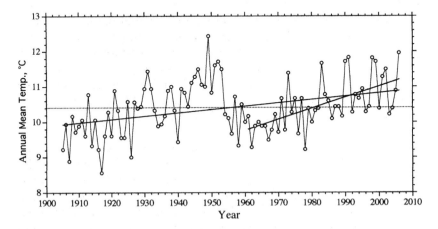

Fig. 2.1 Annual mean temperature at the official Weather Bureau stations for Providence, RI, beginning from 1905. Data are from NOAA (1983, 1971–2006a) and ESSA (1966–1970). Long-term mean temperature from 1905 until 2006 is 10.41 °C. The increase over the record from 1905 to 2006 was 0.094 °C per decade while the increase from 1961 to 2006 was 0.31 °C per decade.

station. Since 1905, the temperatures increased until about 1952, then dropped, and began to rise again after 1960. While the overall trend appears upward, it is noteworthy that the warmest year on record still remains 1949. It must, however, be noted that the weather records were taken mostly within the city of Providence until 1953, when the collecting station was moved to T.F. Green Airport outside the city limits. The existence of a long-term upward trend is provided on the NOAA website, where a similar graph is presented after averaging the data from other stations in RI. No information, however, is available as to which stations, or over which time intervals, are the NOAA data averaged. Data from other adjacent regions—Connecticut and Massachusetts, for instance—also show comparable increases, as does abundant anecdotal information. The increase was about 0.94 °C over the 100 years from 1905 to 2006, and 1.14 °C between 1961 and 2005. The global average increase in temperature reported by the Intergovernmental Panel on Climate Change (IPCC) in February 2007 was 0.74 °C over the same 100-year interval (IPCC, 2007). The New England Regional Assessment (USGCRP, 2002) estimated that the increase for nearly the same interval of time had been 0.95 °C over the region identified as "Coastal New England." These several evaluations support the contention that the increase of nearly 1 °C at the Providence-Warwick station is entirely consistent with what might be expected.

2.3 Precipitation

The Weather Bureau data for Providence and Green Airport in Warwick, RI (Fig. 2.2), suggest a small increase in precipitation (rain + snow) over the past 100 years. The overall increase of about 3 mm yr^{-1} has resulted in a modeled

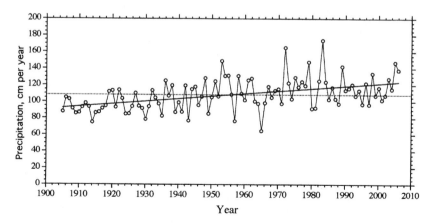

Fig. 2.2 Total annual precipitation (rain + snow) at the weather stations in Providence, RI, from 1905 to 2006. Over the interval reported, the overall mean value was 108.2 cm yr^{-1}. The slope of the simple linear regression is 0.305 cm yr^{-1}.

(a)

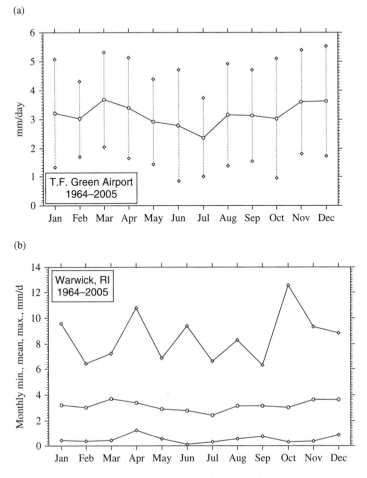

(b)

Fig. 2.3 Daily precipitation at T.F. Green Airport in Warwick, RI. (a) Averaged by month from 1965 to 2005 with error bars showing one standard deviation. The data are presented as daily averages for each month to eliminate the bias due to the different number of days in each month. (b) Averaged by month from 1965 to 2005. The data are presented as daily averages for each month to eliminate the bias due to the different number of days in each month. The middle line is the average during the 40-year interval, and the other two lines are the minimum and maximum monthly daily averages during the interval analyzed.

precipitation increase from 93 cm yr^{-1} in 1905 to 126 cm yr^{-1} in 2006. An early impression that annual precipitation totals have been more variable in the recent decades was not born out after 1990. The overall range is considerable, between 64.6 cm in the drought year of 1965 and 174.0 cm in the wettest year on record, i.e., 1983. The long-term average is 108.2 cm yr^{-1}. Contrary to the common impression, the total precipitation is not markedly greater in the spring than in the fall and early winter (Fig. 2.3a,b). The New England Regional Assessment (USGCRP, 2002) estimated that coastal New England had experienced a 16.3% increase in precipitation between 1895 and 2001. The data shown in Fig. 2.2, however, suggest a 32% increase between 1905 and 2006. The New England Regional Assessment includes several very wet years since 1895 through 1904, and if that is allowed for, the data are roughly comparable. The IPCC report (IPCC, 2007) suggests significant increase in precipitation in this region, but the data are not published yet, and therefore cannot be verified here. The increase is qualitatively in the direction of what should be expected from the evaluations in climate models.

2.4 River Flow

Data on stream flow from nine gauges in the Narragansett Bay watershed maintained by the US Geological Survey were converted to estimate the total river flow to Narragansett Bay (excluding the Sakonnet River) by multiplying the total gauged flow by the ratio of total watershed area to the gauged area. The same procedure was used by Pilson (1985), except that here the total watershed area was as given in Ries (1990), corrected by subtracting the small drainage area of the Sakonnet. The average annual river flow to Narragansett Bay between 1962 and 2004 was 90.9 $m^3 s^{-1}$. As with precipitation, there has been a suggestion (Fig. 2.4a) of a long-term increase amounting to 0.29 $m^3 yr^{-1}$ during the years presented, but year-to-year variability is great enough that one cannot have great confidence in the exact value. The total river flow is decidedly seasonal (Fig. 2.4b) due to the high rate of evapotranspiration during the warmer months of the year. The total fresh water input to the bay also includes the flow of sewage, which is estimated at 5.1 $m^3 s^{-1}$, as well as direct rainfall to the bay surface. These latter inputs are not included in the totals shown in Fig. 2.4.

2.5 Evapotranspiration

An estimate of the annual rate of evaporation and transpiration from the watershed may be obtained from a comparison of the total rainfall on the watershed with the total river flow from the watershed. To make this calculation, data from rainfall records (where available) were averaged from six stations distributed over the watershed and, using the area of the watershed as 4,343 km^2, were converted into units of $m^3 s^{-1}$ and compared with the records of annual total river flow. Given that both sources of data must contain some uncertainty, that there is some lag in river flow after rainfall, or that there may be year-to-year changes in reservoir volumes which are not accounted for, the relationship is remarkably good (Fig. 2.5). The annual evapotranspiration from the Narragansett Bay watershed appears to vary between 40 and 55 cm yr^{-1} depending on the rainfall, and averages about 45 cm yr^{-1}. This is nearly one half of the total precipitation.

(a)

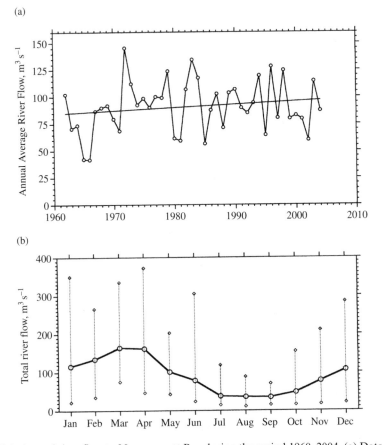

(b)

Fig. 2.4 Annual river flow to Narragansett Bay during the period 1960–2004. (a) Data were obtained by summing the monthly totals (USGS, 1960–1964, 1965–1974, 1975–2004) from each of the gauged rivers draining into the bay and multiplying by the ratio of the total drainage area to the sum of the gauged areas. The procedure is the same as in Pilson (1985, 1989), but the most recent values of the individual gauged areas are used, as well as the total drainage area from Ries (1990) corrected for the drainage area of the Sakonnet River according to Pilson (1985). The rivers included in the calculation were Taunton, Three Mile, Segreganset, Ten Mile, Blackstone, Moshassuck, Woonasquatucket, Pawtuxet, and Potowomut (Hunt) when data were available. The slope of the linear regression line is 0.287 m^3 yr^{-1}. (b) Seasonality of river flow into Narragansett Bay. Data from the gauged rivers, calculated as described for Fig. 2.4a, were averaged by month for the period 1962–2003. The extreme values plotted are the maximum and minimum monthly values observed during this 42-period.

2.6 Wind

Wind speed recorded at T.F. Green Airport in Warwick, RI, is markedly seasonal (Fig. 2.6a). The months of strongest winds are February, March, and April, and the months of weakest winds are July, August, and September.

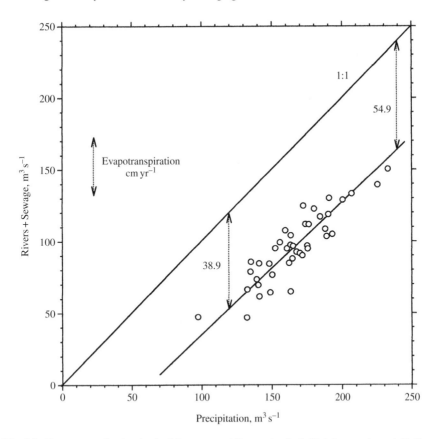

Fig. 2.5 Evapotranspiration in the Narragansett Bay watershed. Total annual precipitation on the watershed of Narragansett Bay (excluding the Sakonnet) was estimated by averaging the precipitation at six or seven stations within the watershed and multiplying by the area of the watershed. Not all stations were continuous during the interval selected (1964–2004). Stations used were (RI): T.F. Green Airport, Woonsocket, North Foster; (MA): Milford, Taunton, Worcester, and Mansfield. The values were converted to units of m^3 s^{-1} and plotted against estimates of total river flow as calculated for Fig.2.4a, added to an estimated flow of sewage (5.1 m^3 s^{-1}). Shown are the 1:1 line (river flow plus sewage equals precipitation) and the functional regression line through the data points. The difference between the two lines at two levels of rainfall (120 and 240 m^3 s^{-1}) was calculated, and presented as an annual loss from the watershed by evapotranspiration in units of cm yr^{-1}.

It appears that there have been remarkable long-term changes in the annually averaged wind speed (Fig. 2.6b). From values averaging near 18 km hr^{-1} in the 1950s, wind speed decreased thereafter to an average of about 14.5 km hr^{-1} in 2004 and 2005. The annual average wind speed appears unrelated to the annual average North Atlantic Oscillation index. Since 1964, the decrease is evident in the time of year both when the winds are strongest, and when they are weakest (Fig. 2.7). At this time, it is not known whether the

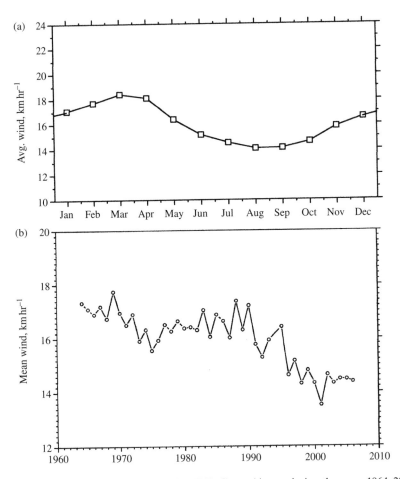

Fig. 2.6 (a) Monthly average wind speed at T.F. Green Airport during the years 1964–2006. Data are from the same sources as for Fig. 2.1. (b) Annual mean wind speed at T.F. Green Airport during the years 1954–2006. Some earlier data are available, but come from stations in downtown Providence where wind speed is not directly comparable to that at the airport, and so are not included here. These calculations were carried out by Steve Granger (personal communication).

apparently decreased wind speed is a local phenomenon, or is supported by similar changes in adjacent regions.

The monthly average winds during the period 1960 and 2005 were separated into vector components directed to the north and to the east, and presented here as the annual averages of each component (Fig. 2.8). The long-term average of the north component is -1.56 km hr^{-1}, that is, this component is generally directed to the south, and there appears to be no secular trend in the strength of this component. The long-term average of the component directed to the east seems to have decreased over this time period, averaging 4.27 km hr^{-1} at present.

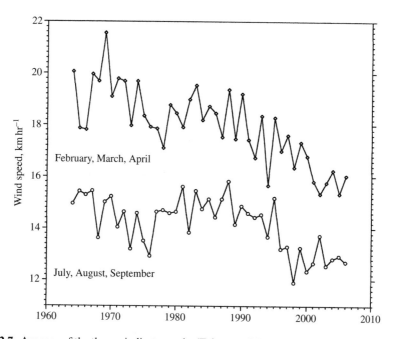

Fig. 2.7 Average of the three windiest months (February, March and April), and the average of the three least windy months (July, August and September) for the years 1964–2006.

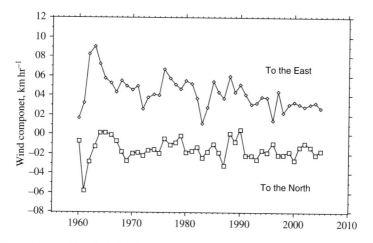

Fig. 2.8 NOAA directional wind information converted to vector-averaged directional components for the years 1959 through 2005 on the monthly averaged data, and then averaged by year. The two components presented are "east-west"—positive to the east, and "north-south"—positive to the north. Winds were on average westerly (positive to the east) and southerly (negative to the north). The northerly component does not show much long-term secular change, but the east-west component appears to have decreased over the interval evaluated . These calculations were carried out by Steve Granger (personal communication).

2.7 Residence Time

The residence time of water in Narragansett Bay (as well as the residence time of dissolved substances and organisms that move with water) is mostly controlled by the input of fresh water, and to a lesser extent by average wind speed (Pilson, 1985). Accordingly, the residence time varies, on average, seasonally, with the longest residence time during the warm summer months of July, August, and September, and the shortest residence time during February, March, and April. Although winds have no major influence on residence time, coincidentally, these are also the times of lowest and greatest strengths of the winds. A seasonal plot of residence time (Fig. 2.9) shows that the range of average residence time is about 20 days in the early spring to about 35 days in the late summer. If there were no fresh water flowing into the bay, the maximum residence time would be some what over 40 days, driven largely by tidal exchange. At the highest flow rates observed, the residence time would be close to 12 days.

Substances that are not metabolized and have the chemical residence time in water less than the water residence time will tend to be trapped in the sediments of the bay, while those that have longer residence times in water will tend to be swept out of the bay. Planktonic organisms with reproduction times longer than the water residence time must have special adaptive mechanisms if they maintain populations in the bay distinct from those in the adjacent coastal waters.

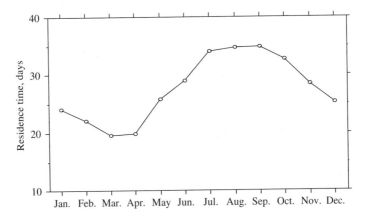

Fig. 2.9 An equation developed from available data on monthly mean salinity, fresh water flow and wind speed (Pilson 1985) was used to predict long-term average residence time of the water according to season. The equation is: $\ln T = 4.053 - 0.00358 FW - 0.0257 W$, where $T =$ residence time in days; $FW =$ monthly average freshwater flow (rivers + sewage) in $m^3\ s^{-1}$ and $W =$ monthly average wind speed at T.F. Green Airport in $km\ h^{-1}$. Data on the input of fresh water (as in Figs 2.4a and 2.5), and wind (as in Fig. 2.6a) are combined to provide long-term monthly average residence time.

2.8 Conclusions

It is generally not a good idea to draw too firm a conclusion about trends in climate using data derived from a single location. Nevertheless, it is apparent that, as with the average for planet earth, the region around Narragansett Bay appears to exhibit a distinct warming trend, which is evident despite the scatter in the data.

The data suggesting a long-term increase of total precipitation in the watershed are less likely to have been affected by the several moves of the recording station. The regression line would imply that, on average, the watershed receives about 30% more precipitation now than it did 100 years ago. Both world-wide evidence and modeling suggest that there has been an increase in precipitation during this time (Evans, 2006; IPCC, 2007), and that the increases are not uniformly distributed around the globe. During the 41 years for which the river flow values are available, the increase in precipitation on the watershed amounts to $16.8 \text{ m}^3 \text{ s}^{-1}$. The increase in river flow during the same period, according to the regression line fit to the data, amounts to nearly $12 \text{ m}^3 \text{ s}^{-1}$. The river flow data are much more scattered, and the fitted regression line is not so compelling. Nevertheless, it is reassuring that the values are as close as they are. Increased precipitation and the consequent increase in stream flow will tend to slightly reduce the residence time of water in Narragansett Bay.

While wind speed has a minor effect on the residence time of water in the bay, the major effect on the ecology of the bay is probably on the vertical mixing that tends to weaken any pycnocline present. The weakest winds are in the warmest summer months, and this is the time when a strong vertical density gradient is most conducive to the formation of low-oxygen water in the bottom layers. Since it appears that the average wind speed may have decreased significantly since good records began to be kept in the early 1950s, it seems that the bay should be more vulnerable to occasional episodes of low oxygen or anoxic water in the bottom layers. The suggestion that the east-west component of wind has decreased also leads to speculation that there may have been changes in the extent to which the flow of water into and out of the bay is apportioned between the East and West Passages. Further exploration of this suggestion requires additional wind data because the station at T.F. Green Airport is not well placed to capture daily changes in wind direction evident in the southern bay, especially in the summer time.

References

Environmental Sciences Service Administration (ESSA). 1966 to 1970. Climatological data, New England. 78–82. Environmental Data Service, US Department of Commerce, Asheville, NC.

Evans, M.N. 2006. The woods fill up with snow. *Nature* 440:1120–1121.

Intergovernmental Panel on Climate Change (IPCC). 2007. Climate change 2007: the physical science basis. *http://www.ipcc.ch*

National Oceanic and Atmospheric Administration (NOAA). 1971–2006a. Local climatological data; Monthly summary for Providence, RI; Annual summary for Providence, RI. Environmental Data Service, US Department of Commerce, Asheville, NC.

National Oceanic and Atmospheric Administration (NOAA). 1998. Local climatological data; Monthly summary for Providence, RI; Annual summary for Providence, RI. Environmental Data Service, US Department of Commerce, Asheville, NC.

National Oceanic and Atmospheric Administration (NOAA). 1971– 2006b. Climatological data, New England. Environmental Data Service, US Department of Commerce, Asheville, NC.

National Oceanic and Atmospheric Administration (NOAA). 1983. Statewide average climatic history. Rhode Island 1895–1982. National Oceanic and Atmospheric Administration, Environmental Data Service, US Department of Commerce, Asheville, NC.

Pilson, M.E.Q. 1985. On the residence time of water in Narragansett Bay. *Estuaries* 8:2–14.

Pilson, M.E.Q. 1989. Aspects of climate around Narragansett Bay. Technical Report, Graduate School of Oceanography, University of Rhode Island, Narragansett, RI. 59 pp.

Ries, K.G. 1990. Estimating surface-water runoff to Narragansett Bay, Rhode Island and Massachusetts. US Geological Survey, Water Resources Investigations Report 89–4164 USGS, Denver, Colorado.

US Geological Survey (USGS). 1960-1964. Surface Water Records of Massachusetts, New Hampshire, Rhode Island and Vermont. US Department of the Interior, Boston, MA.

US Geological Survey (USGS). 1964. Compilation of records of Surface Waters of the United States, October 1950 to September 1960. Part 1-A, North Atlantic Slope Basins, Maine to Connecticut. Water Supply Paper 1721. US Department of the Interior, Boston MA.

US Geological Survey (USGS). 1965 to 1974. Surface Water Records of Massachusetts, New Hampshire, Rhode Island and Vermont. Part 1, Surface Water Records; Part 2, Water Quality Records. US Department of the Interior, Boston MA.

US Geological Survey (USGS). 1975 to 2004. Water Resources Data for Massachusetts and Rhode Island. Water Data Reports MA-RI-75-1 to MA-RI-02-1. US Department of the Interior, Boston MA.

US Global Change Research Program. 2002. New England Regional Assessment. Final Report. *http://www.necci.sr.unh.edu/*.

Chapter 3
Estimating Atmospheric Nitrogen Deposition in the Northeastern United States: Relevance to Narragansett Bay

Robert W. Howarth

3.1 Introduction

Over the past several decades, nitrogen pollution has grown to be perhaps the largest pollution problem in the coastal waters of the United States (NRC, 2000). An estimated two-thirds of the coastal rivers and bays in the country are now believed to be moderately or severely degraded from this pollution (Bricker *et al.,* 1999). The nitrogen comes from many sources, including wastewater treatment plants, agriculture, and atmospheric deposition. Often, the relative importance of these sources for particular estuaries is not well known (NRC, 2000; Alexander *et al.,* 2001; Howarth *et al.,* 2002b). Much of the effort at reducing nitrogen pollution has been directed at wastewater treatment plants, in part because these sources are so obvious. While such point sources are dominant in some estuaries, in most ecosystems the non-point sources of nitrogen from agriculture and atmospheric deposition are more important (Howarth *et al.,* 1996, 2002a,b; NRC, 2000; Alexander *et al.,* 2001). However, in estuaries with high population densities in the watershed, wastewater inputs are sometimes the single largest sources (NRC, 1993). This is the case for Narragansett Bay, as discussed by Nixon and colleagues in Chapter 5 of this volume.

The nitrogen in atmospheric deposition originates both from fossil fuel combustion and from the volatilization of ammonia to the atmosphere from agricultural sources, particularly from animal wastes in confined animal feedlot operations. The importance of this source was virtually unrecognized before the pioneering paper by Fisher and Oppenheimer (1991) noted that the nitrate anion associated with nitric acid in acid rain may be a major source of nitrogen to Chesapeake Bay. Since then, the focus on atmospheric deposition as a source of nitrogen has intensified, and generally, estimates of the importance of this source have tended to increase over time as it has received more attention.

Robert W. Howarth
Department of Ecology & Evolutionary Biology, Cornell University, Ithaca, NY 14853, USA
howarth@cornell.edu

A. Desbonnet, B. A. Costa-Pierce (eds.), *Science for Ecosystem-based Management.*
© Springer 2008

3.2 Atmospheric Deposition as a Nitrogen Source to Coastal Waters

For the United States as a whole, we have estimated that atmospheric deposition of nitrogen that originates from fossil-fuel combustion contributes 30% of the total nitrogen inputs to coastal marine ecosystems, while another 10% of these nitrogen inputs come from ammonia volatized into the atmosphere from agricultural sources (Howarth and Rielinger, 2003). The rest of the nitrogen inputs to coastal waters come from runoff from agricultural sources (44%) and from municipal and industrial wastewater streams (~16%).

Some of the nitrogen from atmospheric deposition is deposited directly onto the surface of coastal waters. This direct deposition to surface waters often contributes between 1% and 40% of the total nitrogen inputs to coastal ecosystems (Nixon et al., 1996; Paerl, 1997; Howarth, 1998; Paerl and Whitall, 1999; Valigura et al., 2000). The direct deposition is most significant in very large systems, such as the Baltic Sea (Nixon et al., 1996) or in coastal systems such as Tampa Bay which have relatively small watersheds in comparison to the area of their surface waters (Zarbock et al., 1996).

In most coastal marine ecosystems, the major route whereby atmospheric deposition contributes nitrogen is not direct deposition onto surface waters, but rather deposition onto the terrestrial landscape with subsequent downstream export in streams and rivers. As discussed below, these fluxes are difficult to measure, leaving significant uncertainty and debate about their magnitude. In the northeastern US as a whole (Gulf of Maine through Chesapeake Bay), our studies have suggested that atmospheric deposition is the single largest source of nitrogen to coastal waters (Howarth et al., 1996; Jaworski et al., 1997; Boyer et al., 2002), while other studies have concluded atmospheric nitrogen deposition is the second largest source after wastewater discharges from sewage treatment plants (Driscoll et al., 2003). Our approach leads to the conclusion that atmospheric deposition of nitrogen onto the landscape—considering only the deposition of oxidized nitrogen compounds that originate from fossil fuel combustion (NO_y)—contributes between 25% and 80% of the nitrogen flux in the different major rivers of New England (Fig. 3.1, Boyer et al., 2002; Howarth and Rielinger, 2003) and approximately 25% of the nitrogen flux in the Mississippi River (NRC, 2000; Howarth et al., 2002b). Using another approach—SPARROW, or Spatially Referenced Regression on Watershed attributes model—Alexander et al. (2001) concluded that atmospheric deposition onto the landscape contributed between 4% and 35% of the nitrogen flux in 40 major coastal watersheds across the United States, with the highest contribution in the northeastern and mid-Atlantic regions. As discussed later in this paper, the SPARROW model may significantly underestimate the role of deposition near emission sources.

The uncertainty over the contribution of atmospheric deposition as a nitrogen source to coastal marine ecosystems stems from two issues: uncertainty over

Fig. 3.1 Percentage of nitrogen in major New England rivers that originates from fossil-fuel derived atmospheric deposition onto the landscape. Reprinted from Howarth and Rielinger (2003), based on data in Boyer *et al.* (2002).

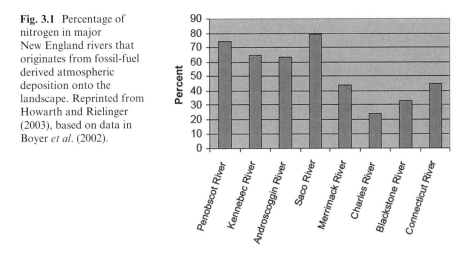

the magnitude of nitrogen deposition onto watersheds, particularly from "dry deposition", and uncertainty over the amount of the deposited nitrogen that is subsequently exported downstream (NRC, 2000; Howarth *et al.,* 2002b). Each of these is discussed in some detail in the following sections.

3.3 Dry Deposition of Nitrogen as a Source

The vast majority of measurements of nitrogen deposition in the United States—including those made by the National Atmospheric Deposition Program (NADP)—measure only "wet deposition" (i.e., nitrogen in rainfall and snow). To estimate wet deposition onto an entire watershed, data at particular monitoring sites are extrapolated statistically considering factors such as local topography and precipitation (Ollinger *et al.,* 1993; Grimm and Lynch, 2005).

Substantial quantities of nitrogen can be deposited from the atmosphere as "dry deposition," which includes aerosols and other particles and uptake of gaseous forms of nitrogen by vegetation, soils, and surface waters. Both in the United States and Europe, the extremely sparse spatial coverage in networks for measuring dry deposition severely limits estimation of this process (Holland *et al.,* 2005). In the United States, dry deposition is routinely estimated only at sites that are part of the CASTNet and AIRMon-Dry programs. At the peak of these programs in the 1990s, these networks consisted of a total of 93 sites across the country, but the number is now down to 70 *(http://www.epa.gov/ castnet/)*. In the watersheds of Chesapeake Bay—an area of 165,000 km^2 that includes land in 6 states—there are only 8 stations for monitoring dry deposition. In New England, there are only 6 stations, with 3 in Maine and only one in southern New England. The vast majority of these dry deposition monitoring stations across the country—and all of them in New England and New York State—are purposefully located far from sources of nitrogen emissions to the atmosphere.

In addition to the limited spatial extent of the dry deposition monitoring networks, these networks do not measure all of the components that can be deposited. For example, particulate NO_3^- and NH_4^+ are routinely measured, as is nitric acid vapor. However, other gaseous nitrogen compounds that may play a significant role in deposition (i.e., NO, NO_2, HONO, peroxy and alkyl based organics, and ammonia gas) are not measured. NO and NO_2 are the major gases emitted from fossil fuel combustion, while ammonia is the major form of air pollution from agricultural sources. Ammonia is also released in vehicle exhaust, although at lesser amounts than for NO and NO_2 (Baum et al., 2001; Cape et al., 2004). To the extent these compounds are deposited, the dry depositional monitoring networks are underestimating total deposition. As currently measured, the dry deposition at the 8 CASTNet sites in the Chesapeake Bay watershed ranges from 23% to 38% of total deposition (T. Butler, pers. comm.), but the actual contribution when all forms of nitrogen gases are considered must certainly be higher.

The manner in which dry deposition rates are calculated—multiplying concentration data obtained at the monitoring sites by "depositional velocities"—may also result in underestimation of this process. For the AIRMon and CASTNet sites, these deposition velocities are estimated as a function of vegetation and meteorological conditions (Clarke et al., 1997). Our knowledge of depositional velocities is based on studies in flat, homogenous terrain; as noted by Bruce Hicks (former Director of the NOAA Air Resources Lab), when estimating dry deposition "we are simulating the world on the assumption that our understanding of [these] special cases applies everywhere. We often display unwarranted confidence" in our estimates (Hicks presentation to the annual meeting of the American Society of Meteorology, October 2005). Complex terrain is likely to substantially increase depositional velocities. Vegetative cover is also important, and different models can vary in their estimates of spatial integrated dry deposition by more than 5-fold depending upon different assumptions of the effect of vegetation (particularly coniferous forests) on depositional velocities (Wesely and Hicks, 1999; Holland et al., 2005).

3.4 Estimation of Total Nitrogen Deposition in the Northeastern US

Boyer et al. (2002) estimated the average deposition of oxidized nitrogen (NO_y) onto the landscape of the major rivers of the northeastern United States (including both wet and dry deposition) following the approach of Ollinger et al. (1993) in using a statistical extrapolation of deposition monitoring data.

They estimated a range of values across these watersheds from $\sim$360 kg N km^{-2} yr^{-1} in the Penobscot River basin in Maine to $\sim$890 kg N km^{-2} yr^{-1} in the Schuylkill River basin in Pennsylvania (Boyer *et al.*, 2002). The average value for this set of watersheds was $\sim$680 kg N km^{-2} yr^{-1}.

Another approach for estimating nitrogen deposition onto the landscape can be obtained from models based on emissions to the atmosphere, with consideration of reaction and advection in the atmosphere, followed by deposition. We used one of these models (the GCTM model; Prospero *et al.*, 1996) to estimate nitrogen deposition in all of the regions—including the northeastern United States—that surround the North Atlantic Ocean (Howarth *et al.*, 1996). The GCTM model predicts depositional patterns globally at a relatively course spatial scale using emission sources as inputs and modeling atmospheric transformations and transport (Prospero *et al.*, 1996). For the northeastern United States, the GCTM model yielded an estimated total NO$_y$ deposition (wet plus dry) of $\sim$1,200 kg N km^{-2} yr^{-1}, a value 80% greater than that derived by Boyer *et al.* (2002) from extrapolation of deposition monitoring data (Fig. 3.2, Howarth *et al.*, in press). A similar, more recent emission-based model (TM3)

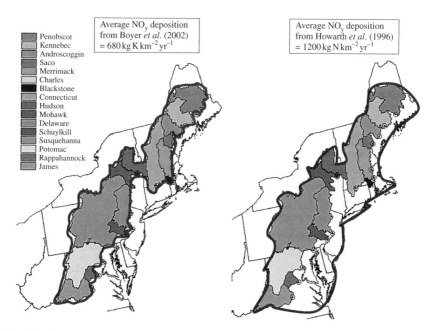

Fig. 3.2 The geographic area considered by Boyer *et al.* (2002) was the area of 16 watersheds in the northeastern United States upriver from the lowest gauging station of the USGS (left). The area considered by Howarth *et al.* (1996) is somewhat larger, and includes the area on the coastal plain (right). Note that the average estimates for deposition of oxidized nitrogen pollution originating from fossil fuel combustion is $\sim$80% greater in the Howarth *et al.* (1996) analysis, probably due to different approaches used for the estimation and/or the different area considered.

developed by Frank Dentener and colleagues, and used by Galloway *et al.* (2004) for their global and regional nitrogen budgets, yields a comparable estimate for the northeastern United States as did the GCTM model (Howarth *et al.*, in press). These emission-based models are attractive, in that at least at very course spatial scales, they are as accurate as the emission data. However, these models are computationally demanding, and until very recently, had not been applied at a spatial scale fine enough to give estimates for the individual 16 northeastern watersheds. A new effort by NOAA/EPA's Atmospheric Sciences Modeling Division uses emissions data and the CMAQ model to estimate nitrogen deposition at a 36-km grid, but the model is still being tested as of late 2006 (presentation by R. Dennis at the National Atmospheric Deposition Program annual Technical Committee meeting, October 2006). This approach shows great promise for the future. Preliminary comparisons of this fine-scale model with the coarser scale output from GCTM and TM3 have shown good agreement (R. Dennis, pers. comm.).

Why is the estimate from the emission-based model (Howarth *et al.*, 1996) so much greater than that from estimates based on extrapolation of the wet deposition monitoring data (Boyer *et al.*, 2002)? There are three possible explanations, which are not mutually exclusive.

First, deposition on the relatively urbanized coastal plain may be much greater than in the watersheds away from the coast. The watershed areas considered by Boyer *et al.* (2002) are upriver from the coast and tend to be more rural than is the coastal plain downstream (Fig. 3.2). Recent studies have found evidence that deposition near emission sources can be much greater than deposition away from emission sources. For example, deposition within New York City was more than twice as high than in more rural areas to the north of the city (Lovett *et al.*, 2000), and deposition in the immediate vicinity of roads was much higher than a few hundred meters away (Cape *et al.*, 2004; presentation by R. Howarth, R. Marino, N. Bettez, E. Davidson, and T. Butler at the National Atmospheric Deposition Program annual Technical Committee meeting, October 2006);

Second, the estimate based on deposition monitoring data (Boyer *et al.*, 2002) may underestimate total deposition. This is of course likely, to the extent that dry deposition is underestimated. As noted above, not all of the important gases that may be deposited are routinely measured by the dry deposition monitoring networks, and depositional velocities may be underestimated in regions with major terrain features. Further, the deposition networks were not designed to measure deposition in the immediate vicinity of emission sources. In fact, most of the NADP wet deposition monitoring sites and most of the CASTNet dry depositon sites are intentionally located far away from urban emission sources.

Third, the estimate from emission-based modeling (Howarth *et al.*, 1996) may overestimate total deposition. This could occur if emissions are overestimated, which may well be true for ammonia emissions, but probably not for emissions of oxidized nitrogen to the atmosphere in the United States (Holland *et al.*, 1999). The difference between the Howarth *et al.* (1996) and Boyer *et al.* (2002) estimates highlighted in this paper is for deposition of oxidized nitrogen (NO_y). Alternatively, emission-based modeling may not accurately capture the spatial pattern of the deposition. These models rely on a mass balance of nitrogen in the atmosphere, so global deposition estimates are as accurate as the emissions data that feed them. However, deposition may be underestimated in some regions and correspondingly overestimated elsewhere.

Obviously, significant uncertainty exists in the overall magnitude of total nitrogen deposition in an area such as the northeastern United States. When considering the differences detailed above, it is important to note that extrapolations based on deposition monitoring (Ollinger *et al.*, 1993; Grimm and Lynch, 2005) do not appear to capture any evidence of higher deposition near urban centers and transportation corridors. For reasons discussed in detail following, I believe it likely that traditional approaches that use deposition monitoring data to estimate total nitrogen deposition result in substantial underestimates, especially for total nitrogen deposition in the urbanized portions of the northeastern United States.

3.5 Using Throughfall to Estimate Total Nitrogen Deposition

The difficulty with measuring dry deposition of N (particularly of gaseous forms such as NO, NO_2, and NH_3) has led some investigators to use tree-canopy throughfall as a surrogate for total N deposition (Lajtha *et al.*, 1995; Lovett *et al.*, 2000; Weathers *et al.*, 2006; Schmitt *et al.*, 2005). Throughfall is the material that falls through the canopy of a forest, and so includes whatever is deposited on the canopy in both wet and dry deposition, plus or minus the net exchange of material with the vegetation. Most studies have found that the assimilation of nitrogen from deposition into leaves of the canopy is generally as great as or greater than the leaching of nitrogen out of leaves (Lindberg *et al.*, 1990; Johnson, 1992; Lovett and Lindberg, 1993; Dise and Wright, 1995; Lajtha *et al.*, 1995). Consequently, many experts on atmospheric deposition have argued that throughfall measurements provide a minimum estimate of total nitrogen deposition (Lindberg *et al.*, 1990; Johnson, 1992; Lovett and Lindberg, 1993; Dise and Wright, 1995; Lajtha *et al.*, 1995; Lovett *et al.*, 2000; Schmitt *et al.*, 2005).

The estimation of total nitrogen deposition from throughfall measurements can yield much higher rates than those inferred from extrapolation of deposition monitoring data. For example, in a forest in Falmouth, MA, on Cape Cod, Lajtha *et al.* (1995) measured wet deposition of 420 kg N km^{-2} yr^{-1} and estimated a total deposition rate of 840 kg N km^{-2} yr^{-1} by assuming that dry deposition equaled wet deposition. This estimate is quite similar to the deposition predicted for that location by the spatial extrapolation of Ollinger *et al.*

(1993). However, from their throughfall data, Lajtha *et al.* (1995) estimated that actual total nitrogen deposition at the site was 1,310 kg N km^{-2} yr^{-1}, or more than 50% greater. In a more recent study, Weathers *et al.* (2006) compared throughfall data with more traditional approaches for estimating nitrogen deposition in Acadia National Park in Maine and in the Great Smoky Mountains National Park in North Carolina. In both locations, they found that total nitrogen deposition rates estimated from their throughfall data were 70% greater than those estimated from NADP and CASTNet wet and dry monitoring data. These throughfall estimates lend strength to the argument that the traditional approaches for estimating total deposition—such as we used in Boyer *et al.* (2002)—yield values that are too small.

3.6 The Fate of Nitrogen Deposited onto the Landscape

Forests are the dominant land cover in the northeastern United States (Boyer *et al.*, 2002), and so much of the nitrogen deposited onto the landscape falls on forests. Only a portion of this nitrogen is exported downstream, with much retained in the forests or denitrified and converted to non-reactive, molecular N_2. Productivity of most forests in the United States is limited by the supply of nitrogen (Vitousek and Howarth, 1991), so as forests receive more nitrogen from atmospheric deposition, production and storage of nitrogen in organic matter can be expected to increase. On average for the northeastern United States, approximately 60% to 65% of the nitrogen inputs to forests through natural nitrogen fixation as well as atmospheric deposition are retained in the forest (primarily accreted in woody biomass) or harvested from the forests in wood (Goodale *et al.*, 2002; van Breemen *et al.*, 2002). A little over 20% is exported from the forest in streams (primarily as nitrate, but also dissolved organic nitrogen), with the rest denitrified (van Breemen *et al.*, 2002). The ability of forests to store nitrogen, however, is limited, and forests can become nitrogen saturated when inputs exceed the needs of trees and the ability for soils to assimilate nitrogen (Aber *et al.*, 1989; Gundersen and Bashkin, 1994; Emmett *et al.*, 1998). Nitrogen export downstream can then increase dramatically (Emmet *et al.*, 1998; Howarth *et al.*, 2002b; Aber *et al.*, 2003).

A recent comparative study suggests that for the forests of northern New England and New York State, the nitrate concentrations in streams and small lakes just downstream increase dramatically as total nitrogen deposition increases above 600 to 800 kg N km^{-2} yr^{-1} (Fig. 3.3, Aber *et al.*, 2003), indicating a substantial increase in nitrogen export from the forests receiving the higher deposition. Figure 3.3 also indicates the estimated average NOy deposition for the northeastern United States in the Boyer *et al.* (2002) and Howarth *et al.* (1996) studies. Note that total deposition, including ammonia, ammonium, and organic nitrogen, would be greater by 20 to 40% (Boyer *et al.*, 2002; Howarth *et al.*, 1996), but is also much more uncertain (Holland *et al.*, 1999; Howarth *et al.*, in press), so I have chosen to illustrate just the NO_y component. Note also that the deposition estimates used in the Aber *et al.* (2003) analysis may also be low, since these are based on extrapolation of monitoring data. On the other hand, all of the data in the analysis of Aber *et al.* (2003) are from fairly rural sites, relatively far from emission sources; their deposition estimates may therefore be fairly reliable. Regardless, Fig. 3.3 suggests that nitrogen deposition onto the landscape on average in the northeastern United States is likely high enough to result in elevated losses of nitrogen from forests, particularly if the higher emission-based estimates used by Howarth *et al.* (1996) are valid.

While forests are often retentive of nitrogen, impermeable surfaces such as roads and parking lots are far less so. While not often studied, nitrogen runoff from these surfaces can be substantial. For example, runoff from highways near Providence, RI, is reported to be 1,700 kg N km^{-2} yr^{-1} of road surface (Nixon *et al.*, 1995). Most if not all of this nitrogen likely originated from atmospheric deposition, much of it from vehicle emissions on the highway.

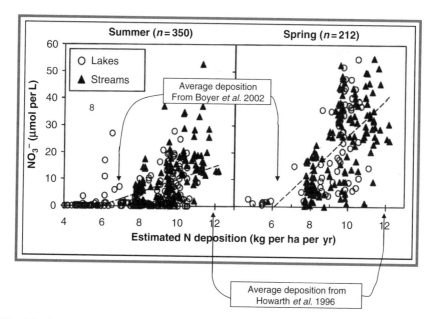

Fig. 3.3 Concentrations of nitrate in small streams and lakes in forested catchments in northern New England in the spring (right) and summer (left) as a function of NO_y deposition onto the landscape. Observe the non-linear response, with nitrate concentrations tending to increase as deposition exceeds 6–8 kg N per hectare per year (600–800 kg N km^{-2} yr^{-1}). Arrows indicate the average deposition rates for oxidized nitrogen compounds (NO_y) estimated for the northeastern United States in Boyer *et al.* (2002) and Howarth *et al.* (1996), respectively. Modified from Aber *et al.* (2003).

3.7 A Closer Look at the SPARROW model

The SPARROW model is one of the best available tools for estimating the sources of nitrogen pollution in particular watersheds (NRC, 2000). The model statistically relates water quality data from US Geological Survey monitoring programs to spatial data on nutrient sources, landscape characteristics such as temperature and soil permeability, and stream properties such as residence time (Smith *et al.*, 1997). As noted previously in this chapter, the SPARROW model has been used to suggest that atmospheric deposition contributes from 4 to 35% of the total nitrogen inputs to a variety of US estuaries (Alexander *et al.*, 2001). One limitation of the SPARROW model as used in the Alexander *et al.* (2001) paper is that it used only wet deposition monitoring data as input for atmospheric deposition as a nitrogen source. Dry deposition data were not used, probably because the sparse spatial coverage of available data would have weakened the statistical analysis too greatly. In the SPARROW approach, the wet deposition data can serve as a surrogate for total deposition, if wet and dry deposition patterns are correlated in space (Howarth *et al.*, 2002b). However, increasingly it seems that wet and dry deposition are not correlated, and dry deposition is proportionately more important in more dry climates (Holland *et al.*, 1999) and in closer proximity to emission sources (presentation by R. Dennis at the National Atmospheric Deposition Program annual Technical Committee meeting, October 2006). This is probably particularly true for nitrogen from vehicle emissions, since relatively reactive gases are released very close to land and vegetation surfaces (Cape *et al.*, 2004; presentation by R. Howarth, R. Marino, N. Bettez, E. Davidson, and T. Butler at the National Atmospheric Deposition Program annual Technical Committee meeting, October 2006). Thus, the atmospheric deposition estimates given by the SPARROW model probably are low since they do not well represent dry deposition near emission sources.

In the version of the SPARROW model used by Alexander *et al.* (2001) to determine the relative importance of various sources of nitrogen inputs to estuaries, one of the identified sources of nitrogen pollution is called "non-agricultural non-point sources." This is nitrogen that is statistically associated with urban and suburban areas, but is not well represented by other nitrogen sources, such as wet deposition as indicated in the NADP monitoring program. Some of this nitrogen may come from home fertilizer use or from general disturbance of the landscape, but I suggest that much of it— perhaps even most of it—may in fact be associated with the dry deposition of nitrogen near vehicle emission sources. If so, the true estimate of the importance of atmospheric deposition as a nitrogen source to coastal systems may be better represented by the sum of the SPARROW estimates for atmospheric deposition and for non-agricultural non-point sources. This combined estimate ranges from 26% to 76% of the total nitrogen inputs to some representative coastal marine ecosystems in the northeastern United States (Table 3.1).

Table 3.1 Estimates from the SPARROW model for the relative importance of atmospheric deposition, "non-agricultural non-point sources," and sewage wastewater as nitrogen inputs to several coastal marine ecosystems in the northeastern United States.

	Atmosphere	Non-agricultural non-point	Wastewater
Casco Bay	22	54	13
Great Bay	9	58	23
Merrimack River	28	43	20
Buzzards Bay	12	14	63
Narragansett Bay	10	19	62
Hudson River	26	21	40
Barnegat Bay	19	28	43
Delaware Bay	22	17	35
Chesapeake Bay	28	22	8

Note that the atmospheric deposition terms are estimated just from wet deposition monitoring data. Note further that the "non-agricultural non-point sources" may include a substantial amount of input from dry atmospheric deposition near emission sources in urban and suburban environments, and this would not be included in the SPARROW estimate of the atmospheric deposition input. See text for further discussion. Based on Alexander *et al.* (2001). Values are percents (%).

3.8 Chesapeake Bay Case Study

Chesapeake Bay is the largest estuary in the United States, and one of the most sensitive to nutrient inputs (Bricker *et al.*, 1999; NRC, 2000). Nitrogen inputs to the Chesapeake have caused widespread loss of seagrasses and have greatly increased the volume of anoxic bottom waters (Boesch *et al.*, 2001). The role of atmospheric deposition as a source of nitrogen to the Chesapeake apparently was not considered until Fisher and Oppenheimer (1991) suggested that it may contribute 40% of the total inputs. Their analysis was simple and preliminary, and was not believed by many scientists who worked on Chesapeake Bay water quality. The most recent analyses by the Chesapeake Bay Program, while giving lower percentages, also suggest that deposition is important, contributing ~25% of the total nitrogen inputs to Chesapeake Bay (7% from direct deposition onto surface waters, and 19% from deposition onto the landscape with subsequent export to the bay ecosystem, using 2003 values; *http://www.chesapeakebay.net/ status.cfm?SID = 126*; see also *http://www.chesapeakebay.net/nutr1.htm*).

Two lines of evidence suggest that the Chesapeake Bay Program model may be underestimating the inputs of nitrogen from atmospheric deposition: 1) the model may be underestimating the magnitude of deposition onto the landscape; and 2) the model may be underestimating the percentage of deposition onto the landscape that is subsequently exported downstream. Each of these is discussed below.

The Chesapeake Bay Program model relies on an estimate of total nitrogen deposition onto the watersheds of 1,210 kg N km^{-2} yr^{-1} (calculated from Fig. A-4 of EPA, 2003). The approach to derive this estimate is very similar to that used by Boyer *et al.* (2002): extrapolation from deposition monitoring data for the 15 NADP wet sites and 8 CASTNet and Airmon dry deposition sties in the watersheds of the Chesapeake (Lewis Linker, Bay Program modeling coordinator, PowerPoint presentation by conference call, January 9, 2006), although the Boyer *et al.* (2002) estimate is in fact somewhat lower (1,010 kg N km^{-2} yr^{-1} for the area-weighted mean for the watersheds of the Susquehanna, Potomac, Rappannnock, and James Rivers up river of the USGS gaging stations). If we assume that the Boyer *et al.* (2002) estimate underestimates by 80% (based on comparison with the global-scale emission-based model used by Howarth *et al.*, 1996), then actual deposition on the Chesapeake watersheds may be as great as 1,550 kg N km^{-2} yr^{-1} (28% greater than assumed for the Chesapeake Bay Program model). This higher estimate is broadly consistent with the preliminary model runs from the CMAQ emission-based model discussed above (R. Dennis, pers. comm.). Note also that locally derived emissions from commercial chicken houses on the Delmarva Peninsula may contribute to the atmospheric deposition load to Chesapeake Bay (Siefert *et al.*, 2004), and this source is not well considered in the Chesapeake Bay Program model.

Perhaps of greater significance is the treatment of nitrogen retention in the landscape by the Chesapeake bay model which assumes on average that 86% to 89% of total nitrogen deposition onto the landscape is retained, and only 11% to 14% is exported downstream to the bay (calculated from Figure A-4, EPA, 2003). Most of this retention is assumed to occur in the 57% of the area of the watershed that is forested, with greater export of deposition onto agricultural lands and urban and suburban areas with impermeable surfaces. The model assumes that most of the forests in the Chesapeake Bay basin are not nitrogen saturated, and therefore leak little if any nitrogen (EPA, 2003).

The average export of nitrogen deposition from all land uses (12%) seems low in comparison with the estimate that average forests in the northeastern United States export over 20% of nitrogen deposition (Goodale *et al.*, 2002; van Breemen *et al.*, 2002). If the deposition in the Chesapeake basin is evenly distributed over land uses, then 43% falls on other land uses where much higher rates of export would be expected. If much of the deposition from nitrogen pollution that originates from vehicles falls near these emission sources (either onto impermeable surfaces or onto vegetation where the rate of deposition would be very high), then very high rates of export might be expected. The preliminary runs of the CMAQ model indeed suggests high deposition—particularly for dry deposition—near heavily populated urban areas. Obtaining better data on nitrogen retention in mixed land-use watersheds has been identified as a high national research need in a multi-agency federal planning document (Howarth *et al.*, 2003). But given current knowledge, it is probably reasonable to assume that the percent export from atmospheric deposition onto the landscape of the Chesapeake Bay basin—including all land uses—is 30% as to assume the 12% used by the Chesapeake Bay model. Ranges from 20% to 40% and even higher can be reasonably inferred from studies of large watersheds (NRC, 2000; Howarth *et al.*, 2002b, in press; Boyer *et al.*, 2002).

Table 3.2 illustrates the sensitivity of nitrogen loading to Chesapeake Bay given various assumptions on the rate of deposition and on nitrogen retention in the landscape. Within this range of reasonable assumptions, the total input of nitrogen to Chesapeake Bay (both directly onto the surface waters and indirectly from deposition onto the landscape and subsequent export downstream) ranges from 34 to 92 thousand metric tons of nitrogen per year, and comprises from 25% to 50% of the total nitrogen load to Chesapeake Bay from all sources. Note that this is similar to the range of 28% to 50% determined from the SPARROW model for Chesapeake Bay (with the upper range including the "non-agricultural non-point sources; Table 3.2). Under the assumptions of greater deposition and lower retention in the landscape, the estimate for total nitrogen load to Chesapeake Bay increases substantially—from 130 to 188 thousand metric tons per year, or 45% greater total nitrogen load. Perhaps surprisingly, monitoring of the load of nitrogen to Chesapeake Bay is not adequate to constrain this total load estimate within this range of uncertainty. As with many other large coastal marine ecosystems, significant portions of the watersheds of Chesapeake Bay are not gaged because of the difficulty in gaging tidal streams and rivers (Valigura *et al.,* 2000; NRC, 2000; Howarth *et al.,* 2002b). These areas of the watershed are therefore not monitored for their nutrient inputs to the Bay. While the fluxes of nitrogen from the watersheds above gaging stations in the Chesapeake Basin are reasonably well known, the fluxes from the watershed in the more urbanized areas on the coastal plain—where nitrogen deposition may

Table 3.2 Importance of atmospheric deposition as a source of nitrogen pollution to Chesapeake Bay under various assumptions. Fluxes are thousands of metric tons of nitrogen per year. Percentage values given in parentheses are percentages of total nitrogen load. The baseline run assumptions are from EPA (2003).

	Total Load to Bay	Input to Bay from Direct Deposition onto Bay Water Surface	Input to Bay from Deposition onto Watersheds	Total Input to Bay from Deposition
Chesapeake Bay model (2000 conditions)	130	9 (7%)	25 (19%)	34 (26%)
Deposition increased to 1,550 kg N km^{-2} yr^{-1} no change in retention assumptions	140	12 (9%)	32 (23%)	44 (32%)
Chesapeake Bay model assumptions on deposition rate; assume 70% retention in landscape	168	9 (5%)	63 (38%)	72 (43%)
Deposition increased to 1,550 kg N km^{-2} yr^{-1}; assume 70% retention in landscape	188	12 (6%)	80 (43%)	92 (49%)

be much greater, and retention of nitrogen in the landscape much less—are estimated only from models and not from empirical monitoring data.

3.9 Application to Narragansett Bay

During the 1980s and early 1990s, Narragansett Bay received an average input of nitrogen of 29 g N m^{-2} yr^{-1} (when normalized over the entire surface area of the Bay; Nixon *et al.*, 1995; note that this corresponds to 29,000 kg N m^{-2} yr^{-1}; in this paper, I express loadings per area of coastal ecosystem water surface in units of g N m^{-2} yr^{-1} and deposition of nitrogen onto the terrestrial landscape in units of kg N km^{-2} yr^{-1} so as to clearly distinguish the two). This estimate includes an input of 1.3 g N m^{-2} yr^{-1} from advection of ocean waters, and the input from land and atmosphere is slightly less than 28 g N m^{-2} yr^{-1}. From the standpoint of the receiving water, this is a moderately high loading, comparable to that for Delaware Bay and the Potomac River estuary and twice that for Chesapeake Bay, but substantially less than the loading to the Hudson River estuary or to Boston Harbor during the 1980s (Nixon *et al.*, 1996; Howarth *et al.*, 2006).

The single largest input of nitrogen to Narragansett Bay is from rivers, estimated to be 17 g N m^{-2} yr^{-1} of surface area of the bay, on average (Nixon *et al.*, 1995). The second largest input of nitrogen to Narragansett Bay is the direct discharge of wastewater treatment plants (7.8 g N m^{-2} yr^{-1}, Nixon *et al.*, 1995). Other inputs are the direct deposition of nitrogen onto the surface of the bay (1.3 g N m^{-2} yr^{-1}) and runoff from urban areas adjacent to the bay (1.6 g N m^{-2} yr^{-1}; Nixon *et al.*, 1995). It is important to note that compared to most estuaries, Narragansett Bay has a low ratio of watershed area to estuarine water surface area (13.2:1; Howarth *et al.*, 2006, LOICZ web site, *http://data.ecology. su.se/mnode/index.htm*). Thus, the loading expressed per area of estuarine area is moderately high, and the flux from the landscape per area of watershed is extremely high (2,000 kg N km^{-2} yr^{-1}, considering wastewater, urban runoff, and river inputs). This is some 20-fold higher than one would expect from such a landscape absent human activity (Howarth *et al.*, 2002b). While such a high flux may not seem surprising given that much of the watershed is heavily urbanized, few other regions show such elevated fluxes. For example, human activity is estimated to have increased the nitrogen flux down the Mississippi River by only 5- to 6-fold (Howarth *et al.*, 2005) and into the Hudson River estuary adjacent to New York City by only 12-fold (Howarth *et al.*, 2006).

Even without the direct wastewater inputs, Narragansett Bay has a very high input of nitrogen from its watershed: ~1,400 kg N km^{-2} yr^{-1} (just considering river inputs and urban runoff). The sources of this nitrogen pollution in the landscape are not well known (Nixon *et al.*, 1995). How much of it might be due to atmospheric deposition onto land surfaces and subsequent export downstream to the bay? For the river inputs, we can evaluate this using the study of Boyer *et al.* (2002), which included the Blackstone River as one of 16 major rivers in the northeastern US; the Blackstone River basin comprises 28% of the entire watershed of Narragansett Bay (Nixon *et al.*, 1995). By assuming that nitrogen exports in large rivers reflect the inputs of nitrogen to their watersheds (regardless of source; Howarth *et al.*, 1996, 2002a,b), the Boyer *et al.* (2002) analysis suggests that atmospheric deposition contributes one third of the nitrogen flux in the Blackstone River basin. While agriculture contributes some to this nitrogen flux, the majority probably comes from wastewater discharges into the Blackstone. As discussed earlier, Boyer *et al.* (2002) may have underestimated the rate of nitrogen deposition. On the other hand, the mass-balance watershed approach of Boyer *et al.* (2002) may underestimate the importance of wastewater inputs in more urbanized watersheds such as the Blackstone (Howarth *et al.*, 2006).

If atmospheric deposition contributes one third of the nitrogen flux from larger rivers into Narragansett Bay, and if most of the direct runoff from urban areas adjacent to the bay originate from atmospheric deposition, then overall atmospheric deposition (directly onto the bay and onto the landscape with subsequent export to the bay) makes up 30% of the total nitrogen inputs to the bay. Note that this is very similar to the SPARROW derived estimate, if the "non-agricultural non-point source" term is indeed associated with near-source deposition of vehicle exhaust (Table 3.1). While significant, atmospheric deposition is clearly less important as a nitrogen input to Narragansett Bay than are the inputs from wastewater treatment plants (Table 3.1).

Prudent management of nitrogen inputs to Narragansett Bay clearly should focus on the wastewater inputs. On the other hand, it may also make sense to further consider the inputs from atmospheric deposition. While there is little evidence of any increase in nitrogen loading from wastewater treatment plants to Narragansett Bay over the past several decades (see Nixon *et al.*, Chapter 5, this volume), atmospheric deposition may well have increased, particularly that in the near-vicinity of vehicles. While the population of Rhode Island grew by only 11% between 1970 and 2000, vehicle miles driven in the state increased by more than 70% (RI Statewide Planning Program, 2001). Improved technology for controlling NO_x emissions from cars since the Clean Air Act Amendments of 1990 has resulted in some decrease in emissions per mile driven for cars, but overall the increase in miles driven, and an increased use of light trucks and SUVs—which are not as stringently regulated—resulted in more NOx emissions from vehicles in the eastern US during the 1990s (Butler *et al.*, 2005). Also, catalytic converters can actually increase the release of ammonia gas in car emissions due to over-reduction of NO_x (Cape *et al.*, 2004).

3.10 Managing Atmospheric Deposition in the United States

Despite the widespread damage to coastal waters from nitrogen pollution, for the most part governments have been slow to systematically apply effective policies for controlling this problem in the United States or elsewhere (NRC,

2000; Howarth *et al.*, 2005). The reasons for this policy failure are many, but one major reason is that management of eutrophication or nutrient pollution often has focused on phosphorus rather than nitrogen since the early 1970s (Howarth and Marino, 2006; Howarth *et al.*, 2005). While this is appropriate for freshwater lakes, nitrogen is the larger problem in most coastal marine ecosystems (NRC, 2000; Howarth and Marino, 2006). Although some local or regional agencies have addressed nitrogen pollution in coastal waters over the past two decades, even today no national standards for coastal nitrogen pollution exist (NRC, 2000; Howarth *et al.*, 2005). Scientific evidence for the necessity of phosphorus control on eutrophication in freshwater lakes and nitrogen control in coastal marine ecosystems has steadily accumulated for many decades, but only in the past 5–10 years has this evidence begun to be fully accepted by water quality managers. Even when managers have recognized that nitrogen is the prime cause of eutrophication in coastal rivers and bays, management practices for non-point sources of nitrogen often have remained focused on those proven effective for managing phosphorus pollution, with insufficient recognition that other practices may be needed for nitrogen because of its much greater mobility in groundwater and through the atmosphere (NRC, 2000; Howarth *et al.*, 2005; Howarth and Marino, 2006).

Both fossil fuel combustion and agricultural practices contribute significantly to atmospheric fluxes of nitrogen but not phosphorus. The magnitude of the contribution of these atmospheric fluxes to coastal nutrient pollution remains uncertain, and understudied. Nonetheless, atmospheric deposition is clearly an important contributor to coastal nutrient pollution in many areas, including Narragansett Bay. This source demands more attention by water quality managers if the goal of reducing coastal nutrient pollution is to be met (NRC, 2000).

Acknowledgments This chapter is heavily based on R. W. Howarth (2006), Atmospheric deposition and nitrogen pollution in coastal marine ecosystems, in D. Whitelaw *et al.* (editors), Acid in the Environment: Lessons Learned and Future Prospects, Springer. I gratefully acknowledge support from grants from the Woods Hole Sea Grant Program, the EPA STAR program, the Coastal Ocean Program of NOAA, the USDA-supported Agricultural Ecosystems Program at Cornell, and an endowment given to Cornell University by David R. Atkinson.

References

Aber, J.D., Nadelhoffer, K.J., Steudler, P., and Melillo, J.M. 1989. Nitrogen saturation in northern forest ecosystems. *BioScience* 39:378–386.

Aber, J.D., Goodale, C., Ollinger, S., Smith, M.L., Magill, A.H., Martin, M.E., and Stoddard, J.L. 2003. Is nitrogen deposition altering the nitrogen status of northeastern forests? *BioScience* 53:375–389.

Alexander, R.B., Smith, R.A., Schwartz, G.E., Preston, S.D., Brakebill, J.W., Srinivasan, R., and Pacheco, P.C. 2001. Atmospheric nitrogen flux from the watersheds of major estuaries of the United States: An application of the SPARROW watershed model. *In* Nitrogen Loading in

Coastal Water Bodies: An Atmospheric Perspective, Valigura, R., Alexander, R., Castro, M., Meyers, T., Paerl, H., Stacey, P., and Turner, R.E. (eds.) *American Geophysical Union Monograph* 57, pp. 119–170.

Baum, M.M., Kiyomiya, E.S., Kumar, S., Lappas, A.M., Kapinus, V.A., and Lord, H.C. 2001. Multicomponent remote sensing of vehicle exhaust by dispersive absorption spectroscopy. 2. Direct on-road ammonia measurements. *Environmental Science and Technology* 35:3735–3741.

Boesch, D. F., R. B. Brinsfeld, and R. E. Magnien. 2001. Chesapeake Bay eutrophication: Scientific understanding, ecosystem restoration, and challenges for agriculture. J. Env. Qual. 30: 303-320.

Boyer, E.W., Goodale, C.L., Jaworski, N.A., and Howarth, R.W. 2002. Effects of anthropogenic nitrogen loading on riverine nitrogen export in the northeastern US. *Biogeochemistry* 57/58:137–169.

Bricker, S.B., Clement, C.G., Pirhalla, D.E., Orland, S.P., and Farrow, D.G.G. 1999. National Estuarine Eutrophication Assessment: A Summary of Conditions, Historical Trends, and Future Outlook. National Ocean Service, National Oceanic and Atmospheric Administration, Silver Springs, MD.

Butler, T.J., Likens, G.E., Vermeylen, F.M., and Stunder, B.J.B. 2005. The impact of changing nitrogen oxide emissions on wet and dry nitrogen deposition in the northeastern USA. *Atmospheric Environment* 39:4851–4862.

Cape, J.N., Tang, Y.S., van Dijk, N., Love, L., Sutton, M.A., and Palmer, S.C.F. 2004. Concentrations of ammonia and nitrogen dioxide at roadside verges, and their contribution to nitrogen deposition. *Environmental Pollution* 132:469–478.

Clarke, J.F., Edgerton, E.S., and Martin, B.E. 1997. Dry deposition calculations for the Clean Air Status and Trends Network. *Atmospheric Environment* 31:3667–3678.

Dise, N.B., and Wright, R.F. 1995. Nitrogen leaching from European forests in relation to nitrogen deposition. *Forest Ecology Management* 71:153–161.

Driscoll, C., Whitall, D., Aber, J., Boyer, E., Castro, M., Cronan, C., Goodale, C., Groffman, P., Hopkinson, C., Lambert, K., Lawrence, G., and Ollinger, S. 2003. Nitrogen pollution in the northeastern United States: Sources, effects, and management options. *BioScience* 523:357–374.

Emmett, B.A., Boxman, D., Bredemeier, M., Gundersen, P., Kjønaas, O.J., Moldan, F., Schleppi, P., Tietema, A., and Wright, R.F. 1998. Predicting the effects of atmospheric deposition in conifer stands: evidence from the NITREX ecosystem-scale experiments. *Ecosystems* 1:352–360.

Fisher, H.B., and Oppenheimer, M. 1991. Atmospheric nitrate deposition and the Chesapeake Bay estuary. *Ambio* 20:102.

Galloway, J.N., Dentener, F.J., Capone, D.G., Boyer, E.W., Howarth, R.W., Seitzinger, S.P., Asner, G.P., Cleveland, C., Green, P.A., Holland, E., Karl, D.M., Michaels, A., Porter, J.H., Townsend, A., and Vorosmarty, C. 2004. Nitrogen cycles: past, present, and future. *Biogeochemistry* 70:153–226.

Goodale, C.L., Lajtha, K., Nadelhoffer, K.J., Boyer, E.W., and Jaworski, N.A. 2002. Forest nitrogen sinks in large eastern U.S. watersheds: estimates from forest inventory and an ecosystem model. *Biogeochemistry* 57/58:239–266.

Grimm, J.W., and Lynch, J.A. 2005. Improved daily precipitation nitrate and ammonium concentration models for the Chesapeake Bay watershed. *Environmental Pollution* 135:445–455.

Gundersen, P., and Bashkin, V. 1994. Nitrogen cycling. *In* Biogeochemistry of Small Catchments: A Tool for Environmental Research, pp. 255–283. Moldan, N., and Cerny, J. (eds.), Chichester, UK: Wiley.

Holland, E., Dentener, F., Braswell, B., and Sulzman, J. 1999. Contemporary and pre-industrial global reactive nitrogen budgets. *Biogeochemistry* 4:7–43.

Holland, E.A., Braswell, B.H., Sulzman, J., and Lamarque, J. 2005. Nitrogen deposition on to the United States and Western Europe: Synthesis of observations and models. *Ecological Applications* 15:38–57.

Howarth, R.W. 1998. An assessment of human influences on inputs of nitrogen to the estuaries and continental shelves of the North Atlantic Ocean. *Nutrient Cycling in Agroecosystems* 52:213–223.

Howarth, R.W. 2006. Atmospheric deposition and nitrogen pollution in coastal marine ecosystems. *In* Acid in the Environment: Lessons Learned and Future Prospects, pp. 97–116. Visgilio, G.R., and Whitelaw, D. (eds.), New York: Springer.

Howarth, R.W., Billen, G., Swaney, D., Townsend, A., Jaworski, N., Lajtha, K., Downing, J.A., Elmgren, R., Caraco, N., Jordan, T., Berendse, F., Freney, J., Kueyarov, V., Murdoch, P., and Zhao-Liang, Z. 1996. Riverine inputs of nitrogen to the North Atlantic Ocean: Fluxes and human influences. *Biogeochemistry* 35:75–139.

Howarth, R.W., Boyer, E.W., Pabich, W.J., and Galloway, J.N. 2002a. Nitrogen use in the United States from 1961–2000 and potential future trends. *Ambio* 31:88–96.

Howarth, R., Walker, D., and Sharpley, A. 2002b. Sources of nitrogen pollution to coastal waters of the United States. *Estuaries* 25:656–676.

Howarth, R.W., and Rielinger, D.M. 2003. Nitrogen from the atmosphere: Understanding and reducing a major cause of degradation in our coastal waters. Science and Policy Bulletin #8, Waquoit Bay National Estuarine Research Reserve, NOAA, Waquoit, MA.

Howarth, R.W., Ramakrishna, K., Choi, E., Elmgren, R., Martinelli, L., Mendoza, A., Moomaw, W., Palm, C., Boy, R., Scholes, M., and Zhao-Liang, Z. 2005. Chapter 9: Nutrient Management, Responses Assessment. In Ecosystems and Human Well-being, Volume 3, Policy Responses, the Millennium Ecosystem Assessment. pp. 295–311, Washington, DC: Island Press.

Howarth, R.W., and Marino, R. 2006. Nitrogen as the limiting nutrient for eutrophication in coastal marine ecosystems: Evolving views over 3 decades. *Limnology and Oceanography* 51:364–376.

Howarth, R.W., Marino, R., Swaney, D.P., and Boyer, E.W. 2006. Wastewater and watershed influences on primary productivity and oxygen dynamics in the lower Hudson River estuary. *In* The Hudson River Estuary, Levinton, J.S., and Waldman, J.R. (eds.), pp. 121–139, UK: Cambridge University Press.

Howarth, R.W., Boyer, E.W., Marino, R., Swaney, D., Jaworski, N., and Goodale, C. in press. The influence of climate on average nitrogen export from large watersheds in the northeastern United States. *Biogeochemistry*.

Jaworksi, N.A., Howarth, R.W., and Hetling, L.J. 1997. Atmospheric deposition of nitrogen oxides onto the landscape contributes to coastal eutrophication in the northeast US. *Environmental Science and Technology* 31:1995–2004.

Johnson, D.W. 1992. Nitrogen retention in forest soils. *Journal of Environmental Quality* 21:1–12.

Lajtha, K., Seely, B., and Valiela, I. 1995. Retention and leaching of atmospherically-derived nitrogen in the aggrading coastal watershed of Waquoit Bay. *Biogeochemistry* 28:33–54.

Lindberg, S.E., Bredemeier, M., Schaefer, D.A., and Qi, L. 1990. Atmospheric concentrations and deposition of nitrogen compounds and major ions during the growing season in conifer forests in the United States and West Germany. *Atmospheric Environment* 24A:2207–2220.

Lovett, G., and Lindberg, S.E. 1993. Atmospheric deposition and canopy interactions of nitrogen in forests. *Canadian Journal of Forestry Research* 23:1603–1616.

Lovett, G.M., Traynor, M.M., Pouyal, R.V., Carreiro, M.M., Zhu, W.X., and Baxter, J.W. 2000. Atmospheric deposition to oak forests along an urban-rural gradient. *Environmental Science and Technology* 34:4294–4300.

Nixon, S.W., Granger, S.L., and Nowicki, B.L. 1995. An assessment of the annual mass balance of carbon, nitrogen, and phosphorus in Narragansett Bay. *Biogeochemistry* 31:15–61.

Nixon, S.W., Ammerman, J.W., Atkinson, L.P., Berounsky, V.M., Billen, G., Boicourt, W.C., Boynton, W.R., Church, T.M., DiToror, D.M., Elmgren, R., Garber, J.H., Giblin, A.E.,

Jahnke, R.A., Owens, J.P., Pilson, M.E.Q., and Seitzinger, S.P. 1996. The fate of nitrogen and phosphorus at the land-sea margin of the North Atlantic Ocean. *Biogeochemistry* 35:141–180.

NRC. 2000. Clean Coastal Waters: Understanding and Reducing the Effects of Nutrient Pollution. National Academies Press, Washington, DC. 405 pp.

Ollinger, S.V., Aber, J.D., Lovett, G.M., Millham, S.E., Lathrop, R.G., and Ellis, J.M. 1993. A spatial model of atmospheric deposition for the northeastern US. *Ecological Applications* 3:459–472.

Paerl, H.W. 1997. Coastal eutrophication and harmful algal blooms: Importance of atmospheric deposition and groundwater as "new" nitrogen and other nutrient sources. *Limnology and Oceanography* 42:1154–1165.

Paerl, H.W., and Whitall, R. 1999. Anthropogenically derived atmospheric nitrogen deposition, marine eutrophication and harmful algal bloom expansion: Is there a link? *Ambio* 28:307–311.

Prospero, J. M., K. Barrett, T. Church, F. Dentener, R. A. Duce, J. N. Galloway, H. Levy, J. Moody, and P. Quinn. 1996. Atmospheric deposition of nutrient to the North Atlantic basin. *Biogeochemistry* 35: 27-76.

Rhode Island Statewide Planning Program. 2001. Part 611–3: Background and trends. 3.1. Historical growth of the transportation system. Transportation Plan 2020, 2001 Update State Guide Plan Element 611, Ground Transportation Plan. *(http://www.planning.ri. gov/humanservices/gtp/pdf/611-3.pdf)*

Schmitt, M., Thoni, L., Waldner, P., and Thimonier, A. 2005. Total deposition of nitrogen on Swiss long-term forested ecosystem research (LWF) plots: comparison of the throughfall and the inferential method. *Atmospheric Environment* 39:1079–1091.

Siefert, R. L., J. R. Scudlkark, A. G. Potter, A. Simonsen, and K. B. Savide. 2004. Characterization of atmospheric ammonia emissions from commercial chicken houses on the Delmarva Peninsula. *Environmental Science and Technology* 38: 2769-2778.

Smith, R.A., Schwartz, G.E., and Alexander, R.B. 1997. Regional interpretation of water quality monitoring data. *Water Resources Research* 33:2781–2798.

US Environmental Protection Agency. 2003. Appendix A, Development of Level-of-Effort Scenarios, Technical Support Document for Identifying Chesapeake Bay Designated Uses and Attainability. *(http://www.chesapeakebay.net/uaasupport.htm)*

Valigure, R.A., Alexander, R.B., Catro, M.S., Meyers, T.P., Paerl, H.W., Stacey, P.E., and Turner, R.E. (eds.). 2000. Nitrogen Loading in Coastal Water Bodies. An Atmospheric Perspective. Coastal and Estuaries Series, No. 57. American Geophysical Union, Washington, DC. 252 pp.

van Breemen, N., Boyer, E.W., Goodale, C.L., Jaworski, N.A., Paustian, K., Seitzinger, S., Lajtha, K., Mayer, B., van Dam, D., Howarth, R.W., Nadelhoffer, K.J., Eve, M., and Billen, G. 2002. Where did all the nitrogen go? Fate of nitrogen inputs to large watersheds in the northeastern USA. *Biogeochemistry* 57/58:267–293.

Vitousek, P.M., and Howarth, R.W. 1991. Nitrogen limitation on land and in the sea. How can it occur? *Biogeochemistry* 13:87–115.

Weathers, K., Simkin, S., Lovett, G., and Lindberg, S. 2006. Empirical modeling of atmospheric deposition in mountainous landscapes. *Ecological Applications* 16:1590–1607.

Wesely, M.L., and Hicks, B.B. 1999. A review of the current status of knowledge on dry deposition. *Atmospheric Environment* 34:2261–2282.

Zarbock, H.W., Janicki, A.J., and Janicki, S.S. 1996. Estimates of total nitrogen, total phosphorus, and total suspended solids to Tampa Bay, Florida. Tampa Bay National Estuary Program Technical Publication #19-96. St. Petersburg, Florida.

Chapter 4
Groundwater Nitrogen Transport and Input along the Narragansett Bay Coastal Margin

Barbara L. Nowicki and Arthur J. Gold

4.1 Overview

The transport of dissolved nitrogen (N) in groundwater has been historically difficult to monitor, model, and predict. Quantitative assessments of its significance to Narragansett Bay and its sub-estuaries suffer from a paucity of information and a lack of direct studies. Two factors exerting the greatest impact on groundwater N delivery to Narragansett Bay are the characteristic surficial geology of the area, and the presence of densely populated un-sewered development in the coastal zone. Local differences in soils, geology, and hydrology are critical in determining the degree of N migration from nonpoint sources to Narragansett Bay. Ultimately, the transport of N via groundwater to Narragansett Bay is governed by the interaction of groundwater flow paths with local soils, geology, and land use.

Rough mass balance calculations suggest that groundwater makes up less than 10% of the direct freshwater input to Narragansett Bay. Nevertheless, localized groundwater seepage to numerous coves and embayments lining the Narragansett Bay shoreline has had a significant impact on these smaller near-shore ecosystems. In areas with individual sewage disposal systems (ISDS or septic systems), groundwater flows provide a direct connection between human waste disposal and nearby marshes, rivers and sub-estuaries. Deteriorating habitat and water quality observed in many of the smaller coves and embayments of Narragansett Bay over the past 20 years have influenced the public's perception of the general health of the coastal waters of "the Ocean State," and have cast a pall over otherwise significant improvements in wastewater and coastal zone management for the bay as a whole.

In this chapter, the interplay of local soils, coastal geomorphology, and land use are described, which in concert with the unique biogeochemistry of nitrogen, act to mediate groundwater N transformation and transport in the coastal

Barbara L. Nowicki
University of Rhode Island School of Education, Chafee Hall, Room 242, Kingston, RI
02881
bnowicki@uri.edu

A. Desbonnet, B. A. Costa-Pierce (eds.), *Science for Ecosystem-based Management.*
© Springer 2008

zone. Studies from other New England coastal areas are drawn upon, in particular, Buzzards Bay and Cape Cod, MA, augmented with insights obtained from work along the Southeastern Atlantic and Gulf coasts. As Rhode Island communities struggle with existing water quality issues, and as the pressure to develop even the most marginal of coastal properties increases, an informed understanding of groundwater N transport and attenuation at the marine coastal margin will be critical to successful environmental management in the coastal zone.

4.2 Groundwater N Inputs—Significant But Difficult to Quantify

While most regional and global estimates of fresh groundwater discharge to the coastal ocean are small (approx. 6% of total; Burnett *et al.*, 2003), groundwater discharge at the land margin is frequently a significant contributor of bacteria, total dissolved nitrogen (TDN), and dissolved organic carbon (DOC) to coastal waters. Fresh groundwater discharge represents a significant fraction of the total water budget for many of the smaller embayments and coastal lagoons characteristic of New England's coastline. For example, groundwater discharge makes up 89% of the water budget for Waquoit Bay, Cape Cod, MA (Cambareri and Eichner, 1998), 60–70% of the freshwater entering Greenwich Bay, RI (Urish and Gomez, 2004), and 10–20% of the total freshwater inflow to Great South Bay, NY (Bokuniewicz, 1980).

Inorganic N concentrations in groundwater from developed coastal areas are frequently orders of magnitude higher than the concentrations of adjoining surface waters (Johannes, 1980; Capone and Bautista, 1985; Valiela *et al.*, 1990; Weiskel and Howes, 1991; Portnoy *et al.*, 1998; Krest *et al.*, 2000; Rapaglia, 2005). Of particular concern is the transport of nitrate-N (NO_3), a readily soluble anion that can travel long distances in groundwater aquifers with little attenuation or removal (Weiskel and Howes, 1991). NO_3 is a common contaminant of surface and groundwaters, arising from natural and agricultural organic matter decomposition, agricultural and residential fertilizer runoff or leaching, atmospheric deposition, and from ISDS discharge, and is of major concern as a drinking water contaminant and as a cause of eutrophication in marine waters (Gold *et al.*, 1990; Valiela *et al.*, 1990; Vitousek *et al.*, 1997; Winter *et al.*, 1998).

The US Geological Survey (USGS) National Water-Quality Assessment Program (NAWQA) has monitored water quality in more than 50 major river basin and aquifer systems, covering about one-half of the land area of the contiguous United States (USGS, 1999). Results of this assessment showed that NO_3 contamination was most prevalent in shallow groundwater (less than 33 m below land surface) beneath agricultural and urban areas. Urban groundwater can become contaminated with nitrogen-rich wastewater from leaky sewer systems. Extensive watershed measurements in the Baltimore Long

Term Ecological Research Project, (Groffman *et al.*, 2004) found that broken sewer lines and disconnected networks can contribute substantial N loading to urban watersheds, typically estimated at 10–15% of total wastewater N loading to standard municipal systems. However, wastewater inputs to groundwater are generally thought to pose a substantial problem only in settings where the water table is lower than the cracked or broken wastewater pipes. When the pipes are below the water table and surrounded by groundwater, leaky pipes tend to serve as conduits transmitting both wastewater and groundwater to treatment or surface discharge locations (Amick and Burgess, 2003). The extent to which sewer line breaks and leaks contribute to groundwater N contamination in the aging infrastructure of the communities at the north end of Narragansett Bay is currently unknown.

Current contamination of shallow groundwater may serve as a harbinger of future contamination of the deeper aquifers commonly used for public drinking water supply (USGS, 1999). In areas where groundwater recharge and flow rates are slow, the groundwater arriving at the coast now may have been formed, and subject to anthropogenic N inputs, decades earlier (Bohlke and Denver, 1995). Thus, current human impacts on underlying groundwater may not be felt for decades, and may last for prolonged periods after source controls are implemented.

In coastal areas of the Northeast, burgeoning population growth and a reliance on septic systems have resulted in dramatic increases in groundwater NO_3 concentrations, with a strong correlation between housing density and groundwater NO_3 contamination (Persky, 1986; Giblin and Gaines, 1990; Weiskel and Howes, 1991; Winter *et al.*, 1998). N budgets for the shallow embayments and salt ponds characteristic of the unsewered coastlines of Massachusetts, southern Rhode Island, Connecticut, and Long Island suggest that groundwater discharge can supply more than 80% of the N inputs to these systems (Bokuniewicz, 1980; Capone and Bautista, 1985; Lee and Olsen, 1985; Valiela *et al.*, 1990; Gobler and Boneillo, 2003). The discharge of N-contaminated groundwater not only alters rates of primary production, but also causes shifts in the types of primary producers that dominate coastal marine systems (Valiela *et al.*, 1990; Gobler and Boneillo, 2003). The resulting phytoplankton and macro-algal blooms cause increased organic loading to bottom sediments, hypoxia, changes in benthic diversity, and losses of valuable sea grass and shellfish habitat (Lee and Olsen, 1985; Valiela *et al.*, 1990;1992; Rutkowski *et al.*, 1999). In fact, sea grass decline has been statistically correlated with the density of unsewered homes in estuarine watersheds (Valiela *et al.*, 1992; Short and Burdick, 1996).

Freshwater discharge to estuaries is rarely distributed uniformly. In many instances, the discharge occurs in areas far removed from the flushing effects of tidal exchange, creating areas of low salinity and high nutrients (Millham and Howes, 1994). Short-term variations in freshwater discharge due to storms, and seasonal variations due to summertime evapotranspiration and dry periods, can cause significant shifts in estuarine salinity and nutrient dynamics resulting in localized blooms and anoxia (Laroche *et al.*, 1997).

Unfortunately, the transport of N to estuaries via groundwater has been extremely difficult to quantify and predict. Valiela *et al.* (1997) presented an excellent overview of this topic based on field data for the Waquoit Bay, MA estuary, and an extensive literature review. Groundwater N originates from diffuse sources that vary in loading rates (mass per area per time) and timing. Major diffuse, or nonpoint inputs within the Narragansett Bay watershed include: atmospheric deposition; fertilized croplands and lawns; animal waste from horses, diary farms, pets and hobby farms; leaky municipal sewer systems, and ISDS. In addition, although NO_3 tends to be the predominant form of N observed in Rhode Island's aquifers, in some settings DON or ammonium can dominate N loads transported to estuaries as well. Because N is subject to a host of biological processes, N loading to groundwater is not a fixed fraction of the input. Instead, the delivery of N to groundwater results from the combination of surface soil characteristics (pore size distribution, permeability, and organic matter content), hydrology, microbial activity, and plant properties. Accurate estimates of groundwater N delivery to the estuarine margin have been elusive, and those that exist appear to vary widely both within and between coastal systems (Table 4.1).

Our understanding of groundwater N transport to estuarine waters has been hindered by the extreme spatial and temporal variability in groundwater fluxes, and by the time and expense required to adequately quantify groundwater flows and pathways at the coastal margin. Comparisons between systems are difficult because there has been no consistent unit of measurement used in reporting groundwater-derived N fluxes, which have been reported per liter of flow, per meter of shoreline, per kilogram of soil, and per square meter or cubic meter of estuarine surface area or volume (Table 4.1). Estimates of fresh groundwater flux have been calculated using a variety of techniques including local measurements of Darcian flow (hydraulic gradient and hydraulic conductivity), salt balances, water budgets, and direct measurements of freshwater flux to individual seepage meters. As each of these methods tends to reflect slightly different time and space scales, estimates of groundwater flux have varied by orders of magnitude even for a single site (Giblin and Gaines, 1990; Table 4.1). Millham and Howes (1994) compared five methods for quantifying fresh water inflow to Little Pond, a shallow coastal embayment on the southwest shore of Cape Cod, MA. They concluded that the most precise estimates of groundwater discharge came from a chloride balance, while a Darcian streamtube approach (based on accurate water table maps and measures of soil hydraulic conductivity) provided the best understanding of the rates and patterns of groundwater flow.

More recently, studies of natural radium isotope enrichments have provided broad-scale integrated measures of groundwater flux to coastal systems on regional scales (Cable *et al.*, 1996; Moore, 1996; Charette *et al.*, 2001; Schwartz, 2003). In particular, ^{222}Rn has been shown to be naturally enriched in groundwater, and is often found in concentrations that are 3 to 4 orders of

Table 4.1 A comparison of groundwater discharge rates for East Coast estuaries and the various methods and units that have been used for measurement.

Site	Soil hydraulic conductivity (md^{-1})	Method of estimation	Groundwater discharge rate	Groundwater-derived N flux	Reference
Nauset Marsh Estuary, Cape Cod, MA (Town Cove)		Seepage chambers, Water budget, Salt balance	$24-72$ L m^{-2} d^{-1}, $4.3 \times 10^4 m^3 d^{-1}$, $0.79-8.9 \times 10^4 m^3 d^{-1}$	35 kg N d^{-1}	Giblin and Gaines (1990)
Nauset Marsh Estuary, Cape Cod, MA		Seepage chambers		$1-3$ mmol NO_3-N m^{-2} h^{-1}	Portnoy et al. (1998)
Buttermilk Bay, MA	29	Water use model, water balance head gradient, hydraulic conduct.	64 $m^3 m^{-2} y^{-1}$ (0.18 $m^3 m^{-2} d^{-1}$)	130 ± 12 mol N $m^{-1} y^{-1}$	Weiskel and Howes (1991)
Little Pond, Falmouth, Cape Cod, MA		Chloride balance, inlet blockage, water budget, water table maps, hydraulic conduct.	$4,800 (\pm 670) m^3 d^{-1}$		Millham and Howes (1994)
Namskaket Marsh, MA, creek bottom		Water mass balance	288 L m^{-2} d^{-1}		Howes et al. (1996)
Waquoit Bay, MA		Radium isotopes	37,000 m^3 d^{-1}	2,100 mol N d^{-1} (DIN)	Charette et al. (2001)
Waquoit Bay, MA		Head gradient, hydraulic conduct.	6,500 m^3 d^{-1}	6.2 kg N d^{-1} (DIN)	Talbot et al. (2003)
Great Sippewissett Marsh, West Falmouth, MA		Radium isotopes	3,900 m^3 d^{-1}		Charette et al. (2003)
Narragansett Bay, RI fringing marsh	0.12 (transition) 0.13 (high marsh) 0.09 (low marsh)	Mini-piezometers, Hydraulic gradient	186 L m^{-1} d^{-1} 372 m^3 d^{-1} (for 2 km shoreline)		Addy et al. (2005)

Table 4.1 (continued)

Site	Soil hydraulic conductivity (m d^{-1})	Method of estimation	Groundwater discharge rate	Groundwater-derived N flux	Reference
Pettaquamscutt Estuary, RI		Radium isotopes	6.4–20 L m^{-2} d^{-1} (summer) 2.1–6.9 L m^{-2} d^{-1} (winter)	61–180 mmol m^{-2} yr^{-1} (DIN)	Kelly and Moran (2002)
Rhode Island Salt Ponds		^{226}Ra	0.1–0.3 cm^3 cm^{-2} d^{-1}		Scott and Moran (2001)
Great South Bay, Long Island, New York		Seepage meters	40 L m^{-2} d^{-1} (within 30 m of shoreline)		Bokuniewicz (1980)
Great South Bay, Long Island, barrier beach		Seepage meters	5–68 L m^{-2} d^{-1}		Bokuniewicz and Pavlik (1990)
Peconic Bay, Long Island, New York		Dye-dilution seepage meters	2–37 cm d^{-1} (10 m seaward of mean tide)		Sholkovitz et al. (2003)
Delaware Estuary		Radium isotopes	14–29 m^3 s^{-1}		Schwartz 2003
Delaware Estuary sandy beaches	75–300 upper beach 30–60 lower beach	Beach, water table topography, Piezometeric sampling, GPR	0.7–3.6 m^3 m^{-1} d^{-1}	0.3–1.6 mol m^{-1} d^{-1} (DIN)	Ullman et al. (2003)
Chesapeake Bay York, James River salt marshes		Hydraulic gradient, hydraulic cond.	5.7–10.4 L m^{-1} d^{-1}		Harvey and Odum (1990)
Chesapeake Ba	0.035–0.43 (marsh)	Darcy's Law	−8.0–80 L m^{-2} d^{-1}		Tobias et al. (2001)
York River Estuary	0.16–1.0 (sand shell layers)	Salt/water balance	0.6–22.6 L m^{-2} d^{-1}		Tobias et al.
Ringfield Marsh		K-Br tracer	16.6 ± 5 L m^{-2} d^{-1}		(2001b)

Table 4.1 (continued)

Site	Soil hydraulic conductivity (md^{-1})	Method of estimation	Groundwater discharge rate	Groundwater-derived N flux	Reference
Chesapeake Bay Wye River Estuary	50–75	Hydraulic gradient and piezometer sampling	0.06–0.5 m^3 m^{-1} d^{-1}	1.1–6.5 g N m^{-1} d^{-1} 1.17 kg N m^{-1} y^{-1} (1993) 1.25 kg N m^{-1} y^{-1} (1994)	Staver and Brinsfield (1996)
Chesapeake Bay (upper limit estimate)		Radium isotopes	200 m^3 s^{-1}		Hussain et al. (1999)
Florida, Florida Keys		Seepage meters Mini-piezometers	8.9 L m^{-2} d^{-1} (<27 m depth) 5.4 L m^{-2} d^{-1} (27–39 m depth)		Simmons (1992)
Florida Bay, Florida Keys		Seepage meters	1–3 cm d^{-1} 13.4–21.2 ml m^{-2}min^{-1} (19-30 L m^{-2}d^{-1})	110 $\pm$ 19 mmol N m^{-2}y^{-1}	Corbett et al. (1999)
Florida Bay, Florida Keys		^{222}Rn, CH$_4$	1.7 $\pm$ 0.25 cm d^{-1}		Corbett et al. (2000)
Apalachicola Bay, NE Gulf of Mexico	36 (3–180)	Conservative tracers, piezometers, water balance	3–9 $\times$ 10^6 m^3 y^{-1}		Corbett et al. (2000)
NE Gulf of Mexico (within 200 m of shore)	0.0015–0.023 (surface aquifer); 8.4 (confined aquifer)	Intercomparison Radon (^{222}Rn) Radium isotopes Seepage Meters	1.1–2.5 m^3 min^{-1}		Lambert and Burnett (2003) Moore (2003) Taniguchi et al. (2003)

GPR – ground-penetrating radar.

magnitude higher in fresh groundwater than in coastal seawater. Using ^{222}Rn, Hussain *et al.* (1999) estimated that the groundwater flux to the Chesapeake Bay may be as much as 10% of the riverine freshwater flux, with a significant portion of the groundwater discharge originating in the Potomac River and other tributaries of the Chesapeake. Unfortunately, many of the newer radium-based estimates still must rely on sometimes poorly quantified estuarine water mass balances and residence times in calculating net groundwater flux (Scott and Moran, 2001). An intercomparison study (Lambert and Burnett, 2003) of methods for measuring submarine groundwater discharge (SGD), including Lee-type manual seepage meters (Lee, 1977), automated seepage meters (Taniguchi *et al.*, 2003), radium isotopes (Moore, 2003), and continuous radon measurements (Lambert and Burnett, 2003), has shown that, when variability in time and space can be carefully controlled (the study was confined to a 5-day period within a 100 m × 200 m area), groundwater discharge estimates are surprisingly consistent between methods. However, the complex geomorphology characteristic of most coastlines, in concert with the dramatic changes to the coastal interface caused by human development, suggests that fine-scale sampling is still required to adequately quantify the processes controlling N transformation and transport in coastal aquifers (Smith *et al.*, 1991; Tobias *et al.*, 2001a,b). Accurately "scaling up" localized groundwater N flux measurements from small-scale individual shoreline estimates to whole estuary estimates, integrated over broader time and space scales, remains a significant problem.

4.3 Local Geomorphology Mediates N Transport

The glacial geology, soils, and pattern of urbanization in the islands and fringing lands of Narragansett Bay have a significant impact on the pathways and removal mechanisms that affect groundwater N delivery to the coastal zone. The lands that border the bay are primarily composed of two distinct types of surficial deposits (Fig. 4.1). Thick stratified deposits of coarse grained sands and gravels (aka "outwash") generally underlie the topsoil along the western sides of Narragansett Bay, while unstratified and unsorted glacial till is found on bay islands and the eastern shore. These deposits represent two extremes of groundwater behavior. The outwash areas have very friable soils (1–2 m deep) above highly permeable, complex arrays of coarse sand and gravel layers that range in thickness from 5 to 35 m (Fig. 4.2). These layers were deposited and sorted by meltwater from retreating glaciers during the late Pleistocene Epoch (see Boothroyd and August, Chapter 1, for a more complete description of the geology of Narragansett Bay). Much of the coastal area of the Northeast, extending from Cape Cod, MA, through the south shores of Rhode Island and Connecticut to Long Island, NY are dominated by glacial outwash, where surficial sediments are characterized by very high permeability. Outwash deposits form the most productive aquifers in New England and specifically, within RI, yield from 100

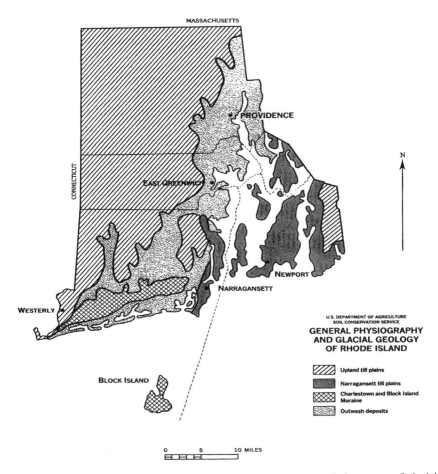

Fig. 4.1 The shores of Narragansett Bay are characterized by two distinct types of glacial geology.

to 700 gal min^{-1} to public water supply wells (Trench, 1991; 1995; Desimone and Ostiguy, 1999). In contrast, areas with glacial till deposits are characterized by a relatively thin layer of friable silt loam overlying a dense, firm layer of compacted material that greatly restricts groundwater flow (Fig. 4.2).

NO$_3$ can move within outwash aquifers for considerable distances (e.g., hundreds of meters), over long time periods (months or years) and can be subject to a variety of transformation processes within the aquifer, at the land–water interface, and in the subterranean estuary. Outwash aquifers are recharged directly from precipitation and also from inflow from adjacent till uplands and from adjacent surface waters. These aquifers are in close hydraulic connection with surface-water systems, and the pumping of wells located nearby stream-aquifer or pond-aquifer boundaries can induce recharge to the aquifer from the stream or pond (Desimone and Ostiguy, 1999).

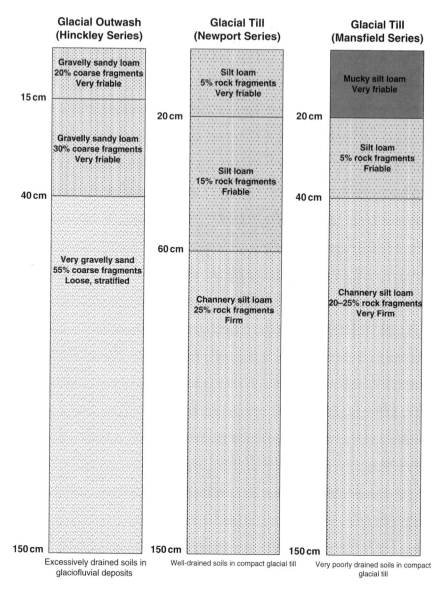

Fig. 4.2 Geologic section for outwash vs till.

Considerable research and monitoring has focused on the fate of groundwater NO_3 in outwash. In these types of aquifers, the patterns of the different layers reflect differences in hydraulic conductivity, and those patterns exert enormous influence on groundwater flow paths. Hydraulic conductivity is a measure of the permeability of the media, essentially the ease with which water can move through the soil. Darcy's Law predicts ground water movement by coupling hydraulic conductivity with hydraulic gradient, which is based on the change in potential energy with distance (a characteristic that equates to slope). Generally speaking (with apologies to groundwater hydrologists), for a given hydraulic gradient, water will move more rapidly through media with greater hydraulic conductivities. Hydraulic conductivity varies greatly between different types of media (Table 4.2). In glacial outwash, the horizontal permeability may be 2 to 20 times greater than vertical (Fetter, 2001). In addition, it is not uncommon to find two to 10-fold differences in horizontal hydraulic conductivity between adjacent layers of outwash media, with low permeability layers conducting little of the groundwater flow.

The results of aquifer tests in the Chipuxet River Basin in southern Rhode Island reflect a generalized hydrogeology that is characteristic of most of the western edges of Narragansett Bay. The Chipuxet River groundwater aquifer resides in stratified drift deposits consisting of "complexly interbedded lenses of sand and gravel and subordinate amounts of silt and silty sand deposited by glacial streams in the Pleistocene Epoch" (Dickerman, 1984; Fig. 4.3). The interbedding of coarser and finer grained materials causes the hydraulic conductivity of the aquifer to be higher for water moving horizontally through the ground and lower for infiltration in the vertical direction. The extreme complexity of the mixed deposits (Fig. 4.3) makes it very difficult to identify the pattern of highly permeable layers, confounding efforts to predict or track NO_3-enriched plumes that follow the twists, dips, and rise of the more permeable layers at a local scale. As groundwater flow paths approach the coastal interface, they tend to be compressed and uplifted (Fig. 4.4a), so that much of the groundwater flow occurs in a narrow margin along the shoreline (Bokuniewicz and Pavlik, 1990; Portnoy et al., 1998).

Rosenshein et al. (1968) used aquifer and specific-capacity tests at 31 wells in the Hunt-Annaquatucket-Pettaquamscutt stream-aquifer system in Rhode Island to estimate the transmissivity (horizontal hydraulic conductivity x saturated thickness) of the soils. They found a wide range of horizontal hydraulic

Table 4.2 Hydraulic conductivity can vary over several orders of magnitude between different types of media.

Material	Hydraulic conductivity (m d^{-1})
Gravel	140
Sand and gravel	60
Very coarse sand	50
Coarse sand	40
Medium sand	30
Fine sand	15
Very fine sand	5
Silt	1
Clay, till	0.03

Note: The values presented here are modified from Dickerman (1984) for the Chipuxet River aquifer in Rhode Island.

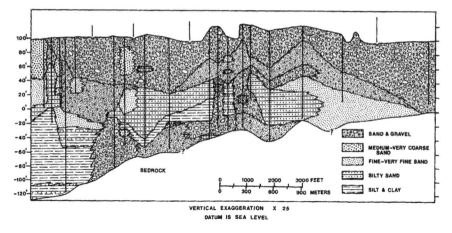

Fig. 4.3 A longitudinal geologic section from the Chipuxet River valley shows the heterogeneity of the complexly interbedded aquifer system (modified from Dickerman, 1984).

conductivities (K_h) ranging from 17 m d^{-1} for fine sand to 157 m d^{-1} for gravel. The largest value of K_h (227 m d^{-1}) was observed in wells adjacent to Secret Lake in North Kingston (Barlow and Dickerman, 2001).

In contrast to the deep aquifers and extended groundwater flow paths of outwash areas, dense till settings are not a major source of drinking water in

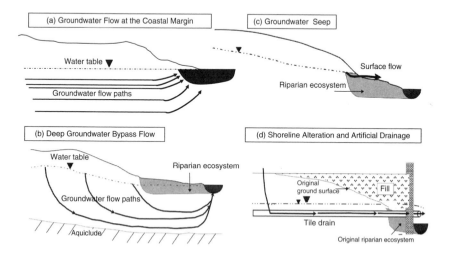

Fig. 4.4 As groundwater flowpaths approach the coastal zone they tend to be compressed and uplifted so that much of the groundwater flow occurs in a narrow margin along the shoreline. Shoreline alterations such as filling, diking, and the installation of tiles and drainfields can cause groundwater flow paths to bypass critical zones of active denitrification and N removal.

Rhode Island due to their low permeability (Trench, 1991; 1995; Desimone and Ostiguy, 1999). The till also restricts groundwater recharge. Surface soils are underlain by a very dense layer with extremely low permeability (e.g., less than 0.1 m d^{-1}) that occurs within 1–2 m below the ground surface (Fig. 4.2). This dense layer acts as a functional aquiclude that prevents substantial downward groundwater movement. Most recharge moves laterally above the dense layer until it seeps out at the surface or is intercepted by storm drains. Groundwater moving through glacial till deposits does not usually encounter subsurface zones with elevated N transformation rates. Rather, it emerges at points where the slope becomes flatter and forms surface flowages (Winter et al., 1998; Fig. 4.4c). Surface seepage often occurs at mineral faces and does not travel as groundwater flow through organic deposits. Warwick and Hill (1988) used field enrichment experiments to examine nitrate removal in surface seeps that flowed across a permanently saturated 20–100 m riparian area with organic soils. They found negligible nitrate depletion and ascribed their results to rapid transit (<1 h) across the riparian zone. In contrast, a host of studies in Rhode Island stream and coastal riparian sites have found groundwater retention times within a 10-m outwash zone to exceed 20 days, or 3–4 orders of magnitude greater than settings with surface seeps (Simmons et al., 1992; Nelson et al., 1995; Addy et al., 2005). In a survey of Rhode Island riparian wetland sites, Rosenblatt et al. (2001) found groundwater surface seeps occurred at 85% of the locations where till intersected riparian wetlands. Less than 20% of comparable outwash sites showed evidence of surface seeps and in most outwash settings, groundwater was subject to extensive biogeochemical processing in wetland soils before it entered streams as baseflow.

Urbanization is another factor that can alter groundwater recharge and groundwater inputs to the bay. We are not aware of studies that have directly measured the effects of urbanization on the magnitude of groundwater flows to coastal waters. However, a robust literature exists on both forest and urban stream hydrology (Satterlund and Adams, 1992; Hornberger et al., 1998; Paul and Meyer, 2001; Dingman, 2002). In forested or vegetated uplands, precipitation that is not removed by evapotranspiration generally leaves a site as groundwater recharge and composes the base flow of streams or estuaries. Stream flow in forested watersheds tends to have relatively stable flow patterns and considerable base flow during dry spells, in response to steady inputs of groundwater. In contrast, streams draining urbanized watersheds typically exhibit extreme fluctuations in discharge. The flashy hydrographs of urban streams provide strong evidence that urban development reduces groundwater inputs to surface waters. Even minor storm events cause rapid increases in stream flow and large increases in discharge compared with pre-storm conditions, signaling rapid transmission on smooth impervious surfaces, like roofs, pavement, and storm drains. Arnold and Gibbons (1996) estimate that the impervious surfaces associated with urban lands can diminish groundwater recharge by 30–70% compared with vegetated landscapes. Hydrologic modifications from urbanization can increase the risk of N delivery to surface waters as storm runoff, bypassing subsurface hot spots of N transformation within the root zone, and often discharging directly into adjacent streams or estuaries.

4.4 Denitrification Provides a Permanent Sink for Groundwater N

Dissolved N transported through coastal soils and aquifers may be diluted, adsorbed, taken up in plant or bacterial biomass, or reduced to ammonium or gaseous nitrogen. Denitrification, the biologically mediated transformation of NO_3 to gaseous nitrogen (N_2 and N_2O), is the primary permanent sink for NO_3 in coastal groundwater systems. A small fraction of soluble N can be converted to gaseous nitrogen during nitrification, and ammonium volatilization can be substantial in areas with extensive surface storage and applications of manure. However, as little animal agriculture occurs within the Narragansett Bay watershed, and NO_3 is the predominant form of N contamination in most of RI's coastal groundwater, denitrification is likely to be the source of gaseous N loss.

In RI aquifers the availability of labile organic carbon generally controls rates of denitrification when NO_3 is abundant and oxygen is limiting or absent. Several studies have reported substantial groundwater denitrification in aquifer zones dominated by pyrite-rich deposits, and in some settings sulfide, iron and iron complexes may serve as the energy source for denitrification, rather than labile C (Kolle et al., 1985; Gayle et al., 1989; Pederson et al., 1991; Postma et al., 1991; Korom, 1992; Joye, 2002); however, this is expected to occur in selected settings and is not likely to be important in most aquifers. Within the lands and islands contributing groundwater directly to Narragansett Bay, we are not aware of extensive anaerobic aquifers that contain a suitable reservoir of electron donor constituents (e.g., sulfide minerals or ferrous iron), thus we do not expect them to contribute substantially to groundwater denitrification in our local aquifers.

Denitrification is most active in the upper portions (upper 30 cm) of the soil, where root turnover and exudates continually replenish labile carbon. It has been measured in groundwater aquifers in a variety of environments including the coastal zone (Slater and Capone, 1987; Capone and Slater, 1990; Howes et al., 1996; Tobias et al., 2001c; Addy et al., 2005) and is frequently inferred from geochemical profiles and tracer studies (Trudell et al., 1986; Nelson et al., 1995; Ueda et al., 2003). Denitrification can remove significant amounts of NO_3 from groundwaters, particularly in sewage or agriculturally contaminated aquifers (Smith and Duff, 1988; Smith et al., 1991; Griggs et al., 2003).

Riparian wetland forests have received much attention as sites of active denitrification, often removing more than 85% of the NO_3 present in groundwater (Correll et al., 1992; Simmons et al., 1992; Correll, 1997; Lowrance, 1998). The "riparian fringe" has been recognized as an important but variable sink for contaminated groundwater and is a frequently used management tool for reducing N loading. Denitrification is common in coastal marine sediments as well, and often removes a significant fraction of the N load to coastal ecosystems (Seitzinger, 1988; Nowicki *et al.*, 1997; 1999). Unfortunately, the rates and efficiency of groundwater N attenuation in the marine coastal fringe are less well known (Addy *et al.*, 2005).

4.5 Labile Organic Carbon–A Key to Groundwater Denitrification

Previous work suggests that patterns in aquifer NO_3 contamination and denitrification are often caused by differences in the bioavailability of organic C (Bradley *et al.*, 1992; Starr and Gillham, 1993; Clay *et al.*, 1996). Translocation of DOC from the soil zone to the water table may be critical to establishing the conditions necessary for denitrification to proceed. Starr and Gillham (1993) hypothesized that under shallow water-table conditions, DOC originating in the soil zone is transported downward by infiltrating water to the aquifer. Once below the water table, oxidation of DOC generates reducing conditions, followed by denitrification. In areas with deeper water tables, most of the labile organic C is oxidized in the vadose zone before reaching the aquifer. As a result, insufficient DOC reaches the aquifer to reduce the oxygen, so oxygen concentrations remain high and denitrification does not proceed. Thus the controlling factor is the residence time of organic C in the unsaturated zone, which itself depends on soil texture and net infiltration rate as well as depth to the water table. A number of studies suggest that denitrification in coastal groundwater is carbon-limited (Slater and Capone, 1987; Giblin and Gaines, 1990; Nowicki *et al.*, 1999), thus the magnitude of N loss from groundwater at the coastal interface may be governed by the degree to which groundwater flow paths interact with organic-enriched transition zones.

4.6 Coastal Transition Zones as Sites for N Removal

The unique geomorphology of coastal transition zones can create opportunities for groundwater flow paths to intersect with zones enriched in labile carbon. Shallow coastal water tables and the compression and uplifting of groundwater flow paths as they approach the coastal zone create the potential for groundwater to interact with surface soils and the plant root zone (Fig. 4.4). Fringing coastal marshes can provide both the organic C source and the anaerobic conditions that favor N removal. In these settings, groundwater flow paths transect buried peat deposits, either freshwater peat overlain by marine deposits or saltwater peat from former salt marshes.

In their analysis of the geology of Rhode Island's coastlines, Boothroyd *et al.* (1985) describe numerous instances where organic matter was buried at depth due to the marine inundation of Pleistocene kettle-hole ponds containing freshwater wetlands, and to the accumulation of fine-grained organic sediments within depressions on the Pleistocene surface. Under present-day conditions, the dynamic processes of accretion and shoreline alterations that shape fluvial and coastal margins often create deposits of labile particulate C in saturated sediments. Coastal lagoons, low-energy back-barrier basins, and protected coves tend to collect organic detritus at the coastal margin. In areas where eelgrass meadows are established, the baffle effect of the grasses further traps fine-grained suspended material and plant debris. Any of these environments may provide the labile carbon and reducing conditions required for groundwater NO_3 reduction. Organically enriched media, capable of fostering substantial denitrification, have been observed in buried horizons and microsites at groundwater depths up to 3 m (Kellogg *et al.*, 2005).

Additionally, a host of processes can create anaerobic micro-sites of particulate labile C that induce denitrification within aerobic soils and aquifers (Gold *et al.*, 1998; Jacinthe *et al.*, 1998). Random "patches" of organic matter can occur irregularly within the soil profile, creating biological "hotspots" of NO_3 removal. The presence of these organic "hotspots" helps to explain the tremendous spatial variability in denitrification rates observed in many studies of groundwater nitrogen dynamics (Tobias *et al.*, 2001a,c; Addy *et al.*, 2005).

NO_3-contaminated groundwater can travel great distances with little apparent loss of N and then undergo considerable rates of denitrification at zones of C enrichment (Jordan *et al.*, 1993; Bowden *et al.*, 1992; Robertson *et al.*, 1991; Smith *et al.*, 1991; Paludan and Blicher-Mathiesen, 1996). Numerous field studies at local scales have found that, where appropriate C sources are lacking, inorganic N can move conservatively (e.g., with minimal transformation) for hundreds of meters through oxygenated outwash aquifers (Giblin and Gaines, 1990; Robertson *et al.*, 1991; Portnoy *et al.*, 1998). Postma *et al.* (1992) tracked septic system plumes on a sandy barrier spit complex at Charlestown Beach, RI. They found little reduction in NO_3 concentrations up to 6 m from the septic field. In contrast, soluble phosphorus concentrations in the plumes were less than 1% of the initial level; demonstrating the more conservative behavior of groundwater N.

Groundwater NO_3 also behaved conservatively in upland outwash aquifers within the vicinity of Greenwich Bay, RI. In a study by Simmons *et al.* (1992), groundwater was amended with NO_3 and Br, a conservative tracer, and the resulting plume was monitored in a series of down gradient wells. Over distances of 5–10 m—corresponding to travel times of 8–37 days—NO_3 declines in many of the wells were similar to the conservative tracer. In contrast, the same study found marked declines in NO_3 compared with Br concentrations when the enriched plumes moved within the sandy deposits underlying wetland soils.

The fact that most fresh groundwater discharge declines with distance offshore (Bokuniewicz, 1980; Harvey and Odum, 1990; Cable *et al.*, 1997; Taniguchi *et al.*, 2002) suggests that coastal groundwater discharge is usually supported by the local water table rather than by deeper, regional aquifers. The local origin of coastal groundwater emphasizes the critical importance of understanding the interplay of local soils, coastal geomorphology, and land use in mediating groundwater N transformation and transport. Nutrient transformation and fate is directly influenced by the path of groundwater flow and whether the groundwater discharges as stream or river flow, as discharge to a fringing marsh or into marsh creek bottoms, through subtidal sediments, or as direct seeps and springs across a beachface. Soil contact times of discharged groundwater can be up to 50–100% longer in marshes than in beaches and subtidal shoreline sediments (Harvey and Odum, 1990; Howes *et al.*, 1996). Pore water drainage, evapotranspiration, and infiltration are important water fluxes that can cause substantial dispersive mixing in tidal marsh soils that does not occur in subtidal sediments or marsh creek bottoms. Ullman *et al.* (2003) have documented significant groundwater discharge across a sandy beachface in Delaware, USA. The beachface serves not only as a zone of mixing between nutrient-rich upland water and groundwater, but also as an efficient particle trap for estuarine organic matter which can subsequently provide remineralized nutrients to the estuarine system.

4.7 Coastal Marshes as Sites of N Transport and Removal

Coastal landscapes are "hydrologically linked" ecosystems forming an uninterrupted continuum from upland forests through coastal plain to tidal marsh and estuary (Correll *et al.*, 1992). Groundwater dynamics in coastal marshes are inextricably tied to perturbations both "upstream" and at their estuarine margin, and can exhibit a rapid response to rain events, drawdown due to evapotranspiration or human water use, diurnal tidal inundation and intrusion, and tidal pumping (Gardner and Reeves, 2002; Gardner *et al.*, 2002). Goni and Gardner (2003) demonstrated the tight linkage between upland, coastal marsh, and estuary when they traced the fate of DOC from overlying forest soils via coastal aquifers to subterranean estuary. Wigand *et al.* (2003; 2004; Chapter 17) have described significant relationships between watershed N loading and denitrification enzyme activity and plant community structure in the salt marshes fringing Narragansett Bay. Groundwater flow can be a significant factor in governing botanical zonation in the salt marsh ecosystem by inhibiting the infiltration and evapoconcentration of saline tidal water, and preventing the formation of hypersaline soils which determine plant distribution (Thibodeau *et al.*, 1998).

As groundwater moves from the upland through the transition zone to high marsh, low marsh and estuary, the hydraulic gradient increases, and deep groundwater tends to upwell at the estuarine margin (Harvey and Odum, 1990). This results in a focusing of groundwater discharge and greater discharge per unit of bottom area close to the marsh-estuarine shoreline (Fig. 4.4; Winter *et al.*, 1998). A number of studies have reported highest soil hydraulic conductivities in test wells furthest inland and lowest conductivities closer to shore (Millham and Howes, 1995; Corbett *et al.*, 2000; Schultz and Ruppel, 2002;

Addy *et al.*, 2005). Lower hydraulic conductivities along the shoreline have been attributed to an increase in organic material of smaller particle size in the sediments at the shore, perhaps due to differential trapping of tidal particulate matter, or to buried horizons created by storms or inundated by sea level rise. Lower K values closer to the estuary further encourage an upwelling of groundwater at the low marsh-estuarine interface.

Groundwater N dynamics are unique in coastal marshes in that the water table can rise and fall daily as the piezometric head changes in response to diurnal tides. These water table fluctuations, as well as the upwelling described above, can bring deeper groundwater flow into contact with DOC from surface soils. Potential sources of DOC include peat deposits, root and rhizome exudates and decomposition, and organic debris accumulated in surface wrack and buried during storm events. An extensive rooting zone in the marsh has great potential to contribute labile carbon to incoming groundwater. Addy *et al.* (2005) found elevated root biomass in a Rhode Island marsh extending to about 75 cm in both high and low marsh, and Turner *et al.* (2004) found live *S. alterniflora* roots down to 1 m depth.

Reports vary in describing groundwater flow below and through salt marsh peat. In areas underlain by unconsolidated coarse sediments, groundwater may bypass the marsh peat (and associated labile C) at the marsh surface and flow rapidly through the more conductive sandy subsoil below. However, macropores due to fiddler crab tunnels, and variable K values for some low marsh soils, can sometimes allow considerable transmission of tidal water downward. Working within a Rhode Island marsh developed over outwash, Addy *et al.* (2005) found that groundwater below the low marsh had significantly higher denitrification capacity in the early spring than the high marsh groundwater, although these zones were less than 10 m apart. They suggest that increased tidal interaction within the low marsh may have increased the transport of labile DOC and other electron donors from the surface marsh sediments into the subterranean estuary. Although the transition zone from fresh groundwater to sea water is frequently depicted in a position directly under the estuary sediments (Burnett *et al.*, 2003), in low gradient New England marshes, with low fresh water flow rates, this brackish area actually occurs within and under the salt marsh itself. Addy *et al.* (2005) found greater groundwater denitrification capacity and lower dissolved oxygen in this area of transitional salinity beneath the low marsh than in groundwater underlying the adjoining high marsh or upland border.

As discussed earlier, the depth and location of groundwater flow paths are critical to determining the efficiency of N removal from groundwater traversing a coastal marsh. Mid-Atlantic coastal plain marshes are frequently subject to very shallow groundwater flow from near surface, hillslope soils. Groundwater flows into the marsh through a high conductivity root zone or onto the marsh surface from direct discharge at the base of the hill slope (Harvey and Odum, 1990; Tobias *et al.*, 2001b,c), providing a relatively long residence time for groundwater in the marsh, and ample opportunity for N removal. For example,

Tobias *et al.* (2001c) followed a high concentration groundwater NO_3 plume confined by a low permeability layer to the surface 10 cm of a mesohaline salt marsh in southeastern Virginia. Under these conditions, denitrification (387–465 μM N d^{-1}; 5,418–6,510 μg N L^{-1}d^{-1}) accounted for 70% of the NO_3 removal from the groundwater plume. NO_3 removal was 90% complete within the 50 cm of marsh nearest to the marsh-upland border. Addy *et al.* (2005) found rates of denitrification ranging from 21 to 538 μg N L^{-1}d^{-1} in the sandy subsoil underlying a fringing salt marsh of Narragansett Bay. However, for many salt marshes in New England, where highly permeable sand aquifers are topped with thin tidal marsh deposits, discharge of groundwater may be preferentially channeled through sandy outwash deposits where residence times are so short that there is little opportunity for NO_3 removal. For example, at Nauset Marsh, MA, groundwater flux was characterized by high-velocity seeps that bypassed the denitrifying potential of low-redox organic marsh and sub-tidal sediments. The marsh's high tidal range and large extent of exposed sediment below the marsh allowed significant amounts of groundwater to bypass marsh peats and discharge directly as seeps and springs across the beachface or into the subtidal zone (Giblin and Gaines, 1990; Portnoy *et al.*, 1998). Sites with the coarsest sediments exhibited the highest groundwater flow rates, but were also characterized by the lowest sediment organic content and negligible NO_3 removal rates via denitrification (Nowicki *et al.*, 1999). When NO_3 concentrations in drinking water wells in the adjacent watershed were compared with those obtained from direct groundwater seepage into the estuary, there was little evidence for NO_3 attenuation or removal. This is in sharp contrast to groundwater dynamics observed in Great Sippewissett Marsh (Falmouth, MA, USA) and Namskaket Marsh (Orleans, MA, USA) where marsh deposits are typically 2–3 m thick, and maximum groundwater discharge occurs at the upland-marsh border and through extensive areas of creek bottom. For these marshes, NO_3 removal via denitrification in the marsh creek bottoms may account for all the groundwater-supplied N to the marsh (Howes *et al.*, 1996).

4.8 The Subterranean Estuary as a Site for N Transport

Willard S. Moore, from the University of South Carolina, first coined the term "subterranean estuary" to describe the zone where fresh groundwater meets seawater at the coastal margin. He used the term to emphasize the importance of the mixing and biogeochemical cycling that can occur at this interface. Estuarine groundwater discharge can include both fresh, seaward flowing groundwater, and saline water recycled within the coastal aquifer (Moore, 1999). Burnett *et al.* (2003) define submarine groundwater discharge (SGD) as a flow of any water at the coastal margin, from the seabed to the coastal ocean, recognizing that it is often a mix of fluids controlled by both land-derived hydraulic gradients, and marine processes such as tidal pumping and current-induced pressure gradients. SGD is commonly characterized by low, diffuse flow rates that make it difficult to detect. However, because it occurs across such a wide area of the coastal zone, SGD flux can be significant. In New England, groundwater flow through the subterranean estuary is often the major mechanism transporting N to estuaries in areas underlain by unconsolidated coarse sediments (Lee and Olsen, 1985; Giblin and Gaines, 1990; Valiela et al., 1990; 1992; Portnoy et al., 1998).

To varying degrees, the subterranean estuary can exhibit the same classic two-layer circulation and mixing pattern that is common in many surface estuaries (Pritchard, 1967). As the tide recedes, fresh groundwater flows out above a layer of salt water, with mixing between the two layers driven by diffusion and dispersion. During incoming tides, salt water enters the aquifer through areas of high permeability and conductivity. Talbot et al. (2003) studied the subterranean estuary at the head of Waquoit Bay, Cape Cod, MA, and found that groundwater flows followed the Ghyben-Hertzberg model of high-density seawater intrusion beneath the lower density fresh groundwater. NO_3 was carried seaward into the estuary as a freshwater plume that mixed little with the intruding saltwater. Ammonium (NH_4) was transported landward into the subterranean estuary through advection of remineralized N through the marine sediments.

Direct measurements of denitrification rates in groundwater flowing in subterranean estuaries are scarce, thus N removal rates must be estimated or assumed in most N transport models (e.g., Weiskel and Howes, 1991; Valiela et al., 1997; 2000; Urish and Gomez, 2004). Currently, few data exist with which to characterize the subterranean estuary of Narragansett Bay, and the magnitude of N removal from groundwater at this interface is unknown.

4.9 Temporal Scales of Coastal Groundwater N Cycling

N transport and transformation in coastal groundwater are subject to unique temporal scales imposed by both tide and season. Riparian studies have shown that soil microbial activity and N removal rates vary seasonally, primarily due to changes in temperature, water table height, and organic supply. Seasonal variability in groundwater depth may cause periodic pulses in denitrification activity by influencing when and where groundwater comes into contact with organic-rich soils and sediments (Groffman et al., 1992; Simmons et al., 1992; Clay et al., 1996). In the Northeast, coastal water tables are highest in the dormant, wet, winter–spring seasons, and lowest during the drier summer growing season. Changes in evapotranspiration losses superimpose a strong seasonal signal on groundwater flow volumes (Millham and Howes, 1994), and pulses of DOC from surface plant materials can occur during fall senescence or at first thaw in the early spring. Periodic coastal storms and high winds can cause temporal pulses of sea-grass wrack and detrital material into back lagoon basins, fringing marsh, and shrub borders. Groundwater–carbon interactions in the marine coastal zone exhibit a temporal complexity and pulsing (Odum *et al.*, 1995) that is more pronounced than at the land–water interface of freshwater riparian zones.

Staver and Brinsfield (1996) have documented the tremendous seasonal and annual variation in groundwater discharge rates to the Wye River estuary, a subestuary of Chesapeake Bay. They found discharge rates ranging from $1.7 \, m^3 \, m^{-1}$ width of shoreline in September, to $16.3 \, m^3 \, m^{-1}$ in March. Almost half of the annual total net groundwater discharge occurred during the three months of February–April, while the summer months of July–September accounted for only 8% of the annual discharge. Annual net discharge was approximately 15% higher in a wet year (1994) than in the preceding year. As average groundwater NO_3 concentrations changed little over time, the variations in the net groundwater flux of N to the Wye estuary were driven by changes in groundwater flow volumes.

Superimposed on these annual and semi-annual patterns is a daily pattern of water table fluctuation caused by tidal forcing. Incoming tides cause coastal groundwater to "back-up" in the coastal zone. Water table heights increase back from the shoreline as the tide comes in, and decrease again as the tide recedes (Urish and Qanbar, 1997; Portnoy *et al.*, 1998). Semi-diurnal tides force coastal groundwater to move up and down and back and forth across various soil and sediment horizons. The freshwater–saltwater interface moves back and forth with each tide, with net movement of fresh water into the estuary on each ebb cycle (Fig. 4.5). Thus there is a clear, tidally driven, temporal pattern to the discharge of fresh groundwater to the coastal zone.

Unfortunately, the high levels of dissolved NO_3 reported in a large percentage of New England's coastal groundwater aquifers suggest that the conditions required for denitrification are either not persistent at the coastal margin, or are overwhelmed by the existing N load. Rising and falling water tables at the coastal interface, due to tides or season, can produce alternating conditions of oxidation and reduction in marsh soils. Microbially mediated N transformation rates (aerobic nitrification/anaerobic denitrification) decline with seasonal low temperatures, requiring longer retention times to produce the same quantity of NO_3 reduction. Nitrification is often limited by insufficient oxygen, while denitrification is often limited by inadequate supplies of labile organic carbon. Aerobic decomposers often compete with denitrifiers for organic substrates. In salt marshes, bacterial sulfate reduction can inhibit denitrification and shift NO_3 reduction from denitrification to dissimilatory nitrate reduction to ammonium (DNRA) (Joye and Hollibaugh, 1995; An and Gardner, 2002). Although denitrification is commonly considered the main NO_3 removal mechanism in groundwater, some studies have shown that DNRA can account for up to 30–50% of NO_3 loss in some systems. Thus the biogeochemistry of NO_3 reduction in coastal groundwaters is governed by both physical and biological factors which vary considerably in both time and space, complicated by a complex geomorphology common at the coastal margin.

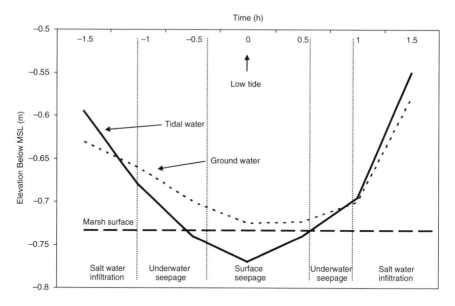

Fig. 4.5 By monitoring groundwater heights and seepage in the intertidal and subtidal areas of Town Cove, Cape Cod, MA, Portnoy *et al.* (1998) determined that maximum freshwater discharge occurred within the 1-h period centered around low tide. As the tide receded, and the piezometric head of groundwater became higher than the tidal head, submarine groundwater discharge occurred. When the tidal height dropped below the MLW mark, the zone of maximum freshwater seepage moved to surface seeps across the exposed beach.

4.10 A Mass Balance of the Fresh Groundwater Input to Narragansett Bay

What is the extent of fresh groundwater flux to Narragansett Bay? There are no direct measurements, so obtaining an estimate requires an understanding of the fate of precipitation on the watershed. A simple mass balance of direct fresh groundwater flow from the watershed to Narragansett Bay is given by:

$$GW_{Flow} = (Precip - ET)\ A - RD - MW - SD;$$

where GW_{Flow} is net fresh groundwater flow to Narragansett Bay; Precip, precipitation; ET, evapotranspiration; A, area of watershed; RD, river discharge to the bay; MW, municipal water withdrawn that enters the bay as point sources; and SD, precipitation that is captured in storm drains.

Although riverine discharge includes substantial inputs of groundwater baseflow, our focus is on the magnitude and fate of direct groundwater N loading to the bay—freshwater N that enters the bay through subsurface saturated media, rather than through rivers, streams, or pipes. We recognize that water balance estimates can be fraught with error due to the temporal, spatial, and methodological variability associated with many of the components of the hydrologic cycle (Winter, 1981). However, a substantial body of field data on river discharge from the watershed suggests that we can provide guidance on the relative extent of direct groundwater flows to the bay.

Ries (1990) conducted an extensive analysis of freshwater inputs to Narragansett Bay. He estimated that the bay's network of river tributaries capture runoff from 90% of the lands within the watershed. The rivers carry runoff waters derived from storm water, overland flow, and groundwater base flow. Long-term estimates of runoff were obtained from three levels of information: (1) 63% of the watershed land area drained into rivers that had continuous flow data; (2) 15% of the watershed land area drained into rivers with partial flow records, and long-term estimates were derived from statistical relationships to gaged rivers with similar watershed features; and (3) 12% of the watershed land area drained into rivers with no flow measurements, and estimates were obtained through a series of statistical and hydrologic techniques.

The 10% of the watershed that did not drain to tributaries included in Ries' analyses (hereafter referred to as the "direct discharge zone") comprise a 1- to 2-km-wide zone immediately adjacent to the bay, as well as the islands of the bay. In these ungaged lands, direct inputs of freshwater are likely to carry a portion of the net precipitation (Precipitation-ET) to the bay. However, the surficial geology and land use features of the direct discharge zone suggest that a substantial portion of the net precipitation will enter the bay as surface runoff, rather than as direct groundwater flow. Roughly one-half of these lands are located over the dense till deposits that characterize the eastern shoreline and islands of the bay. In these settings, numerous small brooks (such as Bailey's Brook on Aquidneck Island) capture groundwater recharge and transmit it as surface discharge to the bay.

The high extent of urban land use (49%; Ries, 1990) in the direct discharge zone is also likely to diminish fresh groundwater inputs from that portion of the watershed. This zone includes much of the shoreline areas of Cranston, Providence, and East Providence. During precipitation events, these areas generate substantial storm water runoff, rather than infiltration and groundwater recharge. Given the surficial geology and urbanization within the ungaged zone, we assume that roughly half of the net precipitation from these areas will enter the bay as direct groundwater inputs.

A firm estimate of the direct contribution of fresh groundwater to Narragansett Bay is beyond the scope of available data. In addition to the 10% of the watershed area that Ries (1990) considered outside of the river tributary network, there is also the potential—as yet undocumented—for fresh groundwater inputs to the bay from deep flow paths beneath the mouths of rivers. Kincaid and Bergondo, in Chapter 10, suggest the existence of such a flow path in the lower bay region. However, given the restrictions on groundwater recharge in the near shore, ungaged flows that result from extensive urbanization, and areas underlain by dense till, we suspect that direct groundwater discharge is likely to range from 5–10% of the total freshwater input to Narragansett Bay.

4.11 The Need to Accurately Quantify Groundwater N loading to Narragansett Bay

The magnitude and significance of groundwater N transport to Narragansett Bay remains elusive. In November, 2004, a Rhode Island Sea Grant-sponsored symposium summarizing the "State of Science on Nutrients in Narragansett Bay" (Carey et al., 2005) concluded that "Groundwater sources of N are thought to be insignificant, but may provide local or seasonal sources of importance." To date, few direct measurements of N transport via groundwater to the bay exist. The resulting symposium document placed primary importance on evaluating the significance of groundwater sources of N to Narragansett Bay.

Of prominent concern for the transport of N to Narragansett Bay is the extensive use of ISDS or "septic systems," by highly populated communities adjoining the bay, where sewered areas have not kept pace with the expanding watershed population. Many ISDS are located on shorelines that have been extensively modified during more than 300 years of human habitation. Natural coastlines and salt marshes have been filled, diked, stabilized, rip-rapped, and drained. Most ISDS use a septic tank for pre-treatment of raw wastewater, discharge the pretreated wastewater into the subsurface environment, and then rely on chemical, physical, or biological processes in subsurface soils for reducing N concentrations and/or loading to groundwater. Thus, in contrast to other sources of nonpoint nutrient pollution, such as atmospheric deposition or agricultural and residential lawn fertilizers where nutrients enter the soil environment via dynamic, biologically active topsoils, ISDS usually release nutrients into low organic subsoils, below the rooting zone of most plants. This minimizes the likelihood of plant uptake and microbial transformation of N, and increases the potential for N losses to surface and groundwaters. The probability and extent of exposure of ecosystems to N emitted from ISDS is controlled by a combination of many factors, including localized site characteristics (soils, hydrology, and slope), population density, system design and maintenance, and proximity to coastal receiving waters (Gold and Sims, 2001). In many cases, ISDS generate concentrated plumes of effluent rather than diffuse inputs (Robertson et al., 1991), potentially overwhelming the capacity of any chemical, physical, or biological nutrient removal mechanisms operative in the subsurface environment.

A number of dense, unsewered residential developments border both the east and western portions of Narragansett Bay; but differences in the surficial geology (Fig. 4.1) generate marked differences in groundwater N dynamics. Greenwich Bay, on the western shore of Narragansett Bay, illustrates the risks of groundwater N loading from ISDS in soils developed over permeable sands and gravels derived from glacial outwash. In contrast, N inputs from unsewered subdivisions in East Bay communities encounter low permeability bedrock or dense till that restricts groundwater recharge and forces wastewater into surface seeps and storm drains, short-circuiting subsurface groundwater flow paths and opportunities for N removal. Although the pathways of movement to coastal waters differ markedly, in both cases high density unsewered residential developments create substantial loading of nitrogen to coastal waters.

Greenwich Bay is unarguably the best-studied subestuary of Narragansett Bay with respect to groundwater. Urish and Gomez (2004) have estimated that 60–70% of the freshwater entering Greenwich Bay from the land is derived from groundwater recharge. The majority (75%) enters the bay as dry weather flow in streams, while the remainder (25%) enters as direct seepage along the coastline. They suggest that 65–75% of groundwater N loadings in the Greenwich Bay watershed originate from ISDS, and that N inputs from the unsewered human population make up 20–40% of the total estimated N input to Greenwich Bay.

The considerable work involved in describing groundwater N transport to Greenwich Bay alone brings into sharp relief the tremendous difficulties involved in quantifying groundwater N transport in the coastal zone. By necessity, Urish and Gomez (2004) made broad assumptions about N travel and attenuation in the Greenwich Bay watershed. Lacking detailed groundwater elevations, they delineated groundwater topography based on surface topography (a reasonable approximation but not always comparable). Groundwater discharge rates were not measured directly, but estimated from a recharge rate of 15 inches per year. Assumptions were made about how much groundwater was discharged per linear foot of shoreline per day, about what proportion was discharged into streams and then into the bay, about how much imported water (from public drinking water wells) ends up as groundwater recharge in unsewered houses, and about how much water is removed from the aquifer through the pumping of private drinking water wells. Most importantly, Urish and Gomez (2004) assumed that no loss or alteration of N occurred after entry into the groundwater system. This follows the work of Giblin and Gaines (1990) who found that groundwater comprised more than 80% of the land-derived N inputs to Nauset Estuary on Cape Cod, with much of this coming from ISDS. As noted earlier, Giblin and Gaines (1990), Portnoy et al. (1998), and Nowicki et al. (1999) found little evidence for N removal during transport in the coarse outwash sediments at Nauset Marsh, while other studies of the New England and Mid-Atlantic coasts have noted that riparian wetlands and coastal marshes can be important sinks for groundwater N (Howes et al., 1996; Gold et al., 2001; Tobias et al., 2001c; Addy et al., 2005; Kellogg et al., 2005). Detailed information about patterns of local geomorphology, land use, and groundwater hydrology is required before the results of these previous studies can be more specifically applied to the Greenwich Bay watershed in predicting groundwater N transport and removal.

Using three different scenarios for N attenuation and transport from ISDS to Greenwich Bay, the authors of the Greenwich Bay Special Area Management Plan calculated that ISDS contribute 47–57 metric tons of N per year to Greenwich Bay. Based on this analysis, ISDS contribute approximately three times the N load of the East Greenwich wastewater treatment facility, and represent the largest land-based source of N to Greenwich Bay (exceeded only by N inputs from the upper West Passage of Narragansett Bay). A planned sewer extension will reduce this amount by approximately half (to 25 metric tons) assuming 100% tie-in to the proposed system (Greenwich Bay SAMP, 2005).

If preliminary calculations are correct, then the areas of Warwick and East Greenwich within the Greenwich Bay watershed that lack sewers also deliver the largest subsidy of groundwater N to adjoining waters. Thus, it is not a surprise that the waters of the estuary immediately adjoining these areas (Brush Neck, Warwick, Apponaug, and Greenwich Coves) have been frequently plagued by nuisance phytoplankton and macroalgal blooms and low dissolved oxygen (<2 mg L^{-1}) in bottom waters (Granger et al., 2000; Chapters 11 (Saarman et al.) and 12 (Deacutis)).

An ongoing struggle by East Bay communities to deal with their groundwater discharge problems further illustrates just how complex and difficult the management of coastal groundwater discharge can be. The town of Portsmouth, RI is characterized by densely developed areas composed predominately of high-density residential housing with a mix of commercial and industrial facilities, some of which are located directly adjacent to the shoreline. The development of beachfront communities in Portsmouth pre-dates the inception of current ISDS regulations. Beachfront cottages were initially constructed for summer occupancy only, and the accompanying ISDS systems were not designed to handle year-round occupancy. Poorly drained soils, and a relatively shallow depth to groundwater result in surface seeps and springs where groundwater actually flows across the surface soils directly to the shoreline. Overflows from failing septic systems, discharges to storm drains from basement sump pumps, and direct piping of gray-water discharge further aggravate problems (RI DEM, 2003).

Thus, the surficial geology characteristic of Rhode Island provides a critical biogeochemical context within which groundwater N is transformed and transported to Narragansett Bay. The western shores of Narragansett Bay are dominated by outwash deposits and here, groundwater N is likely to move to the bay as subsurface flow, potentially altered by conditions of the land–water margin or the bottom sediments. The East Bay is dominated by dense till, and groundwater that emerges as surface seeps here will tend to move as overland flow to the bay and not be subject to subsurface transformations. Urbanization tends to reduce groundwater recharge and transmit water via storm drains, again diminishing the potential for subsurface N removal.

4.12 Considerations for the Future

We are cautiously optimistic that the risks of groundwater N loading to Narragansett Bay will decline over the coming decades. A series of treatment improvements are underway that should reduce N loading from both septic systems and storm water. Cesspools are now widely recognized as a risk for coastal pollution and are likely to be upgraded with advanced treatment in the coastal zone. The Greenwich Bay Special Area Management Plan (Greenwich Bay SAMP, 2005), adopted in May 2005, gives the Rhode Island Coastal Resources Management Council new powers to limit development along the Warwick and East Greenwich shorelines. Specific objectives of the plan include linking 50% of the unsewered residential lots in the Greenwich Bay watershed to public sewers by 2008. In addition, increased monitoring under the RI Department of Health Beach Monitoring Program and the RIDEM Storm Drain Mapping and Monitoring Program continues to target hot spots of untreated wastewater to Greenwich Bay, and to target those areas for treatment. New storm water regulations should also foster improvements in treatment. The effects of these N abatement practices on groundwater N loading to the Narragansett Bay ecosystem are likely to lag behind the timing of the source controls due to the long retention times of groundwater NO_3, so decision makers should not be deterred if immediate improvements do not result from their efforts.

Coastal managers will need to remain vigilant to ensure that continuing pressure to develop coastal land does not result in a relaxing of restrictions that would allow development in "groundwater sensitive areas." The hydrologic connections between uplands and coastal buffer zones should be protected by enhancing or sustaining infiltration on new development, in keeping with practices associated with Low Impact Development (Prince George's County, 1999). Urban watershed managers need to be aware that runoff from impermeable surfaces can alter both the location and quality of recharge to urban groundwater aquifers. Stormwater control structures such as trenches, canals, and retention ponds need to be designed to allow for maximum N attenuation and removal in riparian zones and buffer strips. At the coastal margin, vegetated buffers should be protected and restored to promote NO_3 removal from groundwater, particularly in areas where confining layers near the soil surface prevent groundwater infiltration to deeper strata. Extensive contact between shallow groundwater and the biologically active root zone of wetland vegetation is essential.

In this chapter, we have relied on secondary data about soils, land use, and geology in concert with examples from the literature to gain insight into the locations, timing, and magnitude of groundwater discharges and associated N inputs to Narragansett Bay. Field data on groundwater N loading to Narragansett Bay are sorely lacking, constraining the certainty of our remarks. Although groundwater N inputs from ISDS and storm water may decline, several other potential N sources could increase over the coming decades in response to more intensive land use, additional use of lawn fertilizers, and increased atmospheric deposition. To better assess these future risks, and to improve local management, we suggest that additional work is necessary to identify groundwater discharge zones, and to quantify groundwater N loading and transformations in discharge zones. Future research efforts need to address the development of descriptive, map-friendly attributes for various coastal sites and geomorphic settings that link them to measurable groundwater denitrification potential.

References

Addy, K., Gold, A.J., Nowicki, B., McKenna, J., Stolt, M., and Groffman, P. 2005. Denitrification capacity in a subterranean estuary below a Rhode Island fringing salt marsh. *Estuaries* 29:896–908.

Amick, R.S., and Burgess, E. 2003. Exfiltration in Sewer Systems. US EPA. Cincinnati, OH: National Risk Management Research Laboratory. EPA/600/SR-01/034.

An, S., and Gardner, W.S. 2002. Dissimilatory nitrate reduction to ammonium (DNRA) as a nitrogen link, versus denitrification as a sink in a shallow estuary (Laguna Madre/Baffin Bay, Texas). *Marine Ecology Progress Series* 237:41–50.

Arnold, C.L., Jr, and Gibbons, C.J. 1996. Impervious surface coverage: The emergence of a key environmental indicator. *Journal of the American Planning Association* 62(2):243–259.

Barlow, P.M., and Dickerman, D.C. 2001. Numerical-simulation and conjunctive-management models of the Hunt-Annaquatucket-Pettaquamscutt stream-aquifer system, Rhode Island. US Geological Survey Professional Paper 1636. Published by the U.S. Department of the Interior. USGS Publishing Network. Denver, Colorado. Obtain copies from *(http://water.usgs.gov/pubs/)*. 88 pp.

Bohlke, J.K., and Denver, J.M. 1995. Combined use of groundwater dating, chemical and isotopic analyses to resolve the history and fate of nitrate contamination in two agricultural watersheds, Atlantic coastal plain, Maryland. *Water Resources Research* 31(9):2319–2339.

Bokuniewicz, H. 1980. Groundwater seepage into Great South Bay. *Estuarine, Coastal and Shelf Science* 10:437–444.

Bokuniewicz, H., and Pavlik, B. 1990. Groundwater seepage along a barrier island. *Biogeochemistry* 10(3):257–276.

Boothroyd, J.C., Friedrich, N.E., and McGinn, S.R. 1985. Geology of microtidal coastal lagoons: Rhode Island. *Marine Geology* 63:35–76.

Bowden, W.B., McDowell, W.H., Asbury, C.E., and Finley, A.M. 1992. Riparian nitrogen dynamics in two geomorphologically distinct tropical rain forest watersheds: nitrous oxide fluxes. *Biogeochemistry* 18:77–99.

Bradley, P.M., Fernandez, M., Jr, and Chapelle, F.H. 1992. Carbon limitation of denitrification rates in an anaerobic groundwater system. *Environmental Science and Technology* 26(12):2377–2381.

Burnett, W.C., Bokuniewicz, H., Huettel, M., Moore, W.S., and Taniguchi, M. 2003. Groundwater and pore water inputs to the coastal zone. *Biogeochemistry* 66:3–33.

Cable, J.E., Burnett, W.C., Chanton, J.P., and Weatherly, G.L. 1996. Estimating groundwater discharge into the northeastern Gulf of Mexico using radon-222. *Earth and Planetary Science Letters* 144:591–604.

Cable, J.E., Burnett, W.C., and Chanton, J.P. 1997. Magnitude and variations of groundwater seepage along a Florida marine shoreline. *Biogeochemistry* 38:189–205.

Cambareri, T.C., and Eichner, E.M. 1998. Watershed delineation and groundwater discharge to a coastal embayment. *Ground Water* 36(4):626–634.

Capone, D.G., and Bautista, M.F. 1985. A groundwater source of nitrate in nearshore marine sediments. *Nature* 313:214–216.

Capone, D.G., and Slater, J.M. 1990. Interannual patterns of water table height and groundwater derived nitrate in nearshore sediments. *Biogeochemistry* 10:277–288.

Carey, D., Desbonnet, A., Colt, A.B., and Costa-Pierce, B.A. (eds.). 2005. State of Science on Nutrients in Narragansett Bay: Findings and Recommendations from the Rhode Island Sea Grant 2004 Science Symposium. Narragansett, RI: Rhode Island Sea Grant, 43 pp.

Charette, M.A., Buesseler, K.O., and Andrews, J.E. 2001. Utility of radium isotopes for evaluating the input and transport of groundwater-derived nitrogen to a Cape Cod estuary. *Limnology and Oceanography* 46(2):465–470.

Clay, D.E., Clay, S.A., Moorman, T.B., Brix-Davis, K., Scholes, K.A., and Bender, A.R. 1996. Temporal variability of organic C and nitrate in a shallow aquifer. *Water Research* 30(3):559–568.

Corbett, D.R., Chanton, J., Burnett, W., Dillon, K., and Rutkowski, C. 1999. Patterns of groundwater discharge into Florida Bay. *Limnology and Oceanography* 44(4):1045–1055.

Corbett, D.R., Dillon, K., and Burnett, W. 2000. Tracing groundwater flow on a barrier Island in the North-east Gulf of Mexico. *Estuarine, Coastal and Shelf Science* 51:227–242.

Correll, D.L. 1997. Buffer zones and water quality protection: general principles. *In* Buffer zones: their processes and potential in water protection. Haycock, N.E., Burt, T.P., Goulding, K.W.T., and Pinay, G (eds.). Harpenden, UK: Quest Environmental, pp. 7–20.

Correll, D.L., Jordan, T.E., and Weller, D.E. 1992. Nutrient flux in a landscape: effects of coastal land use and terrestrial community mosaic on nutrient transport to coastal waters. *Estuaries* 15(4):431–442.

Desimone, L.A., and Ostiguy, L.J. 1999. A vulnerability assessment of public-supply wells in Rhode Island. US Geological Survey Water Resources Investigation Report 99-4160. Published by the U.S. Department of the Interior. USGS Publishing Network. Denver, Colorado. Obtain copies from *(http://water.usgs.gov/pubs/)*. 153 pp.

Dickerman, D.C. 1984. Aquifer tests in the stratified drift, Chipuxet River Basin, Rhode Island. US Geological Survey Water Resources Investigation Report 83-4231. Published by the U.S. Department of the Interior. USGS Publishing Network. Denver, Colorado. Obtain copies from *(http://water.usgs.gov/pubs/)*. 39 pp.

Dingman, S.L. 2002. Physical Hydrology, 2nd edition. Prentice Hall. Upper Saddle River, NJ. 646 pp.

Fetter, C.W. 2001. Applied Hydrogeology, 4th edition. Upper Saddle River, NJ: Prentice Hall, 598 pp.

Gardner, L.R., and Reeves, H.W. 2002. Spatial patterns in soil water fluxes along a forest–marsh transect in the southeastern United States. *Aquatic Sciences* 64:141–155.

Gardner, L.R., Reeves, H.W., and Thibodeau, P.M. 2002. Groundwater dynamics along forest-marsh transects in a southeastern salt marsh, USA: description, interpretation and challenges for numerical modeling. *Wetland Ecology and Management* 10:145–159.

Gayle, B.P., Boardman, G.D., Sherrard, J.H., and Benoit, R.E. 1989. Biological denitrification of water. *Journal of Environmental Engineering* 115:930–943.

Giblin, A.E., and Gaines, A.G. 1990. Nitrogen inputs to a marine embayment: the importance of groundwater. *Biogeochemistry* 10:309–328.

Gobler, C.J., and Boneillo, G.E. 2003. Impacts of anthropogenically influenced groundwater seepage on water chemistry and phytoplankton dynamics within a coastal marine ecosystem. *Marine Ecology Progress Series* 255:101–114.

Gold, A.J., and Sims, J.T. 2001. Research needs in decentralized wastewater treatment and management: a risk-based approach to nutrient contamination. National Research Needs Conference Proceedings: Risk-Based Decision Making for Onsite Wastewater Treatment. May 19–20, 2000, St. Louis, MO, Palo Alto, CA, USA: Electric Power Research Institute.

Gold, A.J., DeRagon, W.R., Sullivan, W.M., and Lamunyon, J.L. 1990. Nitrate-nitrogen losses to groundwater from rural and suburban land uses. *Journal of Soil and Water Conservation* 45:305–310.

Gold, A.J., Jacinthe, P.A., Groffman, P.M., Wright, W.R., and Puffer, R.H. 1998. Patchiness in groundwater nitrate removal in a riparian forest. *Journal of Environmental Quality* 27(1):146–155.

Gold, A.J., Groffman, P.M., Addy, K., Kellogg, D.Q., Stolt, M., and Rosenblatt, A.E. 2001. Landscape attributes as controls on ground water nitrate removal capacity of riparian zones. *Journal of the American Water Resources Association* 37:1457–1464.

Goni, M.A., and Gardner, L.R. 2003. Seasonal dynamics in dissolved organic carbon concentrations in a coastal water-table aquifer at the forest–marsh interface. *Aquatic Geochemistry* 9:209–232.

Granger, S., Brush, M., Buckley, B., Traber, M., Richardson, M., and Nixon, S.W. 2000. An assessment of eutrophication in Greenwich Bay. Paper No. 1, Restoring water quality in Greenwich Bay: a whitepaper series. Narragansett, RI: Rhode Island Sea Grant, 20 pp.

Griggs, E.M., Kump, L.R., and Böhlke, J.K. 2003. The fate of wastewater-derived nitrate in the subsurface of the Florida Keys: Key Colony Beach, Florida. *Estuarine, Coastal and Shelf Science* 58(3):517–539.

Groffman, P.M., Gold, A.J., and Simmons, R.C. 1992. Nitrate dynamics in riparian forests: microbial studies. *Journal of Environmental Quality* 21:666–671.

Groffman, P.M., Law, N.L., Belt, K.T., Band, L.E., and Fisher, G.T. 2004. Nitrogen fluxes and retention in urban watershed systems. *Ecosystems* 7:393–403.

Harvey, J.W., and Odum, W.E. 1990. The influence of tidal marshes on upland groundwater discharge to estuaries. *Biogeochemistry* 10(3):217–236.

Hornberger, G.M., Raffensperger, J.P., Wiberg, P.L., and Eshleman, K.N. 1998. Elements of Physical Hydrology. The Johns Hopkins University Press. Baltimore, MD. 302 pp.

Howes, B.L., Weiskel, P.K., Goehringer, D.D., and Teal, J.M. 1996. Interception of freshwater and nitrogen transport from uplands to coastal waters: the role of salt marshes. *In* Estuarine Shores: Evolution, Environments and Human Alterations. Nordstrom, K.F., and Roman, C.T. (eds.). John Wiley and Sons. Hoboken, NJ. pp. 287–310.

Hussain, N., Church, T.M., and Kim, G. 1999. Use of ^{222}Rn and ^{226}Ra to trace groundwater discharge into the Chesapeake Bay. *Marine Chemistry* 65:127–134.

Jacinthe, P.A., Groffman, P.M., Gold, A.J., and Mosier, A. 1998. Patchiness in microbial nitrogen transformations in groundwater in a riparian forest. *Journal of Environmental Quality* 27(1): 156–164.

Johannes, R.E. 1980. The ecological significance of the submarine discharge of ground water. *Marine Ecology Progress Series* 3:365–373.

Jordan, T.E., Correll, D.L., and Weller, D.E. 1993. Nutrient interception by a riparian forest receiving inputs from adjacent cropland. *Journal of Environmental Quality* 22(3):467–473.

Joye, S.B. 2002. Denitrification in the marine environment. *In* Encyclopedia of Environmental Microbiology. Britton, G. (ed.) New York: Wiley Publishers, pp. 1010–1019.

Joye, S.B., and Hollibaugh, J.T. 1995. Influence of sulfide inhibition of nitrification on nitrogen regeneration in sediments. *Science* 270:623–625.

Kellogg, D.Q., Gold, A.J., Groffman, P.M., Addy, K., Stolt, M.H., and Blazejewski, G. 2005. In situ ground water denitrification in stratified, permeable soils. *Journal of Environmental Quality* 34:524–533.

Kelly, R.P., and Moran, S.B. 2002. Seasonal changes in groundwater input to a well-mixed estuary estimated using radium isotopes and implications for coastal nutrient budgets. *Limnology and Oceanography* 47(6):1796–1807.

Kolle, W., Strebel, O., and Bottcher, J. 1985. Formation of sulfate by microbial denitrification in a reducing aquifer. *Water Supply* 3:35–40.

Korom, S.F. 1992. Natural denitrification in the saturated zone: a review. *Water Resources Research* 28:1657–1668.

Krest, J.M., Moore, W.S., Gardner L.R., and Morris, J. 2000. Marsh nutrient export supplied by groundwater discharge: evidence from Ra measurements. *Global Biogeochemical Cycles* 14:167–176.

Lambert, M.J., and Burnett, W.C. 2003. Submarine groundwater discharge estimates at a Florida coastal site based on continuous radon measurements. *Biogeochemistry* 66:55–73.

Laroche, J., Nuzzi, R., Waters, R., Wyman, K., Falkowski, P.G., and Wallace, D.W.R. 1997. Brown tide blooms in Long Island's coastal waters linked to interannual variability in ground water flow. *Global Change Biology* 3:397–410.

Lee, D.R. 1977. A device for measuring seepage flux in lakes and estuaries. *Limnology and Oceanography* 22:140–147.

Lee, V., and Olsen, S. 1985. Eutrophication and management initiatives for the control of nutrient inputs to Rhode Island coastal lagoons. *Estuaries* 8(2B):191–202.

Lowrance, R. 1998. Riparian forest ecosystems as filters for non-point source pollution. *In* Successes, Limitations, and Frontiers in Ecosystem Science. Pace, M.L., and Groffman, P.M. (eds.) New York: Springer. pp. 113–141.

Millham, N.P., and Howes, B.L. 1994. Freshwater flow into a coastal environment: groundwater and surface water inputs. *Limnology and Oceanography* 39(8):1928–1944.

Millham, N.P., and Howes, B.L. 1995. A comparison of methods to determine K in a shallow coastal aquifer. *Ground Water* 33:49–57.

Moore, W.S. 1996. Large groundwater inputs to coastal waters revealed by [226]Ra enrichments. *Nature* 380:612–614.

Moore, W.S. 1999. The subterranean estuary: a reaction zone of ground water and sea water. *Marine Chemistry* 65:111–125.

Moore, W.S. 2003. Sources and fluxes of submarine groundwater discharge delineated by radium isotopes. *Biogeochemistry* 66:75–93.

Nelson, W.M., Gold, A.J., and Groffman, P.M. 1995. Spatial and temporal variation in groundwater nitrate removal in a riparian forest. *Journal of Environmental Quality* 24:691–699.

Nowicki, B.L., Kelly, J.R., Requintina, E., and Van Keuren, D. 1997. Nitrogen losses through sediment denitrification in Boston Harbor and Massachusetts Bay. *Estuaries* 20(3):626–639.

Nowicki, B.L., Requintina, E., Van Keuren, D., and Portnoy, J. 1999. The role of sediment denitrification in reducing groundwater-derived nitrate inputs to Nauset Marsh Estuary, Cape Cod, Massachusetts. *Estuaries* 22(2):245–259.

Odum, W.E., Odum, E.P., and Odum, H.T. 1995. Nature's pulsing paradigm. *Estuaries* 18(4):547–555.

Paludan, C., and Blicher-Mathiesen, G. 1996. Losses of inorganic carbon and nitrous oxide from a temperate freshwater wetland in relation to nitrate loading. *Biogeochemistry* 35:305–326.

Paul, M.J., and Meyer, J.L. 2001. Streams in the urban landscape. *Annual Review of Ecology and Systematics* 32:333–365.

Pederson, J.K., Bjerg, P.L., and Christensen, T.H. 1991. Correlation of nitrate profiles with groundwater and sediment characteristics in a shallow sandy aquifer. *Journal of Hydrology* 124:263–277.

Persky, J.H. 1986. The relation of groundwater quality to housing density, Cape Cod, Massachusetts. US Geological Survey Water Resources Investigation Report, Boston, MA. Published by the U.S. Department of the Interior. USGS Publishing Network. Denver, Colorado. Obtain copies from *(http:// water.usgs.gov/pubs/)*. 22 pp.

Portnoy, J.W., Nowicki, B.L., Roman, C.T., and Urish, D.W. 1998. The discharge of nitrate-contaminated groundwater from developed shoreline to marsh-fringed estuary. *Water Resources Research* 34(11):3095–3104.

Postma, D., Bosen, C., Kristiansen, H., and Larsen, F. 1991. Nitrate reduction in an unconfined sandy aquifer: water chemistry reduction processes, and geochemical modeling. *Water Resources Research* 29:2027–2045.

Postma, F.B., Gold, A.J., and Loomis, G.W. 1992. Nutrient and microbial movement from seasonally-used septic systems. *Journal of Environmental Health* 55:5–10.

Prince George's County. 1999. Low Impact Development Design Strategies: An Integrated Design Approach. Prince George's County, MD., Department of Environmental Resources, Program and Planning Division, 150 pp.

Pritchard, D.W. 1967. What is an estuary: physical viewpoint. *In* Estuaries. Lauff, G.H. (ed.). Washington, DC: American Association for the Advancement of Science, pp. 37–44.

Rapaglia, J. 2005. Submarine groundwater discharge into the Venice lagoon, Italy. *Estuaries* 28(5):705–713.

Rhode Island Coastal Resources Management Council (RICRMC). 2005. Greenwich Bay Special Area Management Plan (SAMP): A Management Program of the Rhode Island Coastal Resources Management Council. Prepared by the Coastal Resources Center/Rhode Island Sea Grant. Narragansett, RI: University of Rhode Island, 475 pp.

Rhode Island Department of Environmental Management (RIDEM). 2003. Total Maximum Daily Load Report: The Sakonnet River, Portsmouth Park and The Cove, Island Park. Providence, RI: Rhode Island Department of Environmental Management, 40 pp.

Ries, K.G., III 1990. Estimating Surface-Water Runoff to Narragansett Bay, Rhode Island and Massachusetts. US Geological Survey, Water Resources Investigations. Report 89-4164. Published by the U.S. Department of the Interior. USGS Publishing Network. Denver, Colorado. Obtain copies from *(http://water. usgs.gov/pubs/)*. 44 pp.

Robertson, W.D., Cherry, J.A., and Sudicky E.A. 1991. Ground-water contamination from two small septic systems on sand aquifers. *Ground Water* 29:82–92.

Rosenblatt, A.E., Gold, A.J., Stolt, M.H., Groffman, P.M., and Kellogg, D.Q. 2001. Identifying riparian sinks for watershed nitrate using soil surveys. *Journal of Environmental Quality* 30:1596–1604.

Rosenshein, J.S., Gonthier, J.B., and Allen, W.B. 1968. Hydrologic characteristics and sustained yield of principal ground-water units Potowomut-Wickford area, Rhode Island. US Geological Survey Water-Supply Paper 1775. Published by the U.S. Department of the Interior. USGS Publishing Network. Denver, Colorado. Obtain copies from *(http://water.usgs.gov/pubs/)*. 38 pp.

Rutkowski, C.M., Burnett, W.C., Iverson, R.L., and Chanton, J.P. 1999. The effect of groundwater seepage on nutrient delivery and seagrass distribution in the northeastern Gulf of Mexico. *Estuaries* 22:1033–1040.

Satterlund, D.R., and Adams, P.W. 1992. Wildland Watershed Management, 2nd edition. John Wiley and Sons. Hoboken, NJ. 436 pp.

Schultz, G., and Ruppel, C. 2002. Constraints on hydraulic parameters and implications for groundwater flux across the upland estuary interface. *Journal of Hydrology* 260:255–269.

Schwartz, M.C. 2003. Significant groundwater input to a coastal plain estuary: assessment from excess radon. *Estuarine, Coastal and Shelf Science* 56(1):31–42.

Scott, M.K., and Moran, S.B. 2001. Ground water input to coastal salt ponds of southern Rhode Island estimated using ^{226}Ra as a tracer. *Journal of Environmental Radioactivity* 54:163–174.

Seitzinger, S.P. 1988. Denitrification in freshwater and coastal marine ecosystems: ecological and geochemical significance. *Limnology and Oceanography* 33(4):702–724.

Sholkovitz, E., Herbold, C., and Charette, M. 2003. An automated dye-dilution based seepage meter for the time-series measurement of submarine groundwater discharge. *Limnology and Oceanography: Methods* 1:16–28.

Short, F.T., and Burdick, D.M. 1996. Quantifying eelgrass habitat loss in relation to housing development and nitrogen loading in Waquoit Bay, Massachusetts. *Estuaries* 19(3):730–739.

Simmons, G.M., Jr 1992. Importance of submarine groundwater discharge (SGWD) and seawater cycling to material flux across sediment/water interfaces in marine environments. *Marine Ecology Progress Series* 84:173–184.

Simmons, R.C., Gold, A.J., and Groffman, P.M. 1992. Nitrate dynamics in riparian forests: groundwater studies. *Journal of Environmental Quality* 21(4):659–665.

Slater, J.M., and Capone, D.G. 1987. Denitrification in aquifer soil and nearshore marine sediments influenced by groundwater nitrate. *Applied Environmental Microbiology* 53:1292–1297.

Smith, R.L., and Duff, J.H. 1988. Denitrification in a sand and gravel aquifer. *Applied Environmental Microbiology* 54:1071–1078.

Smith, R.L., Howes, B.L., and Duff, J.H. 1991. Denitrification in nitrate-contaminated groundwater: occurrence in steep vertical geochemical gradients. *Geochimica et Cosmochimica Acta* 55:1815–1825.

Starr, R.C., and Gillham, R.W. 1993. Denitrification and organic carbon availability in two aquifers. *Ground Water* 31(6):934–947.

Staver, K.W., and Brinsfield, R.B. 1996. Seepage of groundwater nitrate from a riparian agroecosystem into the Wye River Estuary. *Estuaries* 19(2B):359–370.

Talbot, J.M., Kroeger, K.D., Rago, A., Allen, M.C., and Charette, M.A. 2003. Nitrogen flux and speciation through the subterranean estuary of Waquoit Bay, Massachusetts. *Biological Bulletin* 205:244–245.

Taniguchi, M., Burnett, W.C., Cable, J.E., and Turner, J.V. 2002. Investigations of submarine groundwater discharge. *Hydrological Processes* 16:2115–2129.

Taniguchi, M., Burnett, W.C., Smith, C.F., Paulsen, R.J., O'Rourke, D., Krupa, S.L., and Christoff, J.L. 2003. Spatial and temporal distributions of submarine groundwater discharge rates obtained from various types of seepage meters at a site in the Northeastern Gulf of Mexico. *Biogeochemistry* 66:35–53.

Thibodeau, P.M., Gardner, L.R., and Reeves, H.W. 1998. The role of groundwater flow in controlling the spatial distribution of soil salinity and rooted macrophytes in a southeastern salt marsh, USA. *Mangroves and Salt Marshes* 2:1–13.

Tobias, C.R., Anderson, I.C., Canuel, E.A., Macko, S.A. 2001a. Nitrogen cycling through a fringing marsh-aquifer ecotone. *Marine Ecology Progress Series* 210:25–39.

Tobias, C.R., Harvey, J.W., and Anderson, I.C. 2001b. Quantifying groundwater discharge through fringing wetlands to estuaries: seasonal variability, methods comparisons, and implications for wetland-estuary exchange. *Limnology and Oceanography* 46:604–615.

Tobias, C.R., Macko, S.A., Anderson, I.C., Canuel, E.A., and Harvey, J.W. 2001c. Tracking the fate of a high concentration groundwater nitrate plume through a fringing marsh: a combined groundwater tracer and in situ isotope enrichment study. *Limnology and Oceanography* 46(8):1977–1989.

Trench, E.C.T. 1991. Ground-Water Resources of Rhode Island. US Geological Survey Open-File Report. Published by the U.S. Department of the Interior. USGS Publishing Network. Denver, Colorado. Obtain copies from *(http://water.usgs.gov/pubs/)*. 169 pp.

Trench, E.C.T. 1995. Sources of Geologic and Hydrologic Information Pertinent to Ground-Water Resources in Rhode Island. US Geological Survey Open-File Report 93-464. Published by the U.S. Department of the Interior. USGS Publishing Network. Denver, Colorado. Obtain copies from *(http://water.usgs.gov/pubs/)*. 98 pp.

Trudell, M.R., Gillham, R.W., and Cherry, J.A. 1986. An in-situ study of the occurrence and rate of denitrification in a shallow unconfined sand aquifer. *Journal of Hydrology* 83:251–268.

Turner, R.E., Swenson, E.M., Milan, C.S., Lee, J.M., and Oswald, T.A. 2004. Below-ground biomass in healthy and impaired salt marshes. *Ecological Research* 19:29–35.

Ueda, S., Go, C.S.U., Suzumura, M., and Sumi, E. 2003. Denitrification in a seashore sandy deposit influenced by groundwater discharge. *Biogeochemistry* 63:187–205.

Ullman, W.J., Chang, B., Miller, D.C., and Madsen, J.A. 2003. Groundwater mixing, nutrient diagenesis, and discharges across a sandy beachface, Cape Henlopen, Delaware (USA). *Estuarine, Coastal and Shelf Science* 57(3):539–552.

Urish, D.W., and Gomez, A.L. 2004. Groundwater discharge to Greenwich Bay. Paper No. 3. Restoring Water Quality in Greenwich Bay: A Whitepaper Series. Narragansett, RI: Rhode Island Sea Grant, 9 pp.

Urish, D.W., and Qanbar, E.K. 1997. Hydrologic Evaluation of Groundwater Discharge, Nauset Marsh, Cape Cod National Seashore, Massachusetts. Report to the National Park Service. Wellfleet, MA: Cape Cod National Seashore, 69 pp.

US Geological Survey. 1999. The quality of our nation's waters: nutrients and pesticides. US Geological Survey Circular 1225. Published by the U.S. Department of the Interior. USGS Publishing Network. Denver, Colorado. Obtain copies from *(http://water.usgs.gov/pubs/)*. Reston, VA: 82 pp.

Valiela, I., Costa, J., Foreman, K., Teal, J.M., Howes, B., and Aubrey, D. 1990. Transport of groundwater-borne nutrients from watersheds and their effects on coastal waters. *Biogeochemistry* 10:177–197.

Valiela, I., Foreman, K., LaMontagne, M., Hersh, D., Costa, J., Peckol, P., DeMeo-Andreson, B., D'Avanzo, C., Babione, M., Sham, C., Brawley, J., and Lajtha, K. 1992. Couplings of watersheds and coastal waters: sources and consequences of nutrient enrichment in Waquoit Bay, Massachusetts. *Estuaries* 15:443–457.

Valiela, I., Collins, G., Kremer, J., Lajtha, K., Geist, M., Seely, B., Brawley, J., and Sham, C.H. 1997. Nitrogen loading from coastal watersheds to receiving estuaries: new method and application. *Ecological Applications* 7:358–380.

Valiela, I., Geist, M., McClelland, J., and Tomasky, G. 2000. Nitrogen loading from watersheds to estuaries: verification of the Waquoit Bay nitrogen loading model. *Biogeochemistry* 49:277–293.

Vitousek, P., Mooney, H.A., Lubchenco, J., and Melillo, J.M. 1997. Human domination of earth's ecosystems. *Science* 277:494–499.

Warwick, J., and Hill, A.R. 1988. Nitrate depletion in the riparian zone of a small woodland stream. *Hydrobiologia* 157:231–240.

Weiskel, P.K., and Howes, B.L. 1991. Quantifying dissolved nitrogen flux through a coastal watershed. *Water Resources Research* 27:(11):2929–2939.

Wigand, C., McKinney, R.A., Charpentier, M.A., Chintala, M.M., and Thursby, G.B. 2003. Relationships of nitrogen loadings, residential development, and physical characteristics with plant structure in New England salt marshes. *Estuaries* 26(6):1494–1504.

Wigand, C., McKinney, R.A., Chintala, M.M., Charpentier, M.A., and Groffman, P.M. 2004. Denitrification enzyme activity of fringe salt marshes in New England (USA). *Journal of Environmental Quality* 33:1144–1151.

Winter, T.C. 1981. Uncertainties in Estimating the Water Balance of Lakes. *Water Resources Bulletin* 17:82–115.

Winter, T.C., Harvey, J.W., Franke, O.L., and Alley, W.M. 1998. Groundwater and surface water a single resource. US Geological Survey Circular 1139. Published by the U.S. Department of the Interior. USGS Publishing Network. Denver, Obtain copies from *(http://water.usgs.gov/pubs/)*. 79 pp.

Chapter 5
Nitrogen and Phosphorus Inputs to Narragansett Bay: Past, Present, and Future

Scott W. Nixon, Betty A. Buckley, Stephen L. Granger, Lora A. Harris, Autumn J. Oczkowski, Robinson W. Fulweiler and Luke W. Cole

5.1 Introduction

The Industrial Revolution in the United States began in December, 1790, in the headwaters of Narragansett Bay, when Moses Brown used water power from the Blackstone River to spin cotton for the first time in America (Coleman, 1963). But it was not until the War of 1812 and the invention of the power loom and cotton cleaning machinery that the textile industry and its associated base metals and machinery industries began the exponential growth that would turn the Seekonk and Providence River estuaries of upper Narragansett Bay into urban waters of national importance (Fig. 5.1; Nixon, 1995a). Because of its unique history, people have had major impacts on this ecosystem for almost two hundred years. They have filled some places and dredged others, hardened much of the shoreline, introduced species, harvested species, and warmed the waters. For most of the past two centuries they added large amounts of pollutants in the form of organic matter, metals, and nutrients (nitrogen and phosphorus). Within the last thirty years, the inputs of organic waste and metals and phosphorus have all been markedly reduced (Nixon, 1995a; Nixon et al., 2005). Now, it appears that the bay is on the threshold of major reductions in the amount of nitrogen that it receives. These reductions will come largely through the addition of denitrification to urban sewage treatment facilities. The impact of nitrogen reduction during the May–October period is intended to be large, large enough that primary production during summer will be significantly reduced. With less organic matter being formed, it is thought that the consumption of oxygen in the near-bottom waters and sediments of the upper bay and associated estuaries will be sufficiently reduced that dissolved oxygen concentrations there will increase. Because point sources provide a large amount of the nitrogen that currently enters the bay (Nixon et al., 1995; 2005), and because nitrogen has been shown to be the nutrient that most limits

S. W. Nixon

Graduate School of Oceanography, University of Rhode Island, Narragansett, RI 02882
swn@gso.uri.edu

A. Desbonnet, B. A. Costa-Pierce (eds.), *Science for Ecosystem-based Management.*   101
© Springer 2008

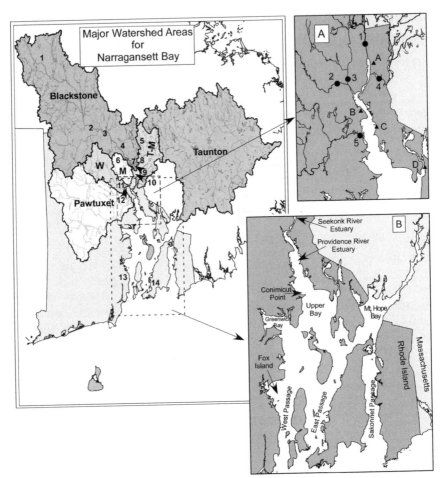

Fig. 5.1 Narragansett Bay and the watersheds of the larger rivers that flow into the bay are shown in the upper left panel. W, M, and T-M denote the watersheds of the Woonasquatucket, Moshassuck and Ten Mile Rivers, respectively. Places mentioned in this chapter are numbered as follows: 1 = Worchester, MA; 2 = Millville, MA; 3 = Woonsocket, RI; 4 = Cumberland, RI; 5 = North Attleboro, MA; 6 = Lincoln, RI; 7 = Valley Falls, RI; 8 = South Attleboro, MA; 9 = Central Falls, MA; 10 = Pawtucket, RI; 11 = North Providence, RI; 12 = Providence, RI; 13 = Graduate School of Oceanography, URI; 14 = Newport, RI. Panel A shows the locations of the sites where we collected river water samples for the nutrient analyses discussed in the text, including the Blackstone (1), Woonasquatucket (2), Moshassuck (3), Ten Mile (4) and Pawtuxet Rivers (5). Upper bay sewage treatment facilities are also shown, including the Narragansett Bay Commission (NBC) facility at Bucklin Point on the Seekonk Estuary (A), the NBC facility at Field's Point on the Providence River Estuary (B), the East Providence facility (C) and the Warren facility (D). Panel B shows major areas of the bay mentioned in the text. Narragansett Bay proper includes the area below Conimicut Point and excludes Mount Hope Bay and the Sakonnet Passage.

summer phytoplankton production in the bay (Kremer and Nixon, 1978; Pilson, 1985b; Oviatt *et al.*, 1995), this effect may be achieved. But the coming nitrogen reductions will impact a bay that is already changing dramatically, presumably in response to warming and reduced surface irradiance (Li and Smayda, 1998; Oviatt, 2004; Borkman, 2002; Fulweiler and Nixon, in press). There may be difficult trade offs between oxygen concentrations in the Providence River and upper bay and the productivity of animals in the mid and lower bay.

Our purpose in this chapter is to provide an overview of nitrogen (N) and phosphorus (P) inputs to Narragansett Bay as they have changed over the historical period, an assessment of the current rates of input from the major sources, and estimates of the possible future inputs with full implementation of probable nitrogen reductions in the major point sources that discharge directly to the bay and in its watershed. Because much of the recent attention of managers, environmental groups, and the public has been focused on other estuaries and coastal marine systems that have become enriched with nutrients from non–point sources (e.g., atmospheric deposition and synthetic fertilizer) since the Second World War, we have included some discussion of the historical factors driving the much earlier fertilization of urban estuaries as exemplified by Narragansett Bay. Digressions into changing waste water treatment technology, the history of water quality monitoring, the importance of waste from urban horses, and the impact of changing human diets may seem a distraction from our main goal, but they are all part of a rich story. While the ecological impacts of N and P in the bay are not the primary focus of this chapter, we can not help touching on this subject. The ecological implications of nutrient addition and removal involve far more than the bottom water oxygen concentrations that have dominated the discussions of nitrogen reduction thus far.

5.2 The Past

5.2.1 What Verrazzano Found

Giovanni da Verrazzano discovered Narragansett Bay for Europe when he sailed into Newport Harbor on April 21, 1524. In the first written description of the bay published several years later, he made numerous detailed observations about the geography of the bay and the people he found living along its shores (Wroth, 1970). Unfortunately, there was no Charles Darwin aboard the *Dauphine* to make and record detailed observations of the water, wetlands, algae, and fish that Verrazzano and his crew encountered. And even if there had been a great naturalist aboard, any ability to make measurements of the nutrients, nitrogen (N) and phosphorus (P), in the bay or its tributaries lay almost 350 years in the future.

In spite of these limitations, it is possible to make some estimates of the amount of nutrients entering Narragansett Bay around the time of Verrazzano's visit (Nixon, 1997). Our estimates are necessarily indirect simplifications and come from recent studies of P exports in streams and rivers draining "minimally disturbed" New England forests (e.g., Likens *et al.*, 1977), N exports from old growth temperate forests in Chile (Hedin *et al.*, 1995), and numerical simulations of pre-industrial atmospheric nitrogen deposition. The focus on nutrients leaving forested watersheds reflects the fact that 90–95% of Rhode Island and Massachusetts was still forested in the early 1600s (Harper, 1918), in spite of the well documented but limited burning of forest undergrowth by the Narragansett Indians (Verrazzano's account in Wroth, 1970; Bidwell and Falconer, 1941; Day, 1953). Despite popular impressions, there is little, if any, evidence that the various groups of people living in the Narragansett Bay watershed practiced extensive maize agriculture and associated widespread land clearing prior to European contact (Bernstein, 1993; Motzkin and Foster, 2002; Nixon, 2004). We should also emphasize that the behavior of nutrients, especially N, in forest ecosystems is complex and the subject of intensive research and a very large scientific literature. Useful recent overviews focusing on New England forests have been published by Aber *et al.* (2002) and Foster and Aber (2004) and on Rhode Island forests by Hooker and Compton (2003). Our purpose here is to develop only a very rough impression of the probable magnitude of N and P inputs to the prehistoric bay.

The reason for turning to studies of N losses from forests in Chile is that virtually all the forests on the east coast of North America, and many of those in Europe, are enriched with a large amount of N that is added in wet and dry deposition from the atmosphere. As a result, they almost certainly export far more N in their streams and rivers than the prehistoric forests around Narragansett Bay would have. Similarly, the direct deposition of N on the surface of the bay itself must be much higher today than it was prior to the Industrial Revolution. The water in the atmosphere is enriched with N largely as a result of fossil fuel combustion and intensive manure production associated with concentrated animal feeding operations (e.g., Galloway and Cowling, 2002; Howarth *et al.*, 2002; Moomaw, 2002; Hamburg *et al.*, Chapter 6, this volume). The former is more important to Narragansett Bay than the latter, with both local and midwestern sources contributing. Howarth, in Chapter 3 of this volume, provides greater detail on this source of N to Narragansett Bay. Anthropogenic N from power facilities can be transported considerable distances between its injection into the atmosphere and its deposition. For this reason (among others), sophisticated numerical simulation models have been developed that can be used to trace N through the atmosphere from its known sources to its probable area of deposition (e.g., Levy and Moxim, 1989; Dentener and Crutzen, 1994). The same models can be run for pre-industrial conditions and, when this is done, they suggest that the deposition of dissolved inorganic nitrogen (DIN) on the Narragansett Bay watershed and on the bay itself would have been about 5 mmol N m^{-2} y^{-1} at the time of

Verrazzano's discovery. Simulations for recent conditions are over ten times higher and compare very well with measured values (78 mmol m^{-2} y^{-1} and 74 mmol m^{-2} y^{-1}, respectively, Nixon, 1997). If we accept the modeled pre-industrial deposition value and assume that the mature old growth forest that Verrazzano found was in steady-state with regard to N, we can calculate an N flux from the watershed to the bay based on a watershed area of 4,708 km^2 (Pilson, 1985a). The export of DIN from old growth evergreen forests in Chile that receive the same amount of atmospheric N deposition is about 1.7 mmol m^{-2} y^{-1}, so we might work with a range of 1 to 5 mmol m^{-2} y^{-1} for the export of DIN from the prehistoric watershed. This compares with a range of 5.4 to 8.4 mmol total N m^{-2} y^{-1} that Howarth *et al.* (1996) described as characterizing total N (TN) losses from "pristine" temperate forests based on an extensive literature review. They cautioned that their range was possibly an overestimate. A recent analysis of N yields from 19 "minimally disturbed" watersheds of the US Geological Survey's (USGS) Hydrologic Benchmark Network found that DIN and TN exports were strongly correlated with runoff across all systems with various types of vegetation (Lewis, 2002). For a watershed with 0.5 m annual runoff (approximately that found for the Narragansett Bay watershed; Pilson, 1985a), the USGS regressions suggest that DIN export would be 5.4 mmol m^{-2} y^{-1} and TN export would be about 20 mmol m^{-2} y^{-1} in the absence of major anthropogenic impact.

Much of the N exported from forests with low human population density is in dissolved organic form (DON). For example, in a recent study of the nutrients exported from a largely forested (by regrowth) watershed of the nearby Pawcatuck River in southern Rhode Island, DIN export amounted to 9.4 mmol m^{-2} y^{-1} and DON export was 8.1 mmol m^{-2} y^{-1} (DIN/DON = 1.2) compared with DIN/DON ratios over 3 in the heavily impacted Blackstone and Pawtuxet Rivers flowing to Narragansett Bay (Fulweiler and Nixon, 2005). In the case of the old growth forests in Chile, DON accounted for 95% of the TN export. Neither the source nor the nature of the DON is well known. In the case of the old growth forests, Hedin *et al.* (1995) felt it was, "...mainly due to the dissolution of highly refractory fulvic acids from soil organic pools."

Studies of the bioavailability of DON in the Delaware and Hudson Rivers have shown that some 40 to 70% of the DON in those systems could be taken up by bacteria (Seitzinger and Sanders, 1997). Of course, neither of those rivers drains old growth forests and the DON they carry may be much more labile than the DON in the prehistoric rivers of southern New England. The forests in Chile also receive over twice the annual rainfall that the Narragansett Bay watershed does, and this may enhance DON export there. It is also true that a significant amount of DON (and some particulate organic N) is deposited directly on the Narragansett Bay watershed and on the bay in wet and dry deposition, but we have no basis for estimating these fluxes under pre-industrial conditions (Neff *et al.,* 2002). As a practical matter, we have excluded DON from the calculation of prehistoric inputs to the bay and note that our estimate may thus somewhat underestimate the amount of biologically available N that entered from the watershed and the atmosphere.

A final consideration is that not all of the DIN exported from the prehistoric forests would have reached the bay, since some would have been denitrified or stored in stream and river sediments. Estimates of the amount of N lost and retained in this way vary considerably in response to a variety of factors (e.g., Seitzinger *et al.,* 2002; Van Breemen *et al.,* 2002; Darracq and Destouni, 2005; Destouni *et al.,* 2006). For example, in their regression analysis of nitrate export from 35 watersheds around the world, Caraco and Cole (1999) obtained a strong correlation with water runoff using a constant retention and removal of 30% based on the results of Billen *et al.* (1991), while a modeling analysis by Seitzinger *et al.* (2002) suggested that in stream removal ranged from 37% to 76% of N input to the rivers in 16 Atlantic coast watersheds of the United States.

Given all of the assumptions and uncertainties, the amount of DIN delivered to the prehistoric stream and river network from the forested watershed may have been between about 5 and 25 million moles per year. If we follow the procedure of Caraco and Cole (1999) and reduce this by 30%, some 3.5 to 17.5 million moles of DIN may have been delivered to the bay from land with another 1 to 2 million moles deposited directly on the surface of the bay. As a rough approximation, the prehistoric Narragansett Bay may have received a total DIN input from land and atmosphere of 5 to 20 million moles per year, or 70 to 280 metric tons.

We began this exercise by noting that Verrazzano provided some detailed comments on the people he encountered, but all of the preceding discussion has been about pristine and minimally disturbed forests. It seems reasonable to ask what impact the people living in the Narragansett Bay watershed might have had on the DIN input estimates given above. Some perspective can be gained by using the surprisingly strong correlations that have been found between human population density and the export of nitrate (and other forms of N) in rivers around the world (e.g., Peierls *et al.,* 1991; Cole *et al.,* 1993; Caraco and Cole, 1999). Numerous studies converge on a population density of some 2–4 persons km^{-2} for the coastal hunter-gatherer populations in southern New England prior to extensive European contact (Nixon, 2004). The empirical regressions suggest that the export of nitrate from watersheds with this population density would be about 20 to 35 kg N $km^{-2} y^{-1}$ (1.4–2.5 mmol $m^{-2} y^{-1}$) for a total nitrate flux to Narragansett Bay of about 7 to 12 million moles of N y^{-1}. Since the regressions are based on modern studies in which synthetic fertilizers, enriched atmospheric deposition, and sewage treatment systems are part of the landscape, and the calculated nitrate flux is essentially equal to or lower than we have estimated, any human influence on N flux to the bay at the time of Verrazzano's visit has been captured in our estimate.

Since little P is transported through the atmosphere, the extensive measurements of dissolved inorganic P (DIP) exported from "minimally disturbed" watersheds in New England may provide a useful model of the DIP export from the forested watersheds that Verrazzano saw (e.g., Likens *et al.*, 1977; 0.06 mmol m^{-2} y^{-1}). While some dissolved organic P (DOP) is also lost from forests, a larger concern in terms of total P (TP) export is particulate phosphorus (PP). For example, in the Pawcatuck River study, DIP accounted for about 30% of the TP while PP accounted for almost 70% (Fulweiler and Nixon, 2005). Some fraction of the PP is almost certainly desorbed from suspended sediments when they first encounter salt water (Froelich, 1988), but the amount of sediment carried by streams draining the prehistoric forests around the bay is unknown and attempting to estimate the amount of biologically available P from this source probably introduces more error than understanding (Nixon, 1997 provides further discussion of this issue). The prehistoric input of DIP to the bay from the watershed may have been on the order of 0.3 million moles per year, or 9.3 metric tons. Again, as with N, it is possible to compare this estimate with an empirical regression relating modern DIP export to human population density in 32 large river systems around the world (Caraco, 1995). The result gives a calculated flux of 0.3 to 0.5 million moles of DIP per year, a result essentially in agreement with our estimate.

5.2.1.1 A Biological Desert?

Taking the total area of Narragansett Bay proper plus Mount Hope Bay (little fresh water drains into the Sakonnet arm of the bay) as 291 km^2 (Chinman and Nixon, 1985), the inputs given above translate into N loadings of roughly 15 to 70 mmol m^{-2} y^{-1} and a P load of 1 mmol m^{-2} y^{-1}. This range of DIN input is much lower than found in such oceanic deserts as the Sargasso Sea and the North Central Pacific Gyre, and would support only very low rates of primary production (Nixon *et al.*, 1996). Such a low primary production appears inconsistent with the great abundance of marine animals found in New England estuaries by early European explorers such as William Wood (1635). The explanation for this remarkable discrepancy may lie in the importance of offshore waters and the nutrients they contained (Nixon, 1997). Admittedly rough estimates of the DIN and DIP brought into Narragansett Bay through gravitational or estuarine circulation suggest that some 80 to 100 million moles of N, and 20 to 30 million moles of P each year, would have entered the bay from offshore (Nixon and Pilson, 1984; Nixon *et al.*, 1995; Chaves, 2004). These fluxes continue today, but their past importance is often not appreciated because of the now large anthropogenic N and P fluxes that enter the bay from land. Under prehistoric conditions, productivity in the bay may have been driven by clear waters, strong tidal mixing, shallow depths, and the entrainment of relatively nutrient rich near-coastal shelf waters. Under such conditions, 80 to 90% of the DIN entering the bay may have been from offshore, while today (or at least until very recently) the offshore contribution may be only about 15% (Nixon, 1997). Comparable values for DIP may have been over 95% from offshore in the past and 50% in the present. Even with the offshore nutrient inputs, primary production by the phytoplankton in the prehistoric bay may only have been about 120–145 g C $m^{-2} y^{-1}$ as calculated using a regression between DIN input per unit area and ^{14}C uptake in some marine systems (Nixon, 1997). But clearer waters with lower nutrient concentrations may have allowed kelp and eelgrass and epibenthic algae to make a much greater contribution to total bay productivity than they do today. A similar increase in the importance of benthic primary production (not including kelp) relative to phytoplankton has been proposed for the prehistoric Chesapeake Bay (Kemp *et al.*, 2005).

5.2.2 What F. P. Webber Saw

While we cannot see the Narragansett Bay that Verrazzano, we do know very well what it (or at least one important part of it) looked like between August 19 and September 7, 1865. We have such a remarkable view because a US Coast Survey party under the command of F.P. Webber spent those few weeks passing back and forth across the Providence River estuary in a cutter off the schooner *Margaret Stevens*. Along with 8,751 soundings in the area between Starve Goat Island (the island no longer exists, but it was then about 750 m south of a much narrower Field's Point) and the uppermost extent of Providence Harbor, the crew made detailed notes on the condition of the bottom. It seems remarkable now, but there were extensive meadows of eelgrass (*Zostera marina* L.) growing in shallows that were 0.6 to 0.9 m deep at mean low water throughout this whole stretch of the estuary and on both sides of the channel (Fig. 5.2). The presence of the eelgrass is important for the history of nutrient inputs to the bay because it establishes that nitrogen inputs to the upper bay were still very low 341 years after Verrazzano. We know this because eelgrass, like other seagrasses, is very sensitive to N loading (Duarte, 1995). There is still some uncertainty about the threshold of N loading that can be tolerated, but the estimates are all less than 1 or 2 mol $m^{-2} y^{-1}$ (Kelly, 2001; Bowen and Valiela, 2001; Nixon *et al.*, 2001). Recent field surveys of eelgrass coverage in southern New England and N loading by the US EPA show that coverage is reduced over 90% at loadings above 0.75 mol $m^{-2} y^{-1}$ (J. Latimer, US EPA Atlantic Ecology Division, Narragansett, personal communication). If we take the 2 mol $m^{-2} y^{-1}$ value and multiply it by the area of the Seekonk (2.03 km^2) and Providence River estuaries (5.48 km^2), the result suggests that the N loading to these systems was probably less than 5.5 million moles per year in 1865.

To put this in perspective, our estimated prehistoric DIN flux from all the watersheds that drain into the Seekonk and Providence River estuaries (the Blackstone, Ten Mile, Moshassuck, Woonasquatucket, and Pawtuxet Rivers, a total watershed of about 2,162 km^2, Ries, 1990) would have provided an input of 1.5 to 7.6 million moles of N per year, a loading consistent with the survival of eelgrass. But, of course, by 1865 those watersheds were no longer covered by

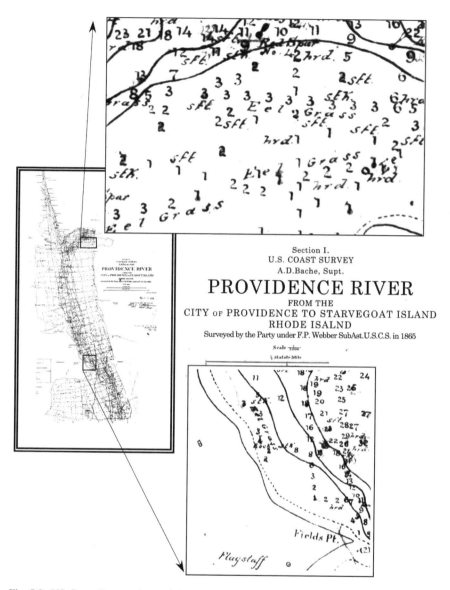

Fig. 5.2 US Coast Survey chart of the Providence River estuary in 1865, with enlargements showing eelgrass. Digitized from a copy provided by the National Archives.

an old growth forest (the maximum extent of cleared agricultural land occurred in 1850, when 80% of Rhode Island was classified as farm land; Hooker and Compton, 2003) and much of the Providence and Seekonk estuaries were surrounded by urban development. The cities of Woonsocket, Lincoln, Cumberland, North Providence, Pawtucket, Central Falls, East Providence,

Providence, and Cranston had an aggregate population of about 105,000 that had been growing at over 3% per year since 1800. The population living in the watershed of the Blackstone River above the Rhode Island state line was over 71,000 by 1870, a population density of 100 persons km^{-2} (Kirkwood, 1876). At the time of the Civil War, Rhode Island was the most industrialized state in the Union (Coleman, 1963) and Providence was a major manufacturing center for textiles and textile machinery, steam engines, armaments, and a wide array of other objects and implements (Gilkeson, 1986; Nixon, 1995a). In fact, it was almost certainly the importance of Providence and the Blackstone River Valley as a manufacturing center that prompted the Coast Survey to dispatch the *Margaret Stevens* for charting Providence Harbor in the midst of war.

The potential N release from the human population was already high by 1865. Measurements of the N and P excreted by a mixed urban population (52% male, 72% adults) in the United States around 1870 by Wolff and Lehman (as reported by Rafter and Baker, 1900) were 0.8 moles (11.2 g) $person^{-1} d^{-1}$ and 68 mmol (2.1 g) $person^{-1} d^{-1}$, respectively. The urban population around the Providence and Seekonk estuaries was thus producing some 32 million moles of N and 2.6 million moles of P each year. Using the regression given by Cole *et al.* (1993), the 100 people km^{-2} in the Blackstone drainage above the state line could potentially have produced a river with over 60 μM nitrate concentrations and a total annual nitrate export of almost 14 million moles.

How can we reconcile the compelling evidence of a very nitrogen sensitive plant growing throughout an area with such a potentially high input of N? Our hypothesis is that little of the N and P generated in human waste (and animal waste; census data show that the cities around Narragansett Bay contained about 5,000 horses in 1865) actually reached the upper bay at the time of Webber's survey. But the timing of his mission was propitious. He captured the Providence River for us just a very few years before it was to change dramatically.

5.2.3 A Celebration with Guns and Bells

On 21 March, 1853, the City Council of Providence appointed a committee to investigate the development of a public water supply. They were motivated by concerns over diseases attributed by some to the increasing juxtaposition of wells and privies and by the increasing hazard of fires in the rapidly growing city. The committee's efforts culminated on Thanksgiving Day, 30 November, 1871, with a 13 gun salute and the ringing of church bells as part of the ceremonies attending the introduction of running water into Providence from the Pawtuxet River (Annual Report of the Providence City Engineer, PCE) (Fig. 5.3). As a fountain was turned on in celebration, an unintended cascade of

Fig. 5.3 The Pettaconsett pumping station on the Pawtuxet River for the Providence public water supply. From the Annual Report of the City Engineer for 1892.

events began that would lead to the intensive pollution and fertilization of Narragansett Bay (Nixon, 1995a).

When the Coast Survey found eelgrass growing in the Providence River, virtually all of the human and animal waste from the urban areas around the bay and in the watershed was deposited on or in the soil (e.g., Kirkwood, 1876; Winsor, 1876; Tarr, 1996). In the cities, the solid waste accumulated in privy vaults or pail systems of various sorts and was periodically collected for use as fertilizer on outlying farms or put in landfills. Per capita water use was 7–11 liters d^{-1} (Tarr, 1996). The system of dry waste disposal largely retained N and P within the soil even when large numbers of people were involved. When running water became available, daily consumption increased to some 190 to 380 liters per person (Tarr, 1996), and the advantages of the "water closet" or indoor flush toilet meant that it was quickly adopted. Increasing amounts of N and P from human waste began flowing over and through the soil. As the urban environmental historian Joel Tarr noted, "…although many cities introduced running water during this period [1800–1880], no city simultaneously made provision for means to remove the water. It was expected that the previous means of water disposal—street gutters or cesspools—would deal with the problem." They were wrong. The water created a mess as soils became saturated and privy vaults overflowed, and Providence was the first of the

Narragansett Bay watershed cities to learn the lesson. The Board of Water Commissioners quickly employed J. Herbert Shedd as chief engineer to design a "water carriage" system of sewerage for the city. He prepared a plan and construction began the same year of 1871.

It is difficult to know how many people were connected to the sewers in the early years. The first report of the number of houses connected was in 1878, when 2,526 were "on line" (PCE, various years). A few years later, in 1884, Chief Engineer Grey made a detailed survey of the numbers of people living in the houses in various neighborhoods around the city and found a range of 7.7 to 8.9, a not unusual density for the time. Applying a mid point of this range to the 1878 house count suggests that the population served may have been about 21,000, or 20% of the city population. This many people would have released about 6 million moles of N and 0.5 million moles of P per year to the Seekonk and upper Providence River estuaries, thus almost doubling our upper estimate of the prehistoric N input to this area. The eelgrass had probably disappeared a few years before. The first official report of the number of people served by the city's sewer system came in the PCE report for 1889, when 54,000 were connected and direct annual N and P discharges to the estuaries would have been about 15 million moles and 1.3 million moles, respectively.

While the new sewers simply collected storm runoff and sanitary wastes and carried them to convenient points on the Seekonk, Moshassuck, Woonasquatucket, or upper Providence Rivers, Shedd's plans provided for the eventual construction of larger intercepting sewers that would carry the wastes below the city for discharge at Field's Point. By 1874, work on the regular sewers had progressed to the point where the Board of Alderman requested a report and further information regarding the plan for sewerage.

According to Rafter and Baker (1900), the report Mr. Shedd submitted was "... of special interest and value, by reason of containing the first thorough analysis of the relation of maximum rainfall to size of sewers to be found in American sanitary literature. In this particular, Mr. Shedd's report of 1874 is an engineering classic, and has been the model upon which nearly all the American sewerage reports since made have been based." There were political objections to the plan, however, and the Mayor of Providence asked the American Society of Civil Engineers to appoint a committee to review the proposal. Among other distinguished members, this group included E.S. Chesbrough, the engineer who had designed the Chicago sewer system in 1855, the first comprehensive sewer system in the United States.

In spite of a very favorable review by the Commission in 1876, it was not until 1882, after conditions in the various rivers had deteriorated from an increasing pollution load, that the Providence City Council directed Samuel Gray, then city engineer, to move forward with plans for construction of the intercepting sewers and to decide on the best method for treating and/or disposing of the sewage. As part of this effort, and at the direction of the Council, Mr. Gray and his assistant, Charles Swan, proceeded to Europe in February of 1884 to inspect the latest developments in sewage treatment technology. On their return, Mr. Gray prepared a detailed report and recommended that the intercepting sewers be completed and carried to Field's Point, where the sewage should be treated with chemicals to facilitate precipitation. The clarified effluent would be discharged into the Providence River on ebb tides. As described in an earlier discussion of the history of sewage treatment in Providence (Nixon, 1995a), these recommendations (similar in their essentials to the earlier Shedd Plan) were followed, and the first sewage (essentially untreated) was discharged at Field's Point in late 1892 (Figs 5.4 and 5.5). It is possible that the discharge of the sewage interacted with a very unusual mix of tide and wind in the late summer of 1898 to produce the most dramatic anoxia and fish kill that Narragansett Bay has ever experienced (Nixon, 1989).

The actual volume of the sewage flowing in the Providence system was first reported in 1897 in the annual report of the City Engineer, and chemical precipitation began in April of 1901 (Fig. 5.6). Treatment produced large volumes of sludge which were first dewatered and used as fill on the site. By 1908, this option had been exhausted and the sludge was taken by scow and dumped in Narragansett Bay in deep water below Prudence Island. This practice continued until 1949, when the city began to incinerate the sludge (Nixon, 1995a). Virtually all of the sewage from the city of Providence was being captured by the sewer system by the 1930s and the number of people served has remained relatively stable since that time (Fig. 5.7). In 1936, the Field's

Fig. 5.4 Building sewers along the Providence River. From the Annual Report of the City Engineer for 1894.

Fig. 5.5 Sewer construction at Field's Point. From the Annual Report of the City Engineer for 1891.

Fig. 5.6 The first sewage treatment facility for the City of Providence at Field's Point, January 1900. Precipitation tanks are in the foreground. From the Annual Report of the City Engineer.

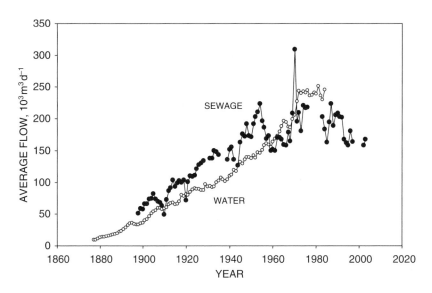

Fig 5.7 The volume of public water supplied and sewage collected in Providence after 1877 and 1900, respectively. From Annual Reports of the City Engineer.

Point facility adopted the activated sludge and aeration process that had been developed at the nearby Lawrence, MA Experiment Station during 1911 and 1912 (MSBH, 1923).

The city of Worcester, near the head of the Blackstone River, also began to develop its water and sewer systems in the early 1870s, and built one of the first treatment facilities in the United States using chemical precipitation in 1889–1890 (Tuttle and Allen, 1924). This facility proved unsatisfactory however, and Worcester replaced it in 1898 with a treatment facility that used sand filtration. Other Rhode Island urban areas surrounding the bay and its tributaries took a decade or more to follow suit, and all of them adopted filtration rather then precipitation when they constructed treatment facilities (Table 5.1; Fig. 5.8). If we assume that running water and sewer systems in the Blackstone, Ten Mile, and Taunton River watersheds followed the pattern for the state of Massachusetts as a whole, 94% of the population was served by water supplies by 1910. By 1900, only one town in the state with a population over 5,000 had not provided a public water supply (Annual Reports of the Massachusetts State Department of Health, MSDH). The population supplied with running water in the ten largest cities and towns on the shoreline of Narragansett Bay (including Woonsocket) increased by about 13,500 people each year between 1870 and 1910, from 0 to over 535,000. The decision by Worcester and the other cities and towns around the bay and in the watershed to adopt trickling sand filters rather than chemical precipitation as in Providence was to have some unfortunate consequences. While the filters reduced particulate and organic loading and nutrients, they required a significant area of land. As the populations of the

Table 5.1 Date when the major construction of public sewers began (though not necessarily a treatment facility) in cities and towns in the Narragansett Bay watershed and along the shoreline of the bay. From Gage and McGouldrick (1922), MDPH (1940) and *http:// www.state.ri.us/dem/programs/benviron/water/permits/wtf/monthly4.htm.*

Rhode Island	Year	Massachusetts	Year
Providence	1871	Worcester	1898
Pawtucket	1883	Hopedale	1900
Newport	1884	Northbridge	1906
Woonsocket	1895	Southbridge	1908
Central Falls	1896		
East Greenwich	1897	North Attleboro	1909
Bristol	1901	Attleboro	1912
East Providence	1910		
Warren	1917	Brockton	1893
West Warwick	1930s	Taunton	After 1940
Cranston	1940s	Swansea	After 1940
Bucklin Point	1954	Fall River [a]	Late 1940s
Smithfield	1978		
Jamestown	1980		

[a] Fall River first built a sewer system in the 1870s, but they constructed a combined storm and sanitary system. *http://www.greenfutures.org/projects/osp/community-d.*

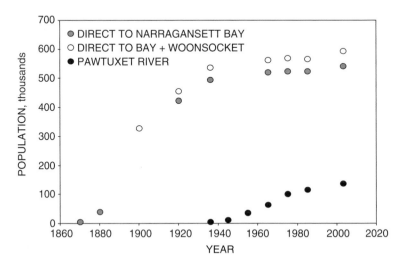

Fig. 5.8 Estimates of the number of people served by sewers in the cities and towns along the shoreline of Narragansett Bay over time. While the City of Woonsocket discharges to the Blackstone River, it does so close to the mouth. The three treatment facilities on the Pawtuxet River are also relatively close to the bay.

towns grew, land became more and more difficult and expensive to acquire. As a result, many of the early filter systems became overwhelmed by the 1920s and 1930s and provided virtually no treatment (e.g., MSDH; Gage and McGouldrick, 1922).

The number of people served by sewer systems in the urban shoreline cities and towns that discharge directly to Narragansett Bay or close to the mouths of the Blackstone and Pawtuxet Rivers rose steadily from 1871 until about 1950, and has remained relatively stable since then (Fig. 5.8). This situation often surprises people and contrasts sharply with many coastal areas now being studied for responses to nutrient enrichment. The explanation is that the addition of people not served by sewers in growing areas of the twentieth century appears to have been largely offset by declines in the populations of the older urban areas with sewers that had experienced such dramatic growth during the nineteenth century (Fig. 5.9). The now large urban areas on the Pawtuxet River only began sewer construction around the middle of the twentieth century, but the numbers of people served by those systems (Cranston, Warwick, West Warwick) have increased only modestly since the 1970s (Fig. 5.8). In contrast to the sewered population, the total population in the Narragansett Bay watershed has continued to increase, especially in the Massachusetts part of the watershed (Fig. 5.10).

With the rapid spread of electrical power in this area between 1910 and 1930, even those people without access to an urban water supply or sewers began to consume much larger amounts of water and to install flush toilets. This must have enriched the ground water with increasing amounts of DIN, but the time required for this N to begin to reach the rivers and the bay is not known. Remarkably however, much of the early imprint of the N pollution of the rivers by the combined influences of direct sewage discharges, groundwater enrichment, manufacturing, and what must have been increasing N deposition from the atmosphere, was captured by chemists measuring the concentrations of nitrogen in the rivers. We have not included fertilizer in the list of important early sources because the intensive use of synthetic fertilizer did not begin until after the Second World War (Smil, 2001), by which time agriculture was only a minor land use in this area (Hooker and Compton, 2003).

5.2.4 What the Chemists Measured

5.2.4.1 Historical Monitoring of the Blackstone River

Some of the earliest measurements of nitrogen in North American rivers were made in the watershed of Narragansett Bay. Like Webber's chart, they allow us to see a period of rapid change that we would otherwise have to reconstruct using proxies and other indirect techniques. Measuring the very dilute concentrations involved was a great challenge to the chemists of the nineteenth

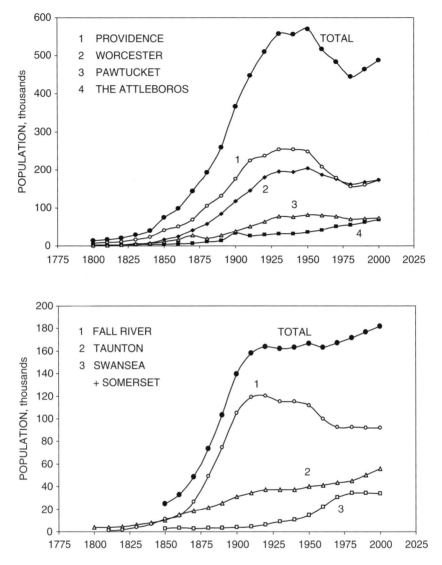

Fig. 5.9 Upper panel: Population in the larger urban areas that discharge sewage to the Blackstone River and the Seekonk and Providence River estuaries. Lower panel: Population in the larger urban areas that discharge sewage to the Taunton River and Mount Hope Bay.

century, and much of the groundbreaking work was done in Britain (an extensive review is given by Hamlin, 1990). The early interest in measuring nitrogen in the rivers in England and in New England had nothing to do with eutrophication, but with developing a method by which sewage pollution could be detected and drinking water supplies could be judged safe or condemned.

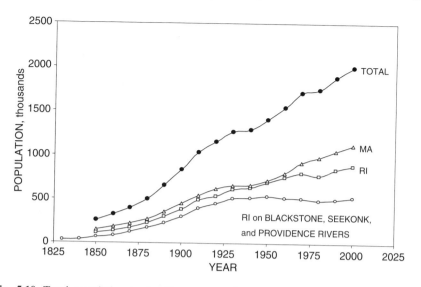

Fig. 5.10 Total population in the Narragansett Bay watershed with separate estimates for numbers in the Rhode Island and Massachusetts portions and in the older urban areas in Rhode Island along the Blackstone River and the Seekonk and Providence River estuaries.

As Hamlin (1990) traces the history, it was the great agricultural chemist and physiologist Justus von Liebig who focused on nitrogen as the best indicator of the decay of organic matter, in his view the indirect cause of disease carried by contaminated water. While Liebig began developing this concept in the 1840s, it was not until the late 1860s that methods were developed to actually measure ammonia and organic N in fresh water and sewage. J.A. Wanklyn's "ammonia process" was the one adopted by William Ripley Nichols in 1871 for the chemical laboratory at the Massachusetts Institute of Technology (Nichols, 1876). Nichols had recently moved from Harvard to MIT, where he became one of the leading water quality chemists in the United States. The same method was later adopted by what was to become the "birthplace of environmental research in America," the Lawrence Experiment Station, founded by the Massachusetts Department of Health (MSDH) on the shore of the Merrimack River in 1887 and still active (McCracken and Matera, 1987). Until at least 1917, nitrates and nitrites were commonly measured in the United States using an alkaline reduction with aluminum (Phelps and Shoub, 1917). Because of analytical difficulties, and because it was not considered an important pollutant in drinking water, DIP was virtually never measured in early monitoring programs (e.g., Wanklyn and Chapman, 1884). The first attempt to standardize analytical methods for drinking water in this country was published by the American Association for the Advancement of Science in 1889. This was superseded by the regular publication of Standard Methods of Water Analysis, the first edition of which appeared in 1905 and the second in 1917. The Lawrence Experiment Station was much involved in all of this, and it is probably safe to assume that the methods used by the MSDH followed the standard methods closely.

In his history, Hamlin (1990) is very critical of the early methods for ammonia and organic N, based largely on a study comparing the three commonly used methods (including Wanklyn's) that was funded by the first US National Board of Health (Mallet, 1882). A close reading of the Mallet Report does not sustain such a pessimistic view however. Wanklyn's "ammonia process" gave measurements of "free ammonia" and "albumenoid ammonia," the latter thought to represent the more easily digested forms of organic N. In his report, Mallet (p. 65) concluded that, "...the figures which probably may fairly be taken to represent the average divergence from the mean of a single determination by the albuminoid ammonia process are—for free ammonia, 2.23%; for albuminoid ammonia, 3.62%." For its time, the analysis was repeatable in the hands of a skilled chemist. It was more difficult to determine the accuracy of any of the methods except to compare results with the amount of N liberated from pure compounds of known composition. In Table XIV of Mallet's report, he showed that the greatest recovery from Wanklyn's technique was 94% for free ammonia and 84% for albuminoid ammonia. Average recoveries were low, but Mallet described the averages as having "no significance" (p. 65). In the end, his conclusion (p. 72) was equivocal: "Of course it's not assumed that the figures [on recovery] really represent either the range of absolute error, or the average amount of absolute error, of the different processes as applied to the doubtless very various forms of organic matter liable to occur in natural waters."

The greatest concern appears to have been how much of the organic N was actually measured by the "albuminoid" N analysis. This can be addressed more satisfactorily than Mallet was able to do because the Lawrence Station carried out thousands of measurements of both "albuminoid" N and Kjeldahl N (a commonly used modern technique employing a strong oxidizing agent to measure organic N) on the same samples of raw and treated sewage effluents and drinking water during the 1920s and 1930s. Comparing many of these analyses suggests that a multiplication of the albuminoid measurements by 1.75 to 2.0 provides an approximation of what a modern measurement of organic N would have found.

As far as we can determine, the first measurements of nitrogen in any of the water flowing into Narragansett Bay were made in the Blackstone River by Nichols (1873) on September 28, 1872 with the following results, all in μM: above Millville, $NH_4 = 2$, $NO_{2,3} = 9$, DIN = 11, Albuminoid N = 14 and below Millville, $NH_4 = 2$, $NO_{2,3} = 17$, DIN = 19, Albuminoid N = 14. Nichols sent E.K. Clarke back to collect more samples on the Blackstone as part of another survey in July of 1875. Samples were taken along the length of the river, and analyses of water from near the lower two stations discussed above gave the following results, all in μM (Nichols, 1876). Unfortunately, nitrate and nitrite were not included. Below Millville, $NH_4 = 12$, Albuminoid N = 13 and near the RI border, $NH_4 = 8$, Albuminoid $N = 9$. As a brief historical footnote, it is likely that some or all of these early Blackstone analyses were actually carried out by Nichols' research assistant, Ellen Swallow Richards, the first female chemist trained in America (Pate, 1992).

It is possible to put these early measurements into a specific historical context because James Kirkwood carried out the first detailed study of the pollution of the Blackstone River for the Massachusetts State Board of Health in 1875 (MSBH, 1876). At the time he estimated that some 55,000 people were connected to sewer systems that discharged to the river and that there were almost 7,000 people working in 44 woolen mills and 27 cotton mills in the Blackstone Valley. The mills were potentially important sources of N for two reasons. First, the wastes from the employees would be discharged into or close by the river during the working day. Second, large amounts of organic compounds containing N were used in textile manufacturing, including dung, urine, and blood. It is even possible to make some estimate of the potential N release from the mills because Kirkwood provided data on the amount of cotton (3,440 mt) and wool (2,230 mt) processed during the year by the Blackstone mills and rough estimates of the amount of blood, dung, and urine used to process a given weight of raw cotton and wool. Assuming that fresh dung, urine, and blood contained 0.5, 0.9, and 2.1% N, respectively (Straub, 1989), Kirkwood's data suggest that the mills may have been discharging a total of almost 3 million moles of N per year when Nichol's made the first measurements of N in the Blackstone. While the textile wastes were often very visible because of the dyes released, the grease removed from wool, and other detritus, they were less important as a source of N than human sewage. The N released from the mills was about 15% of the almost 15 million moles of N released by the sewered population along the river, almost all of which was in Worcester, far above the two stations where Nichols took the measurements shown above. The consumption of wool and cotton only approximately doubled between 1875 and the time it peaked in 1900, and improving technology also reduced the release of nutrients per unit of cloth. For these reasons, the rough exercise with Kirkwood's data suggests that human sewage rather than manufacturing waste was the driving force by far for nitrogen inputs to Narragansett Bay, even at the height of the textile industry.

To put his measurements in the Blackstone in some perspective, Nichols also analyzed the drinking water delivered to his laboratory by the Boston water supply, in which he found 6 μM NH_4, 14 μM NO_3, and 11 μM albuminoid N, essentially the same as he found in the river. Of course, these are single samples, but they have historical value and they become more interesting in the context of the great many samples that would come later. Despite the fact that the Blackstone was far from a pristine system in 1872 and 1875 (Kirkwood described it as "...probably more polluted than any other in Massachusetts", MSBH, 1876; p. 73), the concentrations of N were not yet dramatically elevated, even when compared to the "unpolluted" water being supplied to the citizens of Boston.

With the establishment of the Lawrence Experiment Station, a much more ambitious monitoring program was implemented at various stations on an increasing number of rivers in Massachusetts. Samples were collected monthly from June through November, inclusive, and the results reported by the MSBH (1910; 1920) as six month means. This monitoring began in the Blackstone in summer 1887, in the Taunton in 1898, and in the Ten Mile in 1899. Unfortunately, the results stopped appearing in the MSBH reports after 1919, and the measurement of nitrate and nitrite terminated after 1914. It is possible that some measurements continued at least through the 1920s and 1930s because a brief narrative paragraph on the status of each river continued in the annual reports, but without supporting data. A description of the activities of the Massachusetts Department of Health for the years 1950 through 1956 noted that the department maintained approximately 260 river sampling stations throughout the state (Massachusetts Department of Public Health, 1956), but we do not know what was measured. The same report also says that the Water and Sewage Laboratory, which had operated on the top floor of the State House since 1897, and the old Experiment Station in Lawrence, were both closed during this period and moved into a new Lawrence Experiment Station.

The concentration of DIN in the Blackstone at Millville (near the Rhode Island state line) during the summer low flow period rose almost four fold during the 28 years for which we have quantitative data, largely from increasing

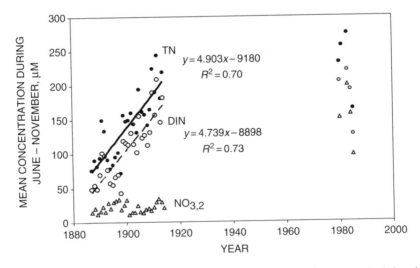

Fig. 5.11 Concentrations of total nitrogen, dissolved inorganic nitrogen, and nitrite plus nitrate in the Blackstone River at Millville between June and November as measured by the Lawrence Experiment Station in the late 1800s and early 1900s, by the USGS in the 1980s. The early TN values are estimated as twice the "albumenoid N" (see text). Almost all the N was ammonia in the early period, while nitrate accounts for most of the N in recent years. The early data are from the Annual Reports of the Massachusetts State Dept. of Health. USGS data are from their annual reports.

ammonia (Fig. 5.11). Even during the first summer of 1887, ammonia levels were far above the 2 μM that Nichols had measured 14 years earlier. The presence of N as ammonia also suggests that sewage was the source of the N. The dramatic increase was the result of the interaction of rapid population growth and the rapid spread of public health infrastructure consisting of water supplies and sewer systems. N concentrations may have continued to increase for perhaps another decade, but almost certainly at a slower rate because of an end to rapid urban population growth and the full build out of water and sewer systems (Figs 5.8 and 5.9). This is consistent with modern (1979–1985) monitoring of N at the same Millville site by the USGS if we average their measurements during the June through November period in years when measurements were obtained during at least five of these months (Fig. 5.11). Comparison between the MSBH results and the USGS also suggests that the earlier analytical techniques were not missing large amounts of N.

Measurements made over annual cycles in the Blackstone above Woonsocket (not far below Millville) in 1913 and 1923 by the RI Board of Purification of Waters (RIBPW) showed a slight decline in average ammonia concentrations (from 68 to 51 μM) and a decline in DIN (from 87 to 75 μM) (Gage and McGouldrick, 1923; 1924). Old data sheets from sampling done by the RIBPW in 1927 that we found in the files of the RI Department of Environmental Management (RIDEM) showed an average of 73 μM ammonia and 23 μM nitrate and nitrite (DIN = 96 μM), and bi-weekly sampling at this site by the MDPH in August 1935 through July 1936 found a mean ammonia concentration of 78 μM. It is hard to know if there were increasing sources of ammonia below Millville after 1923, or if the higher values reflect lower flows or analytical changes. But it is clear that ammonia remained the dominant form of DIN in the river at least through the mid 1930s. Even fewer early measurements appear to have been taken further down the river just before it enters the Seekonk River estuary at Pawtucket. The same old unpublished data sheets from the RIBPW in 1927 showed a mean ammonia concentration of 57 μM and nitrate plus nitrite of 30 μM. These compare with our measurements below Pawtucket in 1982–1983 of 43 μM ammonia and 95 μM nitrate plus nitrite (Nixon et al., 1995). Whatever increases there were in the DIN flux from the Blackstone to Narragansett Bay between the first quarter of the twentieth century and the last appear to be due to sources in Rhode Island.

Our conclusion that N delivery (at least from the Massachusetts portion of the Blackstone) has not changed much from the early decades of the twentieth century appears to contradict a recent report from the USGS (Robinson et al., 2003), but their assessment only considered nitrate concentrations. Comparison of the early and recent data in Fig. 5.11 shows clearly that up until the end of the measurements in 1914, ammonia was the most abundant form of the DIN and the form that was increasing dramatically. The Rhode Island data from above Woonsocket also show ammonia concentrations that were over three times greater than nitrate and nitrite in 1913 and twice the oxidized forms in 1923 (Gage and McGouldrick, 1923; 1924). In measurements of river water collected at Valley Falls, the lowest station sampled by the RI Board of Purification of Waters, N concentrations also changed little, if at all, between 1913 and 1923 and ammonia dominated the DIN pool. By 1979, however, improved sewage treatment and improving oxygen levels in the river resulted in the oxidation of ammonia and a shift to nitrate accounting for almost all of the DIN.

We can also make a very approximate estimate of the amount of N that might have been carried by the Blackstone when the concentration measurements started. Since no long-term flow measurements have been made at Millville, we have to adjust the record from another site. The long-term mean June through November flow of the Blackstone at Woonsocket, RI is just over one million $m^{-3} d^{-1}$ (Ries, 1990). Multiplying this by 0.66, the ratio of watershed above Millville to the area above Woonsocket, and then multiplying by the Millville concentration (TN = 59 µM), the flux of TN in the river may have been about 40 kmol d^{-1} when monitoring started in 1887. Since most of the N was ammonia from point sources, it may be reasonable to apply this flux throughout an annual cycle to derive a minimal annual TN flux down the Blackstone of about 15 million moles, the same amount we estimated for N release in sewage from the Providence sewage system in 1889. By 1914, the TN flux in the Blackstone at Millville was about 53 million moles per year or more. Both the 1887 and the 1914 calculations probably underestimate the total transport of N in the river because monthly measurements of ammonia over an annual cycle in 1897 in the Blackstone at Valley Falls, RI, by the Rhode Island State Board of Health (RISBH) showed that the mean annual concentration was 1.8 times greater than the June through November measurements. Organic N was about the same on an annual basis, and nitrate was about 10% greater. If we apply these findings to the relative proportions of ammonia, organic N, and nitrate at Millville (an admittedly risky extrapolation), the annual estimate for 1887 might be increased by a factor of 1.4 to 21 million moles and the 1914 estimate by a factor of 1.6 to 85 million moles.

It will also prove useful to compare the concentration measurements in the Blackstone with our estimate of prehistoric concentrations based on a DIN export from the forest of 1–5 mmol $m^{-1} y^{-1}$ and an in-stream loss of 30%. With a watershed area above Millville of 717 km^2 (Ries, 1990), the mean summer concentration of DIN might have been 2–10 µM in the prehistoric Blackstone, compared with Nichol's (1873) first measurements in 1872 of about 10–20 µM, the 1887 mean of 45 µM, and the 1914 mean of almost 175 µM. Measurements of "unpolluted" surface water in the late 1800s and early 1900s showed DIN concentrations usually averaging about 5 µM (Table 5.2), well within the range we have calculated for the prehistoric drainage. If we accept a value of 10 µM DIN as reasonable for the Blackstone in 1870, the introduction of running water and sewers to serve a rapidly growing urban population increased this over 17 fold during less than 50 years. By 1903, the USGS described the Blackstone as "...the most polluted river in New England, its name has become synonymous with filth."

Table 5.2 Early analyses of nitrogen concentrations (μM) in "unpolluted" surface water flowing into Narragansett Bay.

		Ammonia	Nitrate + Nitrite	DIN
Pawtuxet River at Hope[a]				
1894–1899	Mean	1.3	4.8	6.1
	Range	0.9–1.5	4.2–5.7	
1902–1906	Mean	1.0	3.7	4.7
	Range	0.8–1.1	2.1–5.0	
Woonsocket water supply[b]				
1907–1912	Mean	1.6	3	4.6
	Range	1.0–2.0	2.8–3.6	
Kickamuit River[c]				
1907–1912	Mean	2.3	3	5.3
	Range	1.7–2.8	2.1–4.3	
Hunt River[d]				
1909–1912	Mean	0.9	4.3	5.2
	Range	0.8–1.1	2.9–5.7	
Clear River[e]				
1908–1912	Mean	1.0	1.9	2.9
	Range	0.8–1.4	0.7–2.9	

Values are means of twice monthly samples. Data from RISBH (1899; 1912).
[a] "... above all sources of pollution from town and mill wastes." RISBH (1899).
[b] "...a large watershed which is owned or controlled by the city...Practically no inhabitants are located in the area...The supply is a sanitary one."
[c] Bristol and Warren public water supply. (1909 very low flow deleted).
[d] East Greenwich public water supply, the watershed described as "...comparatively free from habitation or chance pollution."
[e] sampled at Burrillville, above the state sanatorium, and described as a "good quality" stream.

5.2.4.2 Impact on the Seekonk Estuary

The Blackstone River discharges directly to the Seekonk estuary at the very top of Narragansett Bay, and the changing nutrient delivery by the river must have had a major impact on the estuary. Again we are lucky, because Albert Whitman Sweet, a graduate student at Brown University, studied the chemical and microbiological conditions in the Seekonk River estuary during 1913 and 1914 for his Ph.D. dissertation (Sweet, 1915). His thesis gives a nice overview of the early science of stream pollution and detailed descriptions of the chemical and microbiological techniques then in use. His measurements of ammonia at nine stations at both low and high tide showed mean concentrations of 90 and 100 μM, respectively, with little difference among stations. These levels are twice what is found today, but ammonia today is only about half of the DIN (Doering *et al.,* 1990). Unfortunately, Sweet could not obtain satisfactory nitrate analyses, but his measurements of nitrite were surprisingly high at about 8 μM. These high nitrite levels may reflect low concentrations of dissolved oxygen in the Providence River estuary (Webb, 1981), where much of the ammonia in the Seekonk must have come from on the flooding tides. Oxygen concentrations in the Seekonk were lowest at high tide and ranged, on average, from about 25% of saturation in the lower Seekonk to about 70% just below the dam. On low tides, oxygen in the lower Seekonk averaged about 40% of saturation. The Seekonk estuary was clearly heavily enriched by 1913–1914, but the conditions with regard to dissolved oxygen were better than they must have been in the adjacent Providence River estuary. Of course, the Blackstone Point sewage treatment facility had not yet been constructed (Table 5.1), so the Seekonk was "only" being enriched by the increasing N coming down the Blackstone and Ten Mile Rivers and the N being carried up in the flood tides from Providence.

While Sweet's interest was primarily in the chemistry and microbiology (*B. coli,* now *E. coli*) of the Seekonk estuary, he also made some observations about the ecology of the area. He noted that the Seekonk was at one time "…a very good fishing ground. Fish of marketable variety and size were common, shellfish were very plentiful, particularly the hard and soft shell clam and oysters were found in great numbers." (p.121). But larger fish had not been part of the estuary for "…some time previous to 1908…" and "…fish life in [the Seekonk] has been of no consequence for the past eight years." (p.121). He attributed the decline to the construction of a railroad tunnel and a draw bridge across the estuary below the "Red Bridge," but deteriorating oxygen concentrations in the Providence River estuary may have been the real obstacle preventing fish from moving into the Seekonk. By the time of his study, Sweet found only "Minor fish life, to a slight degree, some shrimp and a few other small types of marine forms…together with marine plant forms, are present in the lower sections." Unfortunately, he does not tells anything more about the plants—except to emphasize that the green seaweed, *Ulva,* "…which feeds upon the free ammonia…was never found in the Seekonk." According to Sweet, however, *Ulva* was "…very prevalent in upper Narragansett Bay." (p.121).

5.2.4.3 Measurements in Other Rivers

The Ten Mile and the Taunton

Monitoring of the Ten Mile River below Attleboro and the Taunton River below Taunton also showed marked increases in ammonia between about 1900 and 1906, but concentrations in these rivers did not get above 20 μM until 1903–1904. Ammonia in the Blackstone already exceeded 30 μM when monitoring began there in 1887. N concentrations in the Ten Mile and the Taunton showed no trend between about 1906 and 1914. As discussed in detail in the next section, our laboratory measured the concentrations of nutrients in the Ten Mile River below Attleboro on a biweekly basis during 2003–2004. The June–November mean DIN concentration in 1914 was 164 μM, most of which was ammonia. In our study it was 115 μM during the same period, with almost all in the form of nitrate. As in the Blackstone, DIN fluxes in the Ten Mile appear to have changed little or declined since the early 1900s. This does not seem to be the case with the Taunton. The most recent nutrient measurements we have found for this river were collected at Taunton by Boucher (1991) in 1988–1989. Based on monthly samples during June through November, she found a mean DIN concentration of 141 µM, 96% of which was nitrate and nitrite. The Lawrence Station reported ammonium of 50 µM and nitrate plus nitrite of 8 µM in 1914. Unfortunately, there appear to be no data available to document the time course of the 2.5 fold increase. The total population of the major urban areas along the Taunton River increased approximately linearly between 1880 and 2000, with rapid increases in Swansea and Somerset between about 1945 and 1975 (Fig. 5.9).

The Pawtuxet

Massachusetts was not the only state to begin to monitor nitrogen in rivers during the late 1800s. The RI State Board of Health (RISBH) paid particular attention to the Pawtuxet River because it served as the source of drinking water for Providence until 1926. Samples were collected bi-weekly at Pettaconsett Pumping Station (Fig. 5.3) between 1876 and 1925 and reported by the Providence Superintendent of Health in 1885 and 1892, by the RISBH (1899) for 1895–1899, and by Gage and McGouldrick (1923; 1924) for 1900–1925. While it was forbidden by law to discharge sewage into the river above the pumping station, there was intensive industrial development. In 1924 there were; 13 large cotton mills; six plants bleaching, dyeing, and finishing cotton; two lace mills; one corduroy mill; and two woolen mills (Gage and McGouldrick, 1923; 1924). Remarkably, however, the long-term monitoring showed no increase in ammonia or nitrate between 1876 and 1920, with an average ammonia concentration during that period of 2.4 µM (range in five year group means of 1.1 to 3.7 µM) and nitrate plus nitrite of 9.8 µM. This is consistent with our analysis of the Blackstone in which textile manufacturing was not a strong source of N. Concentrations in the Pawtuxet rose slightly in the 1920–1925 period to about 4.3 µM ammonia and 16 µM nitrate. Once the water supply no longer depended on the river however, its special protection was lost and N concentrations must have begun to increase. Sewer systems were installed beginning in the late 1950s (see Fig. 5.8), and by the time our laboratory first measured nutrients at the dam in Pawtuxet (the mouth of the river) in 1976, the annual mean ammonia concentration was 95 µM and nitrate plus nitrite was 53 µM. DIN concentration had increased 12 fold over the 1876–1920 background.

The Moshassuck and the Woonasquatucket

In marked contrast to the early Pawtuxet, the small urban rivers in Providence were enriched at least by 1897, when the RISBH carried out bimonthly sampling for ammonia and "total organic nitrogen." Unfortunately, they did not include nitrate and nitrite. Mean annual concentrations of ammonia were 56 and 24 μM at Smith Street on the Moshassuck and at Gaspee Street on the Woonasquatucket, respectively, while the concentrations of organic N were 410 and 130 μM. The latter are extremely high and are due either to analytical problems or to severe organic loading. When we sampled these two rivers in 1982–1983, mean annual ammonia concentrations in both rivers were about 20 μM and total organic N was about 40 μM (Nixon *et al.*, 1995).

5.2.4.4 Early Deposition from the Atmosphere

Our lack of attention to atmospheric deposition of N in causing this enrichment reflects the fact that estimates of NO_x releases to the atmosphere from fossil fuel combustion suggest that the rates of deposition were still very low in the early 1900s (Husar, 1986). With an exponential increase of 4.3% y^{-1} (calculated from the middle of the uncertainty range for the Eastern US given by Husar, 1986), these low rates of deposition would only have doubled between 1870 and 1920. The very low DIN concentrations in surface water not obviously polluted by sewage and animal waste reinforces the relative unimportance of the atmosphere as a source of N before the second half of the twentieth century (Table 5.2). Howarth, in Chapter 3 of this volume, further details atmospheric N contributions to Narragansett Bay.

5.2.4.5 A Note on Horses and Urban Runoff

While nitrogen oxides may have added little N to the rain around 1900, once the rainfall landed on urban areas it would have encountered another source of N that was potentially important. It is hard for us to appreciate today, but large numbers of horses were required in urban areas before the widespread availability of trucks, buses, and cars. The US Census provided counts of "non-farm" horses, and in 1900 there were about 22,000 living in the cities around the shoreline of Narragansett Bay, with 8,000 in Providence alone. A century later, this may seem quaint and charming, but Tarr (1971) provided an amusing and useful reality check on the difficulties presented by large numbers of horses in the cities. He noted that a "normal city horse" produced 10 kg d^{-1} of manure, a total of 80 mt d^{-1} in Providence in 1900. Since horse manure contains about 0.7% N and 0.1% P (Straub, 1989), this would have contained 530 kg N (37.8 kmoles) and 80 kg P (2.6 kmoles). Of course, horses also produce urine, and the daily release of N and P in this form amounts to 80 g N d^{-1} (5.7 moles) for an average full grown horse weighing 630 kg (Altman and Dittmer, 1968). The Providence population would have released 640 kg N d^{-1} (45.7 kmoles). With an N/P ratio in the urine of a grass eating herbivore of about 8 (Altman and Dittmer, 1968), this would suggest a daily release of 80 kg d^{-1} (2.6 kmoles) of P in horse urine. The total production by the Providence horse population in

1900 would have been about 30 million moles of N and 1.8 million moles of P each year. Not all of these large amounts would have been released on the streets, since the animals must have spent most of each day on grass or gravel or dirt in stables and fields. But even one third is 10 million moles of N and 0.6 million moles of P per year in Providence alone. It is very difficult to know how much of this N and P actually reached the bay in storm water runoff. Much of the N in urine must have volatized, and a lot of the P would have been adsorbed on soils and other surfaces. Manure was often collected and removed from the city as well.

We can make a very rough estimate of how much may have been washed off the streets because Metcalf and Eddy (1915) reported the analyses of street runoff from an urban residential area in Lawrence, MA during 1894–1910 which showed an average composition of: Ammonia = 1.6 µM; Nitrate N = 93 µM; Organic N = 0.98 µM; Nitrite N = 11 µM. These concentrations are 10 to 20 times higher than found in urban runoff in this area today (Hanson, 1982; Nixon *et al.*, 1982) and probably reflect the impact of horse wastes. If we assume that the composition of storm runoff in Providence in 1900 was similar to that in Lawrence, we can make a very rough estimate of the flux of N. By 1900, Providence had about 584,500 m^2 of paved (75% with granite block) streets (PCE, 1900). With a long-term mean annual rainfall of 1.06 m (Pilson, 1989) and assuming a runoff coefficient of 1, this would translate into a storm water runoff of almost 620,000 m^3 and a TN flux of about 1.7 million moles or 17% of our calculated TN flux to the streets of Providence. This estimate may be low because runoff from roofs also contributes to storm water and because the concentration data came from a residential area. Rafter and Baker (1900) reported analyses of runoff from heavy traffic areas in London in 1893 showing ammonia concentrations of 2.5–5.0 µM and 3.6–6.1 µM organic N. Even the lower range would increase our calculation by a factor of 2.25 to 38% of the estimated N deposition on the streets of Providence. Perhaps a reasonable estimate is that some 1.7 to 3.8 million moles of N in storm water runoff might be attributed to the horse population of Providence in 1900. Since Providence had about one third of the urban horses around the bay, the total flux to the bay from horse wastes in urban runoff might have been on the order of 5–11 million moles of N y^{-1} at the turn of the last century. The number of urban horses living around the bay was only about 5,000 in 1865 and it peaked in 1900. The numbers dropped sharply after the Model T Ford was introduced in 1908, and urban horses were virtually gone by 1925.

5.2.4.6 Sewage Treatment Facilities and Protein Consumption

The early water quality chemists did not confine their measurements to rivers and drinking water supplies. Once sewage treatment facilities were constructed, they began to measure the concentrations of various pollutants in the effluent and the volume discharged. In some cases, they also made an effort to document the effects of treatment and the efficiency of removal for various pollutants. As part of the latter process, they analyzed raw sewage at various times as well as the treated effluent. As early as 1924, S. Stronganoff working in Russia appreciated that such analyses afforded a unique opportunity to assess the overall nutritional status of an urban population. He used analyses of raw sewage to document the impact of the First World War and the revolution of 1918–1920 on the diet of the inhabitants of Moscow (Tuttle and Allen, 1924). While Stronganoff attempted to estimate caloric intake, we can use the same approach more directly with N. The first sets of measurements of ammonia and albuminoid N in the raw sewage arriving at Providence's Field's Point facility over an annual cycle appear to have been reported for the period 1907 through 1912 by the Rhode Island State Board of Health (1912) (Table 5.3). Since flow data and the number of people served by the system were also reported for 1910 by the City Engineer (PCE, 1910), we can estimate the per capita N excretion by the population. The result is 12 to 12.7 g N d^{-1} (0.86 to 0.91 moles N d^{-1}) for the Providence population of 199,000, depending on whether albuminoid N is increased by 1.75 or 2. In either case, the result is just a bit higher than the 11.8 g person^{-1} d^{-1} reported for a mixed urban population in 1870 (Rafter and Baker, 1900). An excretion of 12 g N d^{-1} suggests a daily average protein consumption of 75 g. This compares with more recent recommended dietary protein intakes of 63 g d^{-1} for males over age 25 and 50 g d^{-1} for females over age 25 (National Academy of Sciences, 1989). The next set of raw sewage analyses that we have found were part of an unpublished correspondence from the Providence Sewage Disposal Works in old files at RIDEM showing a "typical analysis" of raw sewage and treated effluent in 1933 (Table 5.3). Going through the same exercise as before, but for an average daily flow of 137,896 m^3 and a

Table 5.3 Average analyses of raw sewage and treated effluent at the Field's Point sewage treatment facility during 1910 and 1933.

	Raw sewage	Effluent	% Removed
1910			
Free Ammonia	1.73	1.61	7
Albuminoid ammonia	0.72	0.38	47
Total N	2.45	1.99	19
Adjusted Total	3.17	2.37	25
1933			
Suspended solids (ppm)	230	140	39
B.O.D. (5 day, ppm)	228	164	28
Free ammonia	1.3	0.96	26
Albuminoid ammonia	0.49	0.40	18
Total N	1.79	1.36	24
Adjusted total	2.28	1.26	22

Concentrations are in mM of N. Data for 1910 are from the RISBH (1912). Data for 1933 are from unpublished files of the Providence Sewage Disposal Works. Adjusted total is with albuminoid N times 2.0 (see text).

population served of 249,000 (Providence Department of Public Works, 1932), the mean per capita N release comes to 16.7 to 17.7 g d^{-1} (1.19 to 1.26 moles N d^{-1}). This result is essentially the same as found between January 2002 and August 2004 when the Narragansett Bay Commission measured TN in 250 samples of raw sewage at Field's Point using modern techniques (Narragansett Bay Commission, personal communication). Their mean concentration of 1.37 mM ± 0.41 S.D. combined with an average daily flow of 137,869 m^3 and a population served of 208,745 gives an average per capita release of 16.2 g N d^{-1} (1.16 moles N d^{-1}).

These results provide evidence that the average resident in Providence enjoyed an excellent diet in terms of protein at least as early as 1910, that they increased this already generous consumption by about 33% between about 1910 and 1930, and that the average daily protein intake has remained at about 100 g per person per day since then. These results are roughly consistent with the findings of more traditional dietary surveys, though studies of actual protein consumption (compared with protein availability in the food supply) are surprisingly few. The US Census Bureau (1975) reported that detailed national studies of urban households in 1936, during the depth of the Depression, found a range in per capita protein consumption of 66 to 90 g d^{-1} depending on income. A study of over 25,000 US families in the early 1940s found that an average laboring man consumed 88 g of protein each day while business and professional men consumed 96 g (Chancy and Ahlborn, 1943).

The excess consumption of protein relative to nutritional requirements becomes even more striking if we calculate a population mean daily requirement using the National Academy requirements by sex and age group and the age specific Providence census data for 2000 *(http://providence.areaconnect.com/ statistics.htm)*. The resulting size-weighted per capita mean is 49 g protein d^{-1}, about half that being consumed. Put another way, about half of the N discharged to Narragansett Bay from human sewage over the past 75 years has been derived from excess protein consumption.

The analyses summarized in Table 5.3 suggest that the Field's Point treatment facility was removing about 25% of the N from the raw sewage before it was discharged to the Providence River estuary on ebb tides. The facility was using chemical precipitation with lime and chlorination throughout this time (except for a period during the First World War when lime was not available; PCE annual reports). It is possible that the removal was overestimated because no measurements were made of nitrites and nitrates in the effluent, but it is unlikely that this was a large factor at Field's Point. The N that was removed by the treatment would have been associated with the sludge that was dewatered and sent by scow to be dumped in deep water south of Prudence Island (Fig. 5.12). As a result, the treatment did not decrease the N ultimately discharged to the bay, it simply changed the location where it entered. The amount of sludge harvested declined almost from the beginning of the treatment facility operation. During the first ten years, about 200 to 400 g of dry solids were removed from each m^3 treated, but this soon fell to about

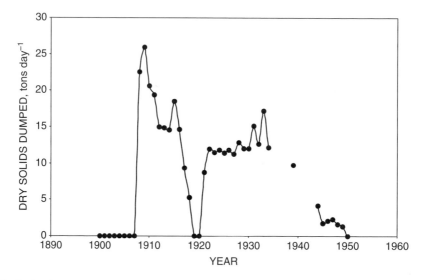

Fig. 5.12 The amount of sewage sludge from the Providence treatment facility at Field's Point that was dumped each year in Narragansett Bay below Prudence Island. The gap just before 1920 was due to a shortage of lime during the First World War. As less sludge was harvested and dumped in the bay, more was released into the Providence River estuary. Data from the Annual Reports of the City Engineer.

100 g m^{-3}, where it remained from 1915 to about 1935. By the early 1940s, only about 25 g m^{-3} were removed. We have not found any measurements of the N contained in the sludge, but modern sludges produced with more advanced technology average about 3% N (National Academy of Science, 1996). Applied to the roughly 12 tons of sludge cake dumped each day in the early 1930s (Fig. 5.12), this would produce an addition of about 325 kg N d^{-1} in the mid bay or 8.5 million moles each year. This falls within our range of the DIN input to the whole of the prehistoric Narragansett Bay and a considerable, but unknown, fraction of the fresh sewage derived PN must have been rapidly decomposed and remineralized. On a per capita basis however, this amount of nitrogen only amounted to about 1.3 g d^{-1} or about 7.5% of the daily excretion by each resident of Providence. If we work back from the treatment facility removal estimates of about 25% reduction in N between raw and treated sewage, the sludge would have to have been closer to 10% N and the input of N to the mid bay would have been three times greater. From the perspective of the bay as a whole of course, virtually all of the N handled by the Providence sewer system reached Narragansett Bay until sludge burning replaced disposal in the bay in 1949. On site incineration was discontinued in December 2005. The sludge harvested at the Field's Point facility today (about 27 dry tons d^{-1}) is taken by tankers to Woonsocket and/or Cranston where it is dewatered and incinerated. A small amount ($< 10\%$) is dewatered on site and sent to a landfill in Maine (C. Walker, Narragansett Bay Commission, personal communication 10/19/06).

As noted earlier, the chemical precipitation facility in Providence was unique in the Narragansett Bay watershed. Other facilities operated various combinations of settling tanks followed by sand filtration. The annual reports of the Massachusetts and Rhode Island health authorities consistently reported very high removal efficiencies by the filtration facilities for ammonia N and organic N (about 80–90%), whether measured by the albuminoid process or Kjeldahl digestion. Fortunately, they also measured nitrates and nitrites, so that we can evaluate the efficiency of TN removal. Data show that the treatment facility in Central Falls, Rhode Island removed an average of 60% of the TN delivered to the facility in 1900, and that the facility serving Pawtucket, Rhode Island averaged the same reduction during 1907 through 1912 (RISBH, 1912). The MDPH annual reports show similar impressive removals of TN in the watershed filtration treatment facilities between 1920 and at least 1940, the last year we examined (Tables 5.4 and 5.5). When they worked well, the filtration

Table 5.4 Average analyses of raw and treated sewage effluent at the Worcester, MA filtration plant.

	Raw Sewage	Treated Effluent	% Reduction
1920			
Ammonia	1.50	1.33	
Organic N	1.37	0.34	
Nitrate		0.76	
Total	2.87	2.43	15
1922			
Ammonia	1.53	1.06	
Organic N	0.95	0.15	
Nitrate		0.61	
Total	2.48	1.82	27
1926–1929			
Ammonia	1.46	1.38	
Organic N	1.15	0.38	
Nitrate		0.13	
Total	2.61	1.89	27
1935			
Ammonia	1.72	0.85	
Organic N	1.15	0.21	
Nitrate		0.69	
Total	2.87	1.75	39
1940			
Ammonia	1.61	0.67	
Organic N	1.41	0.35	
Nitrate		0.54	
Total	3.02	1.56	48

Data are from MDPH annual reports. The system was reconstructed in 1925. Concentrations are in mM of N, mean of day and night samples. Organic N is by Kjeldahl; nitrate includes nitrite.

Table 5.5 Total N removal (%) by some sewage treatment plants using sand filtration in the Narragansett Bay watershed in Massachusetts, as calculated from data in MDPH annual reports.

	1920	1925	1935	1940
N. Attleboro	70	63	59	53
Attleboro	–	63	59	45
Brockton	71	48	21	57

facilities came to remove 25 to 50% of the TN, presumably by denitrification of the nitrate produced in the trickling filters. The larger, modern, full secondary treatment facilities operating in the watershed and on the bay release TN amounting to some 0.8 to 1 mole of N per person each day (Table 5.6). For the Field's Point facility, where we calculated a daily per capita TN inflow to the facility of 1.16 moles, the removal in sludge amounts to about 20%. Assuming that the dietary conditions of other sewered populations is similar, removal commonly falls between 20 and 30% when nitrification and denitrification have not been engineered into the system. It seems possible that the per capita N actually reaching the bay from the sewered population in the watershed and along the shoreline (excluding those served by the Providence system) may have been similar to what it is today, or as little as half that amount. The difficulty in

Table 5.6 Recent (~2001–2003) per capita daily nitrogen and phosphorus discharge rates from different sized sewage treatment facilities in the Narragansett Bay watershed. All facilities are in Rhode Island unless noted.

	Population served	N	P
Field's Point	208,745	0.93	21
Bucklin Point	119,660	0.80	39
Brockton, MA	109,510	0.91	19
Fall River, MA	93,615	0.95	–
Cranston	81,000	0.61	44
Woonsocket	51,370	0.93	116
East Providence	47,835	0.66	34
West Warwick	29,075	0.77	130
Warwick	28,000	0.46	54
Attleboro, MA	18,200	0.68	31
Bristol	16,900	1.07	29
North Attleboro, MA	15,160	1.28	33
Somerset, MA	14,310	0.81	31
Warren	8,000	0.81	16
Burrillville	7,685	0.48	9.4
East Greenwich	2,500	0.99	–
Jamestown	1,720	0.46	59

Nutrient data are from treatment facility monitoring files and estimates of the population served from US EPA (available online at *http://cfpub.epa.gov/cwns/rptffs00.cfm*). Units are moles per person per day for N and mmoles per person per day for P.

making a calculation of historical loading is in knowing how much of the sewage from the various towns and cities was actually treated by the filtration beds at different times. As noted earlier, many of the beds apparently became overloaded, and much of the sewage was almost certainly diverted to rivers and the bay with essentially no treatment.

5.2.5 *What Was Missed*

While the historical monitoring of N in the rivers and sewage treatment facilities discharging to Narragansett Bay provides us with a surprisingly complete early record of N (but not P) inputs to the bay, that situation changed rapidly after the first decades of the twentieth century. Beginning at various times in the 1920s and 1930s, the monitoring programs appear to have been reduced or eliminated or, if maintained, the results were no longer published in the annual reports of public agencies. Doubtless a variety of factors contributed to this, including the Depression and the Second World War. The motivation for monitoring N in rivers and in treatment facility effluents also disappeared during this period as the measurement and interpretation of microbiological indicators of pollution improved (Ashbolt *et al.*, 2001). The last half of the nineteenth century was marked by great debate among water chemists and sanitarians about whether chemical or biological indicators best reflected contaminated water (Hamlin, 1990). But by 1914, the US Public Health Service adopted a bacteriological standard for interstate water supplies, and *B. coli* (now *E. coli*) became the accepted indicator of fecal pollution (Ashbolt *et al.*, 2001).

Since N was not firmly identified as the primary nutrient limiting carbon fixation in most coastal marine ecosystems until the late 1960s (Howarth and Marino, 2006), and marine eutrophication was not commonly perceived as a coastal marine problem until the 1970s or later (Nixon, 1995b; in press), there was little or no motivation among environmental scientists to measure the various forms of N in rivers or sewage treatment facilities discharging to the coast. Even in lakes, there was much debate about the importance of P until the 1970s (Schindler, 1981), and little reason for monitoring P in rivers or treatment facilities discharging to the bay.

However tenuous the link really was between public health and N, it stimulated the monitoring that provided us with a rare opportunity to see the dramatic rise of N inputs to an urban estuary during the period when public health and safety infrastructure in the form of water supplies and sewers was being constructed at a rapid pace. Total N input to Narragansett Bay from land and atmosphere probably increased from less than 50 million moles per year in 1865 to about 500 million moles per year by 1925. Surely, this must have resulted in the eutrophication of the bay and major changes to its ecology. Since hundreds of thousands of people lived around the bay and many used it for recreation and food production, one would think that these changes must have been noticed, remarked upon, and documented. In an attempt to see if and how these changes were perceived, Daniel (2004) examined the Providence Journal newspaper every day during the months of July, August, and September every five years between 1870 and 1935, inclusive. During the fourteen summers consisting of over 1,200 days of sampling, she found almost nothing relating to what we think of as consequences of nutrient enrichment—loss of seagrass, accumulations of nuisance macro algae, intense phytoplankton blooms or water discoloration, fish kills, rotten egg smells, etc. To be sure, there were concerns about sewage and its possible contamination of shellfish beds, low dissolved oxygen in the urban upper bay, oil pollution on beaches, and bad smells in The Cove in downtown Providence (until it was filled in), but almost nothing that would indicate that people were aware of or concerned with the intensive fertilization of the bay. That such things could become newsworthy is shown by the coverage of the dramatic fish kill and phytoplankton bloom of September 1898, which was not captured by Daniel's half decade sampling (Nixon, 1989).

While there was great concern with "pollution" as a public health issue—this concern lead to the construction of the treatment facilities, and finally, in 1920, to the creation of the Rhode Island Board of Purification of Waters—there appears to have been little awareness of the effects of nutrient enrichment *per se*. This is not as surprising as it may first seem. An increase in the primary productivity of the microscopic phytoplankton that support the bay's major food chains would not have been apparent to the public, nor would changes in the species composition of the plankton. Increasing hypoxia due to eutrophication would have been obscured by the much more immediate oxygen demands of the organic matter being added to the upper bay in sewage effluents. And below the Providence River, the lack of strong density stratification (due to the small input of fresh water to this bay compared with many others) and strong tidal and wind mixing make Narragansett Bay relatively safe from the anoxia and fish kills that plague many other estuaries that receive large amounts of nutrients. These conditions are further detailed in Chapters 11 and 12 of this volume. Now, seventy-five years later, the fertilization of the bay has emerged as a major environmental concern. It is a great irony, for as we will show in the next section, the inputs of N have changed relatively little since 1925, and probably not at all during the last quarter century since measurements resumed.

5.3 The Present

5.3.1 The 1970s

While some of the authors of this chapter objected to beginning a discussion of "present" conditions with measurements that were made before they were born, the work of that decade is still vivid and fresh in the experience of their older colleagues. Clearly, the measurements of the 1970s stand apart from the historical work, not only in time, but also in method. After decades with no measurements of N (and no measurements ever of P) in the effluents of the major sewage treatment facilities that discharge to Narragansett Bay or in the major rivers that enter the bay, S.W. Nixon and C.A. Oviatt at the Graduate School of Oceanography at the University of Rhode Island began a program of sampling in June of 1975. Their purpose had nothing to do with public health and little to do with eutrophication. They needed to obtain data that could be used to make monthly estimates of the total inputs of nutrients from the various sources to use in the first numerical ecosystem simulation model of Narragansett Bay that was under development by Rhode Island Sea Grant (Kremer and Nixon, 1978). All forms of N and P were measured approximately monthly at the three largest sewage treatment facilities that discharge directly to the Seekonk and Providence River estuaries (Bucklin Point, Field's Point and East Providence) and approximately biweekly at the mouths of the Blackstone and Pawtuxet Rivers, the two largest rivers that discharge to Narragansett Bay proper (i.e., excluding Mount Hope Bay and the Sakonnet Passage). The rivers were sampled through April 1977. The three treatment facilities measured account for about 70% of the sewage that is discharged directly to the bay (85% of the sewage discharged to Narragansett Bay proper) and the two rivers account for about 38% of the total fresh water drainage (Nixon et al., 1995).

5.3.2 The 1980s

Sponsored by Rhode Island Sea Grant, our laboratory made a more complete inventory of N and P inputs as part of an effort to develop the first annual mass balance of nutrients in Narragansett Bay during 1983–1984. We measured all forms of N and P approximately biweekly in final effluent from the three large treatment facilities measured during the 1970s, as well as biweekly in the Blackstone, Pawtuxet, Moshassuck, and Woonasquatucket Rivers just before they discharged to the bay. Concentrations of all forms of the nutrients were also measured monthly at the mouth of the Taunton River between June 1988 and December 1989 by Boucher (1991), and we used those data to calculate river nutrient fluxes into Mount Hope Bay. We do not know how much of the nutrient load from the Taunton reaches Narragansett Bay proper, though Krahforst and Carullo, in Chapter 13 of this volume, provide estimates for further consideration. About 30% of the river flow is thought to exit from Mount Hope Bay into the Sakonnet River (actually a marine embayment, not a river; Hicks, 1959). With a water residence time of seven days (Saunders and Swanson, 2002), about 18% of the remaining N may be denitrified and buried within Mount Hope Bay (Nixon et al., 1995), leaving just 57% of the Taunton N load to be passed into Narragansett Bay proper. The measured rivers account for almost 75% of the land drainage to the bay.

The USGS also began measuring the monthly concentrations of nutrients in the Blackstone and Pawtuxet Rivers in 1979, but their station on the Blackstone was at Millville, MA, 32 km upstream of the last dam before the river reaches the bay and thus did not capture discharges from Woonsocket and Pawtucket. In a study we carried out for the Narragansett Bay Commission between early April and late August, 2004, the flux of DIN was 28% higher at the Pawtucket dam than at Millville, and DIP was 25% higher (Nixon et al., 2005). The USGS nutrient measurements were reduced to quarterly in 1990, and then eliminated by 2002.

Less frequent sampling of all of the other sewage treatment facilities that discharge directly to the bay was carried out by others during the 1980s as part of The Narragansett Bay Project, so that some estimate of all of the direct discharges could be included. Most of these are very small compared with the three facilities that discharge to the Seekonk and Providence River estuaries, but the Fall River, MA facility that discharges to Mount Hope Bay is about the same size as the Bucklin Point facility, and the Newport, RI treatment facility that discharges to the lower East Passage is about 40% of the Bucklin Point facility. Our inventory of N and P inputs to Narragansett Bay around the mid 1980s has been reported in detail, along with the specifics of sampling locations and analytical methods used in all of the river and sewage studies we have done (Nixon et al., 1995). In reviewing the estimates for the mid 1980s however, we now believe that the values we used for the input from "unmeasured flows" were too high (Nixon et al., 1995; their Tables 14 and 17). At the time the budget for the mid 1980s was developed, we estimated the nutrient flux to the bay from the Ten Mile River (which we did not measure in the 1980s) as well as two small gauged rivers, five small unguaged rivers, and what Ries (1990) called "coastal drainage," by the ratio of the water flow from those sources to the gauged flow in the five rivers for which we had concentration measurements. As discussed in a later section describing our nutrient input estimates for 2003–2004, we believe this procedure inappropriately extrapolated nutrient fluxes in larger urban rivers receiving sewage treatment facility discharges to the smaller suburban and rural watersheds. Recent measurements from a less developed watershed give much lower losses of N and P per unit area and we believe that these provide a more realistic proxy for the less developed small streams that drain to the bay (Fulweiler and Nixon, 2005). Since we now have measurements of N and P fluxes from the Ten Mile River, we can use them assuming that they have not changed significantly from the 1980s. The total adjusted input of N from land and atmosphere of 605 million moles is about 14 to 26% higher than we have estimated here for the mid 1920s. Unfortunately, no earlier estimate of total P input is available for comparison with the adjusted TP input of 36 million moles y^{-1} that we found in the mid 1980s. Of these totals, 72% of the TN was DIN and 61% of the TP was DIP. Direct discharges of sewage to the bay provided 37% of the TN and 38% of the TP entering the bay.

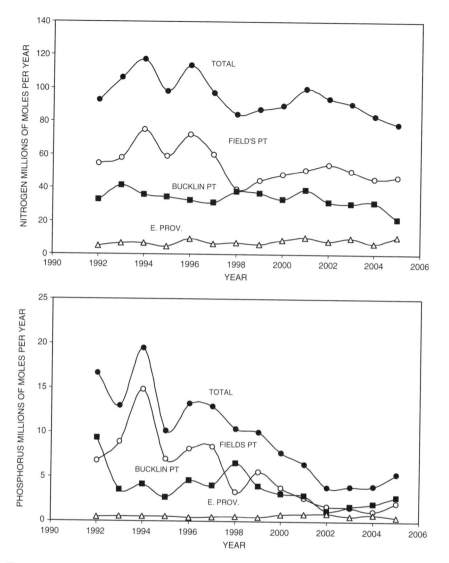

Fig. 5.13 Upper panel: Annual release of dissolved inorganic nitrogen from the three sewage treatment facility that discharge directly to the Seekonk and Providence River estuaries. Data from before 2000 provided by Save The Bay; fluxes from 2000 on calculated from annual monitoring data provided by the treatment facilities. Lower panel: As above except for total phosphorus.

5.3.3 The 1990s

A rising interest in coastal marine eutrophication gave us a chance to make biweekly measurements of nutrient concentrations in the same set of rivers over annual cycles in 1991 and 1992 with the assistance of volunteer employees from Citizens Bank (who collected the water samples) and partial support from the Citizen's Charitable Foundation and Rhode Island Sea Grant (Kerr, 1992). By the early 1990s, many of the treatment facilities were also measuring the concentrations of dissolved inorganic N (DIN) (and often total P) in their effluents. The facilities continued measuring DIN and TP with varying frequencies until 2002, when the larger facilities began to add measurements of total N in their effluents. According to the treatment facility monitoring data which the facilities have provided to us, the total amount of DIN discharged to the Providence and Seekonk River estuaries has remained relatively unchanged or declined slightly since the early 1990s, while TP has declined by about two thirds (Fig. 5.13). The contribution of organic N (mostly DON) to TN varied over time and among facilities, so the DIN release data before 2002 provide an

Table 5.7 Nitrogen and phosphorus inputs to Narragansett Bay proper in the 1970s and 1980s, and in 2002 and 2003 from the three largest sewage treatment plants that discharge directly into the bay.

	1976–77[a]	1983[b]	2002[c]	2003[c]
Narragansett Bay Commission—Field's Point				
Discharge, 10^3 m^3 d^{-1}	151	195	158	168
Total nitrogen	70	83	61	71
Ammonia	22	60	44	39
Nitrite + nitrate	3	3	10	11
Organic nitrogen	45	20	7	21
Total phosphorus	3.3	3.3	1.7	1.6
Narragansett Bay Commission—Bucklin Point				
Discharge, 10^3 m^3 d^{-1}	50	78	79	95
Total nitrogen	20	26	37	35
Ammonia	8.8	20	31	30
Nitrite + nitrate	0.1	0.3	0.6	0.3
Organic nitrogen	11	5.4	6	5.4
Total phosphorus	1.7	2.9	1.2	1.7
Riverside—East Providence				
Discharge, 10^3 m^3 d^{-1}	9.7	22	22	28
Total nitrogen	6.1	12	9.2	11.5
Ammonia	4.0	10	1.0	2.2
Nitrite + nitrate	0.1	0.8	7.0	7.8
Organic Nitrogen	2.1	1.6	1.2	1.5
Total Phosphorus	0.4	0.8	0.9	0.6
PHOSPHORUS TOTAL	5.4	7.0	3.8	3.9
NITROGEN TOTAL	96	121	108	118

* Units are millions of moles per year.
* In 1983 these sources accounted for 66% of the N and 50% of the P from direct sewage discharges to the bay.
[a] Mean discharge on 9–10 days when samples were collected over an annual cycle.
[b] From Nixon et al. (1995).
[c] Data from the Narragansett Bay Commission and East Providence treatment plant.

incomplete picture of the total N discharge from the treatment facilities (Table 5.7). For example, DON accounted for 24% of the TN discharged from Field's Point in 1983, but only 13% at East Providence. While our measurements and the last four years of measurements by the treatment facilities (2002–2005) show that the amount of TN released by the three largest facilities that discharge directly to Narragansett Bay proper did not change significantly between the mid 1970s and 2005, the amount of the N in organic form declined dramatically from 60% of the total in the 1970s to 22% in 1983 and to 17% in 2005. The ecological consequences of this switch from organic N to DIN due to better secondary treatment is not known. Obviously, the organic N that is now decomposed in the treatment facilities is labile N that would have been largely, if not totally, remineralized during its transit of the bay, which has a mean water residence time of 26 days (Pilson, 1985a). But we do not know where in the bay it would have become available to the primary producers. It seems reasonable to suggest that this switch may have increased primary production and the standing crop of phytoplankton in the upper bay and/or Providence River at the expense of the mid and, perhaps, lower bay. Smayda and Borkman in Chapter 15 of this volume further explore the relationship between nutrient availability and primary production. Primary production by the phytoplankton was highest in the Providence River portion of the bay in the early 1970s (Oviatt et al., 1981) and continues to be so today (Oviatt et al., 2002), but the spatial resolution of the data and differences in measurement techniques make it difficult to detect potential changes in the distribution. As discussed in more detail at the end of this chapter, there has been a dramatic long-term decline in chlorophyll concentrations in the mid bay since the early 1970s, but there is no compelling evidence of a corresponding increase in the Providence River estuary. Moreover, the fact that most of the decline in sewage DON occurred by the early 1980s makes it unlikely that it played a significant role in the on-going long-term decline of chlorophyll in the mid bay.

5.3.4 *The Most Recent Measurements*

5.3.4.1 Rivers and Land Drainage

With river nutrient flux measurements available for the 1970s, 80s, and 90s, it seemed compelling to carry out another assessment with the start of a new century. The Narragansett Bay Commission and Rhode Island Sea Grant agreed, and provided support for biweekly concentration measurements at the same sites we had sampled previously, and added the Ten Mile River which flows into the Seekonk River estuary along with the Blackstone. The Ten Mile is a small river with a mean annual discharge of 3.1 m^3 s^{-1} compared with the Blackstone's 24.4 and the Pawtuxet's 11.7 (Ries, 1990). A detailed description of the biweekly river sampling (April 2003 through March 2004) and data analysis, including analytical methods, concentrations, hydrographs, and statistical comparisons among the years studied is given in Nixon *et al.* (2005). The concentration measurements in this study (as well as in all our earlier work) were converted to nutrient fluxes using USGS water discharge data from the nearest gauge and corrected for watershed area below the gauge as done by Pilson (1985a) and Ries (1990). In the case of the Ten Mile River, no water discharge measurements were available during our sampling period. However, we were able to correlate the discharge of the Ten Mile when it was measured by the USGS from 1983 through 2003 with the measured discharge of the Blackstone River at Woonsocket during that period (log Ten Mile discharge, $m^3 d^{-1}$ = 0.83*log Blackstone discharge + 0.23, r^2 = 0.87). We used this regression to estimate discharge from the Ten Mile during our study based on measurements reported by USGS for the Blackstone River at Woonsocket. The "instantaneous" flux measurements were converted to annual flux estimates using Beale's unbiased estimator (Beale, 1962; Dolan *et al.*, 1981) and daily USGS water flow measurements. This estimator provides a flow-weighted annual flux from a set of instantaneous flux measurements that are skewed and/or not normally distributed (Richards, 1999), two features that are characteristic of discrete river sample data sets. The results showed TN flux decreases of 25% in the Blackstone, 12% in the Pawtuxet, and 20% in the Moshassuck since the 1980s survey and a TN increase of 13% in the Woonasquatucket (Table 5.8). Comparable results for the fluxes of TP from each river were decreases of 32%, 41%, and 27%, respectively, for the three rivers that showed TN decreases and an increase of 14% for the Woonasquatucket (Table 5.9). Unfortunately, we have no earlier measurements for comparison with the Ten Mile results except for the historical measurements discussed earlier.

Because the annual flux of nutrients from the rivers varies with water flow (Nixon *et al.*, 1995), it is not easy to recognize longer term trends (or the lack of them) by a simple inspection of the annual flux estimates given in Tables 5.8 and 5.9. For this reason, we carried out a detailed statistical examination of the river flow and concentration data for all of the sampling periods. Since our concentration sampling usually extended several months on either side of a calendar year, we refer throughout this discussion to the sampling by decades (e.g., 1970s, 1980s, etc.) rather than by the specific calendar year used in the flux calculations. We first inspected 1/x and log log transformations of the nitrogen and phosphorus concentration and water discharge data to satisfy conditions of linearity and normality. We found that log transformations did a better job at increasing linearity, while also reducing skewness and kurtosis and producing means and medians that were more comparable. We then tested the fit of a multivariate regression as part of an analysis of covariance using the model:

$$\text{Logconc} = \beta_0 + \beta_1 \log \text{flow} + \beta_{70's} + \beta_{80's} + \beta_{90's} + \beta_{00's} + \epsilon$$

Table 5.8 Annual estimate of nitrogen fluxes into Narragansett Bay from rivers at various times between 1975 and the present.

	1975–1976	1983	1991	1992	2003–2004
Blackstone River					
Mean daily flow	2.47	3.17	2.24	1.99	2.57
Dissolved N					
Inorganic	63.72	98.70	59.14	63.71	68.88
Organic	31.08	28.06	38.36	50.94	23.25
Particulate N		5.04			6.50
TOTAL		131.80			98.63
Pawtuxet River					
Mean Daily Flow	1.06	1.57	1.06	1.10	1.00
Dissolved N					
Inorganic	31.27	46.17	47.70	43.63	44.61
Organic	12.08	17.99	30.04	37.20	11.61
Particulate N		3.41			3.07
TOTAL		67.57			59.29
Woonasquatucket River					
Mean Daily Flow		0.31	0.17	0.18	0.28
Dissolved N					
Inorganic		4.73	2.87	3.80	6.62
Organic		2.39	3.44	3.83	1.67
Particulate N		0.44			0.30
TOTAL		7.56			8.59
Moshassuck River					
Mean daily flow		0.19	0.11	0.12	0.19
Dissolved N					
Inorganic		4.16	1.74	1.82	3.50
Organic		1.40	1.96	1.56	1.01
Particulate N		0.35			0.26
TOTAL		5.91			4.77
Ten Mile River					
Mean daily flow					0.35
Dissolved N					
Inorganic					9.86
Organic					3.30
Particulate N					0.91
TOTAL					14.07

Mean daily flow units are $10^6 \, m^3 \, d^{-1}$. Nitrogen fluxes are in millions of moles per year.

We used a SAS program to carry out this analysis with year variables converted into a binary classification of 1 or 0 for four columns representing the 70s, 80s, 90s, and 00s; the regression was fit to determine if a single line could appropriately describe all the years of data with one slope. When this was the case, an analysis of covariance was completed by fitting this regression to remove the trend and then testing to see if the adjusted means (i.e., intercepts) were significantly different from one another for each of the time groups. Results

Table 5.9 Annual estimate of phosphorus fluxes into Narragansett Bay from rivers at various times between 1975 and the present.

	1975–1976	1983	1991	1992	2003–2004
Blackstone River					
Mean daily flow	2.47	3.17	2.24	1.99	2.57
Dissolved P					
Inorganic	2.62	2.72	2.42	1.05	1.69
Organic	1.89	1.11	1.64	2.04	0.35
Particulate P		1.83	2.34	0.65	1.83
TOTAL		5.66	6.40	3.74	3.87
Pawtuxet River					
Mean daily flow	1.06	1.57	1.06	1.10	1.00
Dissolved P					
Inorganic	2.61	4.45	1.63	1.00	1.96
Organic	0.60	0.93	0.98	1.66	0.32
Particulate P		0.79	1.18	1.11	1.33
TOTAL		6.17	3.79	3.77	3.61
Woonasquatucket River					
Mean daily flow		0.31	0.17	0.18	0.28
Dissolved P					
Inorganic		0.12	0.06	0.07	0.16
Organic		0.06	0.10	0.11	0.04
Particulate P		0.10	0.11	0.13	0.12
TOTAL		0.28	0.27	0.31	0.32
Moshassuck River					
Mean daily flow		0.19	0.11	0.12	0.19
Dissolved P					
Inorganic		0.07	0.05	0.12	0.07
Organic		0.04	0.02	0.04	0.01
Particulate P		0.07	0.05	0.06	0.05
TOTAL		0.18	0.12	0.22	0.13
Ten Mile River					
Mean daily flow					0.35
Dissolved P					
Inorganic					0.24
Organic					0.22
Particulate P					0.35
TOTAL					0.81

Mean daily flow units are 10^6 m^3 d^{-1}. Phosphorus fluxes are in millions of moles per year.

report both the transformed and un-transformed values for adjusted means. The comparison was completed using Tukey's multiple comparisons test to determine which years were different from one another.

Significant ($p < 0.01$) regressions were found for dissolved inorganic nitrogen, dissolved organic nitrogen, and total dissolved nitrogen in the Blackstone and Pawtuxet Rivers, and for dissolved inorganic and total dissolved nitrogen in the Woonasquatucket and Moshassuck Rivers. Statistically significant regressions of dissolved inorganic phosphate with water discharge were found for all the rivers except the Moshassuck. A statistically significant regression between dissolved organic phosphorus concentration and flow was only found for the Pawtuxet River. When the concentration and flow regressions were not significant (DON in the two smaller rivers, DIP in the Moshassuck, DOP in all rivers except the Pawtuxet), we tested the mean concentrations for differences from one another during each sampling period using analysis of variance. Since the Woonasquatucket and Moshassuck Rivers were not sampled in the 1970s, we had a shorter window of time in which to look for changes. No time analysis was done on the Ten Mile River concentration data because we lacked comparable earlier measurements.

Nitrogen

Results of the analyses described above showed that there was no consistent trend in nitrogen concentration in any of the rivers tested. Total dissolved nitrogen was significantly lower in the Blackstone River during the 2003–2004 sampling than in the 1990s, dissolved inorganic nitrogen was significantly lower than it was in the 1980s, and dissolved organic nitrogen was significantly lower than it had been in the 1970s and 1990s (Table 5.10). Total dissolved nitrogen was also significantly lower in the Pawtuxet and Moshassuck Rivers in 2003–2004 than it was in the 1990s (Table 5.10). Overall, dissolved nitrogen (accounting for about 95% of the total nitrogen in the rivers) has not significantly increased in any of the four rivers studied since the 1980s, nor has it increased significantly in the Blackstone or Pawtuxet Rivers since the 1970s (Table 5.10). In fact, total dissolved nitrogen concentrations in the two larger rivers, the Blackstone and the Pawtuxet, were significantly lower in 2003–2004 than they had been a decade earlier. These reductions were substantial and amounted to about 20% in the Blackstone and 25% in the Pawtuxet. The reduction in the Blackstone can be explained entirely by the introduction of N removal at the Woonsocket wastewater treatment facility after 2000. By the time of our measurements, the TN discharge from Woonsocket was down by 28 million moles y^{-1} and TP was down by 4 million moles y^{-1} compared with 2000. Using the average discharge of the river during our study (2.6 million m^3 d^{-1}, Table 5.8), the N reduction amounts to a decrease in average TDN concentration of 30 mmol m^{-3}, exactly what we observed relative to concentrations measured in the 1990s (Table 5.10).

The reduction in the Pawtuxet, despite continued population growth in the watershed, is not as easily explained. In addition, about 13,400 people were connected to the Warwick treatment facility between 1992 and 2003 as part of an aggressive program to reduce pollution inputs to Greenwich Bay from individual on site sewage disposal systems. This is equivalent to adding about 4.5 million moles of N y^{-1} to the facility. At least part of the explanation is that the largest sewage treatment facility that discharges to the Pawtuxet at

Table 5.10 Results of statistical analysis testing differences between decades in nitrogen concentration in the Blackstone, Pawtuxet, Woonasquatucket, and Moshassuck Rivers.

		Blackstone River			
	LOGTDN	**TDN**			
Year	LSMEAN	LSMEAN	70s	80s	90s	00s
70s	2.14	138.49		0.9927	0.9696	0.0336
80s	2.15	141.68	0.9927		0.9997	0.0261
90s	2.15	142.67	0.9696	0.9997		*0.0020*
00s	2.05	112.19	0.0336	0.0261	*0.0020*	
	LOGDIN	**DIN**			
Year	LSMEAN	LSMEAN	70s	80s	90s	00s
70s	1.97	93.83		0.0101	0.4249	0.9567
80s	2.12	130.79	0.0101		*<0.0001*	*0.0043*
90s	1.91	80.96	0.4249	*<0.0001*		0.8351
00s	1.95	88.48	0.9567	*0.0043*	0.8351	
	LOGDON	**DON**			
Year	LSMEAN	LSMEAN	70s	80s	90s	00s
70s	1.66	45.23		0.0230	0.4890	*0.0004*
80s	1.44	27.24	0.0230		*<0.0001*	0.8047
90s	1.74	55.14	0.4890	*<0.0001*		*<0.0001*
00s	1.37	23.33	*0.0004*	0.8047	*<0.0001*	
		Pawtuxet River			
	LOGTDN	**TDN**			
Year	LSMEAN	LSMEAN	70s	80s	90s	00s
70s	2.1625	145.38		0.4378	*0.0001*	0.1589
80s	2.2097	162.06	0.4378		*0.0001*	0.9604
90s	2.3536	225.72	*0.0001*	*0.0001*		*0.0001*
00s	2.2249	167.83	0.1589	0.9604	*0.0001*	
	LOGDIN	**DIN**			
Year	LSMEAN	LSMEAN	70s	80s	90s	00s
70s	2.0389	109.38		0.4573	0.0215	0.0736
80s	2.0863	121.99	0.4573		0.4628	0.6506
90s	2.1288	134.51	0.0215	0.4628		0.9999
00s	2.1264	133.79	0.0736	0.6506	0.9999	
	LOGDON	**DON**			
Year	LSMEAN	LSMEAN	70s	80s	90s	00s
70s	1.5811	38.11		0.9507	*0.0016*	0.5601
80s	1.5311	33.97	0.9507		*0.0002*	0.8869
90s	1.8809	76.02	*0.0016*	*0.0002*		*<0.0001*
00s	1.4649	29.17	0.5601	0.8869	*<0.0001*	
		Woonasquatucket River			
	TDN CONC				
Year	LSMEAN			80s	90s	00s
80s	83.67				0.0829	0.7258
90s	102.56			0.0829		0.3567
00s	91.16			0.7258	0.3567	
	DIN CONC				
Year	LSMEAN			80s	90s	00s

Table 5.10 (continued)

			80s	90s	00s
80s	78.693			*0.0005*	0.6458
90s	50.853		*0.0005*		0.0371
00s	71.002		0.6458	0.0371	
	DON CONC LSMEAN				
Year	ANOVA only		80s	90s	00s
80s	19.85			*<0.0001*	0.9822
90s	55.43		*<0.00001*		*<0.0001*
00s	20.71		0.9822	*<0.0001*	
		Moshassuck River			
	1/y **TDN** CONC	**TDN** CONC			
Year	LSMEAN	LSMEAN	80s	90s	00s
80s	0.0121	82.35		0.9936	0.0158
90s	0.0123	81.63	0.9936		*0.0051*
00s	0.0152	65.62	0.0158	*0.0051*	
	LOGDIN	**DIN** CONC			
Year	LSMEAN	LSMEAN	80s	90s	00s
80s	1.867	73.66		*<0.0001*	*0.0005*
90s	1.672	47.01	*<0.0001*		0.4152
00s	1.718	52.24	*0.0005*	0.4152	
	DON CONC LSMEAN				
Year	ANOVA only		80s	90s	00s
80s	19.85			0.0152	0.9139
90s	55.97		0.0152		*0.0018*
00s	14.16		0.9139	*0.0018*	

ANCOVA/Tukey's Probability Tables (unless noted as ANOVA) with p < 0.01 in bold italic. Nitrogen concentration units are in μM.

Cranston began aerating its effluent during the summer and oxidizing the ammonia to nitrate. Even though the facility was not deliberately providing N removal at the time of our river N measurements, the presence of nitrate apparently stimulated denitrification that removed a large amount of N. The release of DIN from the Cranston facility averaged about 14 million moles y^{-1} during 2003–2004 compared with 27 million moles y^{-1} in 1992–1993. Significant and unexplained declines in nitrate have recently been reported from other streams in the northeast (e.g., Goodale et al., 2003; 2005). Atmospheric nitrogen deposition in the watershed may be declining as it has been in coastal Connecticut (Luo et al., 2002) and perhaps on Cape Cod (Bowen and Valiela, 2000), though Stoddard el at. (2003, p.xi) conclude that "...concentrations of nitrogen in deposition have not changed substantially in 20 years." Howarth (Chapter 3, this volume) provides more details on atmospheric deposition in the coastal northeast, but without strong conclusions regarding recent local trends. Alternatively, there may have been reductions in other sources such as fertilizer use or improved application practices. It is also possible that improvements in wastewater treatment since the early 1990s may have increased denitrification by raising oxygen levels in receiving waters.

Climatic changes may also have played an important role. These concepts are further explored in Chapters in this volume by Smayda and Borkman (15), Sullivan *et al.* (16) and Oviatt (18).

Phosphorus

The total amount of phosphorus brought into the Seekonk and Providence River estuaries by rivers varied by less than a factor of two during the different years of measurement. In contrast to nitrogen, particulate phosphorus is a significant (30–50%) part of the total flux (Table 5.9). Particulate phosphorus was included in all of the sampling except for the first survey in 1975–1976. The dissolved phosphorus was dominated by inorganic phosphate (DIP). Phosphate concentrations in the Blackstone decreased markedly (over 50%) in the 1990s and in the most recent survey compared with conditions in the 1970s and 1980s (Table 5.11). Phosphate in the Pawtuxet River in 2003–2004 was high, and not statistically different from concentrations found earlier. Phosphate in the Woonasquatucket had approximately doubled compared with the 1990s and 1980s. Analysis of variance on phosphate in the Moshassuck showed statistically significant declines (by about two thirds) in the 1990s and in 2003–2004 compared with conditions in the 1980s (Table 5.11). A statistically significant regression between dissolved organic phosphorus concentration and discharge was only found for the Pawtuxet River, where concentrations in 2003–2004 were much lower than they had been in any of the earlier decades back to the 1970s (Table 5.11). Analyses of variance for the other rivers showed large and statistically significant declines in dissolved organic phosphorus in the Blackstone and in the Woonasquatucket in 2003–2004 compared with the 1990s (Table 5.11). No statistically significant changes were found for dissolved organic phosphorus in the Moshassuck River. The relatively large amounts of particulate phosphorus are almost certainly tied to suspended sediment discharge, and the availability of this phosphorus to play a significant role in biological processes in the bay is not known. We have not analyzed the particulate P for time trends.

Since the inspection of statistical tables can be tedious, we also present the total dissolved N and P concentration and water discharge relationships in the two larger rivers graphically, so that the relationships during the different decades can be seen (Fig. 5.14).

5.3.4.2 Seasonal Variation in River Nutrient Fluxes

Because nutrient concentrations vary with season and with water discharge in the rivers, and because there is a strong seasonal variation in the amount of water carried by the rivers, it is useful to examine the seasonal variation in the

Table 5.11 Results of statistical analysis testing differences between decades of phosphorus concentration in the Blackstone, Pawtuxet, Woonasquatucket, and Moshassuck Rivers.

Blackstone River

Year	LOGDIP LSMEAN	DIP LSMEAN	70s	80s	90s	00s
70s	0.601	3.99		0.9437	*<0.0001*	*<0.0001*
80s	0.561	3.64	0.9437		*<0.0001*	*0.0002*
90s	0.265	1.84	*<0.0001*	*<0.0001*		0.9649
00s	0.231	1.7	*<0.0001*	*0.0002*	0.9649	

Year	DOP ANOVA LSMEAN		70s	80s	90s	00s
70s	2.41			*0.0014*	0.9997	*<0.0001*
80s	1.08		*0.0014*		*0.0003*	0.3333
90s	2.44		0.9997	*0.0003*		*<0.0001*
00s	0.44		*<0.0001*	0.3333	*<0.0001*	

Pawtuxet River

Year	LOGDIP LSMEAN	DIP LSMEAN	70s	80s	90s	00s
70s	0.897	7.9		0.8139	*0.0001*	0.1073
80s	0.964	9.2			*<0.0001*	0.0103
90s	0.581	3.81	*0.0001*	*<0.0001*		0.3682
00s	0.707	5.09	0.1073	0.0103	0.3682	

Year	LOGDOP LSMEAN	DOP LSMEAN	70s	80s	90s	00s
70s	0.245	1.76		0.5727	0.1603	*<0.0001*
80s	0.125	1.33	0.5727		*0.0017*	*0.001*
90s	0.436	2.73	0.1603	*0.0017*		*<0.0001*
00s	−0.247	0.57	*<0.0001*	*0.001*	*<0.0001*	

Woonasquatucket River

Year	LOGDIP LSMEAN	DIP LSMEAN	80s	90s	00s
80s	−0.112	0.77		0.8273	*0.0008*
90s	−0.153	0.7	0.8273		*<0.0001*
00s	0.207	1.61	*0.0008*	*<0.0001*	

Year	DOP CONC LSMEAN		80s	90s	00s
	ANOVA only				
80s	0.667			*<0.0001*	0.361
90s	1.57		*<0.0001*		*<0.0001*
00s	0.422		0.361	*<0.0001*	

Moshassuck River

Year	DIP CONC LSMEAN		80s	90s	00s
	ANOVA only				
80s	0.937			*<0.0001*	*<0.0001*
90s	0.214		*<0.0001*		0.6366
00s	0.307		*<0.0001*	0.6366	

Moshassuck DOP—No significant differences between means.
ANCOVA/Tukey's Probability Tables (unless noted as ANOVA) with p < 0.01 in bold italic. Phosphorus concentration units are in μM.

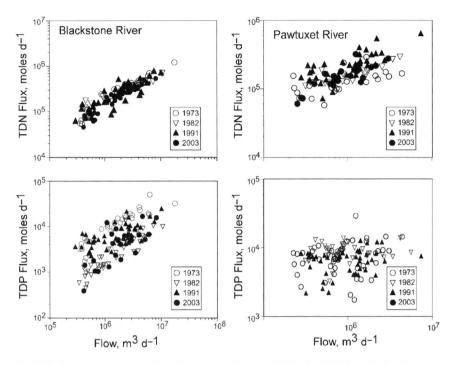

Fig. 5.14 The concentration of total dissolved nitrogen (DIN plus DON) and total dissolved phosphorus (DIP plus DOP) as a function of water discharge in the Blackstone and the Pawtuxet Rivers during our sampling in various decades. Note log-log scales.

delivery of N and P to Narragansett Bay from the landscape. During the spring peak in flow, the flux of total dissolved N to the bay from all the measured rivers was about 0.9 million moles d^{-1}, but this decreased to only 0.15 million moles d^{-1} in late summer (Fig. 5.15). The pattern was similar for DIN. On an annual basis, the DIN flux amounted to 77% of the TN carried by the rivers and DON accounted for about 20%. There was less seasonal variation in P delivery by the rivers, though fluxes of both TP and DIP were lowest in late spring and late summer (Fig. 5.15). Over the year of study, 38% of the TP flux in the rivers was DIP and 8.4% was DOP, with PP providing the remaining 53.6%.

5.3.4.3 Some Loose Ends

Before leaving the discussion of river inputs, it is necessary to deal with the Taunton River, with the smaller rivers where nutrient fluxes have not been measured, with what Ries (1990, p. 32) called the "unguaged area adjacent to the bay," and with the probable lack of importance of ground water.

First the Taunton. A program measuring nutrient concentrations in the Taunton River to compare with those found by Boucher (1991) in the late

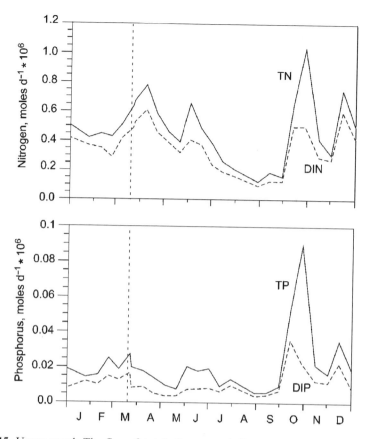

Fig. 5.15 Upper panel: The flux of total nitrogen and dissolved inorganic nitrogen into Narragansett Bay from the Blackstone, Ten Mile, Moshassuck, Woonasquatucket, and Pawtuxet Rivers during our sampling in April 2003 through March 2004. Note the low delivery of N to the bay in summer. Lower panel: As above except for total phosphorus and dissolved inorganic phosphorus.

1980s is in progress (B. Howes, University of Massachusetts, Dartmouth, personal communication), but the results are not yet available to us. This is unfortunate, since the water discharge of the Taunton is actually a bit larger than that of the Blackstone (29.7 vs. 24.4 m^3 s^{-1}; Ries, 1990), though its impact on Narragansett Bay proper is reduced if it is correct that 30% of the river leaves Mount Hope Bay by way of the Sakonnet Passage (Hicks, 1959). Krahforst and Carullo however, in Chapter 13, this volume, show that Narragansett Bay water can have an impact upon Mount Hope Bay. In our mass balance study during the 1980s, the TN flux from the Taunton (based on Boucher's concentration measurements) came out to be 90% of that from the Blackstone, but that was without any reduction since our budget area included Mount Hope Bay (Nixon *et al.*, 1995). If, as discussed earlier, we reduce the Taunton nutrient delivery to Narragansett Bay proper to 57% of that measured at the mouth, the addition from the Taunton to Narragansett Bay proper would have been about 50% of the TN flux from the Blackstone. Apparently without knowing about Boucher's 1988–1989 measurements, Isaac (1997) applied a land use model to the Taunton watershed and estimated a point and non-point TN flux from the Taunton River for 1985 of some 93 million moles, about 80% of what had been measured 3–4 years later. Given all of the uncertainties in the model, the agreement is impressive. Using the same model, Saunders and Swanson (2002) calculated increases of 18% and 6% for non point and point source N loadings, respectively, between 1985 and 2000. Even the point sources were calculated in these exercises based on a assumption of 15 g N m^{-3} times reported flows for each facility. Until the new measurements are available, we will have to work with these estimates and increase the fluxes we calculated from Boucher's (1991) concentration data accordingly.

Second, what about surface flows where no nutrient concentrations have been measured? As mentioned earlier in our discussion of revisions to the published estimates for the mid 1980s, the way these are handled can make a significant difference in the total inventory of nutrient inputs. According to a detailed analysis of the Narragansett Bay watershed by the USGS (Ries, 1990), five small unguaged rivers and almost 440 km^2 of unguaged coastal drainage add a long term mean of 2.6 m^3 s^{-1} of discharge to the bay. To this must be added 4 m^3 s^{-1} that are delivered by two small gauged rivers in which no nutrient measurements have been made. The total of 6.6 m^3 s^{-1} is about 8% of the total land drainage into the bay. Since there are no sewage treatment facilities on these streams and they drain predominantly rural or suburban areas, we will estimate their contribution to the bay using the total drainage area involved (725 km^2) and the TN, DIN, TP, and DIP yields per unit area from the rural/suburban Wood-Pawcatuck watershed in southern Rhode Island and Connecticut recently measured by Fulweiler and Nixon (2005). The result is an annual input to Narragansett Bay (millions of moles) of 15, 6.8, 0.9, and 0.3, respectively. These values are much lower than we estimated in the mass balance for the mid 1980s (Nixon et al., 1995), where we used the ratio of gauged to unguaged water flow to extrapolate the measured nutrient fluxes to the unmeasured sources. Since the larger measured rivers have sewage treatment facilities that discharge into them and they flow through more densely developed areas, we believe that the Wood-Pawcatuck makes a better proxy for estimating the smaller unguaged flows. As noted previously when comparing the N input in the mid 1980s with the estimate for 1925 (Table 5.12), we will work with the adjusted 1980s estimates throughout this chapter. While such reinterpretations are frustrating, they serve a useful purpose in cautioning the reader that many of the numbers that must come together to develop an inventory of nutrient inputs to a system like Narragansett Bay are subject to poorly described and constrained uncertainty as well as to errors of measurement, calculation, and interpretation. While the development of a table of nutrient inputs may be

Table 5.12 Estimates of the input of nitrogen to Narragansett Bay from various sources at various times between 1865 and 1985. Estimates for the first three periods are from data sources discussed in the text while the 1985 data are from Nixon *et al.* (1995) adjusted as described in text.

	1865	1900	1925	1985
Direct deposition from the atmosphere[a]	1.5	4	7	30
Rivers and streams[b]				
DIN	3–17	205	160	255
Organic N	30	95	130	103
Urban runoff	–	5–11	–	37
Direct sewage[c]	0	60	185–235	180
Total	35–50	375–370	480–530	605

Units are millions of moles of N per year (10^6 moles = 14 metric tons).

[a]based on 2.5% per year increase between the pre industrial rate of 1865 and measurements made in late 1980s by Fraher (1991).

[b]pre enrichment DIN discussed in text; 1865 organic N calculated based on 5 µM albuminoid N in "unpolluted" surface waters in late 1800s and early 1900s and organic N = 2x albuminoid. Blackstone measured above Pawtucket at Valley Falls for 1900 and 1925, at Pawtucket in 1980s. Measurements of Pawtuxet by RIDPH at Pettaconsett for 1900 and 1925; at Pawtuxet for 1985. Measurements for Blackstone, Pawtuxet, Ten Mile, Taunton, Moshassuck, and Woonasquatucket. All other fresh waters assigned concentration of "unpolluted" surface water, 5 µM DIN and 5 µM albuminoid N (see Table 5.2). River fluxes calculated from mean annual concentrations measured by RIDPH and MASDH and long-term mean water discharge at river mouths (Ries, 1990).

[c]Providence calculated from PCE annual reports of number of house connections and US. Census data on number of people per house for 1900 and 1920 multiplied by per capita N emission (see text). Other shore line direct sewage calculated from estimates of population served by the various systems (see Fig. 5.7) and per capita N emission reduced by 60% for 1900 and by 25 to 50% in 1925 (see text and Table 5.5) to account for removal in filter treatment systems.

conceptually straightforward, it is also a deeply regressive activity that can be extended almost indefinitely.

Finally, ground water has been conspicuously absent from our discussion of water (and its' associated nutrients) inputs to the bay. While it is clear that groundwater can be an important source of nitrogen in some of the smaller coves or embayments (e.g., within some areas of Greenwich Bay; Urish and Gomez, 2004), the overall water balance for the Narragansett Bay watershed suggests that the volume of groundwater entering the bay as a whole can not be large. As part of his analysis of fresh water inflow to the bay, M.E.Q. Pilson (Oceanography, URI, personal communication) compared the average annual rainfall on the area of the watershed between 1964 and 2000 with the volume of surface runoff reported by the USGS during the same period and corrected for unguaged areas. Only 45 to 55 cm of the annual rainfall is unaccounted for by surface flows, an amount easily consumed by evapotranspiration in this area (e.g., Fennessey and Vogel, 1996). As a result, there is little water left over to support a significant ground water inflow to the bay. Nowicki and Gold (Chapter 4, this volume) reached a similar conclusion that for the bay as a whole, groundwater probably accounted for less than 10% of the fresh water input. This is a fortunate circumstance, since nitrogen inputs through groundwater are extremely difficult to quantify. In the lower bay however, Kincaid and Bergondo (Chapter 10, this volume) suggest groundwater may be an important factor in circulation dynamics, and if so, will need closer scrutiny, at least for that region of Narragansett Bay.

5.3.4.4 Sewage Treatment Facility Discharges

Because most of the sewage treatment facilities that discharge to the bay and in its watershed were making regular measurements of the flow of treated effluent, as well as the concentrations of the various forms of N (often including organic N) and TP during the time of our most recent river concentration measurements, we did not sample the sewage effluents as in previous studies. Instead, we collected annual monitoring results for 2000 through 2003, calculated annual fluxes, then averaged the four years. The only facility of significant size that does not report nutrient concentrations is Newport, and in this case we had to estimate TN and TP releases based on the population served and per capita N and P releases reported by the Field's Point facility (Table 5.6). We also followed this procedure for some of the smaller facilities that did not measure TP or all forms of N in their effluents. In some cases, we also present the individual yearly means (e.g., Fig. 5.13). For the largest three facilities that discharge directly to Narragansett Bay proper, we have also collected the data from 2004 and 2005. We worked with data provided directly by the treatment facilities rather than with state or federal regulatory web sites because this minimizes potential problems of data transmission, formatting, and interpretation, and because most of the larger facilities monitor all year while the compliance monitoring only covers the summer months. While flow measurements are automated and recorded daily, nutrient concentrations are sampled less frequently, and the frequency varied among facilities and over time within each facility. For example, ammonia was measured weekly by the Field's Point

Table 5.13 Nitrogen and phosphorus discharges from 29 sewage treatment plants in the Narragansett Bay watershed.

	Flow	NH_4	NO_2+NO_3	DIN	Organic N	TN	TP
Direct discharges to Narragansett Bay							
NBC Field's Pt.	166,300	44.1	6.57	50.67	14	64.7	2.43
NBC Bucklin Pt.	88,300	31.2	0.42	31.62	5.7	37.3	2.24
Newport[a]	33,800	*Nutrients not monitored*				13	0.49
East Providence	25,200	1.71	6.7	8.41	1.4	9.81	0.77
Bristol	12,100	2.53	2.13	4.66	2.13	6.79	0.18
Warren	7,000	1.53	0.44	1.97	0.8	2.24	0.1
E. Greenwich	4,000	0.55	0.21	0.76	0.43	1.19	0.57[d]
Quonset Pt.[b]	2,300		0.6			0.93	0.08

Table 5.13 (continued)

	Flow	NH$_4$	NO$_2$+NO$_3$	DIN	Organic N	TN	TP
Jamestown[b]	1,600		0.17	0.17		0.29	0.02
Total						136.25	6.88
Fall River[b]	93,500	23.7	2.12	25.82	8.48	34.3	1.19
Total						170.55	8.07
Blackstone River (to Seekonk estuary)							
Worcester	137,400	22.2	10.5	32.7	34.7	67.4	2.25
Woonsocket	27,800	11.6	3. 19	14.79	3.9	18.69	2.17
Smithfield	7,100	3.01	0.23	3.24	1.11	4.35	0.32
Grafton[c]	6,400	1.68	1.12	2.8		2.75	0.12
Millbury[c]	4,600	1.89	0.44	2.33		2.35	0.23
Northbridge[c]	4,250	1.24	0.36	1.6		2.57	0.14
Burrillville	3,000	0.78	0.32	1.1	0.19	1.29	0.03
Hopedale	1,500	0.13					0.02
Leicester	785	0.03					0
Douglas	740	0.08	0.04	0.12	0.05	0.17	0.02
Upton	560	0.05	0.01	0.06	0.03	0.09	0
Total						>99.66	5.3
Ten Mile River (to Seekonk estuary)							
North Attleboro	12,600	0.78				3.76	0.17
Attleborough	18,700	0.16				8.54	0.22
Total						12.3	0.39
Pawtuxet River (to Providence River estuary)							
Cranston	40,600	5.24	6.65	11.89	7	18.9	1.31
West Warwick	21,500	7.03	1.17	8.2		8.19	1.37
Warwick	16,300	1.27	1.92	3.19	1.51	4.7	0.55
Total						31.8	3.23
Taunton River (to Mount Hope Bay)							
Brockton	78,000	15.8	11.9	27.7	8.96	36.7	0.83
Taunton [a]	26,400	2.04				8.29	0.29
Somerset [b]	11,000	2.69	0.76	3.45	0.75	4.2	0.17
Total						49.2	1.29
GRAND TOTAL						**363.5**	**18.3**

Values are the average for 2000–2003, inclusive, unless noted. Units are millions of moles per year for nutrients and cubic meters per day for flow. The values reported here have been calculated from treatment plant records and the frequency of sampling varies among plants and over time.

[a] Estimated assuming 0.9 moles N per person per day and 35 mmoles P per person per day. See Table 5.6.

[b] TP as in footnote a.

[c] TN as in footnote a.

[d] From Granger et al. (2000).

facility between 1998 and 2000, then 61 times during 2001, and 97 times in 2002. We examined how sensitive an annual estimate of N fluxes might be to various sampling schemes by calculating an annual flux for each of the N constituents from the Field's Point facility during the intensively sampled year of 2002 assuming different frequencies of sampling (first measurement each month, mid-month measurement, last measurement of the month, and 25 days selected randomly from the 97 different measurements), and different ways of calculating the annual flux (annual average of all flows and all measurements of each constituent, average of individual fluxes, Beale's estimate). The result is reassuringly consistent, with annual DIN flux estimates only ranging between 43.7 and 44.6 million moles, ammonia fluxes ranging between 33.1 and 35.6 million moles, and organic N ranging between 12.7 and 32.9 million moles. Organic N was measured less frequently and the analytical precision is lower. The coefficient of variation for the 83 organic N measurements made during 2002 was 46% compared with 40% for the 97 ammonia determinations. The coefficient of variation for the daily effluent discharge measurements was 17%.

The 19 sewage treatment facilities in the watershed, and the 10 discharging directly to Narragansett Bay (including Mount Hope Bay), released over 360 million moles of N and over 18 million moles of P per year during the period for which we compiled concentration and flow data (Table 5.13). By grouping the discharges by sub watershed, it is easy to compare the discharge of the treatment facilities to each major river with the flux from that river to the bay (Table 5.14). While treatment facility discharges could account for essentially all of the N delivered to the bay by the Blackstone and Ten Mile Rivers, there are clearly other large sources of N to the Pawtuxet River and the

Table 5.14 The amount of TN and TP discharged by the sewage treatment plants on the rivers draining into Narragansett Bay during 2000–2003 compared with the delivery of TN and TP to the bay from those rivers during 2003–2004.

	Combined Sewage Input to River	Discharge to Bay
Blackstone River		
TN	~100	99
TP	5.3	3.9
Pawtuxet River		
TN	32	59
TP	3.2	3.6
Ten Mile River		
TN	12	14
TP	0.4	0.8
Taunton River[a]		
TN	49	117
TP	1.3	3.9

Units are millions of moles y^{-1}.

[a] River fluxes calculated from concentration measurements in 1988–1989 (Boucher, 1991) as described by Nixon et al. (1995). See text for discussion of the Taunton situation.

Taunton. Of course, a simple mass comparison does not identify the actual source of the N that is being discharged at the mouth of the rivers—that is a much more difficult thing to know. Howarth (Chapter 3, this volume) suggests that about 30% of the N carried by the Blackstone is from atmospheric deposition. This may be correct, but it is an indirect estimate that needs to be verified. As he notes, the regression method used to arrive at this estimate may underestimate the contribution of sewage derived N in urban rivers. An earlier study of oxygen isotope ratios in the nitrate carried by the Blackstone suggested that atmospheric deposition contributed little to the river N load (Mayer et al., 2002). About 25% of the P added to the Blackstone from human sewage appears to be stored within the river, while there appear to be additional sources of P to the Taunton, Pawtuxet, and Ten Mile Rivers (Table 5.14).

For Narragansett Bay proper, the direct discharge of N from sewage treatment facilities along the shoreline is approximately equal to the sewage N discharged to the rivers flowing into that part of the bay (Table 5.14). For TP, the watershed sources are about 40% greater. The three facilities that discharge to the Seekonk and Providence River estuaries account for about 82% of the sewage TN that is discharged directly to Narragansett Bay proper. While we showed earlier that the release of DIN and TP from these facilities has remained approximately constant (DIN) or declined (TP) between 1992 and 2005 (Fig. 5.13), the most recent monitoring data also show that the total TN released from these three facilities has declined since 1983 and is now at the same level we found in our first measurements during the 1970s (Nixon et al., 2005). Combined annual TN release at that time was about 96 million moles, which appeared to increase to 120 million in the 1980s. The apparent increase is almost certainly an artifact that reflects poor management of the Field's Point facility during the 1970s when large amounts of sewage were "bypassed" around the facility during heavy rains (Hoffman and Quinn, 1984). The most recent measurements show combined annual TN releases of 108, 118, 104, and 94 million moles y^{-1} in 2002 through 2005, respectively.

The TN released by the seven other treatment facilities that discharge directly to the bay ($\sim$59 million moles y^{-1}, Table 5.13) has remained the same as in the 1980s, or perhaps decreased by about 5%. The same facilities released about 2.6 million moles y^{-1} of TP during 2000–2003, a decrease of about 60% compared with the estimate of Nixon et al. (1995) for the mid 1980s. These comparisons are more tenuous than for the three upper bay facilities because these discharges were not sampled very often in the 1980s and Newport had to be estimated indirectly for the more recent period.

The inventory of N and P inputs to Narragansett Bay at the start of the twenty-first century suggests that TN inputs have remained constant or declined slightly since the mid 1980s, and that TP inputs have declined by over 25%. The decline in P discharge is due to a combination of improved solids removal and declines in the use of P in detergents (Booman et al., 1987). The pattern for DIN is the same as for TN, while DIP has declined by over 30% (Table 5.15). As described earlier, the estimated N inputs in Table 5.15 include a

Table 5.15. Nitrogen and phosphorus inputs to Narragansett Bay (including Mount Hope Bay) in 2003–2004.

	N	P
Direct atmospheric deposition[a]		
DIN	24	0.13
DON	5.6	
subtotal	29.6	0.13
Measured rivers[b]		
DIN & DIP	133	4.12
DON & DOP	41	0.94
PN & PP	11	3.68
Taunton River[c]		
DIN & DIP	101	3.3
DON & DOP	35	0.6
PN & PP	1	1.4
subtotal	322	14.0
Unmeasured surface drainage[d]		
DIN & DIP	7	0.1
DON & DOP	6	0.03
PN & PP	3	0.25
subtotal	16	0.38
Urban runoff[e]		
DIN & DIP	17	2
DON & DOP	18	0.8
PN & PP	2	1.2
subtotal	37	4.0
Direct sewage discharge[f]		
DIN & DIP	131	5.3
DON & DOP	27	1.6
PN & TP	13	1.2
subtotal	171	8.1
TOTAL DIN & DIP	413	14.9
TOTAL DON & DOP	133	4.0
TOTAL PN & PP	30	7.7
TOTAL	576	26.6

Units are millions of moles y^{-1}.
[a] Assume unchanged from Nixon et al. (1995).
[b] From Tables 5.8 and 5.9.
[c] N fluxes from Nixon et al. (1995) increased by 18% (see text), P left unchanged.
[d] See text.
[e] Assume unchanged from Nixon et al. (1995).
[f] From Table 5.13.

20 million mole y^{-1} increase for the Taunton River over the measurements of the 1980s (increase of 18%) that is based solely on a land-use model. Our measurements in the Blackstone and the Pawtuxet Rivers showed significant declines in nitrogen on the order of 20–25% between the mid 1980s and the most recent

measurements. If measurements subsequently show similar behavior in the Taunton, then the TN input for this river would drop by 45 million moles compared with the value shown in Table 5.15 and TN input to the bay would have declined by over 10% from the adjusted mid 1980s estimate. The ratio of TN/TP entering the bay averaged almost 22 and the ratio of DIN/DIP averaged almost 28. These ratios were about 17 and 20, respectively, in the adjusted inputs for the mid 1980s. The increase in the ratios is due entirely to the decline in P input during the last twenty years.

Those who study nutrient inputs to estuaries are used to seeing plots which show sources or fluxes over time staying monotonously low until the early 1960s, then rising exponentially. The cover of a recent special edition of *Estuaries* (now *Estuaries and Coasts*) dedicated to "Nutrient Over-Enrichment in Coastal Waters" is a perfect example (Rabalais and Nixon, 2002). The rapid growth of interest in nutrient enrichment and eutrophication of coastal marine ecosystems corresponds with the growth of non-point sources of reactive N, including synthetic fertilizer production and fossil fuel combustion. In estuaries where non-point sources of N dominate, scientists have witnessed and studied the recent and often dramatic responses of the environment to these changing inputs (e.g., Turner and Rabalais, 2003; Kemp *et al.*, 2005; Smith *et al.*, 2006; and other papers therein). In this country, many of the estuaries of the mid-Atlantic, southeast, and Gulf coasts fall into this category. In the urban estuaries of the northeast however, point sources are often much more important, and the history of nutrient inputs looks very different (Fig. 5.16). In systems like

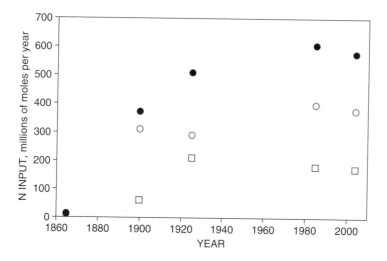

Fig. 5.16 Estimates of the total input of nitrogen to Narragansett Bay from land and atmosphere at various times (solid points). Open circles are inputs from rivers and surface drainage; open squares are direct sewage discharges. The decline in river fluxes around 1925 was due to treatment of sewage by trickling sand filters that stimulated denitrification. See text for sources and assumptions.

Narragansett Bay, the fertilization took place a century earlier and at a much faster rate. This rapid model is almost certainly being repeated in many coastal areas in the developing world (Nixon, 1995b). Unfortunately, the rapid and intense fertilization of Narragansett Bay took place many decades before we had the capacity to make anything but the most superficial observations of its consequences. And even those are scarce and confounded by many other changes taking place in the bay. The question now is if we will remain equally blind and unlearning as nutrient inputs decline rapidly over the coming decade as a result of management interventions that will reduce N discharges from sewage treatment facilities. If so, it will not be due to a lack of tools or curiosity among the scientists.

5.4 The Future

The input of P to Narragansett Bay began to decline at least as early as the mid 1980s or early 1990s and has continued to do so through the present (Tables 5.7, 5.9, 5.11). Using the revised calculations for P input from unmeasured flows discussed earlier, the total input of TP and DIP was about 32 and 22 million moles y^{-1}, respectively, in the mid 1980s. By 2003–2004 it had declined to about 26.6 and 14.9 million moles y^{-1}, respectively. Further reductions are likely with sewage treatment facility upgrades that are currently underway or planned. But the most dramatic change for the coming decade will almost certainly be nitrogen reduction from advanced wastewater treatment (denitrification). Our river N concentration measurements in 2003–2004 captured the leading edge of this development since the Woonsocket treatment facility had already lowered its N discharge to the Blackstone by almost 30 million moles y^{-1} and the Cranston facility released some 13 million moles y^{-1} less to the Pawtuxet River than it did during the early 1990s. Woonsocket is close to the mouth of the river, and, as noted earlier, virtually all of this reduction appears to have been reflected in our measurements. Other treatment facilities have implemented N reduction since our sampling, including the Bucklin Point facility, East Greenwich, Warwick, West Warwick, and Cranston. The first two discharge directly to the bay, while the latter three discharge close to the bay in the Pawtuxet River. It is difficult to know exactly which additional treatment facilities in the watershed or around the bay will be required (or choose) to implement nitrogen reduction, or when, or to what level.

Since N reductions are imposed during the May through October growing season, their potential impact on primary and secondary productivity is great. Because the flux of N into the bay from rivers is strongly correlated with water flow (Nixon et al., 1995), only about 30% of the annual N load delivered by the rivers enters during that six month period (Fig. 5.15). While the rivers (and their upstream sewage treatment facilities) provide most of the N to Narragansett Bay on an annual basis, the treatment facilities that discharge directly to the bay provide twice as much N as the rivers which discharge to that part of the bay in the low flow months of summer and fall.

If we assume that over the coming decade all but the very smallest sewage treatment facilities will move to advanced wastewater treatment, it is relatively straightforward to make a rough estimate of the reduction in N discharges. For this exercise, we assume that Field's Point, Bucklin Point, East Providence, Fall River, Warren, Bristol, and East Greenwich will all reduce their effluents by about 65% to 5 g TN m^{-3}, and that Newport, Jamestown, and Quonset Point will remain unchanged. Since all of these discharge directly to the bay, the loss of N from the effluent translates directly into a reduction in N added to the bay. The situation for the treatment facilities in the watershed is more complicated. With the exception of the large facility in Worcester, near the head of the Blackstone River (which we assume will also go to 5 g m^{-3}), we assume that all of the treatment facilities that discharge to rivers draining to the bay will go to 8 g TN m^{-3}, about a 50% reduction. The difficulty is in knowing how much of a difference this will make to the amount of N that the rivers discharge to the bay. Rivers are not pipes, but complex and dynamic biogeochemical systems that can store N in their sediments or lose it to the atmosphere through denitrification (e.g., Seitzinger *et al.*, 2002). There is still considerable uncertainty about how much N is removed by these processes and how the amount is influenced by factors such as distance, residence time, depth, N concentrations, etc. (e.g., Destouni *et al.*, 2006). Obviously, if half the N carried by a river is being denitrified and the amount of N added to the river is reduced by 50%, only a 25% reduction might be seen in the amount of N reaching the mouth of the river. For this exercise, we have assumed that the in-stream removal is between 10% (Van Breemen *et al.*, 2002) and 30%, the factor used in our earlier historical reconstruction (Caraco and Cole, 1999). We have used functional, bias corrected log-log regressions between total dissolved N (TDN) flux and river flow for all of the rivers we measured in 2003–2004 to estimate average daily TDN fluxes for each month from May through October during a year with median total water discharge for this period during the 75 years between 1930 and 2004. We did the same for a year in which the total water discharge during this period was at the top of the lowest 25% of the years. We assumed that the N delivered during the May–October period in the ungauged surface flow would equal 30% of the annual total and that atmospheric N deposition was evenly distributed throughout the year, as rainfall is (Pilson, 1989). We made the same assumption for urban runoff.

The results of this rough exercise suggest that during a median flow year, the reduction of N input to the bay from land and atmosphere would be on the order of 45 to 55% during June through September. Essentially the same result was found during the drier year, though in extreme droughts even less N would be delivered to the system. This is obviously a significant drop in the input of the major nutrient limiting primary production in the bay (Oviatt *et al.*, 1995; Howarth and Marino, 2006). Even under present high N loading, the concentrations of ammonia and nitrate are near zero in the surface water over most of the bay during summer (Kremer and Nixon, 1978), and, as discussed below, it is not surprising that food limitation appears to be common among bay animals during the late summer.

5.4.1 A "Grand Experiment" or an "Ecological Adventure"?

Some who work on the bay are fond of saying that we are beginning a "grand experiment" to see how the bay will respond as it is weaned from almost a century on a high N diet. But in truth, what we have embarked upon is an ecological adventure. We have no replicates, no controls, and many variables are changing simultaneously. We are also not testing any quantitative predictions, except a general supposition that the concentrations of oxygen in the bottom waters of the upper bay will be less hypoxic for shorter periods and/or that hypoxia may be less frequent and cover a smaller area, and that less nuisance sea weed may accumulate on some of the beaches of the upper bay if N inputs are reduced. As this chapter is written, we are also not making many of the kinds of observations that would be required to document the response of the bay to this ambitious and expensive management intervention. If we do not learn as much as possible it will be a great opportunity missed.

The various numerical models of the bay are not adequate to link changes in bottom water oxygen concentrations to N inputs in a rigorous way (see Chapters 8 and 9, this volume, for further details), so it appears that the approach will be to take out as much N from as many treatment facilities as the political, legal, economic, and engineering constraints will allow, and then see what happens to dissolved oxygen. This is not necessarily a bad approach if one believes that the current oxygen conditions in the bay are having real negative impacts on the ecology of the bay and the environmental services that the citizens want it to provide. And if one believes that reducing primary production will not have undesirable negative impacts on the growth and production of fish and shellfish in the bay.

5.4.1.1 As Blind Men See the Elephant

There is a delightful old Indian legend about six blind men, each examining different parts of an elephant and each proclaiming with great certainty what sort of beast he has encountered (e.g., Saxe, 1963). The late Joel Hedgpeth, a preeminent marine ecologist and biologist of the past century, noted that scientists who study estuaries and bays are in something of the same fix (Hedgpeth, 1978). Like all natural ecosystems, Narragansett Bay is complex and each person can only focus on certain parts of it. And about many (perhaps all) things we have incomplete or imperfect information. The bay's ecology is also changing as we study it. For all these reasons, and doubtlessly others as well, different students of the bay see a lower N future in different ways.

There are at least two reasons why we might set out on the grand ecological adventure with some reservations. First, there is no evidence that the ecology of Narragansett Bay as it has existed for the past seventy-five years or more is in some way "broken" or undesirable, a rare fish kill or occasional rotting sea weed smell notwithstanding. It is not the bay that Verrazzano found or the bay that Webber charted, but it is the bay that more recent generations of people have cherished. Of course, there have been changes in the bay during recent decades, some quite surprising and dramatic (e.g., Sullivan *et al.,* 2001; Oviatt *et al.,* 2003; Fulweiler and Nixon, in press). But those changes have not been due to increasing N inputs because those inputs have been remarkably stable for many years (Fig. 5.16).

Because of the prominence of Chesapeake Bay, the Neuse estuary, the Pamlico estuary, the Gulf of Mexico "Dead Zone" and other systems in which recently increasing non-point nitrogen sources are most important, it often surprises people that the history here has been so different. Oxygen conditions in the Seekonk and Providence River estuaries are actually much improved compared to conditions in the past, before the treatment facilities there went to full secondary treatment (e.g., Sweet, 1915; Gage and McGouldrick, 1923; 1924; Desbonnet and Lee, 1991). When Metcalf and Eddy (1948) surveyed the Seekonk estuary for the Blackstone Valley Sewer Commission on August 13, 1947, the *surface water* was anoxic almost the entire length of the system, and the bottom waters of the upper bay also experienced periods of hypoxia during the weaker neap tides of late summer at least as far back as 1959, when the US Public Health Service made the first intensive measurements (Fig. 5.17; US Department of Health, Education, and Welfare, 1960). The phenomenon has been rediscovered (e.g., Bergondo *et al.,* 2005; Saarman *et al.,* Chapter 11 this volume; Deacutis, Chapter 12 this volume), but it is not a recent development. And we do not know if it has a negative impact on the production of fish and shellfish in the bay. Studies in the 1970s and in 1990 showed that the Providence River estuary, Greenwich Bay, and the upper bay remain the nursery area of the bay, with highest concentrations of fish eggs and larvae (Matthiessen, 1973; Bourne and Govoni, 1988; Keller *et al.,* 1999). The bottom community in the upper bay is dominated by different species than the mid and lower bay, but that does not impact the appreciation of the bay by people, and the "opportunistic" benthic species may make very good food for the fish we like to catch. As for that unwelcome nuisance sea weed, *Ulva,* we know that it has been abundant in the upper bay at least since the early 1900s (Sweet, 1915). The Annual Report of the RI Board of Purification of Waters (1932) noted that during the summer large amounts of *Ulva* had accumulated on the beaches at Buttonwoods in Greenwich Bay and at Narragansett Terrace in East Providence on the Providence River estuary. "At Narragansett Terrace the odors were extremely offensive and in addition the hydrogen sulphide gases produced by the decomposing vegetable matter on the beach caused considerable damage to the paint on two houses. The paint on the interior as well as the

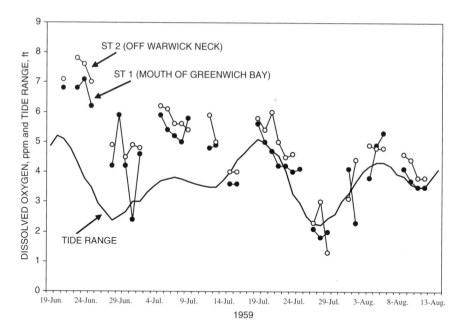

Fig. 5.17 Concentrations of dissolved oxygen in the near-bottom water at two stations in upper Narragansett Bay during the summer of 1959 along with the predicted tide range. Concentrations were much lower in the Providence River estuary. Data from the US Public Health Service (1960) report to the US Army Corps of Engineers, New England Division.

exterior was badly discolored and silver ware and metal fixtures were tarnished..." (p.10). The point of course is not that rare summer fish kills or occasional rafts of rotting *Ulva* are what people have cherished about the bay, but that these things have been part of the urban and heavily fertilized upper bay for a long time; they are not evidence that the bay is dying or in crisis as some have claimed.

The second reason for concern about rushing to N reduction is that we will be imposing a major management intervention on a bay that has been undergoing a dramatic ecological change. Long-term weekly monitoring of the plankton and nutrients in the middle of the bay carried out for many years by T.J. Smayda and his students, and more recently by the Graduate School of Oceanography and the Rhode Island Sea Grant program, shows that the annual mean concentration of phytoplankton chlorophyll has declined by about two thirds since the early 1970s (Fulweiler and Nixon, in press). While most of this decline has been attributed to the loss of the winter-spring bloom (Li and Smayda, 1998), there has also been a marked decline in the mean summer chlorophyll (Fig. 18). Loss of the winter-spring bloom is thought to be due to winter warming and associated cloudiness (Smayda, 1998; Li and Smayda, 1998; Oviatt *et al.*, 2002; Borkman, 2002). The loss or reduction of the

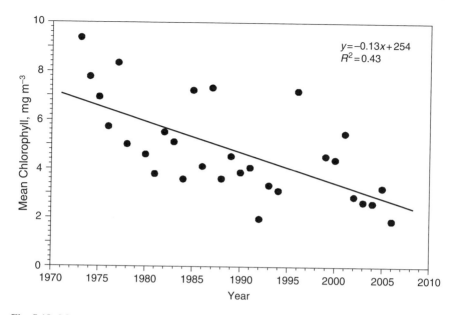

Fig. 5.18 Mean concentrations of phytoplankton chlorophyll during June, July, and August in the near-surface and near-bottom water off Fox Island, in the middle of the West Passage of Narragansett Bay. Data from the early 1970s through 1996 are from T.J. Smayda, Graduate School of Oceanography (GSO), University of Rhode Island, personal communication. Data from 1999 through 2006 are from the plankton monitoring program maintained by the GSO *(http://www.gso.uri.edu/phytoplankton.)*.

winter bloom appears to have caused the decline in the summer standing crop of phytoplankton because there is no large deposition of organic matter on the bottom of the bay in late winter. In the past, this deposited bloom material slowly decomposed as the system warmed in spring and summer, thus releasing nutrients to the overlying water (e.g., Nixon *et al.*, 1976; 1980). Now, the nutrients that accumulate in the water during fall and early winter are flushed from the bay and the fluxes of N and P from the sediments are much lower then they were in the 1970s and 1980s, and oxygen uptake rates are much lower (Fulweiler and Nixon, in press). Because chlorophyll concentrations and primary production are closely correlated in the bay (Keller, 1988), the declines in annual, winter, and summer chlorophyll almost certainly mean that primary production is lower in the mid bay now than it has been during recent decades. Of course, as chlorophyll has declined, the penetration of light through the water column has increased (Borkman and Smayda, 1998), but the decline in annual mean phytoplankton concentration is more important in regulating primary production. A rough calculation by L. Harris using a BZI (biomass, euphotic zone depth, incident light) model approach similar to that of Keller (1988) suggests that mean carbon fixation may now be about half what it was in the mid 1970s in the mid bay. Climate change appears to have caused this oligotrophication of the system even as N inputs have remained high and relatively constant (Nixon, in press). If the N input during summer is now reduced markedly on top of this climate induced oligotrophication, it may have a major negative impact on secondary production. Even under the earlier seasonal cycles of plankton abundance, there was evidence of food limitation in late summer, including field data on benthic infauna in the middle of Narragansett Bay, especially deposit feeders and surface feeders (Rudnick, 1984; Rudnick et al., 1985), and field studies of the dominant winter copepod, Acartia hudsonica, in mid bay during the end of the winter-spring bloom, and of the dominant summer copepod, A. tonsa, in late summer and fall (Durbin and Durbin, 1981). Experiments using the large MERL mesocosms at the Graduate School of Oceanography at the University of Rhode Island have also shown food limitation of the production of dominant copepods in control tanks (Sullivan and Ritacco, 1985), of the abundance and biomass of benthic poly-chaetes and amphipods (Beatty, 1991), and of juvenile Atlantic menhaden growth during summer and fall (Keller et al., 1990). More general correlations between phytoplankton primary production and benthic animal biomass and fisheries landings have been shown by Kemp et al. (2005) and by Nixon (1988) and Iverson (1990), respectively.

Unless the physical factors governing water column stratification in the bay have changed sufficiently that the bottom and surface waters are now more effectively separated for longer periods than they have been in the past, the hypoxia that has been documented in recent years in the upper bay (Saarman et al.; Deacutis, both this volume) may have been more extensive and persistent in the 1970s and 1980s than it is now. The surface waters in the lower bay have warmed slightly since the 1960s (e.g., Nixon et al., 2003; 2004), but we do not know if this extends into the upper bay or if it has been large enough to increase stratification significantly. Unfortunately, there is no long time-series record of chlorophyll in the upper bay or Providence River estuary, nor is the current monitoring of dissolved oxygen of sufficient duration to reveal trends. Chlorophyll was measured in the Providence River estuary by Culver-Rymsza (1988) during 1985–1986, but this was during the infamous "brown tide" and her values may be higher than typical for the time. She sampled on 32 occasions over an annual cycle without regard to the stage of the tide, and chlorophyll in the surface water at six stations averaged 18.1 mg m^{-3}. Monthly surveys at five stations in the Providence River estuary during 1986–1987 showed annual mean concentrations in surface water of 12 mg m^{-3} on low tide sampling and 7.9 mg m^{-3} on high tides (Doering et al., 1988). These results are roughly consistent with surface samples from the same five stations measured biweekly without regard to tide during 1997–1998, when the mean chlorophyll was about 13 mg m^{-3} (Oviatt et al., 2002). Based on this admittedly thin evidence, it does not appear that the levels of phytoplankton chlorophyll in the Providence River estuary have been changing in parallel with those in the mid bay. On the other hand, they do not appear to have increased between the mid 1980s and 1990s either. The lack of a decline in parallel with the mid bay is not as surprising as it may first appear. The traditional winter-spring bloom was a phenomenon that began in the upper West Passage and spread throughout the bay, with a lower manifestation in the Providence River estuary (e.g., Kremer and Nixon, 1978). Because of nutrient inputs from the larger sewage treatment facilities and the rivers, primary production in the Providence River is also much less dependent on nutrients regenerated from the organic matter deposited by a winter spring bloom.

The driving forces behind future N reductions in the bay have included long-term concerns by managers about low dissolved oxygen concentrations during summer in the bottom waters of the Seekonk and Providence River estuaries, recent monitoring that has shown that a larger area of the upper bay is exposed to episodic hypoxia than had generally been realized, and an unexpected (and rare) fish kill in Greenwich Bay in late summer of 2003 (RI Department of Environmental Management, 2003). Changing demographics, housing patterns, and resident expectations have probably also conspired to make occasional rotting sea weed on upper bay shores increasingly unacceptable. These and, doubtless, other considerations have led some environmental groups to exert political pressure to "save" a bay "in crisis" (e.g., Save The Bay, 1998).

The evidence suggests that the 2003 fish kill resulted from an unfortunate and improbable conjunction of physical factors acting in a highly productive and geographically restricted portion of the bay, a situation not unlike the one that caused the great kill in 1898 (Nixon, 1989). But that still leaves the questions of hypoxia in the Seekonk and Providence River estuaries and in the upper bay and the occasional nuisance macroalgae on the beaches. Nitrogen reductions may well alleviate these long-standing conditions, at least to some degree. But at what cost to the productivity and food chains of the mid and lower bay, especially since we now know that the bay has been experiencing a severe climate-induced reduction in the standing crop of phytoplankton during the past few decades? This is an important question for those entrusted with the management of all the bay's resources, and it is a question that needs and deserves input from the scientists who study the bay and from an informed public. As Hedgpeth (1978, p. 5) put it, "When we ask what part of nature our actions affect, without really understanding the entire ecosystem, we find ourselves among those blind men who touched various parts of the elephant without comprehending the nature of the whole animal." Our hope is that this chapter, and the volume of which it is a part, will help to expand the discussion of the impacts of N reduction to all of the bay and to all of its resources and uses.

References

Aber, J.D., Ollinger, S.V., Driscoll, C.T., Likens, G.E., Holmes, R.T., Freuder, R.J., and Goodale, C.L. 2002. Inorganic nitrogen losses from a forested ecosystem in response to physical, chemical, biotic, and climatic perturbations. *Ecosystems* 5:648–658.

Altman, P.L., and Dittmer, D.S. 1968. *Metabolism*. Biological Handbooks, Federation of American Societies for Experimental Biology. Bethesda, MD.

Ashbolt, N.J, Willie, O.K., and Snozzi, M. 2001. Indicators of microbial water quality. *In* Water Quality: Guidelines, Standards, and Health, pp. 289–316. Fewtrell, L. and Bartram, J. (eds.) World Health Organization, London: IWA Publishing.

Beale, B.M.L. 1962. Some uses of computers in operational research. *Industrielle Organisation* 31(1):27–28.

Beatty, L.L. 1991. The response of benthic suspension feeders and their grazing impact on phytoplankton in eutrophied coastal ecosystems. Ph.D. Thesis, Graduate School of Oceanography, University of Rhode Island, Kingston, RI.

Bergondo, D.L., Kester, D.R., Stoffel, H.E., and Woods, W. 2005. Time-series observations during the low sub-surface oxygen events in Narragansett Bay during summer 2001. *Marine Chemistry* 97:90–103.

Bernstein, D.J. 1993. Prehistoric subsistence on the southern New England coast: The record from Narragansett Bay. CA: Academic Press, Inc.

Bidwell, P.W., and Falconer, J.I. 1941. History of agriculture in the Northern United States 1620–1860. New York: Peter Smith.

Billen, G., Lancelot, C., and Meybeck, M. 1991. N, P, and Si retention along the aquatic continuum from land to ocean. *In* Ocean and marine processes in global change, pp. 19–44. Mantoura, R.F.C., Martin, J.M., and Wollast, R. (eds.) Chichester: Wiley and Sons.

Booman, K.A., Pallesen, K., and Berthouex, P.M. 1987. Intervention analysis to estimate phosphorus loading shifts. *In* Systems analysis in water quality management, pp. 289–296. Beck, M.M. (ed.), New York: Pergamon Press.

Borkman, D.G. 2002. Analysis and simulation of Skeletonema costatum (Grev.) Cleve annual abundance patterns in lower Narragansett Bay 1959 to 1996. Ph.D. Thesis, Graduate School of Oceanography, University of Rhode Island, Narragansett, RI.

Borkman, D.G., and Smayda, T.J. 1998. Long-term trends in water clarity revealed by Secchi-disk measurements in lower Narragansett Bay. *ICES Journal of Marine Science* 55:668–679.

Boucher, J. 1991. Nutrient and phosphorus geochemistry in the Taunton River estuary, Massachusetts. Ph.D. Thesis in Oceanography, University of Rhode Island, Narragansett, RI.

Bourne, D.W., and Govoni, J.J. 1988. Distribution of fish eggs and larvae patterns of water circulation in Narragansett Bay, 1972–73. *In* Larval Fish and Shellfish Transport Through Inlets. Weinstein, M.P. (ed.) American Fisheries Society Symposium No. 3.

Bowen, J.L., and Valiela, I. 2000. Historical changes in atmospheric nitrogen deposition to Cape Cod, Massachusetts, USA. *Atmospheric Environment* 35:1039–1051.

Bowen, J.L., and Valiela, I. 2001. The ecological effects of urbanization of coastal watersheds: Historical increases in nitrogen loads and eutrophication of Waquoit Bay estuaries. *Canadian Journal of Fisheries and Aquatic Sciences* 58(6):1489–1500.

Caraco, N.F. 1995. Influence of human populations on P transfers to aquatic systems: A regional scale study using large rivers. *In* Phosphorus in the Global Environment, pp. 235–244. Tiessen, H. (ed) New York: John Wiley and Sons.

Caraco, N.F., and Cole, J. 1999. Human impact on nitrate export: An analysis using major world rivers. *Ambio* 28(2):167–170.

Chaney, M.S., and Ahlborn, M. 1943. Nutrition. Third Edition. Houghton-Mifflin, NY.

Chaves, J.E. 2004. Potential use of ^{15}N to assess nitrogen sources and fate in Narragansett Bay. Ph.D. Thesis, Graduate School of Oceanography, University of Rhode Island, Kingston, RI.

Chinman, R.A., and Nixon, S.W. 1985. Depth–area–volume relationships in Narragansett Bay. NOAA/Sea Grant Marine Technical Report 87. Graduate School of Oceanography, University of Rhode Island, Narragansett, RI.

Cole, J.J., Peierls, B.L., Caraco, N.F., and Pace, M.L. 1993. Nitrogen loading of rivers as a human-driven process. *In* Humans as components of ecosystems, pp. 141–157. McDonnel, M.J., and Pickett, S.T.A. (eds) New York: Springer Verlag.

Coleman, P. 1963. The transformation of Rhode Island, 1790–1860, Providence, Rhode Island: Brown University Press.

Culver-Rymsza, K. 1988. Occurrence of nitrate reductase along a transect of Narragansett Bay. MS Thesis, Oceanography, University of Rhode Island, Narragansett, RI, 135 pp.

Daniel, L. 2004. Narragansett Bay: An Investigative Study of Providence Journal Articles From 1870 –1935. Final report for Oceanography 591, Special Project. 4 pp. with attachments.

Darracq, A., and Destouni, G. 2005. In-stream nitrogen attenuation: Model-aggregation effects and implications for coastal nitrogen impacts. *Environmental Science and Technology* 39:3716–3722.

Day, G.M. 1953. The Indian as an ecological factor in the northeastern forest. *Ecology* 34:329–346.

Dentener, F.J., and Crutzen, P.J. 1994. A three-dimensional model of the global ammonia cycle. *Journal of Atmospheric Chemistry* 9:331–369.

Desbonnet, A., and Lee, V. 1991. Water quality and fisheries-Narragansett Bay. A report to National Ocean Pollution Program Office, National Oceanic and Atmospheric Administration, Rhode Island Sea Grant, Narragansett, RI.

Destouni, G., Lindgren, G.A., and Green, I. 2006. Effects of inland nitrogen transport and attenuation modeling on coastal nitrogen load abatement. *Environmental Science and Toxicology* 40:6208–6214.

Doering, P.H., Weber, L., Warren, W., Hoffman, G., Schweitzer, K., Pilson, M.E.Q., and Oviatt, C.A. 1988. Monitoring of the Providence and Seekonk Rivers for Trace Metals and Associated Parameters. Data Report for Spray Cruises I – VI. Marine Ecosystems Research Laboratory, Graduate School of Oceanography, University of Rhode Island, Narragansett, RI, 2 volumes.

Doering, P.H., Oviatt, C.A., and Pilson, M.E.Q. 1990. Control of Nutrient Concentrations in the Seekonk-Providence River Region of Narragansett Bay, Rhode Island. *Estuaries* 13:418–430.

Dolan, D.M., Yui, A.K., and Geist, R.D. 1981. Evaluation of river load estimation methods for total phosphorus. *Journal of Great Lakes Research* 7: 207–214.

Duarte, C.M. 1995. Submerged aquatic vegetation in relation to different nutrient regimes. *Ophelia* 41:87–112.

Durbin, A.G., and Durbin, E.G. 1981. Standing stock and estimated production rates of phytoplankton and Zooplankton in Narragansett Bay, RI. *Estuaries* 4(10):24–41.

Fennessey, N.M., and Vogel, R.M. 1996. Regional models of potential evapotranspiration and reference evapotranspiration for the northeast USA. *Journal of Hydrology* 184:337–354.

Foster, D.R., and Aber, J.D. 2004. Forests in time the environmental consequences of 1,000 years of change in New England. New Haven, CT: Yale University Press.

Fraher, J. 1991. Atmospheric wet and dry deposition of fixed nitrogen to Narragansett Bay. MS Thesis in Oceanography, University of Rhode Island, Narragansett, RI.

Froelich, P.N. 1988. Kinetic control of dissolved phosphate in natural rivers and estuaries: A primer on the phosphate buffer mechanism. *Limnology and Oceanography* 33:649–668.

Fulweiler, R.W., and Nixon, S.W. 2005. Export of nitrogen, phosphorus, and suspended solids from a southern New England watershed to Little Narragansett Bay. *Biogeochemistry* 76:567–593.

Fulweiler, R.W., and Nixon, S.W. In press. Responses of benthic-pelagic coupling to climate change in a temperate estuary. *Hydrobiologia*.

Gage, S., and McGouldrick, P.C. 1922. Preliminary report of an investigation of the pollution of certain Rhode Island public waters. Annual Report of the Board of Purification of Waters.

Gage, S., and McGouldrick, P.C. 1923, 1924. Report of investigations of the pollution of certain Rhode Island public waters during 1923 and 1924. Presented to the Board of Purification of Waters. Annual Report of the Board of Purification of Waters.

Galloway, J.N., and Cowling, E.B. 2002. Reactive nitrogen and the world: 200 years of change. *Ambio* 31(2):64–71.

Gilkeson, J.S., Jr. 1986. *Middle class Providence, 1820–1940*. Princeton, NJ: Princeton University Press.

Goodale, C.L., Aber, J.D., and Vitousek, P.M. 2003. An unexpected nitrate decline in New Hampshire streams. *Ecosystems* 6:75–86.

Goodale, C.L., Aber, J.D., Vitousek, P.M., and McDowell, W.H. 2005. Long-term decreases in stream nitrate: Successional causes unlikely; possible links to DOC? *Ecosystems* 8:334–337.

Granger, S., Brush, M., Buckley, B., Traber, M., Richardson, M., and Nixon, S.W. 2000. An assessment of eutrophication in Greenwich Bay. Paper No. 1, Restoring water quality in Greenwich Bay: a whitepaper series. Rhode Island Sea Grant, Narragansett, RI. 20 pp.

Hamlin, C. 1990. A science of impurity- water analysis in nineteenth century Britain. Berkeley, CA: University of California Press.

Hanson, L.C. 1982. A preliminary assessment of nutrient loading into Narragansett Bay due to urban runoff. MS Thesis, Marine Affairs, University of Rhode Island, Kingston, RI.

Harper, R.M. 1918. Changes in the forest area of New England in three centuries. *Journal of Forestry* 16:442–452.

Hedgpeth, J.W. 1978. As blind men see the elephant: the dilemma of marine ecosystem research, pp.3–15. *In* Estuarine Interactions, Wiley, M.L.(ed.) New York: Academic Press.

Hedin, L.O., Armesto, J.J., and Johnson, A.H. 1995. Patterns of nutrient loss from unpolluted, old-growth temperate forests: evaluation of biogeochemical theory. *Ecology* 76:493.

Hicks, S.D. 1959. The physical oceanography of Narragansett Bay. *Limnology and Oceanography* 4:316–327.

Hoffman, E.J., and Quinn, J.G. 1984. Hyrdrocarbons and other pollutants in urban runoff and combined sewer overflows. Final report to the National Oceanic and Atmospheric Administration. Graduate School of Oceanography, Narragansett, RI.

Hooker, T.D., and Compton, J.E. 2003. Forest ecosystem carbon and nitrogen accumulation during the first century and after agricultural abandonment. *Ecological Applications* 13(2):299–313.

Howarth, R.W., Billen, G., Swaney, D., Townsend, A., Jaworski, N., Lajtha, K., Downing, J.A., Elmgren, R., Caraco, N., Jordan, T., Berendse, F., Freney, J., Kueyarov, V., Murdoch, P., and Zhao-Liang, Z. 1996. Riverine inputs of nitrogen to the North Atlantic Ocean: Fluxes and human influences. *Biogeochemistry* 35:75–139.

Howarth, R.W., Boyer, E.W., Pabich, W.J., and Galloway, J.N. 2002. Nitrogen use in the United States from 1961–200 and potential future trends. *Ambio* 31:88–96.

Howarth, R.W., and Marino, R. 2006. Nitrogen as the limiting nutrient for eutrophication in coastal marine ecosystems: Evolving views over 3 decades. *Limnology and Oceanography* 51:364–376.

Husar, R.B. 1986. Emissions of sulfur dioxide and nitrogen oxides and trends for Eastern North America. *In* Acid deposition-long-term trends, pp. 48–92. National Research Council Staff (eds.) Washington, DC: National Academies Press.

Isaac, R.A. 1997. Estimation of nutrient loadings and their impacts on dissolved oxygen demonstrated at Mt. Hope Bay. *Environment International* 23:151–165.

Iverson, R.L. 1990. Control of marine fish production. *Limnology and Oceanography* 35:1593–1604.

Keller, A.A. 1988. Estimating phytoplankton productivity from light availability and biomass in MERL mesocosms and Narragansett Bay. *Marine Ecology Progress Series* 45:159–168.

Keller, A.A., Doering, P.H., Kelly, S.P., and Sullivan, B.K. 1990. Growth of juvenile Atlantic menhaden, *Brevoortia tyrannus* (Pices: Clupeidae) in MERL mesocosms: Effects of eutrophication. *Limnology and Oceanography* 35(1):109–122.

Keller, A.A., Klein-MacPhee, G., and Burns, J. St.Onge. 1999. Abundance and distribution of Ichthyoplankton in Narragansett Bay, Rhode Island, 1989–1990. *Estuaries* 22:149–163.

Kelly, J.R. 2001. Nitrogen effects on coastal marine ecosystems. *In* Nitrogen in the environment: sources, problems and management, pp. 207–251. Follet, R.F., and Hatfield, J.L. (eds.) B.V.: Elsevier Science.

Kemp, W.M., Boynton, W.R., Adolf, J.E., Boesch, D.F., Boicourt, W.C., Brush, G., Cornwell, J.C., Fisher, T.R., Glibert, P.M., Hagy, J.D., Harding, L.W., Houde, E.D., Kimmel, D.G., Miller, W.D., Newell, R.I.E., Roman, M.R., Smith, E.M., and Stevenson, J.C. 2005. Eutrophication of Chesapeake Bay: historical trends and ecological interactions. *Marine Ecology Progress Series* 303:1–29.

Kerr, M. 1992. River rescue year one data report. Rhode Island Sea Grant, Narragansett, RI.

Kirkwood, J.P. 1876. Seventh annual report of the State Board of Health of Massachusetts, January 1876. Boston, MA. A special report on the pollution of river waters. New York: Arno Press Inc. pp. 2–408.

Kremer, J.N., and Nixon, S.W. 1978. A coastal marine ecosystem: Simulation and analysis. New York: Springer Verlag.

Levy, H., II and Moxim, W.J. 1989. Simulated global distribution and deposition of reactive nitrogen emitted by fossil fuel combustion. *Tellus* 41B:256–271.

Lewis, W.M., Jr. 2002. Yield of nitrogen from minimally disturbed watersheds of the United States. *In* The nitrogen cycle at regional to global scales, pp. 375–385. Boyer, E.W., and Howarth, R.W. (eds.) The Netherlands: Kluwer Academic Publishers.

Li, Y., and Smayda, T.J. 1998. Temporal variability of chlorophyll in Narragansett Bay, 1973–1990. *ICES Journal of Marine Science* 55:661–667.

Likens, G.E., Bormann, F.H., Pierce, R.S., Eaton, J.S., and Johnson, N.M. 1977. Biogeochemistry of a Forested Ecosystem. New York: Springer-Verlag.

Luo, Y., Yang, X., Carley, R.J., and Perkins, C. 2002. Atmospheric deposition of nitrogen along the Connecticut coastline of Long Island Sound: a decade of measurements. *Atmospheric Environment* 36:4517–4528.

Mallet, J.W. 1882. The determination of organic matter in potable water. *Chemical News*, 18 Aug. pp. 63–66 and 72–75; 25 Aug. pp. 90–92; 1Sept. pp.101–102; 8 Sept. pp. 108–112.

Massachusetts State Department of Health (MSDH). various years. Annual reports. Boston, MA.

Massachusetts Department of Public Health (MDPH). various years. Annual reports. Boston, MA.

Massachusetts State Board of Health. various years. Annual Reports of the Massachusetts State Board of Health, Boston, MA.

Matthiessen, G.C. 1973. Rome Point Investigations. Progress Reports. Marine Research, Inc.

Mayer, B., Boyer, E.W., Goodale, C., Jaworski, N.A., Van Breemen, N., Howarth, R.W., Seitzinger, S., Billen, G., Lajtha, K., Nadelhoffer, K., VanDam, D., Hetling, L., Nosal, M., and Paustian, K. 2002. Sources of nitrate in rivers draining sixteen water sheds in the northeastern U.S.: Isotopic constraints. *In* The nitrogen cycle at regional to global scales, pp. 171–197. Boyer, E.W., and R.W. Howarth (eds.) The Netherlands: Kluwer Academic Publishers.

McCracken, R., and Matera, J. (eds.). 1987. Special Edition Recognizing the 100th Anniversary of the Lawrence Experiment Station. *Journal of the New England Water Works Association* 101(3):349.

Metcalf, L., and Eddy, H.P. 1915. American sewage practice. New York: McGraw-Hill.

Metcalf and Eddy, Engineers. 1948. Condition of the Seekonk River. Report to the Blackstone Valley Sewer District Commission. Boston, MA.

Moomaw, W.R. 2002. Energy, industry and nitrogen: strategies for decreasing reactive nitrogen emissions. *Ambio* 31(2):184–189.

Motzkin, G., and Foster, D.R. 2002. Grasslands, heathlands and shrublands in coastal New England: Historical interpretations and approaches to conservation. *Journal of Biogeography* 29:1569–1590.

National Academy of Sciences. 1989. Recommended Dietary Allowances 10th edition. Washington, DC: National Academies Press.

National Academy of Sciences. 1996. Use of Reclaimed Water and Sludge in Food Crop Production. Washington, DC: National Academies Press.

Neff, J.C., Holland, E.A., Dentener, F.J., McDowell, W.H., and Russell, K.M. 2002. The origin, composition and rates of organic nitrogen deposition: a missing piece of the nitrogen cycle? *In* The nitrogen cycle at regional to global scales, pp. 99–136. Boyer, E. W., and Howarth, R.W. (eds.) The Netherlands: Kluwer Academic Publishers.

Nichols, W.P. 1873. On the condition of certain Massachusetts waters. State Board of Health, Lunacy and Charity.

Nichols, W.P. 1876. Tables of analysis: with remarks on the waters of the different valleys. *In* A special report on the pollution of river waters, pp. 155–173. Kirkwood, J.P. (ed.) New York: Arno Press Inc.

Nixon, S.W. 1988. Physical energy inputs and the comparative ecology of lake and marine ecosystem. *Limnology and Oceanography* 33(4):1005–1025.

Nixon, S.W. 1989. An extraordinary red tide and fish kill in Narragansett Bay. *In* Novel phytoplankton blooms, causes and impacts or recurrent brown tides and other unusual blooms, pp. 429–483. Cosper E.M., Bricelj, V.M., and Carpenter, E.J. (eds.) New York: Springer-Verlag.

Nixon, S.W. 1995a. Metal inputs to Narragansett Bay: A history and assessment of recent conditions. Rhode Island Sea Grant, Narragansett, RI.

Nixon, S.W. 1995b. Coastal marine eutrophication: a definition, social causes, and future concerns. *Ophelia* 41:199–219.

Nixon, S.W. 1997. Prehistoric nutrient inputs and productivity in Narragansett Bay. *Estuarine Research Federation* 20(2):253–261.

Nixon, S.W. 2004. Marine resources and the human carrying capacity of coastal ecosystems in southern New England before European contact. *Northeast Anthropology* 68:1–23.

Nixon, S.W. In press. Eutrophication and the Macroscope. *Hydrobiologia*.

Nixon, S.W., Oviatt, C.A., and Hale, S.S. 1976. Nitrogen regeneration and the metabolism of coastal marine bottom communities. *In* The role of terrestrial and aquatic organisms in decomposition processes, pp. 269–283. Anderson, J., and Macfadyen, A. (eds.) London: Blackwell Scientific.

Nixon, S.W., Kelly, J., Furnas, B.N., Oviatt, C.A., and Hale, S.S. 1980. Phosphorus regeneration and the metabolism of coastal marine bottom communities. *In* Marine benthic dynamics, pp. 219–242. Tenore, K., and Coull, B.C. (eds.) Columbia, SC: University of South Carolina Press.

Nixon, S.W., and Pilson, M.E.Q. 1984. Estuarine total system metabolism and organic exchange calculated from nutrient ratios: an example from Narragansett Bay. *In* The estuary as a filter, pp. 261–290. Kennedy, V.S. (ed.) Florida: Academic Press Inc.

Nixon, S.W., Furnas, B.N., Chinman, R., Granger, S., and Heffernan, S. 1982. Nutrient Inputs to Rhode Island Coastal Lagoons and Salt Ponds. Final report to RI Statewide Planning. University of Rhode Island, Kingston, RI.

Nixon, S.W., Granger, S.L., and Nowicki, B.L. 1995. An assessment of the annual mass balance of carbon, nitrogen, and phosphorus in Narragansett Bay. *Biogeochemistry* 31:15–61.

Nixon, S.W., Ammerman, J.W., Atkinson, L.P., Berounsky, V.M., Billen, G., Boicourt, W.C., Boynton, W.R., Church, T.M., DiToror, D.M., Elmgren, R., Garber, J.H., Giblin, A.E., Jahnke, R.A., Owens, J.P., Pilson, M.E.Q., and Seitzinger, S.P. 1996. The fate of nitrogen and phosphorus at the land-sea margin of the North Atlantic Ocean. *Biogeochemistry* 35:141–180.

Nixon, S.W., Buckley, B., Granger, S., and Bintz, J. 2001. Response of very shallow marine ecosystems to nutrient enrichment. *Human and Ecological Risk Assessment* 7(5):1457–148.

Nixon, S.W., Granger, S., and Buckley, B. 2003. The warming of Narragansett Bay. 41°N, The Magazine of the RI Sea Grant & Land Grant Programs, University of Rhode Island 2:18–20. *(http://seagrant.gso.uri.edu/41N/Vol2No1/baywarming.html)*

Nixon, S.W., Granger, S., Buckley, B.A., Lamont, M., and Rowell, B. 2004. A one hundred and seventeen year coastal water temperature record from Woods Hole, Massachusetts. *Estuaries* 27:397–404.

Nixon, S.W., Buckley, B., Granger, S., Harris, L., Oczkowski, A., Cole, L., and Fulweiler, R. 2005. Anthropogenic nutrient inputs to Narragansett Bay: A twenty five year perspective. A Report to the Narragansett Bay Commission and Rhode Island Sea Grant. Rhode Island Sea Grant, Narragansett, RI. *(www.seagrant.gso.uri.edu/research/bay_ commission_report.pdf.)*

Oviatt, C.A. 2004. The changing ecology of temperate coastal waters during a warming trend. *Estuaries* 27:895–904.

Oviatt, C.A., Buckley, B., and Nixon, S. 1981. Annual phytoplankton metabolism in Narragansett Bay calculated from survey field measurements and microcosm observations. *Estuaries* 4:167–175.

Oviatt, C.A., Doering, P., Nowicki, B., Reed, L., Cole, J.J., and Frithsen, J. 1995. An ecosystem level experiment on nutrient limitation in temperate coastal marine environments. *Marine Ecology Progress Series* 116:171–179.

Oviatt, C.A., Keller, A., and Reed, L. 2002. Annual primary production in Narragansett bay with no bay-wide winter-spring bloom. *Estuarine, Coastal, and Shelf Science* 54:1013–1026.

Oviatt, C., Olsen, S., Andrews, M., Collie, J., Lynch, T., and Raposa, K. 2003. A century of fishing and fish fluctuations in Narragansett Bay. *Reviews in Fisheries Science* 11:221–242.

Pate, S. 1992. Women in science: The lost chapter. Women Chemists Newsletter. *American Chemical Society.* pp 3–4.

PCE. various years. Annual reports of the City of Providence Engineer. Providence, RI.

Peierls, B., Caraco, N.F., Pace, M.L., and Cole, J.J. 1991. Human influence on river nitrogen. *Nature* 350:386–387.

Phelps, E.P., and Shoub, H.L. 1917. The determination of nitrate in sewage by means of ortho-tolidine. *Journal of Industrial and Engineering Chemistry* 9:767–771.

Pilson, M.E.Q. 1985a. On the residence time of water in Narragansett Bay. *Estuaries* 8:2–14.

Pilson, M.E.Q. 1985b. Annual cycles of nutrients and chlorophyll in Narragansett Bay, Rhode Island. *Journal of Marine Research* 43:849–873.

Pilson, M.E.Q. 1989. Aspects of climate around Narragansett Bay. Unpublished report to The Narragansett Bay Project, RI Department of Environmental Management. Graduate School of Oceanography, Narragansett, RI.

Providence Department of Public Works. 1932. Annual Report. Providence, RI.

Rabalais, N.N., and Nixon, S.W. 2002. Preface: Nutrient over-enrichment of the coastal zone. *Estuaries* 25(4b):639.

Rafter, G.W., and Baker, M.N. 1900. Sewage disposal in the United States. New York: Van Nostrand.

Rhode Island Board of Purification of Waters. 1932. Report of the Board of Purification of Waters for the year 1932. Snow and Farnham Co. Providence RI.

RIBPW. various years. Annual Report of the Board of Purification of Waters, Providence, RI.

Rhode Island Department of Environmental Management (RIDEM). 2000. Survey of Wastewater Treatment Facilities in Rhode Island for 1998 and 1999. Providence, RI.

Rhode Island Department of Environmental Management (RIDEM). 2003. The Greenwich Bay fish kill – August 2003 Causes, impacts and responses. Providence, RI. *(www.dem.ri. gov/pubs/fishkill.pdf)*

Rhode Island State Board of Health (RISBH). various years. Annual Reports. Providence, RI.

Richards, R.P. 1999. Estimation of pollutant loads in rivers and streams: A guidance document for NPS programs. Prepared under Grant X998397-01-0 US Environmental Protection Agency.

Ries, K.G., III. 1990. Estimating surface-water runoff to Narragansett Bay, Rhode Island and Massachusetts. US Geological Survey, Water Resources Investigations. Report 89–4164. 44 pp.

Robinson, K.W., Campbell, J.P., and Jaworski, N.A. 2003. Water-quality trends in New England rivers during the 20th Century. USGS Water Resources Investigation Report 03-4012.

Rudnick, D.T. 1984. Benthic Seasonality in Narragansett Bay. Ph.D. Thesis, Graduate School of Oceanography, University of Rhode Island, Narragansett, RI.

Rudnick, D.T., Elmgren, R., and Frithsen, J.B. 1985. Meiofaunal prominence and benthic seasonality in a coastal marine ecosystem. *Oecologia* 67:157–168.

Saunders, W., and Swanson, C. 2002. Comments on US EPA draft permit for Brayton Point station (No. MA 0003654) related to references to Isaac, R.A. (1997), Estimation of Nutrient Loading and their Impacts on Dissolved Oxygen Demonstrated at Mt. Hope Bay. Applied Science Associates, Narragansett, RI.

Save The Bay. 1998. Vital signs: our bay in crisis. Save The Bay, Providence, RI.

Saxe, J.G. 1963. The blind man and the elephant; John Godfrey Saxe's version of the famous Indian legend. New York: Whittensey House.

Schindler, D.W. 1981. Studies of eutrophication in lakes and their relevance to the estuarine environment. *In* Estuaries and Nutrients, pp.71–82. Neilson, B.J., and Cronin, L.E. (eds.) Clifton, NJ: Humana Press.

Seitzinger, S.P., and Sanders, R.W. 1997. Contribution of dissolved organic nitrogen from rivers to estuarine eutrophication. *Marine Ecology Progress Series* 159:1–12.

Seitzinger, S.P., Styles, R.V., Boyer, E.W., Alexander, R.B., Billen, G., Howarth, R.W., Mayer, B., and van Breemen, N. 2002. Nitrogen retention in rivers: model development and application to watersheds in the northeastern U.S.A. *Biogeochemistry* 57/58:199–237.

Smayda, T.J. 1998. Patterns of variability characterizing marine phytoplankton, with examples from Narragansett Bay. *ICES Journal of Marine Science* 55:562–573.

Smil, V. 2001. Enriching the earth. Boston, MA: MIT Press.

Smith, V.H., Samantha, B.J., and Howarth, R.W. 2006. Eutrophication of freshwater and marine ecosystems. *Limnology and Oceanography* 51(1):351–355.

Stoddard, J.L., Kahl, J.S., Deviney, F.A., DeWalle, D.R., Driscoll, C.T., Herlihy, A.T., Kellogg, J.H., P.S. Murdoch, P.S., Webb, J.R., and Webster, K.E. 2003. Response of surface water chemistry to the Clean Air Act Amendments of 1990. EPA 620-R-03-001. US Environmental Protection Agency. Research Triangle Park, NC. *(www.epa.gov/ord/htm/CAAA-ExecutiveSummary-1-29-03.pdf)*

Straub, C.P. 1989. Practical handbook of environmental control. Boca Raton, FL: CRC Press, Inc.

Sullivan, B.L., and Ritacco, P.J. 1985. The response of dominant copepod species to food limitation in a coastal marine ecosystem. *Archiv fuer Hydrobiologie Beih. Ergebn. Limnol.* 21:407–408.

Sullivan, B.K., Van Keuren, D., and Clancy, M. 2001. Timing and size of blooms of the ctenophore *Mnemiopsis leidyi* in relation to temperature in Narragansett Bay, RI. *Hydrobiologia* 451:113–120.

Sweet, A.W. 1915. A sanitary survey of the Seekonk River. Ph.D. Thesis, Brown University, Providence, RI.

Tarr, J.A. 1971. Urban pollution-many long years ago. *American Heritage* 22:65–69.

Tarr, J.A. 1996. The search for the ultimate sink – urban pollution in historical perspective. Akron, OH: University of Akron Press.

Turner, R.E., and Rabalais, N.N. 2003. Linking landscape and water quality in the Mississippi River basin for 200 years. *Bioscience* 53(6):563–572.

Tuttle, A.S., and Allen, K. 1924. The food supply of a city as ascertained from analysis of its sewage. *Sewage Disposal Bulletin* 29:3.

Urish, D.W., and Gomez, A.L. 2004. Groundwater Discharge to Greenwich Bay. Paper No. 3. Restoring Water Quality in Greenwich Bay: A Whitepaper Series, Rhode Island Sea Grant. Narragansett, RI. 9 pp.

US Census Bureau. 1975. Annual reports.

US Department of Health, Education and Welfare Public Health Service. 1960. Effects of proposed hurricane barriers on water quality of Narragansett Bay. Prepared for US Army Corps of Engineers New England Division. New York, N.Y.

US Public Health Service. 1960. Effects of proposed hurricane barriers on water quality of Narragansett Bay. Prepared for US Army Corps of Engineers, New England Division. New York, N.Y.

Wanklyn, J.A., and Chapman, E.T. 1884. Water-Analysis: A Practical Treatise on the Examination of Potable Water, sixth ed., London: Truber & Co.

Webb, K.L. 1981. Conceptual Models and processes of nutrient cycling in estuaries. *In* Estuaries and Nutrients, pp. 25–46. Neilson, B.J., and Cronin, L.E. (eds.) New Jersey: Humana Press.

Wroth, L.C. 1970. *The voyages of Giovanni da Varrazano:1524–1528.* New Haven, CT: Yale University Press.

Wood, W. 1635. New England's Prospect. The Cotes, London. Vaughn, A.T. (ed) reprinted 1977, Amherst, MA: University of Massachusetts Press.

Chapter 6
Nitrogen Inputs to Narragansett Bay: An Historical Perspective

Steven P. Hamburg, Donald Pryor and Matthew A. Vadeboncoeur

6.1 Introduction

The overall alteration of the global nitrogen cycle has been of increasing concern to ecologists and resource managers for several decades (Vitousek *et al.*, 1997), yet most impacts of excess anthropogenic nitrogen are place-specific and cannot be generalized. Thus, the implications of human impacts on the nitrogen cycle can often be best understood in the context of specific ecosystems. While anthropogenic inputs of nitrogen to the Narragansett Bay watershed are broadly representative of the patterns found in other eastern United States estuaries (Castro *et al.*, 2001; Driscoll *et al.*, 2001; 2003), the temporal and spatial patterns are specific to Narragansett Bay, and the detailed patterns need to be understood if effective management of nitrogen is to be undertaken.

Understanding patterns of change in the nitrogen cycle requires knowing spatial and temporal dynamics; how many people and animals lived where, for how long, and with access to what type of technology. To construct temporally and spatially explicit models of nitrogen cycling, it is necessary to develop detailed databases; yet the amount of historical data available on nitrogen dynamics (see Nixon *et al.*, Chapter 5) limits our ability to validate these models. It is possible, however, to use indirect evidence to test the robustness of the assumptions used to model historical N loading to the bay. A quantitative, spatially explicit understanding of changing human population and land use in the watershed over time is essential to the development of effective management strategies that recognize that the Narragansett Bay ecosystem is not in equilibrium with respect to nutrient cycling, climate, or anthropogenic use of the resource. Understanding how nitrogen inputs to Narragansett Bay have changed since the late 1800s, when detailed data on agriculture and the disposal of human waste became available, is a key to interpreting the more detailed data on nutrient dynamics that became available in the later part of the 20th century.

S.P. Hamburg
Center for Environmental Studies, Brown University, Providence, RI 02912-1943
Steven_Hamburg@Brown.edu

A. Desbonnet, B.A. Costa-Pierce (eds.), *Science for Ecosystem-based Management.*
© Springer 2008

178

S.P. Hamburg et al.

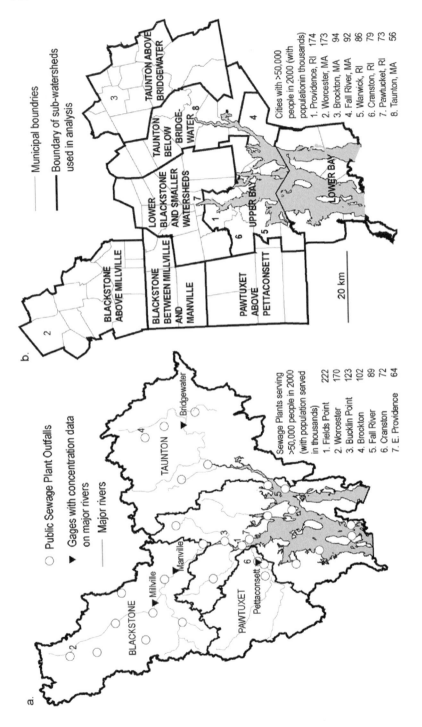

Several researchers have presented reconstructions of the history of anthropogenic nutrient loading to Narragansett Bay based on expected relationships to forcing variables such as population, land use, agricultural practices, and energy consumption (Nixon *et al.,* 1995; Howarth *et al.,* 1996; Jaworski *et al.,* 1997; Roman *et al.,* 2000; Boyer *et al.,* 2002; Moore *et al.,* 2004). To develop a temporal picture of changing nitrogen flux to Narragansett Bay from 1880 to 2000, this chapter expands upon these previous efforts by employing spatially explicit historical data on human populations, domesticated animals, and populations served by sewers. This period was selected as quantitative spatially explicit data are available, and it reaches back to a time prior to the construction of sewers within the watershed. To model changes in N loading over time, historical data on population and domestic animals at town and the subwatershed levels were compiled as shown in Fig. 6.1.

A simple, spreadsheet-based numerical model estimating the N load from human and animal waste between 1880 and 2000 was constructed based on available archival records and a careful review of the narrative history of changing human use of the landscape over that time period. Using this model in conjunction with long-term datasets of river N concentrations, a more complete understanding of how human impacts have altered the N cycle in the Narragansett Bay watershed over the past 120 or so years is developed.

6.2 Historical Changes in the Drivers of Nitrogen Cycling

6.2.1 Changes in Human and Animal Populations

Native populations in the Narragansett Bay watershed are thought to have numbered less than 10,000 before first European settlement in 1634 (Russell, 1980). Total population in the watershed was less than 50,000 (10 persons km^{-2}) through 1790 when the first US Census was taken. Based on an estimate of the nitrogen needs of the human population at that time, and reasoned speculation about the patterns of the presettlement nitrogen cycle, Nixon (1997) estimated that anthropogenic nitrogen loads were less than 20% of the presettlement input through the early 1800s. In the 19th century, a major transformation of the watershed took place, with rapid population increases

---

Fig. 6.1 (a) The Taunton, Blackstone, and Pawtuxet rivers drain the three largest watersheds contributing to Narragansett Bay. There are 30 public sewage systems in the Narragansett Bay watershed, the largest of which (circles) are clustered adjacent to the upper bay. Historical DIN concentration data exist for four points on the major rivers (triangles). (b) As most data on population and agriculture are available only at the town level, towns are aggregated into spatial groups which correspond to the subwatersheds above each of these points on each river. Remaining towns are also separated into spatial groups in order to discern spatial patterns in modeled N trends. The watershed's largest cities are labeled.

of 22–40% per decade between 1840 and 1900 (US Census data), broad deforestation (forest cover hit its nadir in approximately 1850; Harper, 1918) and large increases in domesticated animal populations to a peak of 125,000 animal units in 1910 (US Bureau of the Census, 1922). An "animal unit" (AU) is an agricultural unit of measure corresponding to approximately 454 kg of animal mass (USDA, 1997). By 2000, the watershed's human population had grown to approximately 2 million (more than 460 persons km^{-2}; US Bureau of the Census, 1990), active agricultural land had decreased to 5% of watershed area (USDA, 2002), forest cover had expanded to 72% (USEPA, 1992), and numbers of animal units had declined to approximately 16,000 (USDA, 2002).

These land-use changes are important to shifting patterns of nitrogen cycling in the watershed because of their implications for the quantity of human and animal waste being disposed. Nitrogen is ingested by humans primarily in the form of protein; since the early 1900s, per capita consumption of protein has been approximately stable at 100–110 g N day^{-1} per "man-unit", or typical adult male equivalent (Dirks, 2003; Briefel and Johnson, 2004).

Since European settlement, the population of the Narragansett Bay watershed has gone from being largely self-sufficient in terms of food production, depending on locally grown produce and meat as well as fish from nearby waters, to importing almost all food. Thus, a large quantity of N has been imported (Driscoll *et al.*, 2001; Boyer *et al.,* 2002) as population density increased and trade with more agriculturally productive regions became feasible.

The impacts of anthropogenic activities on the nitrogen cycle are closely tied to how humans secure water, and then dispose of it as wastewater. In Narragansett Bay communities, as in most communities, the time frame of the development of organized provision of these water-based services was staggered (Tarr, 1996; Melosi, 2000). In Providence, the watershed's largest city, a central water supply began providing water in 1871, yet it was not until 1892 that sewage was collected in a centralized system for discharge into the Providence River at Field's Point. In the interim, the sewers served an increasing portion of the city, but disposal was dispersed, presumably into the Providence, Woonasquatucket and Moshassuck Rivers (see Nixon *et al.,* Chapter 5, for more details of the early history of Providence's sewer system). "Water carriage" quickly replaced manual collection of "night soil," with centralized treatment implemented in 1901 (Nixon, 1995), at which time almost 80% of the extant streets of the city were served by sewers (Clapp, 1902). Worcester, the watershed's second largest city, initiated sewer construction by 1870 and began operation of a treatment facility in 1890 (Shanahan, 1994). Today, more than 90% of the watershed population is served by public water supply, though less than 70% are served by public sewers (US Bureau of the Census, 1941).

By the late 20th century, all public sewage systems in the Narragansett Bay watershed provided secondary treatment of wastewater before discharge during dry weather. During wet weather events, the presence of combined storm water and sanitary sewer systems, such as exists in Providence and Worcester, means that the water entering the system exceeds the capacity of the treatment facility to handle it, resulting in the release of untreated sewage. Over the 20th century, sewage treatment evolved first to provide primary treatment (disinfection), then secondary treatment (removal of suspended solids and biological oxygen demand). Although performance varied widely (NRC, 1993), it was not until the 21st century advent of tertiary treatment that significant amounts of N were removed from sewage. Despite the public health benefits of sewers, and the removal of N loading to groundwater and small streams, the net effect of sewage systems is to concentrate waste nitrogen, and often to increase the proportion of total waste N that reaches receiving waters (Valiela *et al.*, 1997; Pryor *et al.*, 2006)

6.2.2 Land-Use Change

The Narragansett Bay watershed, consistent with land-use patterns across the region, has experienced a long-term decline in the amount of agricultural land from the mid-19th to the mid-20th century, with continuing but slower declines to the present day (Fig. 6.2; Whitney, 1994). The bay islands and shoreline, particularly Newport and Bristol (RI) Counties, were the most heavily agricultural, with 85% of Newport County in agricultural use by 1850 (US Census Office, 1893). Bristol County peaked at 72% agricultural land in 1860 (US Census Office, 1850; 1860; Snow, 1867). In contrast, Rhode Island's remaining three counties were all between 40% and 50% agricultural land during the same time period. In general, the pattern of heavier agricultural use in Newport and Bristol Counties continues today (USDA, 2002), despite the fact that the amount of agricultural land has decreased regionally by about 90% since 1860. The rate of decline in agricultural land-uses slowed in the last thirty years of the 20th century (RI Statewide Planning, 2000), though the overall conversion from an agriculturally dominated landscape to a forest-dominated one was largely completed by 1970.

Areas closest to the shoreline of Narragansett Bay have a history of much more intensive agricultural use than do those further inland (Fig. 6.2). This difference has important implications for nitrogen inputs to the bay, as nitrogen is much more likely to be absorbed by terrestrial ecosystems and low order streams when released in the upper reaches of the watershed than if it were released at the land–water margin. In fact, Rhode Island's Newport and Bristol Counties, with 46% and 87% of their land area within 1 km of the bay, respectively, have consistently had the highest densities of animal units in the watershed over the past 150 years (US Census Office, 1850; 1860; 1883; 1902; Snow, 1867; 1877; Perry, 1887; Wright, 1887; Tiepke, 1898; Wadlin, 1899; 1909; Webb and Greenlaw, 1906; US Bureau of the Census, 1913; 1922; 1931; 1941; 1952; 1960; 1971; 1978; 1982; 2000; USDA, 2002). Thus, shorelines of

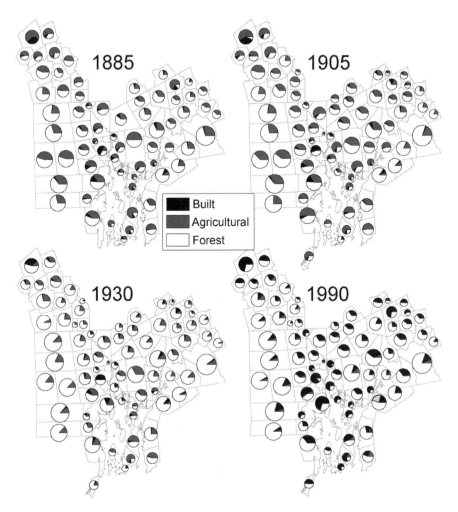

Fig. 6.2 Town-level land-uses (agricultural, forest, and built) in the Narragansett Bay watershed for 1885, 1905, 1930, and 1990. Agricultural lands are those recorded as "improved" (1885–1930) in the census, or in agricultural use in 1990 as identified by remote sensing. The "forest" category likely includes significant amounts of young forests on recently abandoned agricultural land, especially in 1885 and 1905. The "built" category includes all lands that are neither forested nor agricultural (e.g., urban development, low density suburban development, cemeteries, landfills, golf courses, etc.).

Aquidneck and Conanicut Islands, as well as the towns of Bristol and Warren, which together compose most of these two counties, would have experienced some of the most nitrogen-enriched runoff. In 1885, there were approximately 100,000 animal units in the Narragansett Bay watershed; of these, one third were in towns and cities that directly border the bay (Perry, 1887; Wright, 1887), and account for 20% of the watershed's land area.

The fact that the number of animal units peaked in the early 20th century, much later than the peak in the amount of cleared land (US Census Office, 1850; 1860; 1883; 1902; Snow, 1867; 1877; Perry, 1887; Wright, 1887; Tiepke, 1898; Wadlin, 1899; 1909; US Bureau of the Census, 1913; 1922; 1931; 1941; 1952; 1971; 1982; Webb and Greenlaw, 1906; USDA, 2002), means the ratio of manure to cropland and pasture increased in the late 19th century. This more intensive use of manure on the remaining agricultural lands, which were disproportionally near the bay, contributed to an increase in the fraction of manure N entering receiving waters. In fact, the RI State Board of Health (1912) reported that around the turn of the 20th century, "the water-shed of the Newport [public water] supply is still open to pollution from the excreta of cows and fowl which feed along the banks of the brooks supplying the reservoirs." Nitrate concentrations at the "south pumping station" of the Jamestown water supply were observed to be consistently high—between 3 and 5 mg N L^{-1}—at this same point in time (RI State Board of Health, 1912).

In the 19th and early 20th centuries, before the development of modern suburbs, there was very little low-density residential development independent of farms. Nonfarm residences tended to be clustered in villages or cities, which were centered around manufacturing facilities of varying sizes. The Rhode Island towns and cities surrounding the upper bay are, and always have been, the most heavily developed, along with the Massachusetts cities of Worcester, Brockton, Fall River, and Taunton (Figures 6.1 and 6.2). Watershed-wide, forest cover has increased with the decline of agriculture from the late 19th century through the mid 20th century. However, this trend has recently reversed as secondary forests on former agricultural land give way to low-density development (Novak and Wang, 2004). This is significant because aggrading forests are a net N sink (Goodale *et al.,* 2002), while cleared and developed land yield increased runoff and N export relative to forested land (Bowen and Valiela, 2001a).

6.2.3 Agricultural Practices

The practice of agriculture has changed significantly since the mid 19th century—in the early 1800s, there were no synthetic nitrogen fertilizers, and consequently nitrogen in farm systems cycled more tightly than it does in modern farms where nitrogen fertilizer is relatively inexpensive and ubiquitous. However, regardless of how conscientiously animal manure was collected, stored, and used, there would have been significant N leakage and volatilization. Aquidneck and Conanicut Islands may have been particularly affected, as more than 75% of their land area was cleared, mainly for pasturage and to a lesser degree for row crops. Thus, there were probably not adequate vegetated buffers to absorb nitrogen in runoff or shallow groundwater. Forested areas were largely relegated to wood lots required to meet fuel wood needs, and generally located distant to the active agricultural lands of each farm, removing them as effective N sinks (Whitney, 1994).

A comprehensive estimate of commercial fertilizer use was first attempted in 1880. Fertilizer usage was reported as dollars spent, but assuming an average price of two cents per pound (Hamburg, 1984), and that the fertilizer was guano (accumulations of bird and bat excreta mined as fertilizer) with a composition of approximately 5% N (Miller, 1914), we estimate that it represented an exogenous introduction of 180,000 kg N yr^{-1} to the watershed, only a fraction of which would have made it to the bay—Howarth et al. (1996) estimate that 10–40% of fertilizer N applied to loamy soils leaches out in modern agricultural systems. Though it is impossible to know how historical farms in the Narragansett Bay watershed compared with modern farms, the more densely agricultural coastal areas would likely have exported a greater fraction of total fertilizer N to the bay than the less intensively farmed (and more forested) areas in the upper reaches of the watershed.

The use of synthetic fertilizers in the watershed increased rapidly in the late 20th century. County-level fertilizer use data, based on sales, can be used to estimate potential N inputs (Alexander and Smith, 1990; Battaglin and Goolsby, 1994), within the watershed using the mean area weighted county-based average inputs (Boyer et al., 2002). From this, it is estimated that N fertilizer usage in the Narragansett Bay watershed increased from approximately 600,000 kg N yr^{-1} in 1950—more than three times what it is was in 1880—to nearly 2,000,000 kg N yr^{-1} in 1990. (The 1990 value from Plymouth County, MA was more than three times the 1980 value, a far greater increase than seen in any other county, and for this reason we estimated fertilizer use in Plymouth County based on the mean rate of increase between 1980 and 1990 in the other counties within the watershed. Therefore, the 1990 estimate used here may be somewhat lower than the actual value.) The majority of this increase occurred between 1960 and 1970. Fertilizer use was and still is heaviest in Newport County, which remains the most heavily agricultural in the watershed, and which is dominated by the fertilizer-intensive nursery/sod industry (RI DEM, 2003). Although the long-term decline in N from animal waste may be offset somewhat by dramatic increases in the use of fertilizers applied to lawns and golf courses (Bowen and Valiela, 2001a), careful analysis of the nitrogen budget for the Blackstone basin by Boyer et al. (2002) indicated that fertilizer runoff constituted less than 10% of the total N input in the early 1990s. Because fertilizer inputs to the bay are small relative to other inputs, carry high uncertainty, are significant only in the modern period, and have been thoroughly modeled (e.g., Boyer et al., 2002), we have excluded them from our numerical model and address them in a qualitative way in the context of modeled N flux from humans and animals.

6.2.4 Atmospheric Loading

Atmospheric deposition of nitrogen contributes a relatively small fraction of total nitrogen load to Narragansett Bay—10% to 15%—as estimated by Alexander *et al.* (2001), Castro *et al.* (2001), and Moore *et al.* (2004). Howarth, in Chapter 3, further explores atmospheric N deposition to Narragansett Bay and the northeastern US, suggesting that 30% of the N input to the ecosystem is from atmospheric deposition, though his accounting may not fully include urban sewage inputs, which would result in his over-reporting the importance of atmospheric inputs. Bowen and Valiela (2001b) present a historical reconstruction for the northeast; the data are noisy, but show a clear shift in the form of inorganic N deposition over the 20th century. These data show that in the early 1900s, 50–75% of N deposition occurred in the form of ammonium, which was primarily produced by volatilization from livestock waste and natural soil processes. Modern N deposition, in contrast, is dominated by nitrogen oxides, primarily originating from fossil fuel combustion, with nitrate accounting for 60–66% of inorganic N deposition at the nearest NADP monitoring site (Abington, CT, USA) between 1999 and 2005. (*http://nadp.sws.uiuc.edu*).

Historical changes in the form of atmospheric N deposition and in land-use (Fig. 6.2) make it difficult to reconstruct how the transport of atmospheric N to the bay from both local and distant sources has changed. This is particularly true given that during the 19th century sources of atmospheric nitrogen within the watershed would have been dominated by local agriculturally related emissions, making it difficult to separate these inputs from our estimates of domesticated animal waste production. This difficulty and the limited availability of spatially explicit data lead us to exclude atmospheric deposition from our model.

6.3 Observed Historical Patterns of Nitrogen in Tributaries

The combined watersheds of three major tributaries to Narragansett Bay—Taunton, Blackstone, and Pawtuxet—account for 75% of the area of the Narragansett Bay watershed (Fig. 6.1a). In light of the dramatic changes in human population and use of the landscape as described earlier, one can reasonably expect that N loads in these basins have changed significantly over the course of the 20th century. Here, we discuss trends observed in historical and contemporary datasets on nitrogen concentrations in these rivers, providing insight into the influence of changing population and land use on nitrogen loading.

6.3.1 Blackstone River

The Blackstone River is the second largest tributary to Narragansett Bay (Fig. 6.1a), and by 1830 it was known as "the hardest working river in the country," with an average of two dams every three kilometers along the river (Wright *et al.,* 2001).

   The most complete record of nitrogen flux in the Blackstone River, and the entire Narragansett basin for that matter, is from Millville, MA, approximately 11 km upstream of Woonsocket (Fig. 6.1a), where samples were collected on an approximate monthly basis from 1895 to 1923. Approximately 40 km upstream of Millville is Worcester, the Blackstone basin's largest city, so water quality at Millville is heavily influenced by the city's wastewater discharge. Worcester began construction of a sewer system in 1870, and by 1890 the first treatment facility was in operation using chemical precipitation (Shanahan, 1994). The Worcester treatment system began secondary treatment in 1917, and modern sewage treatment began with the construction of the present facility in 1976 (Shanahan, 1994). A number of small treatment facilities are now also located upstream of Millville (Fig. 6.1a), which are reflected in the modern Millville data as well.

   Average dissolved inorganic nitrogen (DIN) concentrations at Millville (Fig. 6.3a) have remained high throughout the 20th century at 1.5–3.0 mg N L^{-1} (Massachusetts State Department of Public Health, 1915; *http://water-data.usgs.gov*). However, the data show that most of the DIN in the early 1900s was in the form of ammonium, which is taken up in the river more readily than nitrates (Bernot and Dodds, 2005). The MA Board of Health monitored nitrogen concentrations at several locations along the river, including just below the treatment works at Uxbridge (24 km downstream of Worcester), and at Millville (16 km farther downstream). Throughout the period from 1892 to 1915, ammonium concentrations showed a strong downstream gradient with approximately 55% reduction by the time it reached Uxbridge, and 80% reduction by Millville (Massachusetts State Department of Public Health, 1915). Data collected by the Rhode Island State Board of Health, in cooperation with the State Board of Purification of Waters (Gage and McGouldrick, 1925), show continued, though diminishing, reductions to the mouth of the river (e.g., Fig. 6.3b; other data not shown). These data also show that nitrate concentrations did not rise proportionally, indicating that DIN was not conserved.

   Data taken at Millville during 2000–2003 showed ammonium concentrations close to those reported in the early 1900s, and decreasing downstream from the treatment facility (Wright *et al.*, 2001, 2004). Mass balance estimates indicated that most ammonium in the upper reaches of the river were taken up by rooted plants rather than nitrified. Thus, upstream ammonium has limited impact on DIN loading to the bay in contrast to nitrate, and DIN loading to the bay is best estimated using data from the lowest segments of the river. Nitrate concentrations in the river, both at Millville and nearer the mouth of the river, have increased significantly over the last century (Robinson *et al.*, 2003). Comparison of the modern (1979–2002) USGS water quality data from Manville (6 km below Woonsocket in Rhode Island; monthly between 1979 and 1996, quarterly until 2002) with that from Gage and McGouldrick's between 1913 and 1923 (Fig. 6.3b) show that DIN concentrations have increased by a factor of roughly 2 over the last century. Based on this analysis,

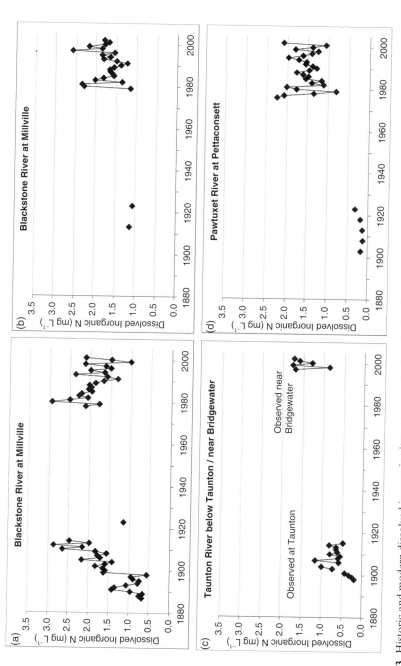

Fig. 6.3 Historic and modern dissolved inorganic nitrogen concentrations in the three main tributaries of Narragansett Bay. These data are taken at locations marked in Fig. 6.1a. All data are annual means.

we conclude that there has been a significant increase of N loading to the bay from the Blackstone River over the course of the 20th century, in contrast to the conclusion drawn by Nixon *et al.* (Chapter 5).

6.3.2 Taunton River

The Taunton River is Narragansett Bay's largest tributary. Because the lower 47 km of the river is tidal, the farthest downstream long-term gaging/monitoring station, near Bridgewater, MA (Fig. 6.1a), captures only about 44% of total river flow to Mount Hope Bay (Ries, 1990). Water quality was monitored from 1898 to 1915 below the present gage at a point which receives only one additional tributary, the Mill River, which adds approximately 15% to the catchment area (*http://waterdata.usgs.gov*). In the years immediately prior to 1900, DIN concentrations in the Taunton River below Taunton were 0.25 mg L^{-1} (3-year mean), yet by the first decade of the 20th century, they had increased to 0.64 mg L^{-1} (5-year mean) (Fig. 6.3c; Massachusetts State Department of Public Health, 1915). At least some of the increase in the early 1900s (Fig. 6.3c) may have been related to the discharge of effluent from wastewater treatment filter beds serving the city of Brockton, the largest city in the watershed. These beds appeared to improve water quality downstream of Brockton, but provided "a marked increase in pollution below Bridgewater" where the effluent from the filter beds was discharged (Massachusetts State Department of Public Health, 1915). Concentration of sewage as a result of the Brockton sewage system, coupled with limited treatment, likely contributed to the increase in N concentrations.

In 2000, the population of the watershed above Bridgewater was 240,000. The only large wastewater treatment facility in the watershed above Bridgewater is in Brockton, with a design capacity of 0.79 m^3 s^{-1}. Smaller facilities at Bridgewater (design capacity 0.06 m^3 s^{-1}) and Middleborough (design capacity 0.10 m^3 s^{-1}; US EPA, 2002a), also discharge to the Taunton, and human waste input now dominates N loading to this river. Overall, DIN in the Taunton at Bridgewater from 1997 to 2002 was more than double the mean value from 1898 to 1915 (Fig. 6.3c).

6.3.3 Pawtuxet River

The Pawtuxet River is the third largest tributary to Narragansett Bay (Fig. 6.1a) with a dam at the mouth of the river preventing tidal influence upstream. In the late 19th and early 20th Centuries, the Providence Water Works drew water from the Pawtuxet River at Pettaconsett, at the approximate location of the present USGS Cranston gage (Gray, 1881; Clapp, 1902). The RI State Board of Health monitored water quality at the intake, including nitrogen concentrations from 1900 to 1925 (Gage and McGoul-drick, 1925; Fig. 6.3d). Data from the RI State Board of Health Annual Reports show generally low levels of nitrate at the junction of the north and south branches between 1901 and 1914. However, concentrations in water from Knight's Spring in what is now the town of West Warwick were consistently high, between 2.5 and 4.5 mg L^{-1}. Historical DIN concentrations observed at Pettaconsett are much lower than observations in the Blackstone or Taunton rivers during the same time period (Fig. 6.3). In 1926, however, recognizing growing difficulties in protecting its water supply, the city of Providence dammed the upper reaches of the river to create the Situate Reservoir. Today, the domestic water supply for 60% of Rhode Island's residents comes from the Scituate Reservoir in the upper Pawtuxet River watershed, while almost all of the wastewater from this population is treated at facilities that discharge either directly to Narragansett Bay or to the lower reaches of the Pawtuxet (Fig. 6.1a). The net effect of this change is that approximately 2.3 m^3 s^{-1} of flow that would otherwise enter the bay at Cranston is instead discharged from several treatment plants on the Providence and Seekonk Rivers (Fig. 6.1a).

6.3.4 Direct Discharges to Narragansett Bay

Twenty-five percent of the Narragansett Bay watershed drains directly to the bay or through small tributaries (Fig. 6.1a). Engineered alterations to drainage, particularly in dense urban areas at the head of the bay, have increased direct drainage. Flows from the three wastewater treatment facilities that discharge to the north end of the bay—Field's Point, Bucklin Point, and East Providence (Fig. 6.1a)—constituted 6% of the total freshwater input to the bay in a very wet month (June of 2001), but 29% of the freshwater input in a very dry month (August of 2002; Pryor et al., 2006; http://waterdata.usgs.gov). Discharge from these three facilities in recent years averaged about 3.1 m^3 s^{-1}, a large proportion of which involves inter-basin transfer, primarily from the Pawtuxet River basin (Pryor et al., 2006; http://waterdata.usgs.gov).

The Field's Point wastewater treatment facility (Fig. 6.1a), the largest in the watershed, was serving nearly 200,000 people by 1906, processing approximately 1.0 m^3 s^{-1} of sewage (Nixon, 1995). This facility, along with the Bucklin Point facility, part of a regional approach to consolidate sewage treatment and constructed in the 1950s, greatly improved water quality in the upper bay, as did the upgrade of the Field's Point treatment facility in 1992 (Patenaude, 2000). The nitrogen output of these facilities in recent years (Pryor et al., 2006) suggests that they remove about 15% of the nitrogen load they collect, assuming 4.4 kg N person^{-1} yr^{-1} and no industrial or nonpoint source N inputs.

6.4 Modeling Historical N Fluxes

A simple spreadsheet-based numerical model of N flux from human and animal waste to Narragansett Bay at decadal time steps from 1880 to 2000 was constructed employing the following historical data: (1) human population served by sewers, (2) un-sewered human population, (3) animal units on farms, and (4) animal units not on farms (Table 6.1). These data exist largely at the town level, and were aggregated by subwatershed (Fig. 6.1b), then converted to N flux into the bay via multiplication by simple weighting factors representing the proportion of N reaching the bay. The weighting factors were derived from the scientific literature—each parameter includes low, median, and high estimates of N flux to the bay in order to reflect the full range of possible loadings (Table 6.2; see Model Parameters, following). Aggregating fluxes from each anthropogenic source allows calculation of the expected N load to the bay from each subwatershed, excluding the contributions of atmospheric deposition and fertilizer. In addition, it allows estimation of the relative importance of each source over time in each subwatershed.

6.4.1 Input Data

6.4.1.1 Sewered and Unsewered Human Populations

Population data are available by town from the United States Census for all years modeled. Sewered population was estimated by town, in 1910 using data from Harris (1913), in 1920 and 1930 for Rhode Island towns based on Gage and McGouldrick (1922; 1925) and RI Department of Health (1936), respectively, on a decadal basis from 1960 (partial) to 1990 by applying the proportion of households served by public sewers in each town (US Census of Housing) to the town's total population in that year, and in 2000 using sewered population data (Patenaude, 2000). To estimate sewered populations prior to 1910, linear interpolation of the proportion of each city's population served by sewers between the first available data point, usually 1910, and the date when sewer construction began, was used (Gage and McGouldrick 1922; 1925). Where data were not available on when sewers were installed, it was assumed that they began in 1900, as only the region's two largest cities had sewers in 1880, while all cities with populations more than 30,000 people in the watershed had sewers by 1910. For all towns for which no data on the origins of their sewer system existed, it was assumed that it was established in 1940, which is the mean date when public sewer construction began in all watershed communities currently served by such systems. This assumption has a minimal effect on the size of the sewered population in each watershed, since all the towns without sewer establishment dates have small populations. Pre-1900 sewered

Table 6.1 Input data used in modeling N flux, by watershed, as shown in Fig. 6.1b.

Sewered population	1880	1890	1900	1910	1920	1930	1940	1950	1960	1970	1980	1990	2000
Mid-Blackstone	0	0	7051	19000	32600	39285	40921	43181	42372	43539	47626	53746	53099
Upper Blackstone	0	42328	105395	126500	161779	177733	187391	198837	185920	177635	174633	196378	201335
Lower Bay	0	1459	3305	6109	9000	17357	24731	37813	47608	57933	39582	48087	44833
Lower Taunton	0	0	4655	10000	13369	15689	21280	25323	30493	41243	54497	68742	73561
Pawtuxet	0	0	0	0	0	0	13459	14131	15861	19987	53831	68240	69973
Other	0	7461	33732	73612	87928	98501	102597	112046	115709	131757	144541	174253	173923
Upper Taunton	0	0	20032	40000	50353	50400	58405	60872	73756	88879	115155	115263	120067
Upper Bay	26000	54000	170732	278234	324564	362334	386175	442843	426631	443738	460936	471748	488440
Unsewered population	1880	1890	1900	1910	1920	1930	1940	1950	1960	1970	1980	1990	2000
Mid-Blackstone	27102	31592	31354	31106	24091	23406	22862	24212	24856	27877	28974	26085	26487
Upper Blackstone	90413	77969	52354	63792	70385	70086	69948	78334	92251	103437	97098	98455	114528
Lower Bay	26926	31374	33615	37210	37245	28819	27281	40092	56218	73226	75827	77878	83287
Lower Taunton	41269	46096	48847	51393	53654	57775	54398	61877	76457	95373	100544	99249	116629
Pawtuxet	16981	19172	21055	24261	25409	27766	16328	21216	29461	48961	40823	45147	48549
Other	66267	82952	75777	65922	75970	94204	93143	103582	128057	148040	147224	141906	162965
Upper Taunton	41016	60175	61104	62379	63698	64076	55637	61581	76068	100226	100662	111279	123653
Upper Bay	164932	193957	163432	138616	126934	134719	124838	91876	111196	96896	45101	34253	34788
Animal units on farms	1880	1890	1900	1910	1920	1930	1940	1950	1960	1970	1980	1990	2000
Mid-Blackstone	4075	4355	4080	4358	3577	2796	2563	2539	1789	1459	845	774	524
Upper Blackstone	11963	15647	17839	17785	15337	12889	11706	12021	11323	7619	7108	4524	3600
Lower Bay	10168	9728	8288	13398	11750	10101	7950	9151	7375	4553	4416	2663	3223

Table 6.1 (continued)

	1880	1890	1900	1910	1920	1930	1940	1950	1960	1970	1980	1990	2000
Lower Taunton	3617	4698	5073	4860	3796	4222	4125	4076	3365	2815	2308	1608	1265
Pawtuxet	9075	9908	9469	9618	8056	6493	6082	5021	3651	2664	1643	1558	1118
Other	11567	14156	14545	14408	11534	8659	8189	8011	6205	5202	3715	2891	2142
Upper Taunton	10032	11027	11279	11125	9330	6045	7941	7184	4803	4584	4443	2521	2666
Upper Bay	7327	8074	7270	7563	6515	5467	5195	4799	3697	3267	2026	1513	1080
Animal units not on farms	1880	1890	1900	1910	1920	1930	1940	1950	1960	1970	1980	1990	2000
Mid-Blackstone	1008	1738	1962	1912	831	0	0	0	0	0	0	0	0
Upper Blackstone	5425	6616	8897	7525	2812	0	0	0	0	0	0	0	0
Lower Bay	2776	1806	1883	2370	992	0	0	0	0	0	0	0	0
Lower Taunton	1083	1020	1031	949	316	0	0	0	0	0	0	0	0
Pawtuxet	471	1088	1111	982	305	0	0	0	0	0	0	0	0
Other	3262	4973	5740	5633	1945	0	0	0	0	0	0	0	0
Upper Taunton	3854	4825	7887	7280	2455	0	0	0	0	0	0	0	0
Upper Bay	10507	13604	15603	15368	5450	0	0	0	0	0	0	0	0

Table 6.2 Data used to calculate loading rates for human populations (Panel A) and for animal populations (Panel B). Parameters used in the model are listed in Panel C.

	Low	Mid	High	Notes on sources
Panel A – Data on N loading from sewered and unsewered populations				
Per person waste N production (kg N person^{-1} yr^{-1})	2.2	4.4	6.2	US EPA (2002a)
	3.3	4.4	5.5	4.4 is a widely cited value from Vollenweider (1968), and is a good consensus value. Valiela *et al.* (1997) say 4.8, and Howarth *et al.* (1996) say 4.2. We take the middle value of 4.4 and use a range of ±25%
Public sewer discharge (% of input N)		75%		Castro *et al.* (2001). Ratio of above consensus N waste per capita and 3.3 kg N yr^{-1} per capita in sewage effluent N per capita (Meybeck *et al.* 1989, similar value found by Van Breemen *et al.* 2002)
Septic tank / cesspool N discharge (% of input N)		95%		Valiela *et al.* (1997)
Leach field discharge (% of input N)		65%		Valiela *et al.* (1997)
First 200 m groundwater discharge (% of input N)		65%		Valiela *et al.* (1997)
Groundwater discharge after 200 m (% of input N)		65%		Valiela *et al.* (1997)
Modern septic system (% of input N)	26%		40%	calculated from above; range reflects systems < 200 m and > 200 m from water bodies
Cesspool (% of input N)	40%		62%	calculated from above; range reflects systems < 200 m and > 200 m from water bodies
Overall non-sewered N loading estimate (% of input)	26%	40%	62%	summary of above calculations
Overall non-sewered N loading estimate (kg N person^{-1} yr^{-1})	0.86	1.76	3.41	
Delivery of input N to mid-high order streams (% of input N)	78%	87%	96%	Low bound from Seitzinger *et al.* (2002) estimate for the Blackstone watershed excluding smallest-order streams, high bound from Alexander *et al.* (2001) estimate for the entire Narragansett watershed. Mid-range estimate is simply the mean of these two

Table 6.2 (continued)

	Low	Mid	High	Notes on sources
Panel B – Data on N loading from animals				
Animal Waste N production (kg N AU^{-1} yr^{-1})	30	50	100	USDA (1997), Boyer *et al.* (2002)
Delivery of input N to mid-high order streams (% of input N)	16%	20%	32%	Johnes (1996)
Export after loss within mid-high order streams (% of input N)	78%	96%	100%	As above, but using, Alexander *et al.* (2001) as the "mid" case, and in the "high" case assuming that export coefficients from Johnes (1996) already account for all in-stream processing
Panel C – Model parameters calculated from above data				
Sewered population loading per capita (kg N $person^{-1}$ yr^{-1})	2.5	3.3	4.1	
Unsewered population loading per captia (kg N $person^{-1}$ yr^{-1})	0.7	1.5	3.3	
Animal waste loading per AU (kg N $person^{-1}$ yr^{-1})	3.7	9.6	32.0	

Note: Estimates for output to streams are for total N.

populations in Providence were based on data from the City Engineers Reports (Gray, 1881; Shedd, 1891).

Un-sewered population was calculated as the difference between total population and sewered population. It is assumed that in the late 1800s and early 1900s all residences not served by sewers dealt with human waste using privies or cesspools that were gradually replaced by modern septic systems. All these systems can contribute to N loading to the bay through groundwater (see Nowicki and Gold, Chapter 4, for a detailed treatment of groundwater N transport to Narragansett Bay). It is possible, particularly in the 1800s, that some houses had "private sewers"—simple pipes or ditches diverting wastes directly to nearby water bodies—though this would be a practical disposal method for only a small minority of all households.

6.4.1.2 Animal Units on Farms and in Urban Areas

State agricultural census data enumerating animals on farms at the town level are available for Rhode Island for 1865, 1875, 1885, 1895, and 1905 (Snow, 1867; 1877; Perry, 1887; Tiepke, 1898; Webb and Greenlaw, 1906) and for Massachusetts for 1885, 1895, and 1905 (Wright, 1887; Wadlin, 1899; 1909). Animal units are calculated from these data using generalized weights

(USDA, 1997): horse, 1.0 AU; cattle, 1.0 AU; swine, 0.25 AU; sheep, 0.1 AU; and goat, 0.1 AU. All other animals, particularly poultry, are reported inconsistently and add less than 10% to the total number of AU in years when they were thoroughly reported (1885–1905 in Rhode Island; 1885 in Massachusetts). These animal data are linearly interpolated for each town to yield 1880, 1890, and 1900 AU estimates. To estimate changes in AUs during the 20th century, decadal, county-level data from the US Census of Agriculture are used.

Rhode Island collected data on nonfarm horses and cows in some years, although not others, and these data were never collected by Massachusetts. The Federal Census of Agriculture reported nonfarm horses and cows for cities of human population more than 25,000 in 1900, 1910, and 1920 (US Census Office, 1902; US Bureau of the Census, 1922). To estimate the number of horses and cows not on farms, nonfarm animal units were regressed against human population at the town level in each year (regression slope = 0.051 AU per capita in 1900, 0.039 in 1910, and 0.012 in 1920; r^2 values range from 0.88 to 0.93). In almost all cases, horses and cattle comprised more than 90% of the total nonfarm animal units in each city. In this analysis, only horses and cattle are included because other categories of animals were reported inconsistently. There are no data enumerating nonfarm animals after 1920, and given the 70% decline in the density of nonfarm AUs per resident between 1910 and 1920, it is assumed that nonfarm animals contributed insignificant N loading to Narragansett Bay after 1920.

6.4.2 Model Parameters

Human waste N production ranges from 2.2 to 6.2 kg N yr^{-1} per capita (US EPA, 2002b), with the most widely cited value being 4.4 kg N yr^{-1} (Vollenweider, 1968), which is consistent with the 4.8 kg N yr^{-1} reported by Valiela et al. (1997) and the 4.2 kg N yr^{-1} reported by Howarth et al. (1996). The consensus value of 4.4 kg N yr^{-1} per capita, with high and low estimates of 125% and 75% of this value, was used in the model (Table 6.2). Following Castro et al. (2001), it was assumed that sewage facilities remove 25% of the N contained in wastewater through sludge removal and denitrification, discharging 2.5–4.1 kg N yr^{-1} per capita in effluent. Meybeck et al. (1989) report a value of 3.3 kg N yr^{-1} in sewage effluent per person served, while Van Breemen et al. (2002) report a value of 3.1 kg N yr^{-1} per capita. A value of 75% N loading for sewage treatment facilities is used in the model (Table 6.2).

The proportion of human waste N entering on-site wastewater systems (cesspools and septic systems) that ultimately reaches the bay is extremely difficult to measure (see Nowicki and Gold, Chapter 4,). Valiela et al. (1997) estimate that 5% is retained in septic tanks and cesspools, 35% of the discharge is retained or lost in leach fields, 35% of the remainder is lost in the first 200 m of travel through the groundwater system, and an additional 35% is lost after 200 m. While these estimates are rough, they allow the development of a range of spatial estimates. For example, a modern septic system that is more than 200 m from the stream into which its water is ultimately discharged will contribute 26% of its wastewater N to that stream. On the other hand, a cesspool system without a leach field that is less than 200 m from a stream will contribute 62% of its wastewater N to the stream. Modern septic systems are a relatively recent development, replacing cesspool systems in the past several decades. In addition, N reaching the bay is discounted as a result of in-stream N loss by a high of 22% in the Blackstone watershed (Seitzinger *et al.,* 2002), excluding the smallest streams, to an average of 4% over the entire Narragansett Bay watershed (Alexander *et al.,* 2001). The resulting loading coefficients range from 0.7 kg N yr^{-1} per capita to 3.3 kg N yr^{-1} per capita. Horsley, Witten, Hegemann, Inc. (1991) report a value of 3.1 kg N yr^{-1} per capita for the watershed of Buttermilk Bay, Massachusetts. Given that this watershed is small enough that all residences are relatively close to receiving waters, it is not surprising that their estimate is on the high side of our range. Table 6.2, Panel A provides the range of model values used to calculate on-site septic system loading to Narragansett Bay.

A mid-range estimate for waste N production per animal unit of 50 kg N yr^{-1} is used (USDA, 1997). Boyer *et al.* (2002) cite late 20th century ranges of 40–140 kg N yr^{-1} for cattle/milk cows and 27–50 kg N yr^{-1} for horses. Historical estimates of animal waste N production fall roughly within this same range, for example, 46–90 kg N yr^{-1} (Roberts, 1907); 68 kg N yr^{-1} per diary cow (Cornell Experiment Station Bulletin, 1891). The model developed here therefore uses a range of 30–100 kg N yr^{-1}, with a mid-value of 50 kg N yr^{-1} on the assumption that only a minority of historical and modern small-scale farms would approach rates of waste production typical of large-scale modern agricultural operations. In the late 19th century, there was an intensification of agriculture associated with increased milk production in parts of the watershed (e.g., Soll, 2006) that would have lead to an increase in N waste production, as diary cattle produce far more waste per animal than do other cattle (Boyer *et al.,* 2002). Diary cattle (milk cows plus heifers) increased slowly as a proportion of total cattle in Rhode Island between 1850 (62%) and 1940 (79%), after which time the proportion began to decrease (US Census Office, 1850; 1893; 1902; Tiepke, 1898; US Bureau of the Census, 1922; 1931; 1941; 1952; 1971; USDA, 2002). Data on milk production per cow therefore can provide a proxy for changes in the rate of waste N production by diary cattle. Milk production per head of diary cattle was approximately stable between 1895 and 1920, but increased by over 50% between 1920 and 1950, and increased an additional 70% between 1950 and 2002. Since total milk production is not reported prior to 1895, it is difficult to estimate increases in milk production during the 19th century noted elsewhere in the region (Soll, 2006). The mid-range estimate of waste production used in the model—50 kg N yr^{-1} AU^{-1}—most certainly underestimates N produced by diary cattle under modern agricultural practices (Boyer *et al.,* 2002), but appears to be reasonable for the early 20th century when diary production per cow was less than half its current value, and when diary cattle were present on the land-scape in much greater numbers than today. Table 6.2, Panel B shows the data values used in the model to calculate N loading from animals in the watershed to Narragansett Bay.

Johnes (1996) estimates that 16–32% of animal waste N is transported to mid- to high-order streams, based on studies of agriculturally dominated watersheds of 46–363 km^2. The variation is primarily a result of the distance between the pasture or field where manure is spread and the first-order streams into which runoff is discharged; the 32% figure is for areas more than 50 m from streams. Each of the 20,000 farms in the watershed disposed of manure and dispersed or concentrated animals on the landscape in unique combinations in relation to perennial and ephemeral surface waters. As an "average" case, it is assumed that the majority of agricultural land is more than 50 m from streams, and so 20% is used as an estimated loading coefficient. The amount of animal N reaching the bay is further discounted using in-stream N removal rates of Alexander et al. (2001) as the mid-range case, and Seitzinger et al. (2002) as the low export case (Table 6.2, Panel B). In the high export case, it is assumed that there is no additional in-stream removal aside from what is already accounted for in the export coefficients from Johnes (1996).

In general, the model may overestimate the contribution of N to the bay from agriculture in the 19th century, and underestimate it in the late 20th century. The range in the parameter representing the mean waste production per animal unit, however, should be large enough to account for differences in the way that animal counts are summed to animal units, and the shifting role of feed imports and animal husbandry (Soll, 2006) over the course of 120 years are modeled.

6.4.3 Validation of Modeling Results

Whole-watershed modeled estimates of N flux are compared with the observed flux values measured by Nixon et al. (1995) in two ways: reported versus modeled N flux from sewage treatment facilities, and measured versus modeled river N flux. Nixon et al. (1995) measured annual output of total N from 10 sewage facilities that discharge directly to Narragansett Bay between 1983 and 1986. We compared these measurements with our modeled outputs using the mean of sewered populations for 1980 and 1990. Nixon et al. (1995) estimated output of 2,560,000 kg N yr^{-1}, and our mid-range estimate for these treatment facilities used here is 2,050,000 kg N yr^{-1} (Fig. 6.4), with a range from 1,550,000 to 2,540,000 kg N yr^{-1}. The high-modeled estimate matches the reported value, and assuming even a modest level of uncertainty in the Nixon et al. estimate would result in substantial overlap between the estimated and observed ranges.

Modeled versus measured total N flux is compared in each of the three major tributaries of Narragansett Bay for the mid-1980s and early 2000s (Fig. 6.5).

Table 6.3 shows that the "middle" case for modeled N flux from human and animal waste for each river accounted for 54–82% of the observed N flux, while the "high" case accounted for 80–124% of the observed flux. If it is assumed that approximately 20% of modern N flux to Narragansett Bay is ultimately attributable to atmospheric deposition and fertilizer runoff (Driscoll *et al.*, 2003), neither of which was included in the model, then the mid- to high-range cases are consistent with the observed fluxes (this analysis used the same aggregations of towns as those employed by Nixon *et al.* (1995), rather those defined in Fig. 6.1b). Though the trends in river DIN concentration data (Fig. 6.3) broadly reflect the long-term changes in total N flux from each subwatershed to the bay, they do not necessarily scale directly with modeled total N flux (Fig. 6.5) in all cases (Table 6.4). Because of the lack of systematic flow data predating 1930 for any of the locations where there are DIN concentration records, DIN flux from the subwatersheds is not calculated. Flow conditions have likely changed in all of these rivers with increased forest cover and groundwater withdrawals, as well as out-of-basin water transfers, all of which would cause the relationship between DIN concentrations and total N flux to change over time. For example, a conservative estimate of the effect of withdrawals from the Scituate Reservoir on flow of the Pawtuxet River is that approximately 25% (2.6 m^3 s^{-1}) of total flow at Pettaconsett is transferred out-of-basin (Providence Water Supply Board; *http://waterdata.usgs.gov*). Nitrogen speciation and the relationship between DIN:TN (organic N was not measured historically) may have also changed

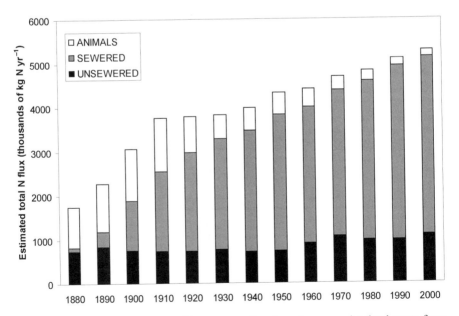

Fig. 6.4 Estimated total N flux to Narragansett Bay from human and animal waste from 1880 to 2000, using the "mid" case parameters from Table 6.2, Panel C.

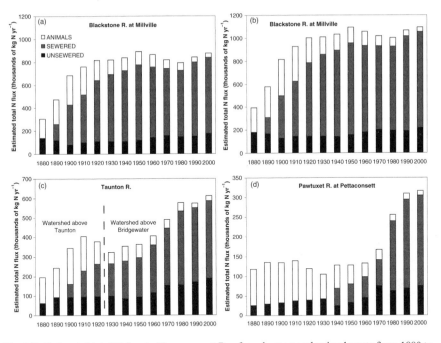

Fig. 6.5 Estimated total N flux to Narragansett Bay from human and animal waste from 1880 to 2000 in four subwatersheds (Fig. 6.1b), using the "mid" case parameters from Table 6.2, Panel C.

over time, as the sources of N have shifted and river conditions (flow, sediment load, temperature, content of industrial wastes) have likely changed.

These changes would affect the N concentrations measured in each river as well as the total N load delivered to the bay, but cannot be accounted for in the model, in which all within-watershed N processes are characterized by constants (Table 6.2). The model does not attempt to account for N derived from atmospheric deposition, fertilizer use, or industrial waste streams, the importance of which has varied across space and time and for which spatially explicit data are not available.

As with any model, it is important to understand the limitations of the input data in evaluating the output. Though only data on animals that are comparable across all datasets are included, changing definitions and methodologies among years, and between census-taking agencies (US Census Bureau, US Department of Agriculture, and state agencies), result in some degree of uncertainty. For example, the US Census Bureau has changed its definition of "farm" slightly with each census, such that the difference in any reported value between consecutive census years must be viewed with some suspicion (Black, 1950). However, the broad trends in the amount of agricultural land and number of animal units at the multi-decadal time scale are relatively unaffected by these variations.

Table 6.3 Comparison of total N flux data observed by Nixon et al. (1995) to modeled estimates of total N flux for each major sub-basin.

Watershed	Year	Observed by Nixon et al. (1995, Chapter 5 of this volume)	Model low estimate	Model mid estimate	Model high estimate	Ratio: mid estimate/observed	Ratio: high estimate/observed
Pawtuxet	1982–3	896	334	486	716	0.54	0.80
	2003–4	830	393	567	814	0.68	0.98
Blackstone	1982–3	1834	679	1036	1659	0.57	0.90
	2003–4	1381	754	1129	1718	0.82	1.24
Taunton	1988–9	1638	619	988	1644	0.60	1.00
	2003–4	1918	662	1064	1775	0.55	0.93

Note: Flux data compiled by Nixon et al. (1995; Chapter 5) for each of the three major watersheds that drain to Narragansett Bay are shown for comparison. All flux data are in units of thousands of kg N yr^{-1}. Driscoll et al. (2003) estimate that around 20% of total N loading is attributable to sources not accounted for in the model. If correct, modeled fluxes should equal approximately 80% of the observed fluxes.

Table 6.4 Trends between observed historical and modern DIN concentration data were compared at four points (Fig. 6.3) with modeled "mid" trends in N flux at those same locations (Fig. 6.5).

	Time Period		Mean DIN (mg L^{-1})		Ratio (modern/historic)		Modeled date ranges
	Historic	Modern	Historic	Modern	Observed DIN data	Modeled TN estimates ("mid" case)	
Blackstone at Manville	1913–1923	1979–2002	1.11	1.74	1.58	1.08	1910–1930; 1980–2000
Blackstone at Millville	1887–1900	1978–2001	1.00	1.93	1.92	1.45	1890–1900; 1980–2000
Blackstone at Millville	1901–1914	1978–2002	1.97	1.93	0.98	1.11	1900–1920; 1980–2000
Pawtuxet at Pettaconsett	1900–1925	1976–2002	0.18	1.56	8.43	2.25	1910–1930; 1980–2000
Taunton at Taunton/ Bridgewater	1898–1915	1997–2002	0.61	1.49	2.44	1.63	1900–1920; 1980–2000

Overall the model provides a solid picture of the changing patterns of N flux into Narragansett Bay and its subwatersheds, limitations not withstanding. While the precision of the historical reconstructions is not as great as that available today, the patterns of input and change with time are robust. The empirical historical data and the modeling result provide a consistent image of change in nitrogen inputs over the past 150 years, though the limits on the precision of this data need to be kept in mind so as to not over interpret the results presented. Our model uses consistent methodology and parameters throughout the entire study period. The ranges given for each estimate are large, and reflect real uncertainty in the model parameterization. It is possible that some of the export parameters (Table 6.2) have changed systematically or in a spatially variable manner within these ranges over the studied time period. Comparison of our modeled, long-term trend estimates with flux values of Nixon *et al.* (1995; Chapter 5) suggest that for the late 20th century, our mid-to-high modeled estimate ranges are realistic. However, Nixon *et al.* (1995; Chapter 5) do not report the uncertainties resulting from their assumptions, which if considered, would likely increase the overlap with our mid-range estimates.

6.5 Model Integration

Modeled trends (Table 6.4) of N flux vary dramatically by subwatershed (Fig. 6.5). In the Blackstone River watershed (Fig. 6.5ab), modeled N loading increased at a roughly constant rate between 1880 and 1920, and has increased more modestly and less consistently since that time as decreases in urban population and in animal units, both in cities and on farms, has been offset by suburban population growth. In the Taunton River (Fig. 6.5c), the increase in modeled N loading has been steadier. The watershed analyzed at each time period was matched to the locations where DIN concentration data have been collected. The modeled watershed was 22% larger from 1880 to 1920 relative to after 1920, including the additions of the city of Taunton and the town of Raynham in 1880–1920. In the Pawtuxet watershed above Pettaconsett, modeled loading remained about level between 1880 and 1960, as increases in population were offset by decreases in the number of animal units in the watershed (Fig. 6.5d). The sudden jump in modeled sewered contribution around 1970 is a result of the City of Warwick's construction of a sewage treatment facility on the Pawtuxet River above Pettaconsett, transferring sewage into the Pawtuxet basin from a population residing primarily outside the basin (Patenaude, 2000).

In the Narragansett Bay watershed as a whole, the modeled rate of increase in N flux was steeper between 1880 and 1920 than after 1920 (Fig. 6.4). In 1880, animal waste N contributed an estimated 53% of all N flux, while in 2000 it contributed 3%. As an increasing number of communities built public sewer systems, the contribution of N from sewers is estimated to have increased from 5% in 1880 to 77% in 2000. This dramatic shift from nonpoint sources of N loading (farm/urban street runoff, on-site wastewater systems releasing N to groundwater) to point sources (sewage facility outfalls) increases the percentage of waste N reaching Narragansett Bay (Table 6.2, Panel C). The increasing proportion of N entering Narragansett Bay through sewage collection systems provides an opportunity for resource managers to reduce total N loading from a very manageable source.

The modeled contribution of each subwatershed (Fig. 6.1b) to N flux per unit area of watershed is shown in Fig. 6.6. The total estimated N flux into upper Narragansett Bay increased more than threefold from 1880 to 1950, and remained relatively stable thereafter. This largely reflects a long-term population trend in the core cities of the Providence metropolitan region, which peaked in population around 1940 (US Bureau of the Census, 1941). The intensity of the loading is four times higher in the upper bay subwatershed than for any other subwatershed.

Data on N fertilizer use by county are only available for the period 1945–1991 (Alexander and Smith, 1990; Battaglin and Goolsby, 1994). Howarth et al. (1996) estimate that on average, 10–40% of fertilizer N applied to loamy soils is exported from temperate drainages. Using a rough figure of 25%, it is estimated that fertilizer N added 150,000 kg N y^{-1} in 1950 (an additional 4% relative to mid-range estimates for human and animal waste). For 1990, it would be an additional 500,000 kg N y^{-1} (10%). For 1880 (assuming the same export rate), the additional 45,000 kg N y^{-1} adds less than 3% to

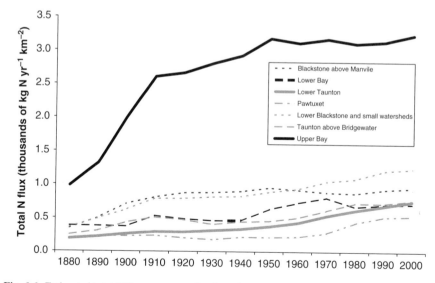

Fig. 6.6 Estimated total N flux per square km from human and animal waste from 1880 to 2000 for each subwatershed (Fig. 6.1b), using the "mid" case parameters from Table 6.2, Panel C.

the estimated flux. It is difficult to assess whether these export estimates are high or low for the Narragansett Bay watershed at any given point in time due to the sensitivity of the fate of fertilizer N to many variables, such as application timing, application rate, weather, land use, and soil texture and chemistry. However, it is clear that the effect of fertilizer N was probably very small until the mid 20th century, and even today it is modest in comparison with N loading from human waste.

Relative to the changes in human population and in agriculture in the Narragansett Bay watershed over the past 120 years, the proportional change in total atmospheric N deposition since 1910 (approximately 20% increase; Bowen and Valiela, 2001b) is quite modest. The shift in the form of atmospheric N deposition from mostly ammonium to mostly nitrate, coupled with dramatic changes in land-cover over the past 120 years (Fig. 6.2) make it difficult to assess whether more or less of the atmospherically deposited N ultimately reaches the bay now than in the past. A landscape increasingly dominated by regrowing forests, however, is likely to retain a greater proportion of incoming N than is one dominated by agricultural land uses (Hamburg, 1984; Goodale *et al.,* 2002; Aber *et al.,* 2003). While it is hypothesized that chronic exposure of terrestrial ecosystems to enriched N deposition will increase N concentrations in drainage waters, there is no evidence that it is occurring in the northeastern US (Aber *et al.,* 2003). On the other hand, increases in the area of impervious surfaces within the watershed likely lead to increased delivery of atmospheric N to surface waters (e.g., Groffman *et al.,* 2004) at least in some localities.

There is no way to accurately estimate N concentrations in the bay resulting from the extensive agricultural activities that dominated the landscape at the end of the 19th century, but the relative spatial distributions of these inputs and the overall scale can be reconstructed. While this nitrogen input may not have been concentrated in the way it is in sewage outfalls, it has potential to have impacted coastal waters, particularly submerged and emergent plant communities (Bowen and Valiela, 2001a). The heavy concentration of farms in close proximity to the lower bay during late 19th century and the early 20th century (Fig. 6.1b), would have resulted in higher N loads than during the later part of the 20th century. Over time, diffuse inputs of N became increasingly concentrated with the introduction of sewage systems and the decline of domesticated animals in the watershed. Given the very large amounts of manure produced in lower bay communities, and lack of concern for impacts of runoff or the presence of buffer strips along streams, it is plausible to envision that these coastal areas were impacted, most likely heavily, by the second half of the 19th century.

Human population in the watershed is likely to continue increasing slowly, though the proportion of the population with access to sewers may decline as unsewered suburban towns are growing far more quickly than cities over the past several decades (US Census). Agriculture is likely to remain a very small component of total N loading to Narragansett Bay. The largest shift in N inputs to Narragansett Bay in the first decade of the 21st century will likely be the declining N in sewage outfalls resulting from mandated denitrification of sewage effluent.

6.6 Conclusions

The pattern of anthropogenic nitrogen discharge to Narragansett Bay has changed over the past 120 years. The modeled input of anthropogenic nitrogen increased rapidly around the turn of the 20th century, more than doubling between 1880 and 1910 (Fig. 6.4), with the simultaneous rise of population centers and commercial farming. These inputs were spread throughout the bay, with major inputs from population centers at the head of the bay as well as substantial inputs from animal agriculture in the lower bay (Fig. 6.6). Nitrogen discharge into the bay from agriculture in the middle and late 19th century was dominated by dispersed, nonpoint sources, including both agricultural sources and decentralized human waste disposal. Unlike the largely concentrated N loading of today, agriculturally derived N would have been discharged to the bay in a spatially diffuse and temporally concentrated pattern.

During the 20th century, modeled total N flux into the bay increased at a relatively steady rate, with an average annual increase of 22,000 kg N yr^{-1}. Our model suggests an increase of 73% in total N flux to the bay from 1900 to 2000, and 39% since 1925, while Nixon *et al.* (Chapter 5) estimate these increases as 53% and 14%, respectively. As discussed earlier, the sources and location of N changed quite dramatically over the course of the century. The increasing concentration of inputs and the increasing intensification of N per unit area to the upper bay has meant that despite slowing population growth, the increase in nitrogen flux has continued (Fig. 6.6), particularly at the very head of the bay where three of the seven largest sewage treatment facilities discharge within 5 km of each other (Fig. 6.1a).

Overall, there have been large amounts of anthropogenic nitrogen entering Narragansett Bay for more than 120 years. The increasing concentration of the N inputs to the upper bay and the relatively stable inputs from most of the rest of the watershed has meant that management of nitrogen entering the bay needs to focus increasingly on the upper bay. The almost four times higher N inputs per unit land area in the upper bay is an opportunity to mitigate the impacts of anthropogenic nitrogen, as well as a burden on the health of the bay ecosystem. Even when mandated denitrification of sewage effluent in the upper bay is implemented around 2010, the rate of anthropogenic N inputs per unit land area will remain higher than in any other subwatershed.

The shifting patterns of anthropogenic nitrogen inputs to Narragansett Bay represent a continuing management challenge; the lower bay has become less impacted while the upper bay is receiving unprecedented nitrogen inputs. With declining inputs of nitrogen from sewage treatment facilities, the challenge moving forward will be defining an acceptable level of nitrogen enrichment in an ecosystem increasingly stressed by climate change and the introduction of exotic organisms.

Acknowledgments Financial support of M.A. Vadeboncoeur provided through Brown University's social science research supplements to S.P. Hamburg.

References

Aber, J.D., Goodale, C.L., Ollinger, S.V., Smith, M.L., Magill, A.H., Martin, M.E., Hallett, R.A., and Stoddard, J.L. 2003. Is nitrogen deposition altering the nitrogen status of northeastern forests? *Bioscience* 53:375–389.

Alexander, R.B., and Smith, R.A. 1990. County-Level Estimates of Nitrogen and Phosphorus Fertilizer Use in the United States, 1945 to 1985. USGS Open File Report 90–130. (*http:// pubs.usgs.gov/of/1990/ofr90130/report.html*).

Alexander, R.B., Smith, R.A., Schwartz, G.E., Preston, S.D., Brakebill, J.W., Srinivasan, R., and Pacheco, P.C. 2001. Atmospheric nitrogen flux from the watersheds of major estuaries of the United States: An application of the SPARROW watershed model. *In* Nitrogen Loading in Coastal Water Bodies: An Atmospheric Perspective, pp. 119–170. Valigura, R., Alexander, R., Castro, M., Meyers, T., Paerl, H., Stacey, P., and Turner, R.E. (eds.) American Geophysical Union Monograph 57. Washington, DC. American Geophysical Union. Any additional required information should be on the book's AGU webpage at: *http://www.agu.org/cgi-bin/agubookstore?memb = stu&topic = CE&book = CECE0572715*.

Battaglin, W.A., and Goolsby, D.A. 1994. Spatial Data in Geographic Information System Format on Agricultural Chemical Use, Land Use, and Cropping Practices in the United States. USGS Water Resources Investigations Report 94–4176. (*http://pubs.usgs.gov/wri/ wri944176/bat000.html*).

Bernot, M.J., and Dodds, W.K. 2005. Nitrogen retention, removal, and saturation in lotic ecosystems. *Ecosystems* 8:442–453.

Black, J.D. 1950. The Rural Economy of New England: A Regional Study. Cambridge, MA: Harvard University Press.

Bowen, J.L., and Valiela, I. 2001a. The ecological effects of urbanization of coastal watersheds: Historical increases in nitrogen loads and eutrophication of Waquoit Bay estuaries. *Canadian Journal of Fisheries and Aquatic Sciences* 58(6):1489–1500.

Bowen, J.L., and Valiela, I. 2001b. Historical changes in atmospheric nitrogen deposition to Cape Cod, Massachusetts, USA. *Atmospheric Environment* 35:1039–1051.

Boyer, E.W., Goodale, C.L., Jaworski, N.A., and Howarth, R.W. 2002. Effects of anthropogenic nitrogen loading on riverine nitrogen export in the northeastern US. *Biogeochemistry* 57/58:137–169.

Briefel, R.R., and Johnson, C.L. 2004. Secular trends in dietary intake in the United States. *Annual Review of Nutrition* 24:401–431.

Castro, M.S., Driscoll, C.T., Jordan, T.E., Reay, W.G., Boynton, W.R., Seitzinger, S.P., Styles, R.V., and Cable, J.E. 2001. Contribution of atmospheric nitrogen to the total nitrogen loads to thirty-four estuaries on the Atlantic and Gulf Coasts of the United States. *In* Nitrogen Loading in Coastal Water Bodies: An Atmospheric Perspective, pp. 77–106. Valigura, R.A., Alexander, R.B., Castro, M.S., Meyers, T.P., Paerl, H.W., Stacey, P.E., and Turner, R.E. (eds.) Washington, DC: American Geophysical Union.

Clapp, O.L. 1902. Annual Report of the City Engineer of Providence for the Year 1901. Providence: Snow and Farnham.

Cornell Experiment Station. 1891. Bulletin 27. Ithaca, NY: Cornell University Agricultural Experiment Station, New York State College of Agriculture.

Dirks, R. 2003. Diet and nutrition in poor and minority communities in the United States 100 years ago. *Annual Review of Nutrition* 23:81–100.

Driscoll, C.T., Whitall, D., Aber, J., Boyer, E., Castro, M., Cronan, C., Goodale, C., Groffman, P., Hopkinson, C., Lambert, K., Lawrence, G., and Ollinger, S. 2001. Acidic deposition in the northeastern United States: Sources and inputs, ecosystem effects, and management strategies. *Bioscience* 51:180–198.

Driscoll, C., Whitall, D., Aber, J., Boyer, E., Castro, M., Cronan, C., Goodale, C., Groffman, P., Hopkinson, C., Lambert, K., Lawrence, G., and Ollinger, S. 2003. Nitrogen pollution in the northeastern United States: Sources, effects, and management options. *Bioscience* 523:357–374.

Gage, S., and McGouldrick, P.C. 1922. Preliminary report of an investigation of the pollution of certain Rhode Island public waters. Annual Report of the Board of Purification of Waters. Providence, RI: State of Rhode Island.

Gage, S., and McGouldrick, P.C. 1925. Report of investigations of the pollution of certain Rhode Island public waters during 1923 and 1924. Presented to the Board of Purification of Waters. Annual Report of the Board of Purification of Waters.

Goodale, C.L., Lajtha, K., Nadelhoffer, K.J., Boyer, E.W., and Jaworski, N.A. 2002. Forest nitrogen sinks in large eastern U.S. watersheds: Estimates from forest inventory and an ecosystem model. *Biogeochemistry* 57/58:239–266.

Gray, S.M. 1881. Annual Report of the City Engineer of Providence. Providence, RI: Providence Press.

Groffman, P.M., Law, N.L., Belt, K.T., Band, L.E., and Fisher, G.T. 2004. Nitrogen fluxes and retention in urban watershed systems. *Ecosystems* 7:393–403.

Hamburg, S.P. 1984. Organic matter and nitrogen accumulation during 70 years of old-field succession in central New Hampshire. Dissertation: Yale University.

Harper, R.M. 1918. Changes in the forest area of New England in three centuries. *Journal of Forestry* 16:442–452.

Harris, W.J. 1913. General Statistics of Cities: 1909. Statistics of sewers and sewage disposal, refuse collection and disposal, street cleaning, dust prevention, highways, and the general highway service of cities having a population of over 30,000. US Department of Commerce, Bureau of the Census. Washington: Government Printing Office.

Horsley, Witten, Hegemann, Inc. 1991. Quantification and Control of Nitrogen Inputs to Buttermilk Bay. Report prepared for the US Environmental Protection Agency, Massachusetts Executive Office of Environmental Affairs, and New England Interstate Water Pollution Control Commission. Barnstable, MA: Horsley, Witten, Hegemann, Inc.

Howarth, R.W., Billen, G., Swaney, D., Townsend, A., Jaworski, N., Lajtha, K., Downing, J. A., Elmgren, R., Caraco, N., Jordan, T., Berendse, F., Freney, J., Kueyarov, V., Murdoch, P., and Zhao-Liang, Z. 1996. Riverine inputs of nitrogen to the North Atlantic Ocean: Fluxes and human influences. *Biogeochemistry* 35:75–139.

Jaworski, N.A., Howarth, R.W., and Hetling, L.J. 1997. Atmospheric deposition of nitrogen oxides onto the landscape contributes to coastal eutrophication in the northeast US. *Environmental Science and Technology* 31:1995–2004.

Johnes, P. 1996. Evaluation and management of the impact of land use change on the nitrogen and phosphorus load delivered to surface waters: The export coefficient modeling approach. *Journal of Hydrology* 183:323–349.

Massachusetts State Department of Public Health. 1915. First Annual Report. Boston, MA: MA Department of Health.

Melosi, M.V. 2000. The Sanitary City: Urban Infrastructure in America from Colonial Times to the Present. Baltimore, MD: Johns Hopkins University Press.

Meybeck, M., Chapman, D.V., and Helmer, R. 1989. Global Freshwater Quality: A First Assessment. World Health Organization/United Nations Environment Programme. Cambridge, MA: Basil Blackwell, Inc.

Miller, C.F. 1914. On the composition and value of bat guano. *The Journal of Industrial and Engineering Chemistry* 6:664.

Moore, R.B., Johnston, C.M., Robinson, K.W., and Deacon, J.R. 2004. Application of Spatially Referenced Regression Models to Evaluate Total Nitrogen and Phosphorus in New England Streams. USGS Water Resources Investigation Reports 2004–5012. Washington, DC: United States Geological Survey. This publication is available online at *http://pubs.usgs.gov/sir/2004/5012/*.

National Research Council. 1993. Managing Wastewater in Coastal Urban Areas. Washington, DC: National Academies Press. pp. 45–47.

Nixon, S.W. 1995. Metal Inputs to Narragansett Bay: A History and Assessment of Recent Conditions. Narragansett, RI: Rhode Island Sea Grant.

Nixon, S.W. 1997. Prehistoric nutrient inputs and productivity in Narragansett Bay. *Estuarine Research Federation* 20(2):253–261.

Nixon, S.W., Granger, S.L., and Nowicki, B.L. 1995. An assessment of the annual mass balance of carbon, nitrogen, and phosphorus in Narragansett Bay. *Biogeochemistry* 31:15–61.

Novak, A.B., and Wang, Y.Q. 2004. Effects of Suburban Sprawl on Rhode Island's Forests: A Landsat View from 1972 to 1999. *Northeastern Naturalist* 11:67–74.

Patenaude, B. 2000. Survey of Wastewater Treatment Facilities in Rhode Island for 1998 and 1999. Rhode Island Department of Environmental Management, Office of Water Resources *http://www.dem.ri.gov/programs/benviron/waste/wwtfrep/index.htm*.

Perry, A. 1887. Rhode Island State Census, 1885. Providence, RI: E.L. Freeman Co.

Providence Water Supply Board—website. *http://www.provwater.com/site_ndx.htm*.

Pryor, D., Saarman, E., Murray, D., and Prell, W. 2006. Nitrogen Loading from Wastewater Treatment Plants to Upper Narragansett Bay. Narragansett Bay Estuary Program Report. Providence, RI: Narragansett Bay Estuary Program.

Rhode Island Department of Environmental Management, Division of Agriculture. 2003. Rhode Island Agricultural Digest 2003. (*htttp://www.dem.ri.gov/programs/ bnatres/ agricult/pdf/digest03.pdf*).

Rhode Island Department of Health. 1936. Annual Report. Providence, RI: H. Beck and Co.

Rhode Island State Board of Health. 1912. Twenty-Ninth Annual Report of the State Board of Health for the year ending December, 1906. Providence, RI: E. L. Freeman Co.

Rhode Island Statewide Planning. 2000. Rhode Island Land Use Trends and Analysis. Technical Paper 149. Providence, RI: RI Statewide Planning. p. 43.

Ries, K.G., III. 1990. Estimating Surface-Water Runoff to Narragansett Bay, Rhode Island and Massachusetts. US Geological Survey, Water Resources Investigations. Report 89-4164. 44 pp. Washington, DC: United States Geological Survey.

Roberts, I.P. 1907. The Fertility of the Land, 10th Edition. New York: The Macmillan Co.

Robinson, K.W., Flanagan, S.M., Ayotte, J.D., Campo, K.W., Chalmers, A., Coles, J.F., and Cuffney, T.F. 2003. Water Quality Trends in New England Rivers During the 20th Century. USGS Water Resources Investigations Report WRIR 03-4012. Washington, DC: United States Geological Survey. This publication is available online at *http://pubs.usgs.gov/wri/wrir03-4012/*.

Roman, C.T., Jaworski, N., Short, F.T., Findlay, S., and Warren, R.S. 2000. Estuaries of the northeastern United States: Habitat and land use signatures. *Estuaries* 23:743–764.

Russell, H. 1980. Indian New England before the Mayflower. Hanover, NH: University Press of New England.

Seitzinger, S.P., Styles, R.V., Boyer, E.W., Alexander, R.B., Billen, G., Howarth, R.W., Mayer, B., and van Breemen, N. 2002. Nitrogen retention in rivers: Model development and application to watersheds in the northeastern U.S.A. *Biogeochemistry* 57/58:199–237.

Shanahan, P. 1994. A water-quality history of the Blackstone River, Massachusetts, USA: Implications for Central and Eastern European Rivers. *Water Science and Technology* 30:59–68.

Shedd, J.H. 1891. Annual Report of the City Engineer of Providence. Providence, RI: Providence Press.

Snow, E.M. 1867. Report upon the Census of Rhode Island, 1865. Providence, RI: Providence Press.

Snow, E.M. 1877. Report upon the Census of Rhode Island, 1875. Providence, RI: Providence Press.

Soll, D. 2006. Milking the Landscape: Reforestation in Norfolk County, MA, 1850–1910. Proceedings of the Massachusetts Historical Society "Remaking Boston" Conference, Boston. (Forthcoming U. of Pittsburgh Press volume).

State Department of Health of Massachusetts. 1915. Annual Report. Boston, MA: Wright and Potter Printing Company.

Tarr, J.A. 1996. The search for the ultimate sink—urban pollution in historical perspective. Akron, OH: University of Akron Press.

Tiepke, H.E. 1898. Census of Rhode Island, 1895. Providence, RI: E.L. Freeman Co.

US Bureau of the Census. 1913. Thirteenth Census of the United States, 1910. Abstract of the Census with Supplement for Rhode Island. Washington: Government Printing Office.

US Bureau of the Census. 1922. Fourteenth Census of the United States, 1920: Volume 5, Agriculture. General Report and Analytical Tables. Washington: Government Printing Office.

US Bureau of the Census. 1931. Fifteenth Census of the United States: 1930: Agriculture Volume 1, Farm Acreage and Farm Values by Township or other Minor Civil Divisions. Washington: Government Printing Office.

US Bureau of the Census. 1941, 1952, 1960, 1971, 1978, 2000. United States Census of Agriculture: 1940 (and preceding years, respectively). Washington: Government Printing Office.

US Bureau of the Census. 1971, 1982. United States Census of Housing: 1970, 1980. Washington: Government Printing Office.

US Bureau of the Census. 1990. United States Census of Housing: 1990: Detailed Housing Characteristics. *http://www.census.gov/prod/cen1990/ch2/ch-2.html*.

US Census Office. 1850, 1860. Agricultural Census Schedules for Rhode Island. Providence: Rhode Island State Archives.

US Census Office. 1883. Tenth Census of the United States: 1880. Washington: Government Printing Office.

US Census Office. 1893. Eleventh Census of the United States, 1890: Agriculture volume. Washington: Government Printing Office.

US Census Office. 1902. Census Reports Volume 5: Twelfth Census of the United States, 1900: Agriculture. Part 1: Farms, Livestock, and Animal Products. Washington: Government Printing Office.

US Department of Agriculture. 1997. Costs Associated with Development and Implementation of Comprehensive Nutrient Management Plans. Part I: Nutrient Management, Land Treatment, Manure and Wastewater Handling and Storage, and Recordkeeping. Washington, DC: US Department of Agriculture.

US Department of Agriculture, National Agricultural Statistics Service. 2002. Census of Agriculture 1987–2002. *http://www.nass.usda.gov/Data_and_Statistics/index.asp*.

US Environmental Protection Agency. 1992. National Land Cover Dataset. *http://www.epa.gov/mrlc/nlcd.html*.

US Environmental Protection Agency. 2002a. Municipally Owned Wastewater Treatment Facilities in New England. Boston, MA: U.S. EPA New England Region, Office of Ecosystem Protection.

US Environmental Protection Agency. 2002b. Onsite Wastewater Treatment Systems Manual. EPA 625/R-00/008. Washington, DC: US Environmental Protection Agency.

Valiela, I., Collins, G., Kremer, J., Lajtha, K., Geist, M., Seely, B., Brawley, J., and Sham, C.H. 1997. Nitrogen loading from coastal watersheds to receiving estuaries: New method and application. *Ecological Applications* 7:358–380.

van Breemen, N., Boyer, E.W., Goodale, C.L., Jaworski, N.A., Paustian, K., Seitzinger, S., Lajtha, K., Mayer, B., van Dam, D., Howarth, R.W., Nadelhoffer, K.J., Eve, M., and Billen, G. 2002. Where did all the nitrogen go? Fate of nitrogen inputs to large watersheds in the northeastern USA. *Biogeochemistry* 57/58:267–293.

Vitousek, P.M., Aber, J.D., Howarth, R.W., Likens, G.E., Matson, P.A., Schindler, D.W., Schlesinger, W.H., and Tilman, D. 1997. Human alteration of the global nitrogen cycle: Sources and consequences. *Ecological Applications* 7:737–750.

Vollenweider, R. 1968. Les bases scientifiques de l'eutrophisation des lacs et des eaux courants sur l'aspect particulier du phosphore et du l'azote comme facteurs d'eutrophisation Rept. DAS/CSI/68-27, P. Paris: OCDE.

Wadlin, H.G. 1899. Census of the Commonwealth of Massachusetts, 1895: Volume 4, The Fisheries, Commerce, and Agriculture. Boston: Wright amp; Potter Co.

Wadlin, H.G. 1909. Census of the Commonwealth of Massachusetts, 1905: Volume 4, Agriculture, the Fisheries, and Commerce. Boston: Wright amp; Potter Co.

Webb, G.H., and Greenlaw, R.M. 1906. Advance sheets of the 1905 Rhode Island State Census. Providence, RI: E.L. Freeman Co.

Whitney, G.G. 1994. From Coastal Wilderness to Fruited Plain, A History of Environmental Change in Temperate North America 1500 to the Present. Cambridge: Cambridge University Press. 451 pp.

Wright, C.D. 1887. The Census of Massachusetts 1885: Volume 3, Agricultural Products and Property. Boston, MA: Wright amp; Potter Co.

Wright, R.M., Nolan, P.M., Pincumbe, D., Hartman, E., and Viator, O.J. 2001. Blackstone River Initiative: Water Quality Analysis of the Blackstone River under Wet and Dry Weather Conditions. Report to EPA Region One, Boston, MA.

Wright, R.M., Viator, O.J., and Michaelis, B. 2004. Dry Weather Water Quality Sampling and Modeling, Blackstone River Feasibility Study; Phase I: Water Quality Evaluation and Modeling of the MA Blackstone River. Report to U.S. Army Corps of Engineers, New England Div. Kingston, RI: University of Rhode Island.

Chapter 7
Anthropogenic Eutrophication of Narragansett Bay: Evidence from Dated Sediment Cores

John W. King, J. Bradford Hubeny, Carol L. Gibson, Elizabeth Laliberte, Kathryn H. Ford, Mark Cantwell, Rick McKinney and Peter Appleby

7.1 Introduction

The organic matter preserved in estuarine sediments provides a number of useful indicators, or "proxies," that can be used to infer paleoenvironmental changes. One type of paleoenvironmental change is anthropogenic eutrophication. Following the operational definition of eutrophication by Nixon (1995a), we define anthropogenic eutrophication as "an increase in the rate of supply of organic matter to an ecosystem that is caused by human activities." The human activity largely responsible for increasing the rate of supply of organic matter in temperate estuaries has been the increased loading of nitrogen (Nixon, 1995a; Vitousek, et al., 1997). Nixon et al. (Chapter 5) and Hamburg et al. (Chapter 6), both in this volume, focus on historical nitrogen inputs from the human landscape, particularly from sewage treatment facilities and agriculture. In this study, we utilize several proxy measurements to analyze the organic matter from radiometrically dated (^{137}Cs, ^{210}Pb, ^{14}C) sediment cores from three sites in the Narragansett Bay ecosystem for evidence of anthropogenic eutrophication.

7.2 Proxy Measurements

We utilize several different proxy measurements of organic matter in sediment cores to make inferences about the evidence of anthropogenic eutrophication within the Narragansett Bay ecosystem. These proxies include the concentrations of carbon and nitrogen—the C/N ratio—nitrogen isotopes, lamination thickness of the biological layer in varved sediments, and the mass accumulation rate (MAR) of chlorophyll a.

John W. King

Graduate School of Oceanography, University of Rhode Island, Narragansett, RI 02882
jking@gso.uri.edu

A. Desbonnet, B. A. Costa-Pierce (eds.), *Science for Ecosystem-based Management.*   211
© Springer 2008

7.2.1 Concentration of Total C and N

The concentration of total C and N represents the fraction of organic matter that remains after remineralization during sedimentation and early post-depositional diagenesis. These concentrations are influenced by other sedimentary components, and, therefore, must be interpreted with caution. However, increasing concentrations of C and N can be an indicator of increased delivery and/or preservation of organic matter (Cornwell *et al.*, 1996). Both may result from anthropogenic eutrophication. This proxy is useful in conjunction with other evidence for anthropogenic eutrophication.

7.2.2 C/N Ratio

The C/N ratio can be a useful indicator of the sources of organic matter. For example, marine algae produce organic matter with C/N ratios of 6–9 (Bordovskiy, 1965), whereas organic matter of terrestrial origin (plants) tend to have C/N ratios greater than 20 (Meyers and Teranes, 2001). In estuarine sediments, the C/N ratio reflects a mixture of marine and terrestrial sources. In general, decreasing C/N ratios can reflect an increase in the importance of marine algae and phytoplankton as a source, whereas increasing C/N ratios indicate an increase in the importance of terrestrial sources. Other studies (e.g., Thornton and McManus, 1994) have shown that C/N ratios can reflect diagenetic changes more than sources in areas subjected to intensive microbial decomposition of organic matter. In general, during these diagenetic processes, the C/N ratio will decrease as particulate N is added to the sediments as detrital biomass.

7.2.3 Nitrogen Isotopes ($\delta^{15}N$)

Recent reviews of the use of nitrogen isotopes ($\delta^{15}N$) to characterize paleoenvironmental changes in the ocean (Altabet, 2005), and in lakes (Meyers and Teranes, 2001; Talbot, 2001), indicate that the interpretation of this proxy is complex. However, because just a few microbially mediated processes tend to dominate the $\delta^{15}N$ pattern recorded in sediments (Talbot, 2001; Altabet, 2005), this proxy can be a very useful indicator of anthropogenic eutrophication in estuaries.

The ratio of ^{15}N to ^{14}N is expressed as:

$$\delta^{15}N(o/oo) = [(R_{sample} - R_{reference})/R_{reference}] \times 1000,$$

where R is $^{15}N/^{14}N$, and the reference is atmospheric N_2 (Peterson and Fry, 1987). To properly interpret $\delta^{15}N$ data of bulk organic matter from sediment cores, it is critical to have knowledge of the major sources and forms of N entering the system, the $\delta^{15}N$ values of these sources of N, and the dominant planktonic and microbial processes operating within the system to modify the original $\delta^{15}N$ of the N entering the system. For example, N derived from groundwater that originates from septic systems tends to be enriched in ^{15}N ($\delta^{15}N$ of $+10$ to $+22$ o/oo) because of denitrification and volatilization of ammonia within leach fields (Kreitler and Browning, 1983). Similarly, the $\delta^{15}N$ values ($+10$ to $+25$ o/oo) of farm runoff and untreated human sewage are high (Teranes and Berasconi, 2000). On the other hand, $\delta^{15}N$ values from groundwater N derived from atmospheric sources tend to be lower ($+2$ to $+8$ o/oo), as does N derived from fertilizer (-3 to $+3$ o/oo) (Kreitler and Browning, 1983). The process currently used (aeration, treatment with activated sludge, removal of most solids, and discharge of liquids and remaining solids) in some of the major wastewater treatment facilities (WWTF) on Narragansett Bay (e.g., Field's Point and Bucklin Point) produces low $\delta^{15}N$ values (-2 to -1 o/oo) for the remaining suspended solids released to the bay (Table 7.1).

The uptake of N by plankton, and dominant microbial processes operating within the ecosystem, can have a significant effect on the $\delta^{15}N$ values of organic matter preserved in cores. The coastal marine plankton produce organic matter with an average $\delta^{15}N$ of $+8.5$ o/oo, whereas land plants using the C_3 metabolic pathway produce average values of $+0.5$ o/oo (Peterson and Howarth, 1987). The relative contribution of these groups impacts on the $\delta^{15}N$ of organic matter preserved in sediment cores within an estuary. For example, increased blooms of coastal marine algae and plankton would tend to increase the $\delta^{15}N$ values observed in sediment cores.

Recent studies have used $\delta^{15}N$ as an indicator of anthropogenic eutrophication in estuaries (e.g., Zimmerman and Canuel, 2002; Bratton *et al.*, 2003; Cole *et al.*, 2004). These studies have shown $\delta^{15}N$ from macrophytes to be a reliable indicator of relative wastewater load to the receiving waters (Cole *et al.*, 2004), and $\delta^{15}N$ from organic matter in the cores to be an indicator of anthropogenic

Table 7.1 $\delta^{15}N$ values for water being discharged by sewage treatment facilities into Narragansett Bay, and in water at various places throughout the bay.

Sewage treatment facility discharge water		
Facility	Average $\delta^{15}N$	Stdev
Bucklin Point	−2.12	0.28
Field's Point	−1.37	0.38
Narragansett Bay water		
Location	Average $\delta^{15}N$	km from Bucklin Point
Blackstone Point	−0.98	2
Sabin Point	2.91	10
Pawtuxet Cove	2.94	11
Colt State Park	5.45	23
Bristol Point	4.96	27
West Passage, mouth	7.15	49

eutrophication and anoxia that allow intense denitrification in both sediments and the water column (Bratton *et al.*, 2003). In both studies, increasing $\delta^{15}N$ values indicate anthropogenic eutrophication within the study areas.

7.2.4 Lamination Thicknesses of the Biological Layer

This proxy is used as an indicator of anthropogenic eutrophication for the annually laminated sediments of the Pettaquamscutt River (Hubeny and King, 2003; Hubeny *et al.*, 2006). The anoxic bottom waters of this site preserve the seasonal pattern of deposition, which consists of a dark spring and summer organic-rich layer produced by biological activity, and a lighter winter layer consisting of more inorganic clastic material eroded from the watershed. An increase in the thickness of the biological layer in conjunction with more intensive human activity (i.e., residential development) in the watershed is a proxy for anthropogenic eutrophication.

7.2.5 Mass Accumulation Rate of Chlorophyll a

Fossil pigments have been used in the recent studies as indicators of anthropogenic eutrophication in estuaries (Hubeny and King, 2003; Hubeny, 2006; Turner *et al.*, 2006). In general, if preservation is not an issue, then increases in fossil pigment accumulation rates are a proxy for increases in phytoplankton production, which may be due to anthropogenic eutrophication.

7.3 Study Sites

Three sites within the Narragansett Bay system, shown in Fig. 7.1 are examined in this study: (1) the lower basin of the Pettaquamscutt River; (2) Potter Cove on Prudence Island; and (3) the Seekonk and Providence Rivers. The general characteristics of these sites are discussed below.

7.3.1 Pettaquamscutt River—Lower Basin

We chose the lower basin of the Pettaquamscutt River estuary for our study for several reasons: (1) the excellent preservation of organic matter and annual layering due to deep anoxic waters; (2) excellent age model provided by radiometric dating and annual layer counts (Hubeny *et al.*, 2006); (3) potential anthropogenic eutrophication caused by extensive residential development and associated failing septic systems in the watershed (Ernst *et al.*, 1999); and

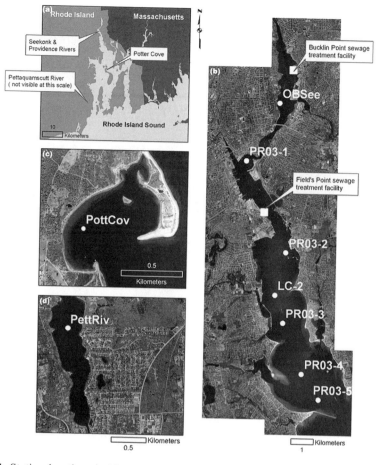

Fig. 7.1 Station locations in Narragansett Bay (a). Northernmost stations (b) are in the Seekonk and Providence Rivers. Mid-bay station (c) is in Potter Cove. Lower bay station (d) is in the Pettaquamscutt River.

(4) potential reversal of anthropogenic eutrophication caused by the replacement of failing septic systems with a sewering project in the 1990s.

7.3.2 Prudence Island—Potter Cove

We chose Potter Cove on Prudence Island for two reasons. First, it is a sheltered site that rapidly accumulates the fine-grained sediments needed to preserve a good record of organic matter accumulation. This type of site is uncommon in the mid-bay region, but has potential to record changes due to nutrient inputs further up the bay. Second, it is intensively used for recreational boating activity, with hundreds of permanently and temporarily moored boats present during the summer season (K. Raposa, personal communication). Therefore, this site is likely to preserve a record that will show anthropogenic eutrophication from nutrient sources either from upper Narragansett Bay or from local recreational boating activity, or both sources.

7.3.3 Seekonk and Providence Rivers

A transect of sediment cores from the Seekonk and Providence Rivers was selected for our study because they are in the direct path of receiving waters of the Field's Point and Bucklin Point WWTFs. These WWTFs became dominant point sources to the upper bay starting in the 1890s (Nixon, 1998). In addition, there has been a clear shift between the 1970s and the 1980s in the dominant form of N discharged from these WWTFs, from organic N to dissolved inorganic nitrogen (ammonia, nitrite, and nitrate) (Carey et al., 2005).

7.4 Methods

7.4.1 Coring

Seven freeze cores 1–1.5 m in length and 20 cm in width (Wright Jr., 1980) were taken from an area approximately 50 m in diameter from the deep hole (approximately 20 m) in the lower basin of the Pettaquamscutt River using an 8.5-m long pontoon boat. Freeze coring was necessary to preserve annual laminations present in the sediments of this basin. A large number of cores were obtained in order to determine the reproducibility of the lamination record and obtain accurate counting statistics. Piston cores of 1–3.5 m in length and 7.5–10.0 cm in diameter were obtained from Potter Cove and the Seekonk and Providence Rivers using the same pontoon boat.

7.4.2 Age Models

The age model for the lower basin of the Pettaquamscutt River was constructed using a combination of annual lamination counts and ^{137}Cs, ^{210}Pb, and ^{14}C dating (Lima et al., 2005; Hubeny et al., 2006). The lamination counts were done after the preparation of thin sections from the sediments using freeze drying and Spurr epoxy resin imbedding technique (Pike and Kemp, 1996; Spurr, 1969), following the approach of Francus et al. (2002). Counting errors were approximately 1% on individual sections (Hubeny et al., 2006).

The varve chronology was validated using ^{137}Cs and ^{210}Pb dating (Lima et al., 2005), pollen dating [the time of European colonization indicated by ragweed pollen (Brugam, 1978)], and by accelerator mass spectrometry (AMS) radiocarbon dating of terrestrial plant fossils done at the Woods Hole Oceanographic Institute National Ocean Sciences AMS facility. The radiocarbon results were converted to calendar ages using the CALIB4.3 program (Stuiver *et al.*, 1998). Overall, error in this age model was approximately ±1%.

The age model for Potter Cove was based on ^{137}Cs and ^{210}Pb dating done using the gamma counting method at Liverpool University Environmental Radioactivity Laboratory. The samples were analyzed using Ortec HPGe GWL series well-type coaxial low background intrinsic germanium detectors (Appleby *et al.*, 1986). ^{210}Pb was determined via its gamma emissions at 46.5 keV, and ^{226}Ra by 295 keV and 352 keV γ-rays emitted by its daughter isotope ^{214}Pb following 3 weeks of storage in sealed containers to allow radioactive equilibration. ^{137}Cs was measured by its emissions at 662 keV. The absolute efficiencies of the detectors were determined using calibrated sources and sediment samples of known activity. Corrections were made for the effect of self-absorption of low energy γ-rays within the sample (Appleby *et al.*, 1992). The errors associated with this age model were less than ±3 years back to 1970, and increase gradually to ±10 years at 1930. Errors were ±15 years at 1900, and ±20 years beyond 1900 (P. G. Appleby, unpublished report).

The age model for the Seekonk River site was based on a combination of ^{137}Cs and ^{210}Pb dating, done using the gamma counting method at the University of Liverpool, and pollen dating (Corbin, 1989). The surface value for the Seekonk River core was from the upper 1 cm of a grab sample obtained from the same location in 1998. The errors associated with this model were less than ±3 years back to 1965, ±10 years by 1930, and ±15 years at 1900 (Corbin, 1989). Errors are ±20 years beyond 1900.

The age model for the Providence River composite was obtained by first correlating a 1-m piston core taken from the dredged channel with a well-preserved sediment-water interface and a basal change in lithology, which corresponds to the last date of channel dredging in 1971, to the upper section of a 3.5-m piston core located 1 km to the north that was missing the interface. After correlation, the surface core was spliced onto the top of the long core. The Providence River composite was then correlated to the well-dated Seekonk River core using down-core variation in copper concentration. Copper is a metal pollutant, which was introduced to the environment by early industrial activity and has a stratigraphy with several major and readily correlateable features (Corbin, 1989). The errors associated with the Providence River composite are comparable to those for the Seekonk River core.

7.4.3 C, N, and $\delta^{15}N$ Analyses

Samples analyzed for C, N, and δ^{15}N were run by continuous flow elemental analysis/isotope ratio mass spectrometry (EA/IRMS). Samples were dried at 100°C, large shell fragments removed, and then ground, weighed, and placed in tin capsules, and analyzed using VG Optima IRMS with Carlo Erba EA Model #NA-1500. The precision of $\delta^{15}N$ measurement was $\pm0.3\%$ o/oo, and the precision of total C and N measurements was $\pm0.05\%$. The organic C results were corrected for the effects of decomposition with time using the model of Middleburg (1989).

7.4.4 Fossil Pigment Analyses

Samples were extracted in cold acetone, filtered with a 0.45 μm PTFE membrane filter, and analyzed by high performance liquid chromatography (HPLC) (Wright et al., 1991; Bianchi et al., 1996). The HPLC system comprised of a Waters Z690 Alliance separation module with a 996-photodiode array detector and a 474 scanning fluorescence detector with excitation set at 410 nm and emission at 660 nm. The pigments were identified and quantified by comparing retention times and PDA spectra to the pigment standards. Concentrations (μg L^{-1}) were converted to mass accumulation rates (μg cm^{-2} yr^{-1}) using dry bulk density measurements and annual lamination sedimentation rates (Hubeny et al., 2006).

7.4.5 Biological Lamination Thickness

The annual lamination thin sections were scanned on a flat-bed scanner with transparency capabilities under cross-polarized films to produce high-resolution (1440 dots per inch) TIFF images (De Keyser, 1999). The images were analyzed using Adobe Photoshop^{TM}, lamination boundaries marked with the path tool, and thicknesses measured using the algorithm of Francus et al. (2002) (Hubeny et al., 2006).

7.4.6 Trace Metal Analyses

Subsamples were obtained from cores at approximately 5–20 cm intervals (depending on the total length of the core and sedimentation rate) for trace metal analyses. At each sample interval, approximately 5 g of wet sediment was subsampled with acid-washed plastic spatulas from a 2-cm slice (e.g., 0–2 cm, 5–7 cm, etc.), and transferred to an acid-washed polypropylene centrifuge tube. The samples were freeze-dried for at least 48 hours, or until they were dry.

The Seekonk River samples were prepared using a partial digestion technique (Corbin, 1989). Approximately 2.0 g of dried sediment was treated with 50 ml of 2 N nitric acid, placed in a heated (55°C) water bath, and periodically agitated and vented over a 48-hour period. After the digestion was complete, the samples were agitated and centrifuged at 2500 RPM for 10 min, and decanted into an acid-cleaned bottle. The samples were analyzed for copper using an ARL model 3410 inductively-coupled argon plasma atomic emission spectrometer (ICP-AES) (King et al., 1995).

All other samples were prepared using a total digestion technique in 2003. Different digestion methods were used, because the standard method in trace metal geochemistry used for sample digestion has changed during the 14 years between the two studies. At each sample interval, approximately 0.200 g of freeze-dried sediment was transferred to Teflon centrifuge tubes, and treated with concentrated hydrochloric (1 ml), nitric (5 ml), and hydrofluoric (4 ml) acids, and covered and placed in a heated sonicator for 48 h. After cooling, the samples were combined with 30 ml of 5% boric acid solution to neutralize the HF, and then brought up to volume in 50 ml volumetric flasks using deionized water. The samples were stored in acid-cleaned bottles until analysis. Copper concentration for total digestions was determined using a Perkin Elmer 4100ZL graphic furnace atomic absorption spectrometer (GFAA) with Zeeman background correction.

7.5 Results

7.5.1 Pettaquamscutt River—Lower Basin

The results of our studies of $\delta^{15}N$, biological lamination thickness, and chlorophyll a mass accumulation rate (MAR) are plotted against calendar age (Fig. 7.2). The age model from this site is discussed in detail in Lima et al. (2005) and Hubeny et al. (2006). The $\delta^{15}N$ values began to shift to more positive values around 1930, but did not exceed the range of natural historical values observed since 1720 AD until the 1950s. The shift from the mean pre-1950 values of $\delta^{15}N$ to the mean post-1950 $\delta^{15}N$ was +1.7 o/oo. A single point spike in chlorophyll a occurred in the late 1930s; an increasing trend started in the 1950s, which continues until the present. Biological lamination thickness first increased beyond the range of natural variation between 1955 and 1990, and then increased dramatically between 1990 and the present.

7.5.2 Potter Cove—Prudence Island

The age model for Potter Cove shown in Fig. 7.3a was constructed using the CRS model (Appleby et al., 1986) to estimate the age from ^{210}Pb data. The dated sediment record from Potter Cove first shows evidence of anthropogenic influences starting in the late 1800s (Fig. 7.3). Copper concentration increased above natural background, and the C/N ratios decreased between 1840 and

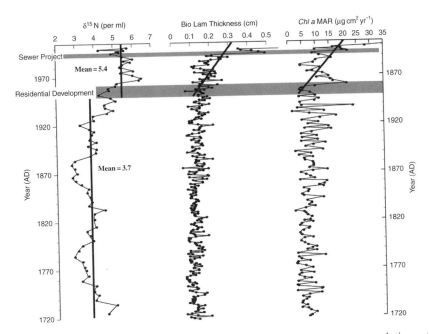

Fig. 7.2 Variation in $\delta^{15}N$, biological lamination thickness, and the mass accumulation rate of chlorophyll *a* during the last 300 years for the Pettaquamscutt River Lower Basin.

1900. In addition, the C and N concentrations, and nitrogen isotope values, increased between 1880 and 1900.

Between 1900 and 2002, a number of changes occurred. Since 1900, copper concentration increased to a peak of 100 μg g^{-1} in 1950, decreased to about 60 μg g^{-1} in 1980, and then increased to maximum values of 100 μg g^{-1} or higher in 2002. C and N concentrations were fairly constant between 1900 and 1960, but increased between 1960 and 1980. Maximum values were observed between 1980 and 2002. The C/N ratios decreased from values above 10 to 8.5 from the early 1900s to 2002. Nitrogen isotope values increased from +4.7 to +7.4 between 1900 and 2002. The overall change in nitrogen isotope values from the mid-1800s to 2002 was an increase of +2.5 o/oo.

7.5.3 Seekonk and Providence River

A radiometrically dated (^{137}Cs and ^{210}Pb) record of copper concentration (input) to the Seekonk River (Corbin, 1989) was used to develop an age model for mid-Providence River cores by matching the copper concentration curves of the cores (Fig. 7.4). The copper curve contains a range of distinctive features that reflect the history of industrial and sewage inputs—an initial increase in 1750 AD, a major increase in 1840 AD as the industrial revolution

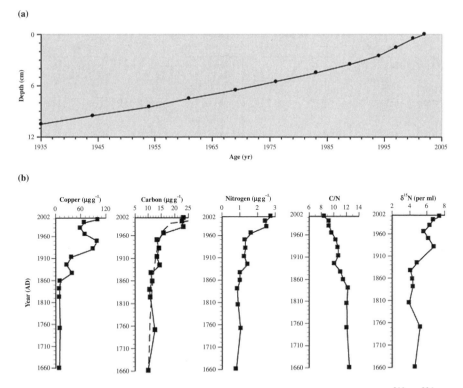

Fig. 7.3 (a) Age–depth model for Potter Cove, Prudence Island, derived from $^{210}Pb/^{226}Ra$ analysis with a constant rate of supply (CRS) model interpretation (P. G. Appleby, unpublished data). (b) Variation in copper, carbon and nitrogen concentrations, C/N ratio, and $\delta^{15}N$ values during the last 400 years for Potter Cove. The dashed line on the carbon curve indicates the concentrations expected by diagenesis of surface carbon concentrations with time estimated using the model of Middleburg (1989).

intensified, a decrease during the Depression in 1935—and the impact of environmental regulations and decreasing industrial activity, which showed a decreasing trend between 1965 to present (Corbin, 1989). The general shape and trends of the features are comparable, although industrial-age Seekonk River values are higher due to the proximity of historical discharges from a wire manufacturer (Washburn Wire) to the coring site. The correlation points used to construct the age model and the resulting copper concentration versus age curve for the Providence River cores are shown in Fig. 7.4

The construction of the age model allowed the study of historical changes in nitrogen isotope ratios in the Providence River. In addition, studies of nitrogen isotope ratios of suspended solids collected from sewage treatment facility effluent and from a number of stations in a north–south transect in the Narragansett Bay system (Table 7.1), and from the sediments of a N–S transect of short piston cores from the Providence River dredged channel (Fig. 7.5), facilitate the interpretation of the long core record.

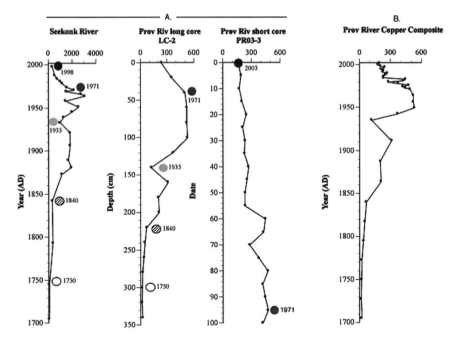

Fig. 7.4 Construction of the Providence River age model using correlation to the radiometrically dated copper concentration ($\mu g\, g^{-1}$) record from the Seekonk River (Corbin, 1989). Data from PR03-3 are spliced onto the top of PC03-2 to construct the Providence River copper composite curve.

The short piston cores obtained from the dredged channel prior to the recently completed dredging contain a high resolution record from the interval of approximately 1971–2003 (i.e., from the time of previous channel dredging to the time of collection). The short core locations shown in Fig. 7.1 are within the receiving waters of the Narragansett Bay Commission (NBC) Bucklin Point WWTF and Field's Point WWTF. The average $\delta^{15}N$ values of the cores increased by +4 o/oo between the north end to the south end of the Providence River. In addition, three of the four cores, 03–1, 03–3, and 03–5, showed an increase in $\delta^{15}N$ values between approximately 1971 and 2003.

The $\delta^{15}N$ values obtained from cores PR03-3 and LC-2, and the age assignments used to construct the mid-Providence River $\delta^{15}N$ composite, are shown in Fig. 7.6a A similar methodology was used to construct the composite curves for C and N concentration and C/N ratio (Fig. 7.6b). The $\delta^{15}N$ record shows values fluctuating around approximately +7.7 until the late 1800s, decreasing to approximately +6.5 in the 1960s, then increasing to approximately +8.3 in the late 1970s, and remaining above +8.0 between 1980 and 2003. The C concentration showed an increasing trend between approximately 1900 and 2003, whereas the N concentration showed an increasing trend between approximately 1930 and 2003. The C/N ratio fluctuated around a value of 12 between 1800 and 1940,

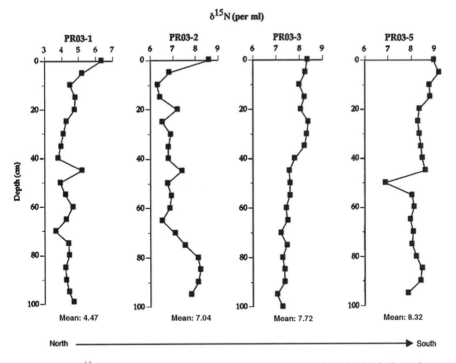

Fig. 7.5 The δ^{15}N record of a core transect in the Providence River dredged channel (see Fig. 7.1 for location). Mean values of δ^{15}N for each core are shown, and increase down river.

and then decreased from approximately 13.3 to 10 between 1940 and 2003. The δ^{15}N and N concentration composite records are compared to the reconstructed record of total N input to Narragansett Bay (Nixon, 1995a; see Fig. 7.6c). The total N record largely reflects sewage inputs from 1890 to 1985 (Nixon, 1995b). The N concentration record is similar to the total N input record between the early 1900s and the end of the input record around 1985, whereas there is no relationship observed with the δ^{15}N record.

7.6 Discussion

7.6.1 Pettaquamscutt River—Lower Basin

Previous studies of the Pettaquamscutt River watershed (Urish, 1991; ASA et al., 1995) have found that 65% of the total freshwater inputs are from groundwater. This region is anomalous with respect to the general importance of groundwater N input (Nowicki and Gold in Chapter 4). The fjord-like structure of this system is atypical of the overall Narragansett Bay ecosystem.

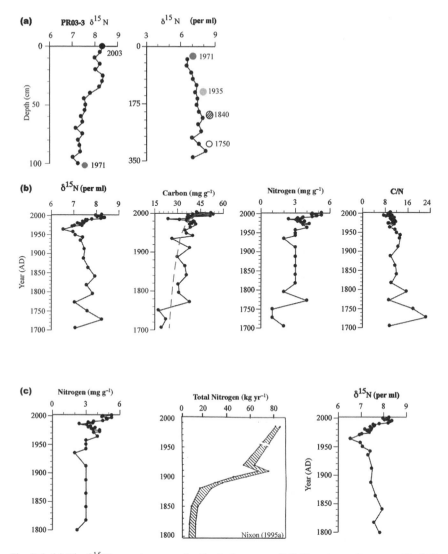

Fig. 7.6 (a) The δ^{15}N records versus depth of a long core (LC-2) and a surface core (Pro3-3) from the mid-Providence River. Ages are derived from Fig. 7.4 and used to construct a composite record of δ^{15}N shown in Fig. 7.6(b) by splicing the surface core on the top of the long core at the 1971 AD datum. The same approach is used to construct the carbon, nitrogen, and C/N curves in Fig. 7.6(b). (b) The composite δ^{15}N, C, and N concentration, and C/N ratio records versus calendar age for the mid-Providence River. The dashed line on the carbon curve indicates the concentration expected by diagenesis of the observed surface sediment carbon concentrations with time estimated using the model of Middleburg (1989). (c) Comparison of the nitrogen concentration and δ^{15}N composite records from the mid-Providence River to the total nitrogen input to Narragansett Bay record of Nixon (1995a).

Furthermore, groundwater is a major conduit of nutrients from failing septic systems within the watershed (ASA *et al.*, 1995). Major residential development of the watershed began in the 1950s (Ernst *et al.*, 1999), and all homes utilized septic systems. The use of septic systems in an area of active groundwater flow is likely to have quickly increased N inputs to the watershed. N derived from septic systems has enriched $\delta^{15}N$ values of $+10$ to $+22$ o/oo, primarily due to denitrification and volatilization of ammonia in these systems (Kreitler and Browning, 1983; Aravena *et al.*, 1993; Macko and Ostrom, 1994).

The sedimentary record from the Pettaquamscutt River (Fig. 7.2) provides clear evidence of anthropogenic eutrophication. Higher values of $\delta^{15}N$ are observed during the 1930s and 1950s, which may indicate higher anthropogenic inputs of N. In conjunction with major residential developments that occurred during 1950–1970, there is a shift of approximately $+2.2$ o/oo in $\delta^{15}N$, which may reflect increased inputs to the groundwater from septic systems. It is unlikely that an increase in sediment column and/or water column denitrification in the basin contributed to the shift in $\delta^{15}N$ because the sediment column and most of the water column within the basin are permanently and stably anoxic (Gaines and Pilson, 1972) and have been for almost a millennium (Hubeny *et al.*, 2006). In recent decades, the $\delta^{15}N$ values have decreased from the peaks observed in the 1970s, but remain well above the 1950s values.

Increases in biological lamination thickness and chlorophyll *a* MAR (Fig. 7.2) occur in conjunction with residential development beginning from the 1950s through approximately 1980. Due to the highly anoxic bottom waters of the basin, little degradation of organic matter and fossil pigments is observed (Hubeny *et al.*, 2006). Therefore, changes in preservation during remineralization and burial are not an issue at this site. These increases are interpreted as due to enhanced phytoplankton productivity caused by increased N loading within the watershed (Turner *et al.*, 2006). After a period of relative stability in the 1980s, both proxies increased again in the 1990s. A single point increase in chlorophyll *a* MAR in the late 1930s may indicate increased nutrient inputs due to the hurricane of 1938.

The sewer project that replaced septic systems within the watershed during the early 1990s (Fig. 7.2) does not appear to have resulted in any significant improvement in the system. To date, the $\delta^{15}N$ values have not decreased significantly, whereas biological lamination thickness and chlorophyll *a* MAR have actually increased, indicating further eutrophication. One possible explanation for the lack of change in $\delta^{15}N$ values is the lag in system response due to long residence time (several years) of N in groundwater. The apparent increase in eutrophication may reflect residential development in the 1990s on the very steep western slope of the lower basin of Pettaquamscutt River, which has increased soil erosion rates, reflected by increased clastic lamination thickness and sediment thickness (Lima *et al.*, 2005; Hubeny, 2006). Increased soil erosion and the recent developments may increase nutrient inputs to the system, and the major increases in biological lamination thickness and chlorophyll *a* MAR in the 1990s may indicate that anthropogenic eutrophication in the upper basin has actually increased, despite anticipated decreases on the completion of the sewer project.

7.6.2 Prudence Island—Potter Cove

The sedimentary record from Potter Cove reflects anthropogenic impacts (Fig. 7.3). The record of copper concentration reflects inputs from both local and regional sources. We interpret the initial increase above natural background values between 1860 and 1880 and the trend of increasing values until 1950, followed by a decreasing trend between 1950 and 1980, as primarily reflecting comparable trends in the upper bay. The trends observed in the upper bay primarily reflect trends in industrial and sewage treatment plant inputs (Corbin, 1989; Nixon, 1995b). The trend of increasing copper concentrations observed between 1980 and 2002 in Potter Cove contradicts a decreasing trend observed in upper Narragansett Bay (see Fig. 7.4) during this interval. For this reason, we interpret the recent increase in copper concentration in Potter Cove as reflecting inputs from local sources. Boating activity has increased dramatically in Potter Cove in the last 20 years (Raposa, personal communication), and the use of copper ablation anti-fouling bottom paints has simultaneously become standard practice for recreational boaters. We interpret the source of observed recent increase in copper as bottom paints used in recreational boating.

The observed increases in C and N concentrations, the decrease in C/N ratio, and the increase in $\delta^{15}N$ values, which are initiated in the late 1800s shortly after the initial increase in copper concentrations (Fig. 7.3b), are interpreted as evidence of anthropogenic eutrophication in Potter Cove. The surface sediments from the upper 20 cm (last 130 years) are very dark grey, indicating the likely presence of iron sulfides, either pyrite or monosulfide, and either low sedimentary or water column oxygen conditions during this interval, whereas sediments below this depth are light brown, indicating more oxygenated conditions. Therefore, changes in organic matter preservation during remineralization and burial probably do not control the temporal trends observed since the late 1800s.

To test this hypothesis, we calculated the curve expected from diagenesis of organic C with time (Middleburg, 1989) for Potter Cove that is shown in Fig. 7.3b. Intervals of observed C values in excess of predicted values may be interpreted as times of excess C production (Cornwell *et al.,* 1996; Zimmerman and Canuel, 2002) or eutrophic conditions. Two intervals of excess C were observed; 1900–1940 and 1970–2002.

The C and N concentration variations reflect the intervals of increased loading of organic matter to Potter Cove during the last century. We interpret the initial phase of excess concentrations between 1900 and 1940 as reflecting baywide increases in nutrient loading, whereas the marked increase between

1970 and 2002 is interpreted as reflecting increased nutrient loading from local recreational boating discharges. The decreasing trend in C/N ratios from the late 1800s to the present is interpreted as a shift in the proportion of organic matter derived from terrestrial sources toward higher inputs from marine sources (Bordovskiy, 1965; Meyers and Teranes, 2001). The C/N trend is interpreted to reflect increased marine algae production in response to increased nutrient loading. The +3 o/oo increase in $\delta^{15}N$ values observed between the late 1800s and the present is interpreted to reflect a shift to a heavier N source— human sewage from the upper bay and recreational boating (Teranes and Berasconi, 2000).

Recent studies of water column hypoxia and anoxia in Narragansett Bay have identified Potter Cove as an area of low oxygen conditions during the summer months (C. Deacutis, personal communication; K. Raposa, personal communication). Overall, the sedimentary record of Potter Cove is interpreted to reflect anthropogenic eutrophication beginning from the late 1800s and increasing to the present due to increased nutrient loading from both upper bay and local sources. The increase in $\delta^{15}N$ since 1970 may reflect the increase in water column denitrification due to low oxygen conditions within the cove (e.g., Bratton et al., 2003), although no direct evidence of denitrification is available.

7.6.3 Seekonk and Providence River

We believe that the Seekonk and Providence River sedimentary records of copper concentration shown in Fig. 7.4 reflect the history of anthropogenic inputs primarily from industrial sources and WWTFs (Corbin, 1989; Nixon, 1995b). Industrial sources date to the 1790s in Rhode Island, and WWTFs were constructed in the 1890s (Nixon, 1995b). The major features include a slow increase above natural background values from the late 1700s until about 1840 during early industrialization of the area, and a rapid increase occurring between 1840 and 1880 reflecting increased industrial inputs. An increasing trend between 1890 and the late 1920s reflects increased inputs from both industry and WWTFs. The decrease between the late 1920s and mid-1930s reflects the decrease in economic activity during the Great Depression. The increase between the late 1930s and 1950s indicates high economic activity and high inputs during World War II and post-war years. Finally, the major decline between the late 1960s and mid-1970s indicates the decline in regional industrial activity and the onset of strict environmental regulations (e.g., Clean Air and Clean Water Acts). The concentrations continued to decline from the 1970s to 2003, the time of core collection. To a first order, the observed sediment copper concentrations record input changes as described by Nixon (1995b).

The $\delta^{15}N$ records of a N–S transect of surface cores from the dredged channel of the Providence River (Fig. 7.5) reflect changes between 1971 (last time of dredging) and 2003 (time of core collection). The $\delta^{15}N$ data obtained from suspended solids in Bucklin Point and Field's Point effluent are slightly negative. In addition, suspended solids from the water column have a strong down-bay increase in $\delta^{15}N$ (Table 7.1). The N in organic matter from terrestrial sources tends to have low positive $\delta^{15}N$ values, whereas coastal marine plankton produces organic matter with average $\delta^{15}N$ values of $+8.5$ o/oo (Peterson and Howarth, 1987). For these reasons, we interpret the approximately $+4.0$ o/oo down-river increase in average $\delta^{15}N$ values observed in the surface cores to reflect the decreasing influence of N from WWTFs and terrestrial sources, and the increasing influence of marine algae and plankton.

It is also noteworthy that three of the four cores (PR03-1, 3-3, and 3-5), as shown in Fig. 7.5, have trends of increasing $\delta^{15}N$ between 1971 and 2003. Bratton et al. (2003) have interpreted similar results from Chesapeake Bay as evidence of increasing seasonal water column denitrification due to increased nutrient loading and decreased oxygen concentrations. A similar mechanism may be operating in the Providence River although no direct evidence is currently available. Saarman et al. (Chapter 11) and Deacutis (Chapter 12), both in this volume, indicate hypoxic conditions in the upper bay, lending further support to this hypothesis.

An example of the approach used to construct composite records of $\delta^{15}N$ for the mid-Providence River is shown in Fig. 7.6a. The $\delta^{15}N$, C and N concentration, and C/N ratio are shown in composite records (Fig. 7.6b). The C and N concentrations increased significantly around 1770. The comparison between C concentrations expected after diagenesis with observed values (Fig. 7.6b) indicate that the system was producing excess organic matter between 1770–1980, which could readily be advected down bay as suggested by Smayda and Borkman in Chapter 15. The C and N concentration records show major increasing trends for most of the 20th century. Overall, the C/N ratio has a decreasing trend since 1700, indicating a possible shift from terrestrial to marine sources of organic matter. The decreasing trend in C/N between 1940 and 2003 can be interpreted as the increasing importance of marine sources of organic matter over this interval. The overall evidence from these three proxies is interpreted as indicating anthropogenic eutrophication during the 20th century. The $\delta^{15}N$ record has two major trends—a decrease between approximately 1900 and 1970, and an increase between 1970 and 2003, which cannot be readily explained by the change in total N inputs shown in Fig. 7.6c. However, a consideration of historical changes in the sewage treatment process, the effectiveness of sewage treatment, and the dominant forms of N discharged from Field's Point WWTF may provide insight into these trends.

Toward the late 1800s, the $\delta^{15}N$ record varied in the range of $+7.0$–8.0 o/oo, and reflected a system dominated by a marine signal with some terrestrial inputs. The construction of Field's Point WWTF in 1890 is interpreted to have increased the influence of input of terrestrial organic matter, resulting in a slow decrease of $\delta^{15}N$ until 1940. In 1936, Field's Point was converted to the activated sludge process (Nixon, 1995b). Today, this process produces suspended solids with low $\delta^{15}N$ values (Table 7.1), and we interpret the shift to lower values between 1940 and 1970 as reflecting the influence of organic matter inputs derived from WWTFs utilizing the activated sludge process. During the 1970s, the Field's Point WWTF began to fail, and the efficiency of solids removal decreased until the facility became little more than a chlorination facility in the late 1970s (Nixon, 1995b). The trend of increasing $\delta^{15}N$ values during the 1970s probably reflects increasing $\delta^{15}N$ in solids produced by the much less-efficient sewage processing. Lightly processed sewage tends to have higher $\delta^{15}N$ values (Teranes and Berasconi, 2000; Cole *et al.*, 2004).

Repairs and upgrades to both Field's Point and Bucklin Point WWTFs in the early 1980s have resulted in major changes in the composition of N in the discharge (Carey *et al.*, 2005). Ammonia discharged by WWTFs into estuaries tends to have high $\delta^{15}N$ values (+10–15 o/oo, or higher) (Cifuentes *et al.*, 1989). The increased discharge of ammonia since approximately 1980, and the high bioavailability for algal uptake, is a probable cause for the increase in $\delta^{15}N$ values for sedimentary organic matter observed between 1980–2003. Overall, the history of sewage processing methods and processing efficiency of major WWTFs within the study area appear to exercise more control over the $\delta^{15}N$ record observed in the mid-Providence River than the total sewage and total N inputs. Anthropogenic inputs and sewage processing methods clearly affect the $\delta^{15}N$ record.

The trend of C and N concentrations and C/N ratios provide strong evidence of anthropogenic eutrophication during the last century. The trend of increasing $\delta^{15}N$ values since 1970 may also be interpreted as evidence of anthropogenic eutrophication, decreased oxygen concentrations, and increased water-column denitrification (Bratton *et al.*, 2003). However, sorting out whether the observed changes in $\delta^{15}N$ are due primarily to the characteristics of N input to the system over time, or post-input processes (e.g., denitrification by bacteria or differential uptake by plankton and algae) will require additional sediment studies. Compound-specific $\delta^{15}N$ studies of biomarkers from cores should be illuminating.

7.7 Conclusions

Strong evidence exists for anthropogenic eutrophication due to residential development from 1950 onward for the lower basin of the Pettaquamscutt River based on the proxies $\delta^{15}N$, biological lamination thickness, and chlorophyll *a* MAR. In the 1990s, a sewer project aimed at mitigating N inputs from septic systems has had little impact to date on anthropogenic eutrophication. In fact, recent developments on the western shore of the basin appear to be accelerating eutrophication.

Strong evidence exists from the proxies C and N concentrations, C/N ratios, and $\delta^{15}N$ for anthropogenic eutrophication in Potter Cove. Both upper bay regional WWTF and local sources—recreational boating—of nutrients are implicated, primarily upper bay WWTF between 1900 and 1980 and recreational boating between 1980 and 2002.

The C and N concentrations and C/N ratio records provide strong evidence for anthropogenic eutrophication during the last century in upper Narragansett Bay. The N and C records are similar to the N input record for the Narragansett Bay system as estimated by Nixon (1995a). Nutrient inputs from Field's Point and Bucklin Point WWTFs appear to have strongly impacted the δ^{15}N record of Providence River sediments. Changes in sewage processing methods and the forms of N entering the system over time appear to be the dominant control over the record. The δ^{15}N proxy records changes in anthropogenic inputs and activities.

References

Altabet, M.A. 2005. Isotopic Tracers of the Marine Nitrogen Cycle: Present and Past. Handbook of Environmental Chemistry. Volume 2, Part N: 1-x. Berlin Heidelberg: Springer-Verlag. 43 pp.

Appleby, P.G., Nolan, P.J., Gifford, D.W., Godfrey, M.J., Oldfield, F., Anderson, N.J., and Battarbee, R.W. 1986. ^{210}Pb dating by low background gamma counting. *Hydrobiologia* 141:21–27.

Appleby, P.G., Richardson, N., and Nolan, P.J. 1992. Self-absorption corrections for well-type germanium detectors. *Nuclear Instruments and Methods in Physics Research Section B-Beam Interactions with Materials and Atoms* 71:228–233.

Applied Science Associates (ASA), URI Watershed Watch, SAIC Engineering, Inc., and Urish, Wright and Runge. 1995. Narrow River stormwater management study: problem assessment and design feasibility Appendices. Applied Science Associates. Narragansett, Rhode Island, USA. 321 pp.

Aravena, R., Evans, M.L., and Cherry, J.A. 1993. Stable isotopes of oxygen and nitrogen in source identification of nitrate from septic systems. *Ground Water* 31:180–186.

Bianchi, T.S., Demetropoulos, A., Hadjichristophorou, M., Argyrou, M., Baskaran, M., and Lambert, C.D. 1996. Plant pigments as biomarkers of organic matter sources and coastal waters of Cyprus (eastern Mediterranean). *Estuarine, Coastal, and Shelf Science* 42:103–115.

Bordovskiy, O.K., 1965. Accumulation and transformation of organic substances in marine sediment, 2. Sources of organic matter in marine basins. *Marine Geology* 3:5–21.

Bratton, J.F., Colman, S.M., and Seal, R.R., II. 2003. Eutrophication and carbon sources in Chesapeake Bay over the last 2700 yr: human impacts in context. *Geochimica et Cosmochimica Acta* 67(18):3385–3402.

Brugam, R.B. 1978. Human disturbance and the historical development of Linsley Pond. *Ecology* 59(1):19.

Carey, D., Desbonnet, A., Colt, A.B., and Costa-Pierce, B.A. (eds.) 2005. State of science on nutrients in Narragansett Bay: Findings and recommendations from the Rhode Island Sea Grant 2004 Science Symposium. Rhode Island Sea Grant, Narragansett, RI. 43 pp.

Cifuentes, L.A., Fogel, M.L., Pennock, J.R., and Sharp, J.H. 1989. Biogeochemical factors that influence the stable nitrogen ratio of dissolved ammonium in the Delaware Estuary. *Geochemica et Cosmochimica Acta* 53:2713–2721.

Cole, M.L., Valiela, I., Kroeger, K.D., Tomasky, G.L., Cebrian, J., Wigand, C., McKinney, R.A., Grady, S.P., and Carvalho da Silva, M.H. 2004. Assessment of a $\delta^{15}N$ isotopic method to indicate anthropogenic eutrophication in aquatic ecosystems. *Journal of Environmental Quality* 33:124–1332.

Corbin, J. 1989. Recent and historical accumulation of trace metal contaminants in Narragansett Bay sediments. M.S. Thesis, University of Rhode Island, Kingston, RI. 295 pp.

Cornwell, J.C., Conley, D.J., Owens, M., and Stevenson, J.C. 1996. A sediment chronology of the eutrophication of Chesapeake Bay. *Estuaries* 19:488–499.

De Keyser, T.L. 1999. Digital scanning of thin sections and peels. *Journal of Sedimentary Research* 69:962–964.

Ernst, L.M., Miguel, L.K., and Willis, J. 1999. The Narrow River Special Area management plan for the watershed of the Narrow River in the towns of North Kingstown, South Kingstown, and Narragansett. Prepared for the Rhode Island Coastal Resources Management Council. 10 chapters.

Francus, P., Keimig, K., and Besonen, M. 2002. An algorithm to aid varve counting and measurement from thin sections. *Journal of Paleolimnology* 28:283–286.

Gaines, A.G., and Pilson, M.E.Q. 1972. Anoxic water in the Pettaquamscutt River, *Limnology and Oceanography* 17:42–49.

Hubeny, J.B. 2006. Late Holocene climate variability as preserved in high-resolution estuarine and lacustrine sediment archives. PhD Dissertation, Graduate School of Oceanography, University of Rhode Island. Narragansett, Rhode Island, USA. 239 pp.

Hubeny, J.B., and King, J.W. 2003. Anthropogenic eutrophication as recorded by varved sediments in the Pettaquamscutt River Estuary, Rhode Island, USA. *GSA Abstracts with Programs* 35:282.

Hubeny, J.B., King, J.W., and Santos, A. 2006. Subdecadal to multidecadal cycles of Late Holocene North Atlantic climate variability preserved by estuarine fossil pigments. *Geology* 37(7):569–572.

King, J., Corbin, J., McMaster, R., Quinn, J., Gangemi, P., Cullen, D., Latimer, J., Peck, J., Gibson, C., Boucher, J., Pratt, S., LeBlanc, L., Ellis, J., and Pilson, M. 1995. A Study of the Sediments of Narragansett Bay, Volume 1—The Surface Sediments of Narragansett Bay. Final Report submitted to the Narragansett Bay Project, Graduate School of Oceanography, University of Rhode Island. Narragansett, Rhode Island, USA. 29 pp.

Kreitler, C.W., and Browning, L.A. 1983. Nitrogen-isotope analysis of groundwater nitrate in carbonate aquifers: Natural sources versus human pollution. *Journal of Hydrology* (Amsterdam) 61:285–301.

Lima, A.L., Hubeny, J.B., Reddy, C.M., King, J.W., Hughen, K.A., and Eglinton, T.I. 2005. High-resolution historical records from Pettaquamscutt River basin sediments: 1. ^{210}Pb and varve chronologies validate record of ^{137}Cs released by the Chernobyl accident. *Geochemica et Cosmochimica Acta* 69:1803–1812.

Macko, S.A., and Ostrom, N.E. 1994. Pollution studies using stable isotopes, *In* Stable Isotopes in Ecology and Environmental Science, pp. 45–62. Lajtha, K., and Michener, R. (ed.) Oxford: Blackwell Scientific Publications.

Meyers, P.A., and Teranes, J.L. 2001. Sediment organic matter, *In* Tracking Environmental Change Using Lake Sediments Volume 2: Physical and Geochemical Methods, pp. 239–269. Last, W.M., and Smol, J.P. (eds.) The Netherlands: Kluwer Academic Publishers, Dordrecht.

Middleburg, J.J. 1989. A simple rate model for organic matter decomposition in marine sediments. *Geochimica et Cosmochimica Acta* 53:1577–1581.

Nixon, S.W. 1995a. Coastal marine eutrophication: a definition, social causes, and future concerns. *Ophelia* 41:199–219.

Nixon, S.W. 1995b. Metal inputs to Narragansett Bay: A history and assessment of recent conditions. Narragansett, RI: Rhode Island Sea Grant.

Nixon, S.W. 1998. Enriching the sea to death. *Scientific American Present* 9(3):48–53.

Peterson, B.J., and Fry, B. 1987. Stable isotopes in ecosystem studies. *Annual Review of Ecology and Systematics* 18:293–320.

Peterson, B.J., and Howarth, R.W. 1987. Sulfur, carbon, and nitrogen isotopes used to trace organic matter flow in the salt-marsh estuaries of Sapelo Island, Georgia. *Limnology and Oceanography* 32:1195–1213.

Pike, J., and Kemp, A.E.S. 1996. Preparation and analysis techniques for studies of laminated sediments, *In* Palaeoclimatology and Palaeoceanography from Laminated Sediments. Kemp, A.E.S., (ed.) Special Publication of the Geological Society of London, 116:37–48.

Spurr, A.R. 1969. A low-viscosity epoxy resin embedding medium for electron microscopy. *Journal of Ultrastructure Research* 26:31–43.

Stuiver M., Reimer, P.J., Bard, E., Beck, J.W., Burr, G.S., Hughen, K.A., Kromer, B., McCormac, G., van der Plicht, J., and Spurk, M. 1998. INTCAL98 radiocarbon age calibration, 24,000–0 cal BP. *Radiocarbon* 40:1041–1083.

Talbot, M.R. 2001. Nitrogen isotopes in palaeolimnology, *In* Tracking Environmental Change Using Lake Sediments Volume 2: Physical and Geochemical Methods, pp. 401–439. Last, W.M., and Smol, J.P. (eds.) The Netherlands: Kluwer Academic Publishers, Dordrecht.

Teranes, J.L., and Bernasconi, S.M. 2000. The record of nitrate utilization and productivity limitation provided by δ^{15}N values in lake organic matter – A study of sediment trap and core sediments from Baldeggersee, Switzerland. *Limnology and Oceanography* 45:801–813.

Thornton, S.F., and McManus, J. 1994. Application of organic carbon and nitrogen stable isotope and C/N ratios as source indicators of organic matter provenance in estuarine systems: evidence from the Tay Estuary, Scotland. *Estuarine, Coastal and Shelf Science* 38:219–233.

Turner, R.E., Rabalais, N.N., Fry, B., Atilla, N., Milan, C.S., Lee, J.M., Normandeau, C., Oswald, T.A., Swenson, E.M., and Tomasko, D.A. 2006. Paleo-indicators and water quality change in the Charlotte Harbor estuary (Florida). *Limnology and Oceanography* 51(1/2):518–533.

Urish, D.W., 1991. Freshwater inflow to the Narrow River. Narragansett, RI: Maritimes.

Vitousek, P.M., Aber, J.D., Howarth, R.W., Likens, G.E., Matson, P.A., Schindler, D.W., Schlesinger, W.H., and Tilman, D. 1997. Human alteration of the global nitrogen cycle: sources and consequences. *Ecological Applications* 7:737–750.

Wright, H.E., Jr. 1980. Cores of soft lake sediments. *Boreas* 9:107–114.

Wright, S.W., Jeffrey, S.W., Mantoura, R.F.C., Llewellyn, C.A., Bjornland, T., Repeta, D., and Welschmeyer, N. 1991. Improved HPLC method for the analysis of chlorophylls and carotenoids from marine phytoplankton. *Marine Ecology Progress Series* 77:183–196.

Zimmerman, A.R., and Canuel, E.A. 2002. Sediment geochemical records of eutrophication in the mesohaline Chesapeake Bay. *Limnology and Oceanography* 47(4):1084–1093.

Chapter 8
Circulation and Transport Dynamics in Narragansett Bay

Malcolm L. Spaulding and Craig Swanson

8.1 Introduction

This chapter provides a synthesis of our current understanding of the circulation and pollutant transport dynamics of Narragansett Bay and several of its major sub-sections—Providence River, Mount Hope Bay, Greenwich Bay, and the Sakonnet River. The synthesis draws on field observations, analytic studies, and the application of various two- and three-dimensional hydrodynamic and pollutant transport models to the system(s). The chapter covers the period from the late 1950s when Hicks published his summary of the physical oceanography of Narragansett Bay (Hicks,1959a and b) to the present. It represents an update of Spaulding's (1987) earlier summary. The reader should see Chapter 1 (Boothroyd and August) for a general introduction to the physical and geological layout of Narragansett Bay. This chapter provides an overview of circulation in the bay, divided into sub-sections on tidal-, wind-, and density-induced flows, highlighting some unique features of bay circulation. A summary of flushing times of the bay and various major sub-areas, as a function of freshwater inputs, is provided next. A review of major efforts to understand pollutant transport in the bay, including thermal discharges and dredging, concludes the chapter.

8.2 Setting

Bay bathymetry varies significantly between the various passages and the confluence of these in the upper bay and Mount Hope Bay (Fig. 8.1). The East Passage is relatively deep (37 m) at the mouth. The depth decreases with distance up the passage, reaching mean depths of 12 m at the entrance to Mount Hope Bay and 8 m at the entrance to the Providence River. The West Passage

Malcolm L. Spaulding
University of Rhode Island, Department of Ocean Engineering, Box 40, 223 Sheets Laboratory, Narragansett, RI 02882
spaulding@oce.uri.edu

A. Desbonnet, B. A. Costa-Pierce (eds.), *Science for Ecosystem-based Management.*
© Springer 2008

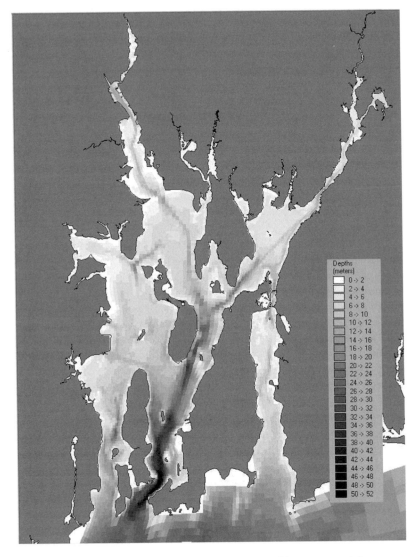

Fig. 8.1 Narragansett Bay bathymetry from NOAA. Legend in meters on right.

and lower Sakonnet River are considerably shallower, with mean depths of 10 and 7.5 m, respectively. At the confluence of the West and East Passages, north of Patience Island, the water depths are approximately 8 m and merge at the entrance of the Providence River. The Providence River is quite shallow, with a mean depth of 4.0 m [relative to mean low water (MLW)]. A channel has been dredged along the central axis of the river, approximately 200 m in width and 12.2 m in depth (MLW). The dredged channel extends from south of Conimicut

Point to the head of the river. The river width decreases sharply at the head of the bay, and the dredged channel width becomes a significant portion of the total river width. Much of Mount Hope Bay is shallow, with mean depth of 4.5 m on the northwestern side. A channel has been dredged from the mouth of the bay—connecting to Narragansett Bay—to the lower reaches of the Taunton River. The channel is about 12 m deep and 150 m wide.

Sixty percent of the bay watershed is in Massachusetts, and 40% in Rhode Island (Pilson, 1985). The watershed includes seven river sub-drainage basins— Blackstone, Moshassuck, Pawtucket, Taunton, Ten Mile, Warren/Barrington, and Woonasquatucket Rivers, plus the Narragansett Bay basin—defined as the land that is directly drained into the bay (Ries, 1990; Fig. 1.8 in Chapter 1). Flows in the Warren and Barrington Rivers are sufficiently small so that they are not gauged and are normally affiliated with the Narragansett Bay basin. Table 8.1 provides the annual mean flow rate for each system. The data were obtained from the US Geological Survey web site (*http://ma.water.usgs.gov/ basins/narrag.htm*) and include records from the beginning through 2003. It should be noted that the flow rates are based on data from gauges that are located several kilometers upstream of the river mouths. The principal freshwater discharges to the upper bay are the Blackstone and Pawtucket Rivers, while the principal discharge to Mount Hope Bay is the Taunton River (see Chapter 2, for details on freshwater input to the bay). River flows display a strong seasonal cycle with peak flows in spring and minimum values in early fall (Ries, 1990). As an example, the maximum values for the Blackstone River are observed in March and April and the minimum in August and September. The highest values are about 4.8 times greater than the smallest values (Ries, 1990). Similar variations are observed for other river basins.

As shown in Table 8.1, the annual mean runoff is consistent among the basins at about 0.02 $m^3 s^{-1}$ per km^2 of drainage area. Using this estimate, the total freshwater input to the entire drainage basin is estimated at 94.3 $m^3 s^{-1}$. In addition, the Providence River receives discharges from the Field's Point

Table 8.1 Summary of freshwater input to Narragansett Bay from major rivers.

River name	Mean annual flow rate ($m^3 s^{-1}$)	% of Total	Drainage area (km^2)	Yield flow rate/area ($m^3 s^{-1}$) km^{-2}
Blackstone	22.00	42.74	1076.97	0.020
Moshassuck	1.14	2.21	59.80	0.019
Pawtuxet	9.87	19.17	517.78	0.019
Taunton	13.48	26.19	675.69	0.020
Ten Mile	2.90	5.63	137.47	0.021
Woonasquatucket	2.09	4.06	99.15	0.021
TOTAL	51.48	100.00	2566.86	0.020
Total Drainage Area			4081.00	
Narragansett Bay Drainage Area			1514.14	

Data source: *http://ma.water.usgs.gov/basins/narrag.htm*

sewage treatment plant at 1.84 m^3 s^{-1}, and the Seekonk River from the Bucklin Point treatment plant at 0.96 m^3 s^{-1} (Pilson, 1985). These are equivalent to the inputs from smaller rivers. Finally, freshwater input by direct rainfall on the bay (annual average rainfall of 104.8 cm yr^{-1}), corrected for evaporation, is estimated at 6.52 m^3 s^{-1} (Pilson, 1985). The total input of freshwater is, therefore, about 103.62 m^3 s^{-1}. This is in excellent agreement with Pilson's (1985) earlier estimate of 104.8 m^3 s^{-1}.

8.2.1 Oceanic Forcing

Narragansett Bay is primarily forced by water level variations, currents, salinity, and temperature through its three connections to Rhode Island Sound—the entrance to the Sakonnet River and the East and West Passages of Narragansett Bay proper. Tidal forcing in the adjacent Rhode Island Sound is described by Spaulding and Gordon (1982), Petrillo (1981), and Spaulding and Beauchamp (1983). Kincaid and Bergondo (in Chapter 10) provide detailed insight into the relationship of Narragansett Bay with Rhode Island Sound.

8.3 Circulation Dynamics in Narragansett Bay

8.3.1 Overview and Estuarine Type Classification

Circulation in the bay is dominated by lunar semi-diurnal (M$_2$) tides with a period of 12.42 h. Wind-driven flows are relatively modest and principally serve to enhance mixing in the bay. During strong wind events, however, the normal tidal-induced flows in various passages of the bay can be arrested or even reversed (Spaulding et al., 1999; Weisberg, 1976). In the relatively narrower upper reaches of the bay (Providence and Seekonk Rivers; Lower Taunton River), where there is significant freshwater input and the presence of dredged channels, density-induced flows become important and result in classic estuarine circulation, with up-bay transport at depth and down-bay transport at the surface.

The strength of the residual flows is directly related to the freshwater input rate. In these upper reaches, the water column is typically stratified, and the degree of stratification depends on the freshwater input rate. In the mid-section of the bay, the water column is either well mixed or very weakly stratified depending on tidal and wind conditions. Neap tides and weak winds favor stratification, while spring tides and strong winds favor well-mixed conditions. These conditions are important to the onset of hypoxic events, as discussed in detail in Chapters 11 and 12. The lower bay is typically well mixed. Based on Hansen and Rattray's (1966; 1970) method, which evaluates the salinity difference between surface and bottom values divided by the average value versus the ratio of net surface circulation divided by cross-sectional averaged freshwater flow rate, Narragansett Bay is a Type 2 system. In this estuarine type, the net flow reverses with depth, and both tidal diffusion and net circulation contribute to up-estuary transport of salt. The Providence and Taunton Rivers are characterized as Type 2b, because they both are stratified, and the mid and lower bay are Type 2a, because they tend to be well mixed.

8.3.2 Tidal Circulation

The circulation in Narragansett Bay and its various sub-bays is dominated by tidal forcing, making up more than 85% of the current variance (Turner, 1984; Spaulding and White, 1990; Ward and Spaulding, 2002). Sub-tidal frequency currents, as a result of wind- and density-induced forcing are of secondary importance, but may dominate during storms or hurricanes, or when mean flows are of primary interest (e.g., river input) (Turner, 1984).

Water level data have been collected by NOAA/NOS at two permanent benchmark stations, Newport and Providence, since the early 1900s. In addition, NOAA/NOS has made long-term water level measurements at Quonset, Fall River, and Conimicut Light and short-term (for several months) measurements at 20 stations throughout the bay in 1977 (Turner, 1984). Recently, NOAA, through its Physical Oceanographic Real Time System (PORTS) initiative, is making real-time water level measurements at Newport, Providence, Quonset Point, and Fall River via the internet.

NOAA, Turner (1984), Spaulding and White (1990) and several others have performed harmonic analyses of data to determine the magnitude and phase of principal tidal constituents in the bay. Table 8.2 summarizes the amplitudes and phases for principal semi-diurnal (M_2, S_2, N_2), diurnal (K_1, O_1, P_1), and tidal harmonic (M_4, M_6) components for Newport, Fall River, Quonset, Conimicut Light, and Providence based on NOAA's analysis (NOAA PORTS). The harmonic components are over-tides of the semi-diurnal tides, and caused by quadratic frictional dissipation as the tide propagates up Narragansett Bay. The speed for each component, in degrees per hour, is also provided. The phases are given in degrees, relative to Greenwich Mean Time (GMT). Local standard times (LST), however, are 5 h earlier than GMT. The phase shift depends on the speed of the individual constituent. The analysis shows that over 85% of the variance in surface elevation records is a result of the tides (Turner, 1984). The Defant Number $(M_2 + S_2)/(K_1 + O_1)$ is 5.4, and clearly shows that water levels in the bay are predominantly semi-diurnal. The M_2 amplitude is by far the largest (0.518 m), followed by S_2 (0.11 m) and N_2 (0.12 m). The M_4 (0.055 m), K_1 (0.065 m), and O_1 (0.052 m) components are considerably smaller but have comparable magnitudes. The values given here are for Newport; however, other stations show a similar scaling of magnitudes.

Table 8.2 Tidal constituent speeds (degrees per hour), amplitude (m) and phases (degrees GMT).

Tidal Constituent		Newport		Fall River		Quonset Point		Conimicut Light		Providence	
Name	Speed (degrees per hour)	Amp (m)	Phase (deg GMT)	Amp (m)	Phase (deg GMT)	Amp (m)	Phase (deg GMT)	Amp (m)	Phase (deg GMT)	Amp (m)	Phase (deg GMT)
M2	28.984104	0.518	2.2	0.614	8.7	0.538	5	0.593	7.2	0.643	9.5
S2	30	0.11	24.3	0.131	31.1	0.116	27.1	0.13	30	0.138	33.6
N2	28.43973	0.123	346.2	0.151	353.5	0.133	349.4	0.147	351.9	0.152	354.6
K2	30.082137	0.032	28.4	0.036	30.2	0.031	25.8	0.036	28.6	0.038	29.7
K1	15.041069	0.065	166.6	0.07	168.9	0.064	165.1	0.068	165.7	0.073	169.4
O1	13.943036	0.052	200.4	0.052	200.3	0.048	198.2	0.051	198.6	0.056	202.2
P1	4.958931	0.023	81.7	0.026	185.1	0.023	185.7	0.025	183.2	0.025	182.3
M4	57.968208	0.055	36	0.101	64.8	0.066	48.9	0.092	58.3	0.103	62.1
M6	86.952313	0.005	221.7	0.022	335.4	0.008	270.9	0.018	304.4	0.027	312.7

Amp = Amplitude. Data source: *http://co-ops.nos.noaa.gov/station_retrieve.shtml*

The semi-diurnal tides are responsible for the dominant 12.42-h tidal cycle observed in the bay, with two tides every 24 h 50 min. The diurnal components serve to modulate the tides, giving neap and spring tides every 14.8 days and the associated diurnal inequality. The tides are amplified by a factor of about 1.24 from Newport to Providence for semi-diurnal components, 1.1 for diurnal components, and 1.85 and 5.4 for the M_4 and M_6 components, respectively. Turner (1984) plots the amplitude versus up-bay distance of all the major constituents and shows that amplification increases progressively with distance. Table 8.3 provides a summary of the observed mean and spring tidal ranges and mean tidal level at stations progressing from the mouth to the head of the bay. The mean level is referenced to MLLW. Tidal amplification is seen to progressively increase with distance up the bay. Spring tidal range at each station is approximately 25% higher than the mean range.

The phase lags between Providence and Newport are 7–9° for semidiurnal tides, 2–3° for diurnal tides, and 26° and 91° for the M_4 and M_6 components, respectively. For the dominant M_2 component, high tide at Providence is 15 min later than at Newport. Turner (1984) plots the phase lag versus up-bay distance

Table 8.3 Mean and spring tidal range for various locations in Narragansett Bay (*http:// co-ops.nos.noaa.gov/station_retrieve.shtml*).

Location	Latitude/Longitude	Mean Tidal Range (m)	Spring Tidal Range (m)	Mean Tidal Level (m)
Narragansett Pier	41° 25.3′ 71° 27.3′	1.16	1.43	0.61
Sakonnet	41° 27.9′ 71° 11.6′	0.95	1.13	0.52
Anthony Point, Sakonnet River	41° 38.3′ 71° 12.7′	1.16	1.46	0.64
Beavertail Point	41° 27.1′ 71° 24.1′	1.07	1.31	0.58
Castle Hill	41° 27.8′ 71° 21.7′	1.01	1.25	0.55
Newport	41° 30.3′ 71° 19.6′	1.06	1.34	0.57
Conanicut Point	41° 34.4′ 71° 22.3′	1.16	1.43	0.61
Wickford	41° 34.3′ 71° 26.7′	1.16	1.43	0.61
Prudence Island, (south end)	41° 34.8′ 71° 19.3′	1.16	1.46	0.61
East Greenwich	41° 39.9′ 71° 26.7′	1.22	1.52	0.64
Bristol Ferry	41° 38.2′ 71° 15.3′	1.25	1.55	0.67
Bristol, Bristol Harbor	41° 40.1′ 71° 16.7′	1.25	1.55	0.67
Bristol Highlands	41° 41.8′ 71° 17.6′	1.28	1.59	0.73
Fall River, Massachusetts	41° 42.2′ 71° 09.9′	1.34	1.68	0.73
Providence, State Pier no.1	41° 48.4′ 71° 24.1′	1.34	1.72	0.73
Pawtucket, Seekonk River	41° 52.1′ 71° 22.8′	1.40	1.77	0.76

for all the major constituents, and shows that the phase lag increases progressively with distance. Turner's analysis shows that the amplitudes and phases along the East-West Passages, which merge in the Providence River and the lower Sakonnet River, do not vary significantly across the width of the bay. The co-amplitude (constant amplitude) and co-phase (constant phase) lines are, therefore, normal to the central axis of the bay. The exceptions are the tides in Mount Hope Bay and Fall River. In this case, the principal path for tidal waters to reach Mount Hope Bay and the lower Taunton River is via the passage between Bristol and Aquidneck Island. Flows directed north up the Sakonnet River are restricted by a very narrow passage at the inoperative railroad bridge causeway system. If the flow path distance up the East Passage to Fall River is used, rather than the distance northward from the mouth of the bay, then the Fall River tidal constituent data are consistent with those for similar distances up bay—1.2 amplification, 13.5-min phase lag for M_2 relative to Newport.

The amplification of tide in the bay is primarily due to its standing wave nature and to the decrease in the width and depth from the mouth to the head of the bay. The small phase lag for semi-diurnal and diurnal components is a direct result of the fact that bay length (40 km) is small (1/10) as compared to the tidal (semi-diurnal) wave length (400 km).

Based on extensive current observations in the Providence River (Turner, 1984) and Mount Hope Bay (Spaulding and White, 1990) and more limited observations near Quonset (Berger-Maguire, 2004) and Sakonnet River (Kim and Swanson, 2001), it may be concluded that 80–95% of current variance is dominated by the tides, with most of the remainder explained by sub-tidal frequencies. In the upper reaches of the Providence and Taunton Rivers, the tidal current energy is reduced to 50–60% due to the standing wave nature of the tides, with the sub-tidal and higher frequencies comprising 15–20% each.

Figure 8.2 shows the estimated upper level maximum flood and ebb tidal currents for select stations throughout the bay based on NOAA observations (see also NOAA, 1971). Flood and ebb currents are rectilinear at most stations. The tidal currents are observed to vary considerably, and are strongly dependent on the location. On a broader scale, currents are observed to be stronger in the deeper, lower East Passage than in the shallow Sakonnet River and West Passage, and appear weaker up the Providence River. In the central portions of the bay, the peak currents are in the range of 15–20 cm s^{-1}. Stronger currents are observed in narrower passages of the bay, such as the Sakonnet River Railroad Bridge (1 m s^{-1}) and the entrance to the East Passage (0.75 m s^{-1}), and weaker currents are observed in Wickford Cove, Greenwich Bay, and the upper reaches of the Providence River due to the standing wave nature of the tides. The current directions typically follow the large-scale topography of the bay, but can be modified by local bathymetry.

Table 8.4 provides the amplitudes and phases for the principal tidal constituents of currents at select stations around the bay. The list is illustrative, not exhaustive. The constituent analyses shown are for mid-depth locations and for the currents along the local major axis of current time series. The tidal currents

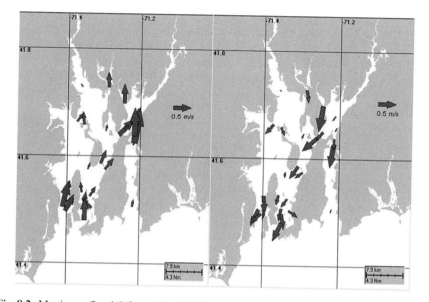

Fig. 8.2 Maximum flood (left panel) and ebb (right panel) tidal currents in Narragansett Bay based on NOAA/NOS data.

are dominated by the M_2 component with speeds in the range of 15–23 cm s^{-1}. Other semi-diurnal components (N_2, S_2) are significantly smaller at 3–7 cm s^{-1}, while the diurnal components (K_1, O_1) are 2 cm s^{-1}. The tidal harmonics (M_4, M_6) are comparable to the smaller semi-diurnal components. Strong tidal harmonics combined with the principal lunar semi-diurnal tides result in double-peaked flood and single-peaked ebb flows in the bay (see below). The currents at the Sakonnet River Railroad Bridge stand out from the rest due to the much stronger flows and the very large tidal harmonics. This can result in flows that ebb, which should otherwise be flood conditions during neap tides. This phenomenon is described later in this chapter.

A comparison of lunar semi-diurnal phases for surface elevations and currents shows currents preceding elevation changes by about 80°. This is consistent with standing wave theory, which predicts a 90° phase lag between the surface elevation and the currents (Officer, 1976).

8.3.3 Model Predictions and Validation

The earliest effort to model tidal circulation in the entire Narragansett Bay system, including the Sakonnet River, was performed by Swanson and Spaulding (1975; 1977) using a two-dimensional, vertically-averaged hydrodynamic model with square grid spacing of 370 m. In an earlier study, Hess and White (1974) performed tidal simulations for the main part of the bay, excluding

Table 8.4 Tidal constituent speeds (degrees per hour), major axis current amplitudes (m s⁻¹) and phases (degrees GMT).

Tidal constituent		Gaspee point[1]		Conimicut point[1]		Quonset point[2]		Sakonnet Railroad bridge[3]		Mount Hope bridge[4]	Brightman bridge[4]
Name	Speed (degrees h⁻¹)	Amp (m s⁻¹)	Phase (deg GMT)	Amp (m s⁻¹)	Phase (deg GMT)	Amp (m s⁻¹)	Phase (deg GMT)	Amp (m s⁻¹)	Phase (deg GMT)	Amp (m s⁻¹)	Amp (m s⁻¹)
M2	28.9841	18.3	155	16	140	16	140.8	92	244	20.3	22.6
S2	30	6.9	150	3.7	143	3	133	23	266	2	2
N2	28.4397	6.6	181	5	84	2.8	116.6				
K1	15.0411	2	296	2.1	33			1.9	50.6		
O1	13.943	1.7	97	1	66			9	124		
M4	57.9682	5.3	70	3.4	67	4.2	50.7	50	318	8	8
M6	86.9523	2.6	167	1.9	172			33	228	4	4

[1]Turner (1984); data set E1D2—Gaspee Point East, mid-depth; data set E1D6—Conimicut Point, mid-depth.
[2]Berger-Maguire (2004), Quonset Point PORTS station ADCP mid-depth (5 m).
[3]Kim and Swanson (2001), Sakonnet Railroad Bridge, mid-depth.
[4]Spaulding and White (1988).
Amp = Amplitude.

the Sakonnet River and Mount Hope Bay. Swanson and Spaulding's model was forced with tidal time series based on estimated amplitudes and phases of the principal tidal constituents at the mouth of the bay. Model predictions were summarized in the form of tidal current and amplitude charts at hourly intervals through one complete tidal cycle referenced to high water at Newport. These charts were designed to be comparable to the Narragansett Bay tidal current charts produced by NOAA (1971). The model predictions were compared against the NOAA tidal current atlas data and water level data collected by NOAA in Providence. The principal findings of the study were that, as the tidal wave propagates up the bay, the model correctly predicts the amplification and phase lag for the tide at Providence and Fall River, and shows that tidal current at any location in the bay scales linearly with the tidal amplitude at the mouth of the bay. One very interesting finding was that the delay in the time of high water, relative to Newport, at a given location in the East Passage was shorter than that in the other two passages at similar distances from the bay mouth. This difference was due to the fact that the speed of tidal propagation is faster in the deeper, lower, and mid-East Passage than in the shallower West Passage and Sakonnet River. This time difference may be progressively removed as the tide progresses up the bay by cross-bay transport among the islands.

A recent bay-wide modeling effort was the development of a real-time now-casting and forecasting system (Opishinski and Spaulding, 2002) developed by Ward and Spaulding (2002). This system used a three-dimensional, boundary-fitted hydrodynamic model (Muin and Spaulding, 1997a) to predict tidal circulation in the bay. The model domain covered Long Island Sound, Block Island Sound, Rhode Island Sound, and Buzzards Bay at coarse resolution (several kilometers), and Narragansett Bay at high resolution (200 m). The model was forced with data for the amplitudes and phases of the seven largest tidal constituents based on tidal co-range and co-phase lines for the southern New England Bight. Model predictions were compared with water level observations at eight locations in the bay—Newport, Quonset, Conimicut, Providence, Fall River, and three locations in the Sakonnet River—and Acoustic Doppler Current Profiler (ADCP) current measurements at four stations—Sakonnet, Fall River, Quonset, and Providence—for the period September to December in 2001. Model predictions of the surface elevation were compared to the observations in terms of the amplitudes and phases of principal tidal constituents.

For M_2 tides, model-predicted amplitudes of sea surface levels were generally within 0–2 cm (maximum difference of 3%), and 2–3 min difference in phase. For tidal constituents with smaller initial amplitudes, take M_4 as an example, the amplitude differences were similar to that for M_2 at 1.4 cm (maximum difference of 12%) and 5–10 min difference in phase. Model-predicted and observed water levels were regressed against one another and showed correlation coefficients of 0.992 or higher, and root mean square (RMS) errors of 2–3.9%. Model-predicted tidal currents along the major axis of flow were compared against observed values at ADCP stations, which gave correlation coefficients of 0.84 or higher for all stations except one at the Providence River. For this location, the correlation coefficient was very low (0.32). This was a result of weak currents at the measurement site near the hurricane barrier. The RMS errors were 8–17% for all sites except at Providence, which had much larger errors (24.5%). Model-predicted maximum ebb currents at mid-depth are presented in Fig. 8.3. The flood currents (not shown) are of same speed, but in opposite directions. The predictions show that tidal currents are rectilinear with flood and ebb in opposite direction currents, and are strongest in the lower East Passage, the narrows at Sakonnet, and the lower Taunton River. Peak currents are in the range of 15–20 cm s^{-1} at most other locations. Model predictions are available from an on-line digital atlas giving hourly predictions referenced to high tide *(www.narragansettbaymap.com)*. The user simply enters a date and time, and the system determines the time of high tide for that day, and then provides an appropriate field from the atlas.

The most recent modeling study of bay-wide circulation was reported as an application in the presentation of a new model by Sankaranarayanan and Ward (2006). It is a three-dimensional, orthogonal, coordinate semi-implicit prognostic hydrodynamic model in spherical coordinates driven by tides, winds, and density differences. After presenting the model equations and solution approach, along with the comparison of model predictions against analytic solutions, the model was applied to Narragansett Bay. The model predicted surface elevations, three-dimensional instantaneous and mean currents, salinities and temperatures, and compared them to the observations. The model accurately captured the spring neap and low-frequency fluctuations in surface elevation, as well as the ebb tide dominance and double flood characteristics of horizontal currents and the vertical structure of both instantaneous and residual (mean) currents. The model reproduced the temperature and salinity structure at both tidal and lower frequency time scales.

A three-dimensional, finite-volume model application to Narragansett Bay was performed by Zhao et al. (2006). This model used very high horizontal resolution to capture transient eddies generated by irregular coastlines, and showed that model-predicted eddies were qualitatively consistent with limited observations.

Muin and Spaulding (1994; 1997b) applied a high-resolution, boundary-fitted, three-dimensional hydrodynamic model (Muin and Spaulding, 1996, 1997a) to predict tidal- and wind-driven circulation in the Providence River. The vertical eddy viscosity was parameterized in terms of an empirical relationship, which is dependent on model-predicted turbulent kinetic energy, Richardson number, and turbulent length scale. The model grid size varied between 50 and 250 m, depending on the location, and allowed the resolution of dredged channel and narrowest sections of the river. The model was forced by lunar semi-diurnal tides specified at the open boundary to the south. No river forcing was assumed. Model-predicted currents were in good agreement with Turner's (1984) observations and gave rectilinear tidal flows with maximum values of 17 cm s^{-1} at Conimicut Point, 16 cm s^{-1} at Gaspee Point, 10 cm s^{-1} at Field's

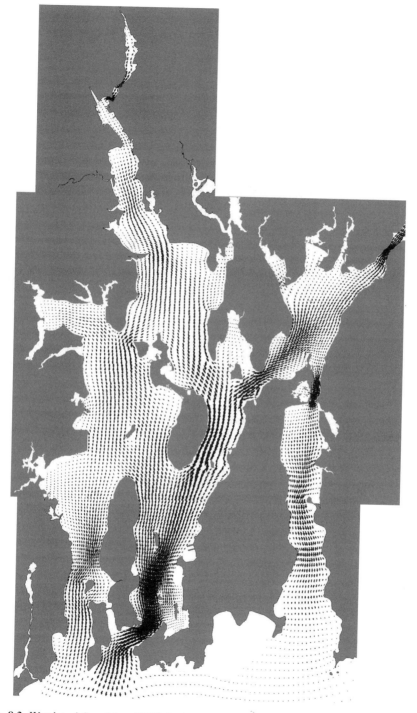

Fig. 8.3 Ward and Spaulding (2002) hydrodynamic model predicted maximum ebb mid-depth tidal currents for Narragansett Bay.

Point, and 25 cm s^{-1} at Cold Spring Point. The strength of the currents was correlated with the width of the channel and percentage of width that is represented by the dredged channel. The flow pattern at the bottom was similar to that at the surface, but reduced by about 20–30% due to enhanced bottom frictional dissipation. Muin and Spaulding (1997b) studied the vertical structure of flow at deep (Field's Point) and shallow locations (Cold Spring Point). At both stations, turbulent kinetic energy was generated from flow-induced bottom stress and diffused up into the water column. This led to decreasing current speeds with distance from the surface. At the deep water site, the thickness of the tidal boundary layer (5 m) was about half the water depth (10 m), and hence the currents at the bottom lagged those at the surface. This was particularly noticeable at slack tide. For a shallow station (3m), the boundary layer was greater than the water depths, and there was no phase lag between surface and bottom currents.

Swanson et al. (1999, 2006) and Spaulding et al. (1999) report on the application of a three-dimensional, boundary-fitted hydrodynamic model to predict circulation in Mount Hope Bay. The focus of the application was to study the impact of the intake and discharge through cooling water from the Brayton Point power plant. The plant generates 1600 MW and uses 57 m^3 s^{-1} of water for condenser cooling. The model included the central and upper portion of Narragansett Bay and Sakonnet River at coarse resolution, and Mount Hope Bay at a much higher resolution (200–300 m). In the immediate vicinity of the power plant, a 50–100-m grid was used. The horizontal and vertical eddy viscosities were held constant at 5 m^2 s^{-1} and 30 cm^2 s^{-1}, respectively. Similar values for eddy diffusivities for heat and salinity were 5 m^2 s^{-1} and 1–4 cm^2 s^{-1}. The model employed 11 sigma layers to represent the vertical structure. The model was forced with observed Taunton River flows, winds derived from Green Airport, and tidal constituent-based estimates of the water level at open boundaries in Narragansett Bay and upper Sakonnet River. Information on the flow rates and temperature of discharge from the power plant was based on plant records. Solar radiation and air temperature were obtained from a station on Prudence Island operated by the Rhode Island Department of Environmental Management for the National Estuarine Research Reserve. Salinity, temperature, currents, and water level data were collected 1 m below the sea surface and 1 m above the bottom at Gardners Neck, off Brayton Point, Borden Flats, and at the center of the Mount Hope Bridge in the month of August in 1996 and 1997.

The model predicted that the tidal currents were strongest in the deeper areas in the lower Taunton River and along the central axis of the dredged channel from the lower Taunton River to the Mount Hope Bay Bridge. Currents were substantially weaker in the shallow area on the west side of the dredged channel and in the vicinity of Brayton Point. Currents generally flooded to the northeast and ebbed to the southwest, following the central axis of the dredged channel and the bay. The influences of the power plant intake and channel discharge were clearly observed in the vicinity of Brayton Point. The model suggested that the power plant discharge was affected by flows exiting the lower Taunton River on the ebb tide, but the river discharge had little influence on the flows at flood tide.

A comparison of model predictions against observations showed an excellent ability to represent sea surface variations (within 2%, $R^2 = 0.97$), which were dominated by the tides (85–95% of current variance). The model also correctly predicted the amount of tidal energy in the semi-diurnal, diurnal, and the two principal tidal harmonic components. At the Brayton Point location, the model predicted that the tidal currents at the bottom were slightly lower than at the surface, which is consistent with observations. The model underpredicted the maximum current speeds (19.6 cm s^{-1} observed and 14.9 cm s^{-1} predicted at surface; 14.1 cm s^{-1} observed and 10.5 cm s^{-1} predicted at bottom). A comparison of model predictions against observations at other stations showed similar results. In general, the stronger the currents, the better the model data comparisons.

Hydrodynamic modeling has been performed in various parts of the bay for specific studies. These include studies of the tidal circulation of the Providence River by Muin and Spaulding (1994; 1997b), Swanson and Mendelsohn (1994), Mendelsohn et al. (1995), and Swanson et al. (1993; 1996); of Mount Hope Bay by Huang (1993), Huang and Spaulding (1994; 1995a,b), Spaulding et al. (1999), and Swanson et al. (2006); of Greenwich Bay by Spaulding et al. (1998); of the Sakonnet River and Mount Hope Bay by Kim and Swanson (2001); of the central section of the West Passage in the vicinity of Quonset Point by Swanson et al. (1987) and Berger-Maguire (2004); and of the central East Passage and lower Providence River by Swanson and Mendelsohn, (1993). Bergondo (2004) and Bergondo and Kincaid (2003) performed additional modeling investigations on the bay. Collectively, these studies have improved our understanding of specific areas, but have not changed our fundamental understanding of the bay's tidal circulation.

8.3.4 Impact of Tidal Harmonics (Double-peaked Flood, Single-peak Ebb)

In analyzing current observations on Narragansett Bay, Haight (1936) noted that tidal currents throughout the bay displayed a distinct double-peaked flood and single-peaked ebb. He noted that this feature was more pronounced as one moved further up the bay, and was largest in the vicinity of the Sakonnet River Bridge. His analysis of the current data showed that this effect was due to the significant magnitude of M_4 and M_6 tidal harmonics, and the fact that both M_4 and M_6 tidal currents were ebbing at the time M_2 tides were flooding. He attributed the increase in the magnitudes of M_4 and M_6 components with distance up the bay to the fact that the bay's principal longitudinal resonant period (5.72 h) was between the periods of these two components (M_4: 6.21 h; M_6: 4.14 h). Gordon (1982) simulated the response of the bay to tidal forcing using a linear damped system, and found a resonant period between 3.8 and 5.0 h, depending on the value of frictional damping assumed. These results were consistent with Haight's (1936) analysis. Turner (1984) performed an extensive analysis of current data that he collected in the Providence River, and his results confirmed the presence of double-peaked flood and single-peaked ebb at both surface and bottom stations in the dredged channel. As a proxy for tidal currents, Turner (1984) plotted the amplitude and phase of principal tidal constituents derived from surface elevation records collected throughout Narragansett Bay, normalized by their magnitudes at the mouth of the bay, as a function of distance from the mouth of the bay. His results clearly showed strong amplification of M_4 and M_6 components by three and five times, respectively, at the head of the Providence River relative to the mouth. This compares to an amplification of the M_2 tide by only 1.28.

The recent hydrodynamic modeling studies of the bay, or portions of it, and more recent current measurement programs (Berger-Maguire, 2004), have shown the ubiquitous nature of double-peaked flood and single-peaked ebb currents. This is due to the phase changes of M_4 and M_6 components relative to the M_2 tide moving up the bay. However, there is no strong asymmetry in flood versus ebb current speed. Since the total flood tidal flux must equal the ebb tidal flux, there is no direct effect on net pollutant transport.

8.3.5 Tidal Residuals

Simulations of bay circulation have been performed with and without the inclusion of non-linear convective acceleration terms in the governing equations. These cases have shown that tidal- induced residual flows in the bay are generally very small. Both non-linear bottom friction and convective terms are relatively insignificant since velocities are small—and the velocity products even smaller—except in narrow passages like Sakonnet River. In these locations, the residuals sometimes show eddy patterns, but the effects of these patterns on mean pollutant transport are small. This is consistent with current meter observations collected in the Providence River (Turner, 1984), in Mount Hope Bay (Spaulding and White, 1990), lower Taunton River (Swanson et al., 2006), and in the Quonset shipping channel (Berger-Maguire, 2004)

8.3.6 Wind-driven Circulation

8.3.6.1 Impact on Circulation and Intra-passage Transports Paths

Haight (1936) noted that winds could temporarily exert a profound influence on circulation in Narragansett Bay, but his records were too short to quantify the relationship. Weisberg (1972; 1976) obtained a 51-day-long record of bottom currents in the shipping channel just south of the entrance to the Providence River. Weisberg's analysis of data showed that 45% of the energy was in the sub-tidal frequencies with winds dominating this portion of the record. Wind-induced bottom currents varied from 25 cm s^{-1} up bay to 10 cm s^{-1} down bay. Winds from the south or southeast tended to force bottom currents down the bay, while winds from the north had the opposite effect.

Turner's (1984) analysis of his current measurements at surface and bottom stations in the Providence River showed that 8–25% of current variance occurred at sub-tidal frequencies. Turner low-pass filtered his current observations and compared them to simultaneous wind observations. He found a strong relationship with wind-forced currents at both the surface and bottom stations flowing in the direction opposite to the wind and proportional to wind strength. Currents from the bottom meter were generally stronger than that from the surface meter (3 m below the surface). These results indicated that both meters were in the return portion of the flow, and that the flow in the direction of the wind was restricted to several meters of the upper water column. In two very strongly forced, short-term events, the upper level currents were observed to flow in the direction of wind while the bottom meter showed a flow opposite to wind direction. It appeared that, for strongly forced events, the wind-generated turbulence penetrates deeper into the water column, and deepens the level of no motion. Turner found that, while wind-induced flows could be comparable or larger than both tidal- and density-induced flows for short periods of time, there was no evidence that it significantly altered the mean flow characteristics of the Providence River.

Gordon and Spaulding (1987) applied a three-dimensional hydrodynamic model to study tidal- and wind-driven circulation in Narragansett Bay. The model employed a Galerkin spectral method (Legendre polynomials) in vertical and explicit finite differences in horizontal space and time. The model covered the entire bay, including the Sakonnet River and Mount Hope Bay, using a 926-m^2 grid system. This resolution was adequate to represent the main passages in the bay, but not the smaller sub-sections of the bay. The model, when first applied to predict the tidal dynamics of the bay, was able to predict the lunar semi-diurnal tides within 2% of the observed amplitude and 2° in phase.

Simulations with Gordon and Spaulding's model were run with steady wind forcing of 1 dyne cm^{-2} ($\sim$7.5 m s^{-1}) along (north-south) and across (east-west) the axis of the bay. After steady-state conditions were achieved, an approximately linearly-varying setup was predicted in the bay, reaching 4 cm in the Providence River. For the easterly directed winds, a set down of -2 cm was predicted on the western side of the bay and a set up of 1.5 cm on the eastern side in Mount Hope Bay. For the northerly directed winds, the depth mean currents show an inflow into the bay in the West Passage and Sakonnet River and a return flow out of the bay in the East Passage. The West Passage inflow was diverted into the East Passage south and north of Prudence/Patience Islands, and returned to Rhode Island Sound via the East Passage. In the Providence River, the mean flows are up-river in the shallows and down-river in the dredged channel. When looking at the vertical structure of the velocity, it was observed that the flows were in the direction of the wind at the surface. The bottom currents were in the direction of winds in shallow areas (West Passage, Sakonnet River), but opposed to the wind direction in deep waters (East Passage, dredged channel in the upper bay, and Providence River). In deeper waters, the pressure gradient dominated the vertical momentum diffusion in the local momentum balance throughout the water column. Peak surface and bottom currents were 20 and 15 cm s^{-1}, respectively. For the easterly directed winds, the mean flows were into the bay in the West Passage, and out of the bay in the Sakonnet River. The flows were very small in the East Passage. In the upper bay and Providence River, the mean flow was once again up river in the shallows and down bay in the dredged channel. In addition, the surface currents were in the direction of wind, suitability modified by channel geometry, and bottom currents in the direction or opposed to winds depending on the depth of the passage. Maximum currents for the easterly directed winds were about half those for the northerly directed winds. Additional simulations were performed with and without tides, which showed that a 25% reduction in current speeds is possible with tides. This reduction was possible due to the enhanced effective frictional dissipation of the combined (tidal and wind) flow at the seabed.

The impact of wind direction on net flows between passages was also assessed. Gordon and Spaulding (1987) predicted steady-state flows for eight wind directions with constant 1 dyne cm^{-2} ($\sim$7.5 m s^{-1} wind speed) forcing. They explored a variety of values of vertical eddy viscosity, and whether tidal forcing was employed. The results were presented in terms of volume flux through nine transects across the various passages in the bay. These passages are as follows: West (1) and East (2) Passage, Sakonnet River (3) at the bay mouth, Jamestown to Prudence Island (4), southern tip of Prudence to Aquidneck Island (mid-East Passage) (5), entrance to Mount Hope Bay (6), Patience Island to Greenwich Bay (upper West Passage) (7), Prudence Island to Bristol (upper East Passage) (8), and Conimicut Point (entrance to Providence River) (9). Model-predicted flows for the major passages are shown in Table 8.5 for five separate simulation cases where wind direction, vertical eddy viscosity, and inclusion of tidal forcing were varied. The tidal flood and ebb fluxes for each transect are also provided for reference. The predicted transports were found to be symmetrical with wind direction. Flows into (out of) the mouth of the West Passage are balanced by flows out of (into) the mouth of the East Passage, with flows through Sakonnet River giving only a minor contribution in either case. There is no mean transport into/out of the Providence River. The predicted transport paths are complicated, but, in general, winds with a northerly (southerly) component result in inflows (outflows) in the West Passage and outflows (inflows) in the East Passage. Simulations show that tidal transports in the various sections range from 1.5 m^3 s^{-1} (entrance to the Providence River at Conimicut Point) to 13 m^3 s^{-1} (East Passage at the entrance to the bay). Wind-driven flows for constant wind-forced cases have magnitudes about

Table 8.5 Calculated transport (10^3 m^3 s^{-1}) at selected cross sections for the constant wind forcing cases and peak tidal flow conditions. Values on second line are percentages relative to flood tide transport (Gordon and Spaulding, 1987).

Section	Case number					Tidal flood	Tidal Ebb
	1	2	3	4	5		
1	1.48	0.98	3.05	1.31	0.85	6.93	−6.66
	21%	14%	44%	19%	12%		
2	−1.96	−1.3	−4.25	−0.86	−0.49	13.04	−12.22
	15%	10%	33%	7%	4%		
3	0.47	0.31	1.2	−0.46	−0.37	5.1	−4.93
	9%	6%	23%	9%	7%		
4	−0.84	−0.47	−1.68	−0.86v	−0.51	3.41	−3.26
	25%	14%	49%	25%	15%		
5	−1.14	−0.84	−2.56	−0.01	0	6.26	−5.74
	18%	13%	41%	0%	0%		
6	0.49	0.32	1.2	−0.45	−0.35	−1.19	2.07
	25%	17%	63%	24%	19%		
7	0.66	0.52	1.37	0.47	0.35	2.76	−2.57
	24%	19%	50%	17%	13%		
8	−0.66	−0.52	−1.37	−0.47	−0.35	1.86	−1.78
	35%	28%	73%	25%	19%		
9	0	0	0	0	0	1.46	−1.36
	0	0	0	0	0		

Case summary

Number	Wind stress	Wind direction* (dyne cm^{-2})	Vertical Eddy viscosity (cm^2 s^{-1})	Tidal forcing
1	1	N	65	Yes
2	1	N	325	Yes
3	1	N	65	No
4	1	E	65	Yes
5	1	E	325	Yes

*Toward which wind blows.

10–35% of the tidal fluxes, and are lowest for the passages with largest transport and highest for the passages with lowest transport. Simulation results show that transport increases with increasing wind stress. The scaling is slightly less than linear due to quadratic bottom friction losses. There has been no verification of the above predictions with field observations, although one short-term wind event, summarized below, is consistent with these predictions.

On May 13, 1995, two MetOcean GPS tracking buoys were deployed in the West Passage approximately 100 m south of the Jamestown Bridge (Spaulding *et al.*, 1996). The buoys were designed to track near-surface waters (about 1 m) and had minimal sail area to minimize the effects of wind. The buoys were followed from approximately 10:00–22:30 local time (Fig. 8.4). High tides at Newport for the day were predicted at 07:08 and 19:33 with ranges of 1.37 and 1.58 m, respectively. The mean tidal range at Newport was 1.08 m, and hence conditions on the day of experiment corresponded to spring tidal conditions. The early winds were weak, but soon reached 9.4 m s^{-1} from the northeast and were sustained at this level throughout much of the tracking period. The buoys initially moved southward in the West Passage driven by the spring-enhanced ebbing tides, and exited the mouth of the bay by 12:02. The transport rate of the buoy was well in excess of the tidal current speeds. The tide should have turned

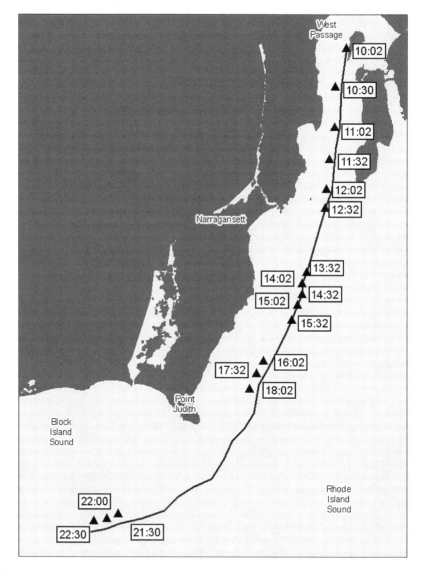

Fig. 8.4 Comparison of observed (triangles, with times noted) and model predicted (solid line) trajectory for a drifting surface buoy on May 13, 1995 shown in center panel (from Spaulding *et al.*, 1996).

at about 13:00 when the buoys were off Black Point. Instead, the buoys continued on their shore parallel, southerly trajectory and eventually rounded Point Judith. They were retrieved south of the Point Judith breakwater at 22:30 (Fig. 8.4).

Simulations were performed using a two-dimensional, vertically-averaged hydrodynamic model of Narragansett Bay and nearby Rhode Island and Block Island Sounds. Four different simulations were performed using the following forcing: C1—lunar semi-diurnal forcing at the open boundary, C2—58 tidal constituents forcing at the open boundary, C3—observed water level forcing for the experimental period, and C4—observed water level forcing and constant winds from the northeast at 9.4 m s^{-1}. Trajectory simulations were performed by varying the drift factor and angle and the hydrodynamic forcing field. Trajectory predictions with a 3.5% drift factor and C4 currents were in excellent agreement with observations. Simulations with any other current forcing and drift factor or angle gave trajectories that were confined to the bay. A close inspection of the C4 hydrodynamic model predictions showed that the winds were sufficiently strong to effectively arrest the flood tide currents throughout the water column (based on the vertically averaged model results) in the West Passage and nearby areas directly to the south along the coastline of Narragansett. These predictions explain the fact that the buoy trajectories continued to move to the south even during what should have been flood (northerly) currents. As predicted by the simulations of Gordon and Spaulding (1987) and Spaulding *et al.* (1996), the flood tide currents were predicted in the East Passage, however, weakened by the opposing winds. This is a dramatic illustration of the impact of wind forcing on the circulation in the bay.

Weisberg and Sturges (1976) collected current data at four locations in the vertical (2, 3.9, 7.3, and 9 m below the sea surface, water depth of 12 m) at a station in the center of the West Passage located off Rome Point from approximately October 20 to November 25, 1970. An analysis of the resulting data showed that tidal currents dominated the record and displayed a phase advance of 0–3 h with depth. After removing tidal variations by filtering the record, they found non-tidal currents to be an order of magnitude less than the tidal currents that were well correlated with the winds. For typical weather period forcing (2–3 days), they found the coherence between the longitudinal winds and surface currents to be high (0.8), with current in the direction of, and lagging the wind by, 3 h. They observed numerous instances where net flows over the entire water column were either into or out of the bay, with typical wind-induced transports of 500 m^3 s^{-1}. Averaged over a 1-month-long period, they estimated a mean surface current of 1.2 ± 1.6 cm s^{-1}, with the large error bounds attributed to large uncertainty in the net currents.

Gordon and Spaulding (1987) applied their hydrodynamic model to predict wind-forced flows observed in the study by Weisberg and Sturges (1976). The model was forced by tides at the open boundary and by winds as reported at Green Airport (*http://www.ncdc.noaa.gov/oa/ncdc.html*). The observations included two strong wind-forcing events, with winds of 7.5 m s^{-1} toward the south during the first event of October 23, 1970 and of 8 m s^{-1} toward the southwest during the second event of October 26, 1970. During these strongly forced events, currents were in the same direction as the wind at all depths, but strongest at the surface (7–9 cm s^{-1}) and decreasing toward the bottom. The model-predicted currents were in good agreement with observations at the surface, but overpredicted the current strength at depth. This may have been because the vertical eddy viscosity used in the model (65 cm^2 s^{-1}) was too large. For more weakly forced flows, and periods after strongly forced southerly flows when water was returning to equilibrium conditions, the model predictions were in poor agreement with the observations. This may be partially attributed to the fact that non-local wind forcing (through wind-induced water level fluctuations at the bay mouth) was not available, and hence not used to force the model.

Muin and Spaulding (1997b) used the three-dimensional model described earlier to perform wind-forced simulations of the Providence River. The model was forced with both lunar semi-diurnal tides and along- (toward northwest) and across- (toward northeast) axis winds (2.3 m s^{-1}) in order to ensure that non-linear interactions between the tides and winds were represented in the predictions. Simulations were run until steady-state conditions were achieved. The model predicted currents in the direction of wind (up river) in shallow areas on both sides of the dredged channel and opposed to the wind at all depths in the dredged channel. Currents at the surface were predicted to be stronger than at depth, except in the dredged channel where bottom currents were stronger. The pattern is most pronounced where the channel is narrow as compared to the width of the river (e.g., Conimicut Point to Field's Point). At Field's Point, where dredged channel width and width of the river are comparable, the surface current is considerably weaker since up-river wind stress is almost balanced by the down-river pressure gradient. Peak currents are about 3 cm s^{-1}. Simulations for the northeast winds show a very complex pattern with no clearly discernable trends, and peak currents of 1–2 cm s^{-1}.

8.3.7 Stratification, Mixing, and Density-induced Circulation

Poor water quality in the Seekonk and Providence Rivers had been evident from Hicks' (1959a,b) early work, and reconfirmed by several subsequent investigations (Oviatt, 1980; Doering et al., 1988; Desbonnet and Lee, 1991; Granger, 1994; Kester et al., 1996; Kincaid, 2001; Bergondo et al., 2005). The water column in the upper bay is principally stratified by freshwater discharges from the major rivers; the higher the freshwater discharge, the stronger the stratification. Figure 8.5 shows the observed surface and bottom salinity from the head of the Seekonk River with distance down bay. The upper reaches are highly stratified (16% difference between surface and bottom) but decrease with distance down the bay (2% at Conimicut Point).

A result of this stratification is that, in summer, the weaker winds and lower saturation levels for dissolved oxygen, combined with the decay of phytoplankton from prior blooms and benthic oxygen demand, result in the depression of bottom dissolved oxygen levels. Chapters 11 (Saarman *et al.*) and 12 (Deacutis) of this volume provide critical detail and analysis of contemporary summer dissolved oxygen conditions in upper Narragansett Bay. In winter, the near-bottom waters remain well oxygenated due to reduced system metabolism, higher dissolved oxygen saturation due to lower temperatures, reduced thermal stratification, and enhanced vertical mixing caused by stronger winds. Similar patterns of enhanced vertical stratification in summer followed by reduced stratification in winter are also evident in the lower Taunton River (Spaulding and White, 1990).

Recently, Kester *et al.* (1996), Deacutis (1999), Bergondo (2004), and Bergondo *et al.* (2005) have shown that, in summer, bottom water dissolved oxygen levels appear to move much further down the bay than previously considered. These authors have also shown that low dissolved oxygen events are strongly correlated with neap tidal conditions and thus with periods of reduced vertical mixing.

This observation can be explained by noting that the mean value of vertical mixing scales approximately as the horizontal velocity squared (Munk and Anderson, 1948; Bowden and Hamilton, 1975; Officer, 1976). Effects of stratification are parameterized by multiplying this basic formulation by $(1 + \alpha Ri)^{-n}$, where Ri is the Richardson number, defined as the ratio of potential energy to turbulent energy produced by shear stresses; α is a scaling coefficient;

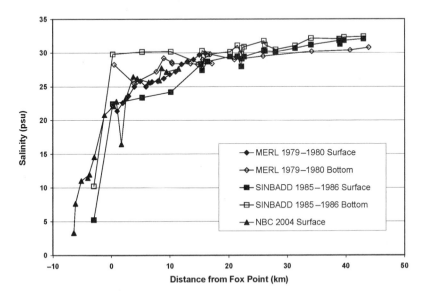

Fig. 8.5 Observed surface and bottom salinity versus distance from the head of the Seekonk River.

and n ranges from $1/2$ to $7/4$. In the mid-section of the bay, where water column stratification is limited, vertical mixing is dominated by the velocity square term, and hence the mixing is substantially reduced with a reduction in tidal range. As an example, if the vertical mixing coefficient is on the order of 15 cm^2 s^{-1} for mean tidal range, the value is reduced to 9.6 cm^2 s^{-1} for neap case (21.6 cm^2 s^{-1} for spring range). This represents a reduction by a factor of 1.56. This neap-spring variation in vertical mixing is important in many estuarine systems, and if sufficiently strong, can alter salinity intrusion. Mendelsohn *et al.* (2000) show an excellent example for the Savannah River, where the upper reaches of the river change from well mixed to strongly stratified conditions over the spring-neap cycle. At spring (neap) tides, salinity intrusion is reduced (increased) due to enhanced (reduced) vertical mixing. The impact of tidal range (current strength) on less strongly stratified systems shows similar behavior, but has not been as widely appreciated. It has been highlighted in Narragansett Bay by its impact on dissolved oxygen levels.

8.3.8 Two-layer Flow in Providence River and Mount Hope Bay and Multi-layer Flow in East Passage

Hess (1974; 1976) developed and applied a rigid-lid, steady-state, three-dimensional hydrodynamic model to Narragansett Bay (not including the Sakonnet River and Mount Hope Bay; square grids–1.85 km) to predict the estuarine and salinity distribution. This was an extension of his earlier work on tidal circulation in the bay (Hess and White, 1974). The vertical eddy viscosity/diffusivities were parameterized using a density-gradient-dependent Richardson formulation, while the horizontal values were approximated by Elder's velocity shear approach. The values of vertical eddy viscosity ranged from 1 to 10 cm^2 s^{-1}, and horizontal values ranged from 10 to 1,000 m^2 s^{-1}. The model was forced with freshwater input from the rivers and observed salinity at the mouth. Simulations were performed until the model reached steady-state conditions (several months), and model-predicted salinities were compared to the data collected from non-synoptic, short-term field measurements at nine stations throughout the bay. The model accurately predicted a decrease in salinity for both the surface and bottom values from the mouth to the head of the bay and the progression of the bay from well mixed to stratified conditions (salinity difference of 9.5 %) at the head of the bay. The predictions were within 2 ppt or less of the observations. The model predictions showed the expected two-layer estuarine flow pattern with outflows at the surface and inflows at depth. The flows were strongest in the East Passage. The model predicted several inflows of several cm s^{-1} at the entrance to the Providence River, which was consistent with Weisberg's (1976) observations. No other data were available at the time of the study to compare model predictions. The model grid system was too coarse to represent the Providence River, thus precluding comparisons to Turner's (1984) measurements (summarized below). Hess (1976) presented maps of transverse circulation in the middle lower portion of the West Passage and at the entrance to the Providence River. At a site in West Passage, the model predicted a cyclonic transverse circulation with down-welling currents on the western side of the channel. The maximum vertical velocity was predicted to be 4.7×10^{-3} cm s^{-1}. This led to a slight tilting of the isohalines downward (upward) on the western (eastern) side of the channel. At the Providence River entrance transect, the transverse circulation was anti-cyclonic, while the isohalines showed a similar tilting on the western side as in the West Passage. Hess (1976) explains the force balances responsible for each of these cases.

Turner (1984) performed large-scale field measurements in the Providence River from September 1981 through November 1983. He deployed surface and bottom current meters along the side of the dredged channel in the Providence River at Gaspee Point (east and west side), Conimicut Point, and off the Pawtuxet River. The deployments typically lasted for one or several months. Salinity and temperature measurements were made at each station about 3 m below the surface (to avoid vessel traffic) and 2 m above the bottom (to minimize bottom effects). Water depths at various sites were typically 8.5–9 m, and hence vertical separation of the meters was limited to 3–4 m. In addition, Turner collected surface elevation data at Warwick Point, the State Pier in Providence, and at Newport; river flow data for the Blackstone and Pawtuxet Rivers; and wind data from Green Airport.

To develop an understanding of the long-term sub-tidal flow, Turner extracted tidal signals from the record and then low-pass filtered the data (30 h cutoff) to develop a progressive vector diagram for each deployment and meter site. The observations show mean flows up the Providence River over the duration of the experiment, showing small fluctuations with periods of several days, which appeared to be correlated with local winds. The mean current speeds were in the range of 2.5–7.2 cm s^{-1} (mean value 4 cm s^{-1}), and all directed up river. The current directions were highly variable for the upper level observations, but were generally consistent with the local orientation of the dredged channel for bottom observations. Speeds were strongest in the spring. These observations are consistent with estuarine residual circulation patterns where density-induced flows are up estuary near the bottom, and are in good agreement with Hansen and Rattray's (1966; 1970) model.

The observations, however, did not show the normal down-estuary transport near the surface. This surprising result was reconciled by the fact that the upper observation meter was located at the depth, or just slightly below the depth, of shallow areas on either side of the dredged channel (3.5 m), and hence represented an area where the vertical structure of flow was in transition from up-estuary at the bottom to down-estuary at the top. The two sampling stations measured up-estuary transport in the bottom of dredged channel.

Spaulding et al.'s (1999) application of a three-dimensional hydrodynamic model to Mount Hope Bay predicts that the water column in the vicinity of the mouth of Taunton River is strongly stratified as freshwater exits from the lower river on ebb tide. The freshwater plume is predicted to follow the dredged channel. As the Taunton River plume enters Mount Hope Bay proper, it is mixed with bay waters, and the water column becomes partially well mixed. On the flood tide, these partially mixed waters were transported up the river, and hence the water column at the river mouth had limited stratification. Model predictions were consistent with Spaulding and White's (1990) current and hydrographic observations. Spaulding and White (1990) showed mean flows of about 10 cm s^{-1} from Mount Hope Bay to the West Passage of Narragansett Bay, and a two-layer flow—inflow at the surface (2.3 cm s^{-1}) and outflow at the bottom (1.8–3.8 cm s^{-1})—in the lower Taunton River. The model gave similar two-layer flows in the Taunton River but with lower magnitudes.

8.3.9 Some Unusual Features

8.3.9.1 Tidal Harmonics at the Sakonnet River Bridge

Toward the head of the Sakonnet River, which connects Mount Hope Bay to Rhode Island Sound, there is a narrow natural passage that has been further modified by the construction of causeways for two railroad bridges, north (current) and south (old), creating an hour-glass-shaped system. Kim and Swanson (2001) studied the tidal flow dynamics of the area using water level (north and south of the two bridges) and current (between the two bridges) measurements, and a tidal hydrodynamic model of Narragansett Bay. This application necessitated a highly refined grid system for the northern end of the Sakonnet River, particularly in the vicinity of the bridges. Model predictions and observations were in very good agreement, and show a strong double-peaked flood and single-peaked ebb, a characteristic of Narragansett Bay. The double-peaked flood was sufficiently strong that the flow actually reversed and ebbed for what otherwise was a portion of the flood tide. This behavior was a result of the abnormally large magnitudes of the harmonics of M_2 tides (M_4, M_6). As an example, the amplitudes of along-channel M_2, M_4, and M_6 tidal constituent current speeds were 0.924, 0.502, and 0.332 m s^{-1}, respectively. The magnitudes of over-tide currents were about 54% (M_4) and 36% (M_6) of M_2. This compared to about 30% (M_4) and 10% (M_6) for Narragansett Bay proper. Kim and Swanson (2001) have shown that the principal mechanism causing the double flood pattern is the magnitude of M_4 and M_6 components and their phase lag relative to the lunar semi-diurnal tide. They have further attributed the enhanced magnitude of the M_4 component to non-linear bottom frictional losses, particularly in the vicinity of the narrowest portions of the causeway. DeLeo et al. (1997) and DeLeo (2001) investigated these phenomena. It is interesting to note that the current speeds at the Railroad Bridge are among the strongest observed in Narragansett Bay.

8.3.9.2 Phase Lags Near Slack Water between Dredged Channels and Shallows (Providence River)

Muin and Spaulding (1997b) applied a high-resolution, three-dimensional hydrodynamic model to predict tidal- and wind-driven circulation in the Providence River. The model grid size ranged from 250 m in the lower river to 50 m in the upper river. The model grid system was sufficiently fine to resolve the dredged channel that is a key bathymetric feature of the river. The model was exhaustively compared to observations collected by Turner (1984). In areas of the river that were wide compared to the width of the dredged channel (Gaspee Point, for example), the model predicted that the currents in the shallow water on either side of the channel led the currents in the dredged channel by about 1 h. The effect was most noticeable at slack water, where currents in the shallows flowed in the opposite direction to those in the channel.

8.3.9.3 Impact of Quonset and Davisville Channels on Flows and Water Quality

As part of larger investigation on the potential impact of re-development of Quonset Point as a container port, researchers at Applied Science Associates, Inc., and URI Ocean Engineering, led by Berger-Maguire (2004), performed a 1.5-year long field observation program (August 2002 to December 2003). A hydrodynamic modeling study was performed to assess the impact of Quonset and Davisville shipping channels (with and without deepening) on circulation and water quality in the area. The field observation program included the deployment of two long-term moorings equipped with YSI water quality sensing systems (salinity, temperature, dissolved oxygen, turbidity, and pH); one in the center of the West Passage in the Quonset Shipping channel (sensors at surface mid-depth and bottom), and the other just north of Gould Island (sensors at mid-depth and bottom). The latter mooring was used to characterize the potential water source that might enter a deepened Quonset channel. In addition, ADCP measurements were available from the NOAA PORTS system in the Davisville shipping channel. ASA/URI also conducted three, month-long deployments, either in or adjacent to the shipping channel. Finally, vertical profiling of water quality parameters was performed at nine stations along the central axis of the Quonset Channel, and four stations along the Davisville Channel, approximately once per month from September 2002 to December 2003.

The observations showed that the currents were dominated by semi-diurnal tides and their harmonics (Fig. 8.6). The currents at the surface were observed to be slightly larger than those at depth. At sub-tidal frequencies, the currents at the surface are in the direction of the wind and rotate clockwise with depth. Wind-driven flows to the south (north) generate return flows near the bottom to the north (south), but follow topographic steering to the northwest (southeast) at the bottom of the Quonset Channel. There is no evidence of any mean

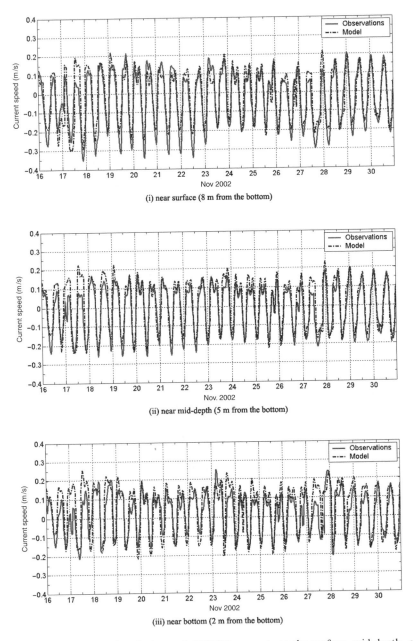

Fig. 8.6 Model predicted and observed (ADCP) currents at the surface, mid-depth, and bottom in the Quonset Shipping Channel, November 16–30, 2002 (Berger-Maguire, 2004).

circulation at any of these stations. The salinity measurements show a mean 2.3 ppt difference between surface and bottom stations, with higher values in summer and lower values in winter. The vertical structure of salinity and temperature is very similar for stations along the Quonset Channel. There is no evidence in either the density fields or the current observations to indicate any significant baroclinic-driven flow along the channel bottom. The dissolved oxygen measurements show a seasonal cycle (at all depths) with highest values in winter when wind mixing is strongest, saturation levels are the highest, and biological activity is the lowest, and lowest values in summer when saturation levels decline, wind mixing is less vigorous, and biological activity is the highest. During several two- or three-day periods in summer, the bottom waters were found to be more oxygenated than waters at the surface or mid-depth, and consistent with source water at depth in the East Passage. The shipping channel appears to provide a path for more oxygenated bottom water to be transported from the East to the West Passage.

Hydrodynamic model simulations show results that are consistent with other bay-wide modeling results, and are in very good agreement for tidal time scales (Fig. 8.6) and reduced performance for wind-driven events. Model predictions show little impact from deepening the Quonset Channel from 10 to 14.5 m on tidal circulation, salinity, or dissolved oxygen levels in the study area, and only minor changes in the channel proper.

8.4 Flushing Dynamics of Bay and Sub-bay Areas

Pilson (1985) employed the fraction of freshwater method to estimate the flushing time for Narragansett Bay. He divided the bay into nine segments and estimated the mean volume, surface, and depth of each segment. He then estimated the freshwater input to the bay from principal river discharges, input from sewage treatment facilities (Field's and Bucklin Points), runoff from ungauged areas, and direct freshwater input to the bay surface from rainfall for six different years for which he made salinity observations over the bay. He estimated the volume-weighted salinities in each segment of the bay for every month during the years for which he had salinity data. He then estimated the flushing time by dividing the volume of freshwater in the bay by the estimated freshwater input. He developed the following relationship between the predicted flushing time, T (days), and the freshwater input, Q_f (m^3 s^{-1}).

$$T = 41.8e^{-0.00435Q_f}$$

This relationship had a correlation coefficient of 0.841. Pilson extended the correlation analysis to include the effects of mean wind speed, but this did little to improve the predictions because of very low correlation. For the annual average freshwater input of 105 m^3 s^{-1}, the flushing time was predicted to be 26.5 days. For low freshwater inputs (50 m^3 s^{-1}), the flushing time increased to 33.6 days, while for high input rates (250 m^3 s^{-1}), the value decreased to 14.1 days. Hence, we can conclude that, as the freshwater flow increased, the flushing time decreased exponentially. This is consistent with an enhancement of estuarine circulation in the bay with increasing freshwater input. It further suggests that exchanges between the bay and Rhode Island Sound are primarily dependent on the freshwater flow into the bay. Kincaid and Bergondo address this in greater detail in Chapter 10.

For comparison, a tidal prism analysis by Hicks *et al.* (1953) gave a flushing time of 9.8 days. They assumed the whole bay as one basin and hence neglected the fact that water at the head of the bay cannot be transported along the length of the bay in one tidal cycle. Tidal prism analyses are expected to predict minimum flushing times since they assume complete mixing, and that no water that exits the bay on ebb tide returns on the following flood.

Wang and Spaulding (1985) developed a modified, segmented tidal prism flushing model for the bay. In this approach, the bay was divided into segments consistent with the tidal excursions in each portion of the bay, and a vertical stratification parameter was incorporated to account for stratification in the corresponding section. The model was applied to the same data set used by Pilson (1985), and predicted a flushing time that decreased approximately exponentially with freshwater input. The predicted value at low flow (50 m^3 s^{-1}) was 31.5 days, and at high flow (250 m^3 s^{-1}) was 13.53 days, both in excellent agreement with Pilson's independent estimates. The one important difference is that the tidal prism model showed an asymptotic behavior to flushing times (13 days) for flows above 300 m^3 s^{-1}, whereas Pilson's analysis predicts a continuing decrease in the flushing time.

Asselin (1991) and Asselin and Spaulding (1993) performed an extensive investigation into the flushing time of the Providence River, including the upstream, tidally-influenced Seekonk River. This investigation was motivated by the desire to better understand the flushing of contaminants from the Providence River combined sewer overflows (CSO) in the upper bay. They employed one time-dependent and three steady-state methods to estimate the flushing time. The time-dependent method was used to estimate the decrease in concentration following the release of dyes into the bay during three CSO (wet weather) events in October 1988, May 1989, and June 1989. A steady-state fraction of freshwater method, using the observed salinities, was applied to the May and June 1989 wet-weather experiments to determine the flushing time. The same method was applied to five SPRAY cruises (Doering *et al.,* 1988) (October 11, 1986; December 15, 1986; March 11, 1987; April 22, 1987; June 27, 1987; and August 12, 1987) in the upper bay, which collected salinity, temperature, nutrient, and trace metal concentrations at 10 stations in the Providence and Seekonk Rivers. Finally, Swanson and Jayko (1988) applied a box model (Swanson and Mendelsohn, 1995) to predict the flushing time in the Providence River using data collected for four bay-wide SINBADD cruises (Pilson and Hunt, 1989). SINBADD cruises were performed (4 days each) in October 1985, November 1985, April 1986, and May 1986 to obtain estimates of the input and distribution of nutrients and trace metals at 20 stations within the bay, and two at the mouth of the bay. The box model (Swanson and Jayko, 1988) was two layered and allowed for both advective and diffusive transports in the bay.

For the Providence River, the model showed that the advective and diffusive transport processes were equivalent in transporting salt. The flushing times from each of these methods were regressed against the freshwater flow rates corresponding to each experimental period, which gave:

$$T = 9.02e^{-0.02167Q_f}$$

where T (days) is the flushing time and Q_f $(m^3 s^{-1})$ is the freshwater input to the Providence and Seekonk Rivers. This relationship had a correlation, $R^2 = 0.889$, to the data. For the mean freshwater flow of 42.3 $m^3 s^{-1}$, the flushing time was estimated to be 3.6 days. The flushing time ranged from 1.3 days for high flow (90 $m^3 s^{-1}$) to 8 days for low flow (5 $m^3 s^{-1}$). The flushing time for the Providence and Seekonk Rivers at mean flows was, therefore, about 13.6% of that for the whole bay. A separate analysis was performed to estimate the flushing times for the Seekonk River, which gave:

$$T = 3.05e^{-0.03372Q_f}$$

where $R^2 = 0.794$. The flushing time of the Seekonk River for mean freshwater input (26.7 $m^3 s^{-1}$) was 1.2 days and ranged from 0.4 day at high flow (58.6 m^3 s^{-1}) to 2.9 days at low flow (1.9 $m^3 s^{-1}$). The flushing time for the Seekonk River is, therefore, about one-third of that for the entire Providence and Seekonk River system.

Asselin and Spaulding (1993), noting that exponential fit to flushing time versus freshwater flow data predicted a lower flushing time than observed, fitted the data using an equation with Q_f^{-1} dependence. They determined the flushing times as:

Providence River:

$$T = 89.68/Q_f \quad R^2 = 0.910$$

Seekonk and Providence River:

$$T = 129.19/Q_f \quad R^2 = 0.980$$

Seekonk River:

$$T = 39.51/Q_f \quad R^2 = 0.790$$

These relationships provide a better fit to the data than the exponential form, giving 3.05 days as compared to 3.6 days for the Providence and Seekonk Rivers, and 1.47 days as compared to 1.2 days for the Seekonk River for mean freshwater input rates. This approach, however, has a deficiency that flushing time approaches infinity as freshwater flow approaches zero.

Erikson (1998) used a two-layered box model to perform estimates of the flushing times for Greenwich Bay and associated coves. She divided the bay into seven separate boxes—four representing the coves, two the central portion of the bay, and one the transition between the central bay and Greenwich and Apponaug Coves. The freshwater inputs from stream flow, runoff, and ground-water were estimated for each box. The mean volume and surface area and depth were estimated from the NOAA bathymetric charts. Mean salinities were derived for each box using data collected at 14 different stations for 27 different intensive measurement programs from August 1995 to May 1997 (S. Granger, personal communication). Erikson (1998) curve-fitted the model predictions with an exponential relationship and determined the flushing time versus fresh-water input for each box in the bay, and for the bay in its entirety:

$$T = 178.32e^{-7.96Q_f} + 4.34e^{-0.08Q_f}$$

The fit had an $R^2 = 0.87$. For a mean freshwater input of 1.19 m^3 s^{-1}, the model predicted a flushing time of 3.96 days. The flushing time ranged from 2.57 days (6.54 m^3 s^{-1}) to 182 days (no freshwater input). The flushing times for individual sections of the bay varied strongly depending on the volume of freshwater input to that segment.

Spaulding et al. (1998) applied a two-dimensional, vertically-averaged, boundary-fitted hydrodynamic model to predict tidal circulation in Greenwich Bay. The model was validated with surface elevation data in Greenwich and Apponaug Coves. A particle-based trajectory model was linked with the hydrodynamic model and employed to estimate the flushing times for dye release that had been performed in Greenwich Cove (model—1.2 days, observed—1.1 days) (Turner, 1986) and Apponaug Cove (model—0.7 day, observed—0.5 day), the flushing of fecal coliforms from Greenwich Bay following a rainfall-induced release (model—3.35 days, observed—3–4 days) (FDA, 1993), and compared to Erikson's (1998) box model predictions (model—3.5 days, observed—3.35 days). Model predictions were generally in good agreement with the experiments, and Erikson's (1998) flushing model, except in very low flow conditions when flushing times were estimated to increase substantially.

Turner et al. (1990) performed a continuous dye release from City Pier and River View outfalls, located in the central eastern side of Mount Hope Bay, for a 3-h period on September 22, 1990. The dye was tracked in the water column with measurements every 4–6 h over the next few days using standard fluoro-metric techniques. The dye was observed to disperse horizontally, but remained in the upper water column (upper few meters) during the measurement program. The analysis of decrease of dye concentration at select stations yielded a flushing time of 1.9–2.3 days. Huang (1993) and Huang and Spaulding (1995b) applied a three-dimensional hydrodynamic and water quality model to simulate the circulation and dispersion of dye release. The model was forced by tides (at Mount Hope Bridge), Taunton River flow, and observed winds. The model employed a hyperbolically stretched grid system in the vertical that allowed the grid cells to be concentrated in the vicinity of the sea surface. The model-predicted dye concentration patterns were in good agreement with observations of dye concentrations, as was the model predicted flushing time of 1.8–2 days.

Abdelrhman (2005) applied two-dimensional, vertically averaged finite element hydrodynamic and pollutant transport models to predict tidal circulation and flushing time in Greenwich Bay and its sub-embayments, as well as the Providence and Seekonk Rivers and Narragansett Bay. The hydrodynamic model was forced only by tides, except for the Providence and Seekonk Rivers where river flow was included. He estimated flushing times from the integrated loss rate for the water body, and assumed that no constituent returned on subsequent tides. Because the goal of his study was to present a simplified modeling approach, he did not calibrate or verify either the tidal circulation or constituent flushing response. The results of his analysis found that Narragansett Bay had a simulated flushing time of 12.6 days, the Providence/ Seekonk River—4.8 days, Greenwich Bay—9.2 days, Apponaug Cove—0.9 day, Greenwich Cove—0.7 days, and Warwick Cove—1.9 days. These predictions generally compare favorably to other analyses using the fraction of freshwater or tidal prism method, except for Warwick and Greenwich Coves where one was lower and the other higher.

8.5 Pollutant Transport and Fate

8.5.1 Impact of Thermal Discharges from the Brayton Point Power Station on Mount Hope Bay

Brayton Point power station, a 1600-MW power generating station, is located in Somerset, Massachusetts, on Mount Hope Bay between the confluence of the Taunton and Lee Rivers. The station, for its through cooling system, draws subsurface water from Mount Hope Bay at two sites: the Taunton River (eastern) side of Brayton Point for coal-fired Units 1, 2 and 3, and the Lee River (western) side for oil-fired Unit 4. Water is pumped through the station and then released via an open channel discharge, which is a canal with a venturi structure at its mouth to enhance exit velocities. The outfall generates a discharge jet, which enhances mixing and dissipates its momentum relatively close to the outfall. The subsequent thermal plume is advected with tidal oscillations and wind-induced circulation in the area. The station discharges a maximum of $57 \, m^3 \, s^{-1}$ of heated effluent when its four generating units are in operation (Swanson *et al.*, 2006). The station currently operates in accordance with a discharge permit allowing a 12.2°C temperature rise between intake and outfall temperatures during regular operations, with a maximum effluent temperature limit of 35°C.

The USEPA permits this discharge, but was, however, concerned about its biological impacts on Mount Hope Bay. In connection with the renewal of station's permit application, its owner, then New England Power, commissioned a series of studies in Mount Hope Bay, including the development of a hydrothermal model to simulate the thermal structure of the bay. To develop this model, a two-component study was performed (Swanson *et al.*, 2001, 2006): (1) physical measurements of Mount Hope Bay were collected to determine the spatial and temporal distribution of circulation and water temperature during different seasons; and (2) a simulation model was constructed to predict the hydrodynamics and thermal structure of Mount Hope Bay under different thermal loads and environmental conditions. The model was calibrated and verified with data collected during the field program.

An extensive field program to map the thermal structure in space and time in Mount Hope Bay was conducted (Rines, 1998; 1999), and major surveys were conducted during the four seasons. Thermistor strings were generally laid out in arcs of concentric circles centered on the station, and consisted of buoyed lines with self-logging thermistors attached at different depths. The thermistor surveys show that, in the Taunton River, events were driven mostly by tides, weather, and river flows, with no effect from the Brayton Point station plume. In Mount Hope Bay, the temperature pattern varied with depth, distinctly tidal at the deepest thermistor where weather influences were mitigated. The shallowest thermistor recorded the largest temperature peaks, due to the effects of daily warming and cooling at the air/water interface. Close to the outfall, strong temperature peaks with tidal periodicity were common. No significant vertical stratification was seen in the observations. At mid-depths during winter, these temperatures peaked, representing the movement of the Brayton Point power station thermal plume below the surface. This can exhibit a stable condition, since the surface discharge consisted of warm, more saline water into colder and fresher receiving water. It should be noted that the intake was located at the bottom (hence draws more saline water), while the discharge was located at the surface. During summer, the plume was consistently at the surface. There was no upwelling or downwelling, because the vertical profiles from month-long moorings did not display any bending of the thermoclines.

A hydrodynamic (or hydrothermal) model (Muin and Spaulding, 1997a,b) was used to simulate the effects of tide, river flow, air temperature, solar radiation, and wind-induced environmental forcing on circulation and thermal energy balance. An environmental heat transfer sub-model (Swanson *et al.*,1998) at the water surface closely follows Edinger *et al.* (1974), and contains an explicit balance of short-wave solar radiation, long-wave atmospheric radiation, long-wave radiation emitted from the water surface, convective (sensible) heat transfer, and evaporative (latent) heat transfer between water and air.

The model successfully simulates the multiple time scales (tidal, daily, and weekly) seen in data during the period as well as the vertical structure. Thus, the model captures the plume distribution and persistence, and ultimately its dissipation. The model was not designed, however, to capture the finer detail of the plume as indicated by the analysis of remotely sensed imagery by Mustard *et al.* (undated).

Numerous scenarios were run and presented during the course of the studies in support of the Brayton Point station's permit application. These included existing conditions, historical conditions, no plant operation, and various options reducing discharge flow and temperature to assess the effects on the plant's thermal discharge characteristics.

Figure 8.7a,b shows typical spatial distributions of temperatures in Mount Hope Bay at maximum ebb and maximum flood for a representative hydrothermal model run. The inserts on the figures show the sectional view of the plume from the outfall. At maximum ebb, the thermal plume extends in the southwest direction (Fig. 8.7a) since the ebb current moves in that direction. As the tide turns, the plume generated during the previous ebb moves northward, while the plume generated during the flood moves toward the mouth of the Taunton River. At maximum flood (Fig. 8.7b), the remnants of the ebb plume are seen to the west of the station discharge, while the flood plume just discharged is seen to the east.

8.5.2 Impact of Dredging in the Upper Bay on Circulation and Water Quality

The Federal Navigation Channel is a 27-km-long channel that begins near the head of the Providence Harbor south of the Hurricane Barrier and follows the Providence River south to deep water near Prudence Island (USACE, 2001). The northern 4 km comprise the main harbor of the Port of Providence. The channel is 180 m wide, except between Field's Point and Fox Point, where it has varying widths up to 520 m. The channel has an authorized depth of 12.2 m MLW, but shoaling has reduced controlling depths to as little as 9.2 m MLW.

Restoring the authorized dimensions involved the removal and disposal of $3.3 \times 10^6 \text{ m}^3$ of sediment. Some of the sediment was unsuitable for unconfined open-water disposal because of contaminants; hence a series of confined aquatic disposal (CAD) cells were established in the Providence Harbor portion of the channel. The volume of these CAD cells required an additional $1.2 \times 10^6 \text{ m}^3$ of material to be removed.

A major concern expressed by the regulatory agencies to USACE over the dredging of Providence Channel was the potential for sediment plumes to disrupt or harm marine resources in the area. These agencies recommended that dredging be allowed only at certain times (dredging windows) to minimize the potential impact. The concern was to assess the impact of implementation of

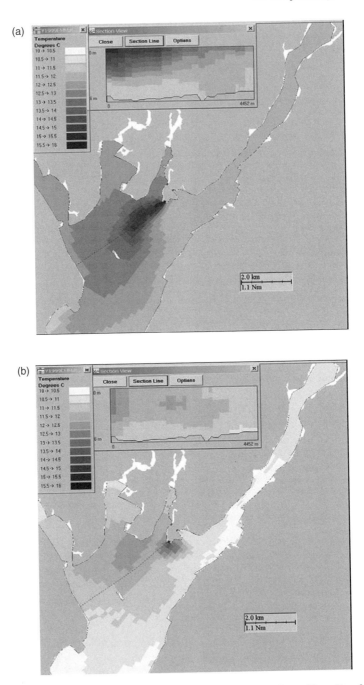

Fig. 8.7 (a) Plan view of distribution of surface temperatures in Mount Hope Bay for EMM model run at maximum ebb. Vertical cross-sectional view of the plume from the outfall along the transect is shown as a dotted line in the insert. (b) Plan view of distribution of surface temperatures in Mount Hope Bay for EMM model run at maximum flood. Vertical cross-sectional view of the plume from the outfall along the transect is shown as a dotted line in the insert.

dredging windows on the dredging schedule. USACE conducted additional studies to address interagency discussion of dredging windows. This included the modeling of releases generated from the dredging process to determine the extent and duration of dredged plumes. Total suspended solids (TSS) and copper (Cu) were the constituents modeled. A series of dredging sites were selected as representatives of locations along the length of the dredged channel from the Fox Point Reach in Providence Harbor to the Rumstick Neck Reach in the upper bay south of Conimicut Point. The model simulated the release of material from one, two, or three sites simultaneously with a variable release rate in the vertical. Three release rates were used to simulate the effects of different types of buckets and barge filling protocols.

The WQMAP hydrodynamic model was applied to the Providence River Harbor and Channel areas to generate a representative set of currents in the river and upper Narragansett Bay. The model included the Seekonk and Providence Rivers, as well as the upper and mid-bay portions of Narragansett Bay. The model was calibrated in several recent studies, particularly examining other problems related to dredged materials: velocities at mid-bay sites for USACE (Swanson and Mendelsohn, 1998) and velocities at proposed Rhode Island Coastal Resources Management Council disposal sites (Swanson and Ward, 1999a,b).

The WQMAP currents were used as input to SSFATE (Johnson et al., 2000). This model simulated the transport, both vertically and horizontally, of dredged material released at select dredging sites using a particle-based modeling approach. Particle movement was based on an advective velocity calculated from interpolated hydrodynamic model results as well as a diffusive velocity that vectorially adds a random component based on a typical estuarine dispersion. The particle model allows the user to predict the transport and fate of classes of settling particles (e.g., sands, silts, and clays). The fate of multicomponent mixtures of suspended sediments is predicted by linear superposition. The WQMAP pollutant transport model also used the currents as input to calculate the transport of dissolved copper, a constituent of the dredged material.

Figure 8.8 shows the near-surface (0–2 m) extent of the sediment plume at the Fox Point Reach Central (FPRC) dredging site with a release rate of 1.5% of the amount dredged. The lower panel shows the down-channel extent of the plume at low slack water, while the upper panel shows the up-channel extent at high slack water. At low slack water, the plume (at a concentration >20 mg L^{-1}) extends approximately 1,000 m down-channel of the dredging site. There is some residual material up channel of the site generated during the previous flood tide. At high slack water, the plume extends approximately 780 m up channel, again with a small residual down channel of the site. The tidal circulation tends to move the plume up and down the channel, away from the release site. Thus, any location away from the immediate dredging site sees elevated concentrations for only a portion of the tidal cycle.

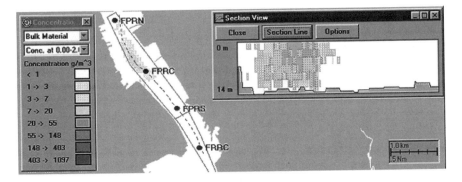

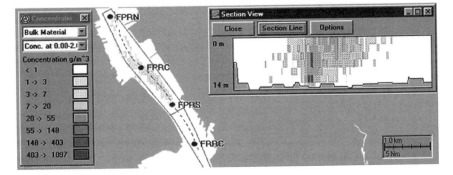

Fig. 8.8 Extent of surface (0–2 m) sediment plume concentrations (g m^{-3} or mg L^{-1}) with bucket release rate of 1.5% for Fox Point Reach Central (FPRC) site over a typical tidal cycle. Bottom panel shows maximum down-channel extent at low slack water and top panel shows maximum up-channel extent at high slack water. Vertical section shows distribution of concentrations (g m^{-3} or mg L^{-1}) over depth along dashed section line shown in plan view. Scale shown is for plan view; section view is scaled to the length of the dashed line. Concentration contours correspond to ranges in the Newcombe-Jensen (1996) fisheries impact model.

8.6 Summary and Conclusions

8.6.1 Tides

Our understanding of the tidal circulation of the bay is generally very good. The standing wave character of the tides and the short length of the bay (40 km) relative to the tidal wave length (400 km) dominate the basic tidal current and elevation patterns in the bay. This results in amplification of the tide with distance up the bay, and a relatively limited phase difference between the time of high (low) tide at the mouth and the head of the bay. The system is only marginally sensitive to bottom friction dissipation, since current speeds are very modest. The differential depths of the major passages alter this basic pattern, with faster tidal wave propagation in the deeper East Passage than in the shallow Sakonnet River and West Passage. The tidal amplitudes are enhanced in the upper reaches of the system (head of Providence River and Taunton River) by the decrease in cross-sectional areas with distance up bay. The longitudinal (along axis) resonant frequency of the bay, located between the frequencies of the harmonics of the lunar semi-diurnal tide (M_4 and M_6), results in the enhancement of these overtides (harmonics) and concludes in the characteristic double-peaked flood and single-peaked ebb tide. This interaction is prevalent throughout the bay, but most dramatic at the Sakonnet River Railroad Bridge, where the current reverses direction in the middle of the flood during the neap portion of the tidal cycle. For shallow areas in the bay, tidal currents are observed to be in phase throughout the water column, while in deeper waters (>10 m) a distinct phase difference is noted between the surface and bottom currents. This is particularly noticeable around slack water. In areas of the bay where shallow water exists in close proximity to either natural or dredged channels, there are clear phase differences between the flows in shallow vs. dredged channels, notably during the change of tides.

The application of a wide variety of numerical models has shown that tidal elevations and currents can be predicted throughout the bay with differences of several percentage for elevations (2–3 cm, 3–7° in phase for the lunar semi-diurnal tides) and 20% for currents. The models are able to capture all the principal tidal dynamics of the bay, including tidal amplification, phase lags between surface and bottom currents, phase lags between deep channels and adjacent shallow areas, differential tidal wave propagation speeds, the role of overtides in generating double-peaked flood and single-peaked ebb current patterns, and the impact of small geometry and bathymetry on local current patterns.

8.6.2 Wind-forced Flows

Few studies have been performed to investigate wind-forced flows in the bay or its major sub-areas. Modeling and field observation programs have shown that wind forcing is significant only during strongly forced events that last for 2 to 3 days. During these events, wind-induced currents can reach magnitudes comparable to those induced by the tides, and can induce significant transport in the various passages of the bay. Northerly directed winds feature inflows in the West Passage and Sakonnet River, and outflow in the East Passage. About half the water that enters the West Passage is transported to the East Passage north of Jamestown, and the remainder to the north of Patience Island. Water that enters the Sakonnet River (less in comparison to that entering via the West Passage) is transported via Mount Hope Bay to the East Passage and exits the bay. For easterly directed winds (predominant for the area), inflows are predicted for the West Passage, and outflows for both the Sakonnet River and East Passage. Wind-driven flow patterns are symmetric with wind forcing, and so the reversal of wind direction can result in corresponding reversal of wind-induced flows. From the fact that mean winds vary in direction from the southwest (summer) to the northwest (winter), we understand that intra-passage flow patterns vary primarily as a result of the strength of the north-south component of the winds. In the Providence River, northerly (southerly) directed winds result in up (down) river transport in the shallows and down (up) river transport throughout the water column in the dredged channel. The currents are strongest at the surface in the shallows and at the bottom in the dredged channel. Wind-induced setups are largest at the head of the bay and are typically 10 cm. Because of the intra-passage transport, wind-induced flows can readily mask the underlying density-induced flows in the lower and mid-bay areas.

8.6.3 Density-induced Flows

Data and model predictions show that the bay is typically weakly stratified to well mixed in the lower and mid-sections of the bay. The degree of stratification in the mid to upper portion of the bay is dependent on freshwater input and the strength of tidal and wind-induced mixing. Neap tides, and hence weaker currents and tidal mixing, favor partial stratification. The upper bay (Providence and Seekonk Rivers) is typically stratified due to freshwater discharges from the Blackstone and other rivers discharging into the head of the bay. In the Providence River, the dredged channel plays a central role in guiding the saltier bottom water up river. The return flow at the surface, however, is guided by river geometry with a level of no motion at less than 3.5 m below the surface. For the Taunton River, stratification is due to discharge from the river. The strength of stratification is dependent on the freshwater input rate, with greatest stratification under higher freshwater flow. Stratification results in a classic estuarine flow pattern with up river currents of 3–7 cm s^{-1} at the bottom, and downriver flow at the surface of comparable magnitudes. The models currently available are able to reproduce the large-scale density-induced circulation patterns of the bay, but the small-scale flow features in areas of high stratification are less well modeled.

A variety of methods have been applied to estimate the flushing time of Narragansett Bay and its major sub-areas. All the methods show that flushing time is principally determined by freshwater input to the system, with limited dependence on the winds. For Narragansett Bay proper, flushing time ranges between 14 and 34 days for low to high freshwater input rates, with a mean value of 26.5 days. For the Providence and Seekonk Rivers, the values range from 1.3 to 8 days, with a mean value of 3.6 days. It is interesting to note that, for small embayments like Greenwich Bay, the flushing times are quite large, given the small sizes of the system (4 days—Greenwich Bay). This is a result of low freshwater input rates and very weak tidal currents and mixing. There is no evidence in any of the flushing studies, field observation programs, or numerical modeling studies of areas in the bay, where pollutants might be trapped, that residence times greatly increased by the presence of large-scale eddies or similar features. Chen *et al.,* in Chapter 9, further explore eddy persistence and its effects on circulation of the bay.

8.6.4 Pollutant Transport

8.6.4.1 Thermal

An extensive field program to map the thermal structure of the Brayton Point thermal plume in space and time showed that events in the Taunton River were driven mostly by tides, weather, and river flows, with no effect from the plume. In Mount Hope Bay, the temperature pattern was distinctly tidal in nature, with the most regular cycles seen at the deepest thermistor where weather influences were mitigated. The shallowest thermistors recorded the largest temperature peaks, probably due to the effects of daily warming and cooling at the air/water interface. Close to the outfall, strong temperature peaks with tidal periodicity were common. No significant vertical stratification was seen in the observations. Temperatures at mid-depths peaked during winter, representing the submerged plume passing the thermistors below the surface. During summer, the plume was consistently at the surface.

Model predictions showed that dynamic plume locations were consistent with the thermal mapping field program. At maximum ebb, the thermal plume extends from the south to southwest direction since the ebb current moves in that direction. As the tide turns, the plume generated during the previous ebb moves northward while the plume generated during the flood moves toward the mouth of the Taunton River. At maximum flood, the remnants of the ebb plume are seen to the west of the station discharge while the flood plume just discharged is seen to the east.

8.6.4.2 Dredging

A series of studies were conducted to evaluate the environmental effects of channel dredging in the Providence River and its subsequent disposal, the potential change in currents at candidate disposal sites within Narragansett Bay, and the effects on water quality of the potential dredged material plume, containing both suspended sediment and dissolved copper, generated during dredging.

USACE assessed the candidate dredged-material disposal sites around Narragansett Bay. As part of that study, a hydrodynamic model was used to assess the expected maximum tidally induced current speeds at each site. Different levels of filling were examined to determine the sensitivity of each site to changes in depth. The variation by site was larger than the variation by fill condition (minimum, mid, or maximum).

Total suspended solids and Cu were the constituents modeled for environmental impacts. At low slack water, the TSS plume (at a concentration >20 mg L^{-1}) extends approximately 1,000 m down channel of the dredging site. There is some residual material up channel of the site generated during the previous flood tide. At high slack water, the plume extends approximately 780 m up channel, again with a small residual down channel of the site. The tidal circulation tends to move the plume up and down channel away from the release site. Thus, any location away from the immediate dredging site sees elevated concentrations for only a portion of the tidal cycle.

References

Abdelrhman, M.A. 2005. Simplified modeling of flushing and residence times in 42 embayments in New England, USA, with special attention to Greenwich Bay, Rhode Island. *Estuarine Coastal and Shelf Science* 62:339–351.

Asselin, S. 1991. Flushing times in the Providence River based on tracer experiments, MS Thesis, Ocean Engineering, University of Rhode Island, Narragansett, RI.

Asselin, S., and Spaulding, M.L. 1993. Flushing times for the Providence River based on tracer experiments. *Estuaries* 16(4):830–839.

Berger Group, Inc. and Maguire Group, Inc. 2004. Circulation: Pre EIS application consulting, engineering, and environmental sciences-Port Services at Quonset Point, RI, report prepared for State of RI Governor's Office, Providence, RI, April, 2004.

Bergondo, D.L. 2004. Examining the processes controlling water column variability in Narragansett Bay: time-series data and numerical modeling. Ph.D. Dissertation, University of Rhode Island, Narragansett, RI, 187 pp.

Bergondo, D., and Kincaid, C. 2003. Observations and modeling on circulation and mixing processes within Narragansett Bay, EOS Transactions of the American Geophysical Union, Fall Meeting, 2003.

Bergondo, D.L., Kester, D.R., Stoffel, H.E., and Woods, W. 2005. Time-series observations during the low sub-surface oxygen events in Narragansett Bay during summer 2001. *Marine Chemistry* 97:90–103.

Bowden, K.F., and Hamilton, P.L. 1975. Some experiments with a numerical model of circulation and mixing in a tidal estuary. *Estuarine, Coastal, and Marine Science* 3:281–301.

Deacutis, C.F. 1999. Nutrient impacts and signs of problems in Narragansett Bay, *In* Proceedings of a Workshop on Nutrient Removal from Wastewater Treatment Facilities. Nutrients and Narragansett Bay, Kerr, M. (ed.) Rhode Island Sea Grant, Narragansett, RI. pp. 7–23.

DeLeo, B. 2001. Investigation of the physical mechanisms controlling exchange between Mount Hope Bay and the Sakonnet River, MS Thesis in Oceanography, University of Rhode Island, Narragansett, RI. 190 pp.

DeLeo, W., Kincaid, C., and Pockalny, R. 1997. Circulation and exchange within Sakonnet River Narrows, Estuarine Research Foundation Meeting Abstracts, Fall, 1997.

Desbonnet, A., and Lee, V. 1991. Water quality and fisheries- Narragansett Bay. A report to National Ocean Pollution Program Office, National Oceanic and Atmospheric Administration, RI Sea Grant, Narragansett, RI.

Doering, P.H., Weber, L., Warren, W., Hoffman, G., Schweitzer, K., Pilson, M.E.Q., and Oviatt C.A. 1988. Monitoring of the Providence and Seekonk Rivers for Trace Metals and Associated Parameters. Data Report for Spray Cruises I–VI. Vol. 2. Marine Ecosystems Research Laboratory, Graduate School of Oceanography, University of Rhode Island, Narragansett, RI.

Edinger, J.E., Brady, D.K., and Greyer, J.C. 1974. Heat exchange and transport in the environment. Report No. 14, Cooling Water Res. Project (RP-49), Electrical Power Research Institute, Palo Alto, CA.

Erikson, L. 1998. Flushing time of Greenwich Bay; Estimates based on freshwater inputs, MS Thesis, Ocean Engineering, University of Rhode Island, Narragansett, RI.

Gordon, R.B. 1982. Wind-driven circulation in Narragansett Bay, Doctoral Dissertation, Ocean Engineering, University of Rhode Island, Kingston, RI. 166 pp.

Gordon, R. and Spaulding, M.L. 1987. Numerical simulation of the tidal and wind driven circulation in Narragansett Bay. *Estuarine, Coastal, and Shelf Science* 24:611–636.

Granger, S.H. 1994. The basic hydrography and mass transport of dissolved oxygen in the Providence and Seekonk Rivers estuaries, MS Thesis, University of Rhode Island, Narragansett, RI. 230 pp.

Haight, F.J. 1936. Currents in Narragansett Bay, Buzzards Bay and Nantucket and Vineyard Sounds, Special Publication No. 208, US Department of Commerce, Coast and Geodetic Survey, Washington, DC.

Hansen, D.V., and Rattray, M. 1966. New dimensions in estuarine classification. *Limnology and Oceanography* 9(3):319–326.

Hansen, D.V., and Rattray, M. 1970. Gravitational convection in straits and estuaries. *Journal of Marine Resources* 23(2):104–122.

Hess, K.W. 1974. A three dimensional model of steady state gravitational circulation and salinity distribution in Narragansett Bay, Doctoral Dissertation, Ocean Engineering, University of Rhode Island, Kingston, RI.

Hess, K.W. 1976. A three dimensional numerical model of the estuary circulation and salinity distribution in Narragansett Bay. *Estuarine, Coastal, and Shelf Science* 4:325–338.

Hess, K.W., and White, F.M. 1974. A numerical tidal model of Narragansett Bay, University of Rhode Island, Marine Technical Report No. 20, University of RI, Narragansett, RI. 141 pp.

Hicks, S. 1959a. The seasonal distribution of certain physical and chemical oceanographic variables for Narragansett Bay, Appendix B, Sustaining data summary on fishery resources in relation to the hurricane damage control program for Narragansett Bay and vicinity RI and MA, US Fish and Wildlife Service, US Department of the Interior, Boston, MA.

Hicks, S.D. 1959b. The physical oceanography of Narragansett Bay. *Limnology and Oceanography* 4:316–327.

Hicks, S.D., Wehe, T.J., and Campbell, R. 1953. Inshore survey project: Final Harbor Report, Narragansett Bay and its approaches; physical oceanography, Narragansett Bay Marine Laboratory, University of Rhode Island, Narragansett, RI. Ref No 53–12.

Huang, W. 1993. Three dimensional numerical modeling of circulation and water quality induced by combined sewage overflow discharges, PhD Thesis, Ocean Engineering, University of Rhode Island, Kingston, RI.

Huang, W., and Spaulding, M.L. 1994. Three dimensional modeling of circulation and salinity in Mt Hope Bay and the lower Taunton River, *In* Proceedings of the 3rd International Conference on Estuarine and Coastal Modeling. Malcolm L. Spaulding and Keith Bedford (eds.), Oak Bridge, IL, September 8–10, 1993, pp. 500–508.

Huang, W., and Spaulding, M.L. 1995a. A three dimensional model of estuarine circulation and water quality induced by surface discharges. *Journal of Hydraulic Engineering* 121(4):300–311.

Huang, W., and Spaulding, M.L. 1995b. Modeling of CSO induced pollutant transport in Mt. Hope Bay. *Journal of Environmental Engineering* 121(7):492–498.

Johnson, B.H., Anderson, E., Isaji, T., and Clarke, D.G. 2000. Description of the SSFATE numerical modeling system. DOER Technical Notes Collection (TN DOER-E10). U.S. Army Engineer Research and Development Center, Vicksburg, MS.

Kester, D.R., Fox, M.F., and Magnuson, A. 1996. Modeling, measurements and satellite remote sensing of biologically active constituents in coastal waters. *Marine Chemistry* 53:131–145.

Kim, H.S., and Swanson, J.C. 2001. Modeling of double flood currents in the Sakonnet River, *In* Proceedings of the 7th International Conference on Estuarine and Coastal Modeling (ECM 7), Malcolm L. Spaulding (ed.), St. Pete Beach, Fl, November 5–7, 2001.

Kincaid, C. 2001. Results of hydrographic surveys on the Providence and Seekonk Rivers: summer period, Report submitted to the Narragansett Bay Commission, Providence, RI. 43 pp.

Mendelsohn, D., Howlett, E. and Swanson, J.C. 1995. WQMAP in a Windows Environment, *In* Proceedings of the 4th International Conference on Estuarine and Coastal Modeling. Malcolm L. Spaulding and Ralph Cheng (eds.), San Diego, CA. American Society of Civil Engineers. October 26–28, 1995, pp. 555–569.

Mendelsohn, D.L., Peene, S., Yassuda, E., and Davie, S. 2000. A hydrodynamic model calibration study of the Savannah River estuary with an examination of factors affecting salinity intrusion, *In* Proceedings of 6th International Conference on Estuarine and Coastal Modeling. Spaulding, M.L. and Butler, L. (eds.), New Orleans, LA, November 3–5, 1999, pp. 663–685.

Muin, M., and Spaulding, M.L. 1994. Development and application of a three dimensional boundary fitted circulation model in the Providence River, *In* Proceedings of 3rd International Conference on Estuarine and Coastal Modeling. Malcolm L. Spaulding and Keith Bedford (eds.), Oak Bridge, IL, September 8–10, 1993, pp. 432–446.

Muin, M., and Spaulding, M.L. 1996. Two-dimensional boundary fitted circulation model in spherical coordinates. *Journal of Hydraulic Engineering* 122(9):512–520.

Muin, M., and Spaulding, M.L. 1997a. A three dimensional boundary fitted coordinate hydrodynamic model. *Journal of Hydraulic Engineering* 123(1):2–12.

Muin, M., and Spaulding, M.L. 1997b. Application of three-dimensional boundary fitted circulation model to Providence River. *Journal of Hydraulic Engineering* 123(1):13–20.

Munk, W.H., and Anderson, E.R. 1948. Notes on the theory of the thermocline. *Journal of Marine Research* 7:276.

Mustard, J.F., Swanson, C., and Deacutis, C. 2006. Final report for project: Narragansett Bay from space: a perspective for the 21st century. Prepared by Brown University, Providence under NASA Contract NAG-13-39.

Newcombe, C.P., and Jensen, J.O. 1996. Channel suspended sediment and fisheries: a synthesis of quantitative assessment of risk and impact. *North American Journal of Fisheries Management*, Vol.16, pp. 693–727.

NOAA, 1971. Tidal current charts of Narragansett Bay, NOAA National Ocean Service, Rockville, MD.

NOAA   PORTS   *http://tidesandcurrents.noaa.gov/station_retrieve.shtml?type   =   Harmonic+Constituents.*

Officer, C.B., 1976. Physical Oceanography of Estuaries. John Wiley and Sons, New York.

Opishinski, T., and Spaulding, M. L. 2002. Application of an integrated real time monitoring and modeling system to Narragansett Bay and adjacent coastal waters incorporating internet based technology, *In* Proceedings of 7th International Conference on Estuarine and Coastal Modeling. Malcolm L. Spaulding (ed.), Sponsored by the University of Rhode Island Conference Office, St Pete, Florida, November 5–7, 2001, pp. 982–1002.

Oviatt, C. 1980. Some aspects of water quality in and pollution sources to the Providence River, prepared by Graduate School of Oceanography, University of Rhode Island, Narragansett, prepared for EPA Region I, Boston, MA.

Petrillo, A. 1981. Southern New England coastal sea bottom pressure dynamics, MS Thesis, Ocean Engineering, University of Rhode Island, Narragansett, RI.

Pilson, M.E.Q. 1985. On the residence time of water in Narragansett Bay. *Estuaries* 8:2–14.

Pilson, M.E.Q., and Hunt, C.D. 1989. Water quality survey of Narragansett Bay, A summary of results from the SINBADD cruises: 1985–1986. Marine Ecosystem Research Laboratory, Graduate School of Oceanography, University of Rhode Island, Narragansett, RI.

Ries, K.G., III. 1990. Estimating surface-water runoff to Narragansett Bay, Rhode Island and Massachusetts. US Geological Survey, Water Resources Investigations. Report 89-4164. 44 pp.

Rines, H. 1998. Mapping temperature distributions in Mount Hope Bay with respect to the Brayton Point Station outfall. Report to NEPCo. ASA, Narragansett, RI. April, 1998.

Rines, H. 1999. Mapping temperature distributions in Mount Hope Bay, Fall 1998 deployment. Report to NEPCo. ASA, Narragansett, RI. September, 1999.

Sankaranarayanan, S., and Ward, M.C. 2006. Development and application of a three-dimensional orthogonal coordinate semi-implicit hydrodynamic model. *Continental Shelf Research* 26:1571–1594.

Spaulding, M.L. 1987. Narragansett Bay: Issues, resources, status, management, Section on Circulation dynamics, NOAA Estuary of the Month, Seminar Series, US Department of Commerce, NOAA, Estuarine Programs Office, Washington, DC.

Spaulding, M.L., and Gordon, R. 1982. A nested numerical tidal model of the southern New England Bight. *Journal of Ocean Engineering* 9(2):107–126.

Spaulding, M.L., and White, F.M. 1990. Circulation dynamics in Mt. Hope Bay and the lower Taunton River, Narragansett Bay, *In* Coastal and Estuarine Studies, Vol. 38, Cheng, R.T. (ed.), Springer-Verlag, New York.

Spaulding, M.L., Opishinski, T., and Haynes, S. 1996. COASTMAP: an integrated monitoring and modeling system to support oil spill response, *Spill Science and Technology Bulletin* 3(3):149–169.

Spaulding, M.L., Sankaranarayanan, S., Erikson, L., Fake, T., and Opishinski, T. 1998. COASTMAP, an integrated system for monitoring and modeling of coastal waters: Application to Greenwich Bay, *In* Proceedings of the 5th International Conference on Estuarine and Coastal Modeling. Malcolm L. Spaulding and Alan Blumberg (eds.), American Society of Civil Engineers, Reston, VA, May 1998, pp. 231–251.

Spaulding, M.L., Mendelsohn, D., and Swanson, J.C. 1999. WQMAP: an integrated three-dimensional hydrodynamic and water quality model system for estuarine and coastal applications. *Marine Technology Society Journal Special Issue on State of the Art in Ocean and Coastal Modeling* 33(3):38–54.

Swanson, J.C., and Jayko, K. 1988. A simplified estuarine box model of Narragansett Bay, Report to the Narragansett Bay Project, Rhode Island Department of Environmental Management, Providence, RI. ASA #85-11.

Swanson, J.C., and Mendelsohn, D. 1993. Application of WQMAP to Upper Narragansett Bay, Rhode Island, *In* Proceedings 3rd International Conference, Estuarine and Coastal Modeling. Malcolm L. Spaulding and Keith Bedford (eds.), American Society of Civil Engineers, Oak Brook, IL, September 8–10, 1993, pp. 656–678.

Swanson, J.C., and Mendelsohn, D.L. 1994. Application of a water quality modeling, mapping and analysis system to evaluate effects of CSO abatement alternatives on upper Narragansett Bay, Rhode Island. 1994 Water Environment Federation CSO Specialty Conference, Louisville, Kentucky, July 1994.

Swanson, J.C., and Mendelsohn, D. 1995. BAYMAP: A simplified embayment flushing and transport model system, *In* Proceedings of the 4th International Conference on Estuarine and Coastal Modeling. Malcolm L. Spaulding and Ralph Cheng (eds.), San Diego, CA. American Society of Civil Engineers, October, 26–28, 1995, pp. 570–582.

Swanson, J.C., and Mendelsohn, D. 1998. Velocity estimates for candidate dredged material disposal sites in Narragansett Bay. Prepared by Applied Science Associates, Inc,

Narragansett, RI, for Science Applications International Corporation, Newport, RI and the New England Division, US Army Corps of Engineers, Waltham, MA, ASA Project 97-059, March 12, 1998.

Swanson, J.C., and Spaulding, M.L. 1975. Tidal current and height charts for Narragansett Bay, Marine Technical Report No. 35, University of Rhode Island, Narragansett, RI.

Swanson, J.C., and Spaulding, M.L. 1977. Generation of tidal current and height charts for Narragansett Bay using a numerical model, Marine Technical Report No. 61, University of Rhode Island, Narragansett, RI.

Swanson, J.C., and Ward, M.C. 1999a. Improving coastal model predictions through data assimilation, In Proceedings of the 6th International Conference on Estuarine and Coastal Modeling (ECM6). Malcolm L. Spaulding and H Lee Butler (eds.), New Orleans, LA, November 3–5, 1999, pp. 947–963.

Swanson, J.C., and Ward, M. 1999b. Extreme bottom velocity estimates for CRMC dredged material disposal sites in Narragansett Bay. Submitted to SAIC, Newport, RI. Submitted by Applied Science Associates, Narragansett, RI, ASA Project 98-044.

Swanson, J.C., Turner, A.C., and Jayko, K. 1987. CSO Study Area B, receiving water quality analysis. Appendix I in combined sewer overflow mitigation study, CSO Area B, Moshassuck River interceptor drainage basin. Report by O'Brien amp; Gere Engineers, Syracuse, New York and Narragansett Bay Commission, Providence, RI.

Swanson, J.C., Mendelsohn, D.L., Wright, S., Turner, C., Rines, H., Galagan, C., and Isaji, T. 1993. Receiving water quality model for Narragansett Bay Commission combined sewer overflow facilities. Report to Louis Berger amp; Associates, Inc., Providence, RI, ASA #91-47.

Swanson, J.C., Grgin, J., and von Zweck, P. 1996. The integration of receiving water impacts in the evaluation process of alternative designs for CSO abatement in Providence, RI, In Proceedings of the North American Water and Environment Congress 1996. Chenchayya Bathala (ed.), Sponsored by ASCE, Anaheim, CA, June 22–28, 1996, pp. 1537–1542.

Swanson, J.C., Mendelsohn, D., and Isaji, T. 1998. Receiving water quality modeling of combined sewer overflow impacts to the Providence River. Submitted by Applied Science Associates, Inc., Narragansett, RI to Louis Berger amp; Associates, Inc., Providence, RI. ASA Report 96-105, April, 1998.

Swanson, J.C., Rines, H., Mendelsohn, D., Isaji, T., and Ward, M. 1999. Mt. Hope Bay winter 1999 field data and model confirmation. Report to PG amp; E Gen, Somerset, MA. ASA, Narragansett, RI. October, 1999.

Swanson, J.C., Kim, H.S., Isaji, T., and Ward, M. 2001. Summary of hydrodynamic model results for Brayton Point Station simulations. Report to PG amp; E National Energy Group, Somerset, MA. ASA, Narragansett, RI. January, 2001.

Swanson, C., Kim, H.S., and Sankaranarayanan, S. 2006. Modeling of temperature distributions in Mount Hope Bay due to thermal discharges from the Brayton Point Station. Northeast Naturalist Special Issue 4:145–172.

Turner, A.C. 1984. Tidal and sub-tidal circulation in the Providence River, MS Thesis, Ocean Engineering, University of Rhode Island, Kingston, RI.

Turner, A.C. 1986. Dye study at Greenwich Cove, Narragansett Bay, prepared for the Narragansett Bay Project and EPA Region I. ASA Report 85-11, Applied Science Associates, Inc., Narragansett, RI.

Turner, A.C., Asselin, S., and Feng, S. 1990. City of Fall River combined sewer overflow facilities: receiving water impacts field measurement program. Report No, 99-024, Applied Science Associates, Inc., Narragansett, RI.

USACE, 2001. Providence River and Harbor maintenance dredging project, Final Environmental Impact Statement. US Army Corps of Engineers, New England District, Concord, MA, August.

US Food and Drug Administration (FDA). 1993. Greenwich Bay, Rhode Island shellfish growing area survey and classification considerations. US Public Health Service. Food and Drug Administration, Davisville, RI.

Wang, X., and Spaulding, M.L. 1985. A tidal prism flushing model of Narragansett Bay, Applied Science Associates, Inc., Narragansett, RI.

Ward, M., and Spaulding, M. 2002. A nowcast/forecast system of circulation dynamics in Narragansett Bay, *In* Proceedings of 7th International Conference on Estuarine and Coastal Modeling. Malcolm L. Spaulding (ed.), Sponsored by the University of Rhode Island Conference Office, St. Pete, Florida, November 5–7, 2001, pp. 1002–1022.

Weisberg, R.H. 1972. Net circulation in Narragansett Bay, MS Thesis, Graduate School of Oceanography, Narragansett, RI. 90 pp.

Weisberg, R.H. 1976. The non-tidal flow in the Providence River of Narragansett Bay: a stochastic approach to estuarine circulation. *Journal of Physical Oceanography* 6:721–734.

Weisberg, R.H., and Sturges, W. 1976. Velocity observations in the west passage of Narragansett Bay: a partially mixed estuary. *Journal of Physical Oceanography* 6:345–354.

Zhao, L., Chen C., and Cowles, G. 2006. Tidal flushing and eddy shedding in Mount Hope Bay and Narragansett Bay: an application of FVCOM. *Journal of Geophysical Research* 111, C10015, doi:10.1029/2005JC003135.

Chapter 9
Critical Issues for Circulation Modeling of Narragansett Bay and Mount Hope Bay

Changsheng Chen, Liuzhi Zhao, Geoffrey Cowles and Brian Rothschild

9.1 Introduction

Narragansett Bay is a medium-sized estuary located along the coast of the northeast United States with shoreline in both Massachusetts and Rhode Island. The bay covers 380 km², has an average water depth of 7.8 m, and a maximum depth of 56 m. Narragansett Bay contains three large islands—Aquidneck, Conanicut, and Prudence, all of which are oriented roughly north–south, and divide the bay into three interconnected channels—the West Passage, the East Passage, and the Sakonnet River (Fig. 9.1). The narrow linkages between these waterways control the water exchange among the various sectors of the bay. The connection to the sea is found in the southern reaches of the bay, where it opens onto the inner New England Shelf via Rhode Island Sound. In the northeast corner of the bay lies a semi-isolated shallow estuary called Mount Hope Bay. It is connected to the greater portion of Narragansett Bay through a narrow, deep channel of about 800 m in width and 25 m in depth. In view of water exchange dynamics, Narragansett Bay and Mount Hope Bay are an integrated inter-bay complex.

In recent decades, intensive short- and long-term field measurements have been made in Narragansett Bay. These observations show that regional warming has caused a dramatic increase in the stratification of the water column (Hicks, 1959; Nixon et al., 2004). Annual mean water temperatures in the Narragansett Bay–Mount Hope Bay system underwent an increase of 2°C from 1985 to 2001, following a decrease during 1972–1984 (Fig. 9.2a, dashed line), with a net increase of ~1.1°C overall (Fig. 9.2a, solid line). The warming trend was also observed in Woods Hole, Massachusetts (Nixon et al., 2004), and in coastal waters of the northeast United States (Oviatt, 2004). The bay, which remained vertically well mixed throughout the year in 1954–1955 (Hicks, 1959), has been strongly stratified since the summer of 1990.

Changsheng Chen
The School for Marine Science and Technology, University of Massachusetts at Dartmouth, 706 South Rodney French Blvd., New Bedford, MA 02744
c1chen@umassd.edu

A. Desbonnet, B. A. Costa-Pierce (eds.), *Science for Ecosystem-based Management.*   281
© Springer 2008

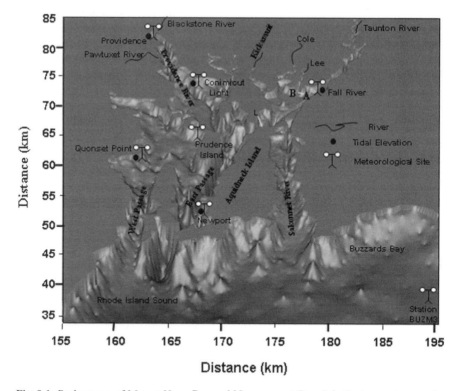

Fig. 9.1 Bathymetry of Mount Hope Bay and Narragansett Bay. L is the transect across the Narragansett Bay–Mount Hope Bay channel used to estimate water transport.

The observed increase in stratification is believed to have a negative impact on the Narragansett Bay ecosystem (Karentz and Smayda, 1998; Li and Smayda, 1998; Keller *et al.*, 1999; 2001). There is an associated reduction in vertical mixing, which leads to a reduction in oxygen exchange between the atmosphere and deeper waters of the bay. Direct observations of hypoxic (<2 mg L^{-1}) dissolved oxygen concentrations have been made in the northern bay (Bergondo *et al.*, 2005). An enlargement of the observed hypoxic area in summer from 2001 to 2002 was consistent with a significant increase in observed water stratification in the Providence River, Greenwich Bay, and the adjacent regions of Narragansett Bay (Deacutis *et al.*, 2006; Fig. 9.2c). A strong hypoxic event occurred on August 20, 2003, which caused significant mortality of menhaden, as well as many finfish, eels, crabs, soft-shell clams, and grass shrimp in Greenwich Bay, and subsequent closure of a large number of beaches in the area due to deteriorating environmental conditions (Deacutis *et al.*, 2006).

The warming tendency is also thought to advance the timing of the peak abundance of resident marine species (Sullivan *et al.*, 2001; Sullivan and Keuren, 2004), diminish eelgrass (Sullivan *et al.*, 2001), and dramatically decrease commercial fishery stocks such as winter flounder (Jeffries, 2002;

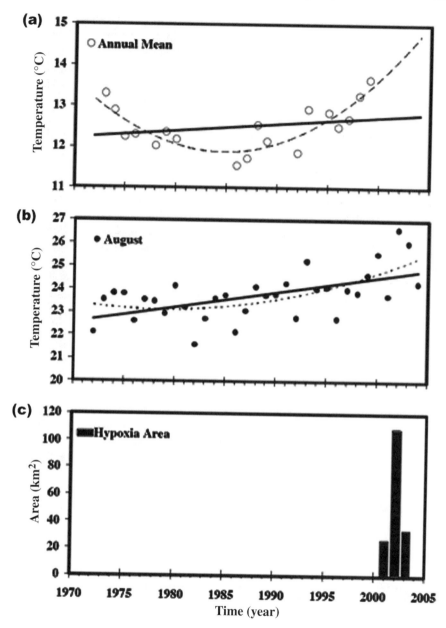

Fig. 9.2 Annual mean (upper) and August (middle) near sea surface water temperatures measured in Narragansett Bay–Mount Hope Bays from 1970–2001 and the area of hypoxia observed from 2001 to 2003. The warming trend shown in this figure is the same as that measured at Woods Hole by Nixon *et al.* (2004).

Nixon *et al.*, 2004). The winter-spring phytoplankton bloom, which normally occurs in Narragansett Bay, has not occurred in the recent warm winters (Oviatt, 2004), implying that warming in coastal waters may cause a permanent change in the phytoplankton seasonal cycle. This can, in turn, cause significant changes at higher trophic levels. These hypoxic events will lead to better evaluation of the effects of potential long-term temperature trends on the bay's ecosystem.

A direct anthropogenic warming influence on Mount Hope Bay is the Brayton Point power station, located at the northern end of the bay. Water is drawn from Mount Hope Bay, used for cooling at the facility, and discharged back into Mount Hope Bay at a higher temperature. The inflow temperature ranges from $7°C$ in winter to $26°C$ in late summer, and the discharge, about $40 \, m^3 \, s^{-1}$, has a temperature range from $12°C$ in winter to $31°C$ in late summer.

Hypoxia has been more serious in upper Narragansett Bay and the Providence River than in Mount Hope Bay (a region with higher heating) in summer. The reason for this discrepancy lies in differences in the relative balance of physical and biological processes that lead to hypoxia in these two shallow bays. From the power station cooling system, water is discharged into Mount Hope Bay as a strong jet; it is unclear if this may lead to an increase in vertical stratification through the introduced buoyancy flux or a decrease through enhanced mixing.

In the interest of evaluating and monitoring the potential impacts of ecosystem changes in Narragansett Bay, an observational network, which includes moorings and monthly ship surveys, has been developed through a cooperative effort from the Rhode Island Department of Environmental Management, Rhode Island Division of Water Resources, The Narragansett Bay National Estuarine Research Reserve, the Narragansett Bay Commission, the University of Rhode Island, and Roger Williams University. The *in situ* water temperature, salinity, nutrients, dissolved oxygen, and chlorophyll *a* data recorded from this network help to monitor the large-scale physical and ecological conditions in the bay. While there is a clear need to focus on monitoring ecosystem change, due to the temporal and spatial sparseness of these observations, there is a limitation on the physical features that these measurements are able to resolve (Zhao *et al.*, 2006). If, however, a coastal ocean model is used in conjunction with the observation network, a comprehensive analysis of the complex, temporally varying, coupled physical and biological dynamics in Narragansett Bay can be made. The underlying physical causes of ecosystem variability will be better understood, increasing the effectiveness of management and planning strategies.

This chapter focuses on the capabilities required by such a model to resolve the physical processes underlying change in the Narragansett Bay–Mount Hope Bay ecosystem. We first review the physical forcing that generates local circulation and stratification, and then present some results of modeling process studies to elucidate the numerical capabilities required to resolve detailed circulation and water exchange, focusing on transport between Narragansett Bay and Mount Hope Bay.

9.2 Physical Forcing

Short-term circulation variability in Narragansett Bay is driven primarily by tides, local sea breezes and synoptic winds, and seasonally modified river discharge. Long-term variability in the physical environment derives from changes in freshwater runoff and wind patterns linked to decadal scale atmospheric variability, including the well-known North Atlantic Oscillation (NAO), as well as long-term trends in regional water temperature.

Tidal forcing in Narragansett Bay–Mount Hope Bay is dominated by the semidiurnal M_2 tidal constituent, which accounts for 70–80% of the total energy in bay currents, and generates most of the near-shore vertical mixing (Gordon and Spaulding, 1987; Spaulding and White, 1990; Kincaid, 2006; Zhao et al., 2006). For fortnightly and monthly variations of tidal motion, N_2, S_2, K_1, and O_1 tidal constituents need to be included, although ratios of N_2 to M_2 and S_2 to M_2 are only 0.25 and 0.2, respectively, and ratios of K_1 to M_2 and O_1 to M_2 are less than 0.15 (Zhao et al., 2006). Tidal waves in this region propagate into Narragansett Bay–Mount Hope Bay from the inner New England Shelf. The M_2 amplitude of the wave is about 46–48 cm at the entrance of the bay, and about 58–59 cm at the northern end (Fig. 9.3). Co-phase lines are oriented northeast–southwest, with a phase difference of about 8° between the entrance and the northern reaches. Due to Coriolis effects, tidal elevation is slightly higher on the right side coast (east) than on the left side coast (west). Tidal variation in Mount Hope Bay is mainly controlled by wave propagation through the narrow Narragansett Bay–Mount Hope Bay channel. For a given latitude, the tidal phase in Mount Hope Bay lags that of upper northwest Narragansett Bay by only 1–2°.

The amplitude of the tidal currents in Narragansett Bay–Mount Hope Bay varies significantly with location due to variable bathymetry and local acceleration driven by narrow passages between the islands in the bay. Tidal ellipses are generally oriented parallel to local isobaths (Fig. 9.4). Eddies are generated by the separation of tidal currents in diverging channels and around islands and coastal headlands. In deep channels and passages, the maximum speed of currents may exceed 100 cm s^{-1} (Zhao et al., 2006).

Freshwater discharge into the bay ecosystem derives primarily from three major rivers: the Taunton River at the northeastern head of Mount Hope Bay, and the Blackstone and Pawtuxet Rivers at the northwestern head of Narragansett Bay. These rivers drain approximately 4,500 km^2 of adjacent watershed in Massachusetts and Rhode Island (Pilson, 1985). The annual average discharge rate (based on outflow data from 1929–2003) is about 14 m^3 s^{-1} for the Taunton River, 22 m^3 s^{-1} for the Blackstone River, and 10 m^3 s^{-1} for the

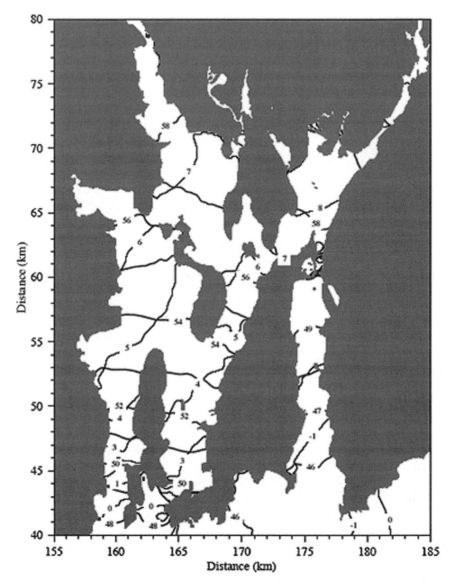

Fig. 9.3 Co-tidal chart of the model-predicted M_2 tidal elevation. Heavy solid line: Co-amplitude (cm) and thin solid line: co-phase (°G). This figure is adopted directly from Zhao *et al.* (2006).

Pawtuxet River (Fig. 9.5). Peaks in river discharge are generally found in December and March, with a monthly-averaged volume flux of approximately $40 \ m^3 \ s^{-1}$. There is significant inter-annual variability in the timing and magnitude of the peak discharge rate. In 1972, for example, the maximum discharge rate in the Blackstone exceeded $60 \ m^3 \ s^{-1}$ in December and $80 \ m^3 \ s^{-1}$ in March.

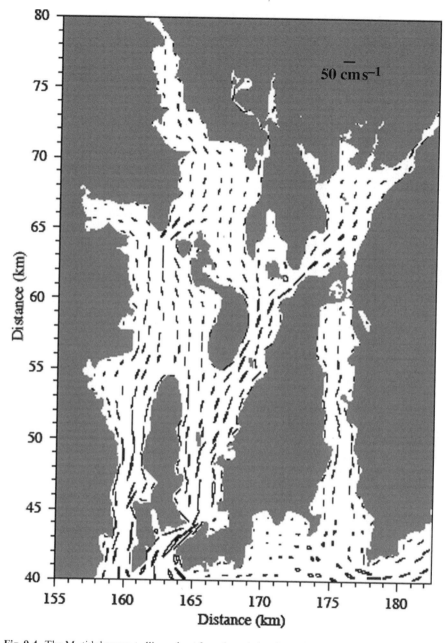

Fig. 9.4 The M_2 tidal current ellipse chart for selected sites from the model results. This figure is adopted directly from Zhao *et al.* (2006).

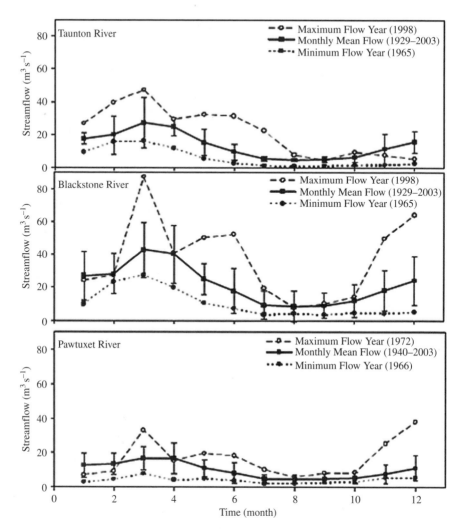

Fig. 9.5 Monthly freshwater discharges for the Taunton, Blackstone, and Pawtuxet Rivers. The data used for this analysis were downloaded from the USGS website.

That year also featured an additional late spring peak in June exceeding 40 m^3 s^{-1}. In 1965, a year of anomalously low discharge, the maximum discharge rate in the Blackstone was only about 20 m^3 s^{-1}.

River discharge has a direct impact on the seasonal variation of near-surface stratification and nutrient loading into the bay (Weisberg, 1976; Weisberg and Sturges, 1976; Kremer and Nixon, 1978). Buoyancy-induced flow in Narragansett Bay–Mount Hope Bay is driven mainly by freshwater discharge from rivers.

Wind forcing in Narragansett Bay–Mount Hope Bay is highly variable, both temporally and spatially. While synoptic-scale (2–5 days) fluctuations are

Table 9.1 Correlation coefficients of the wind velocity at six meteorological measurement stations in the Narragansett Bay–Mount Hope Bay system.

u v	Providence	Conimicut Light	Potter Cove	Quonset Point	Newport	Buzzards Bay
Providence	1	0.64	0.63	0.57	0.49	0.56
Conimicut Light	0.88	1	0.73	0.47	0.40	0.66
Potter Cove	0.85	0.86	1	0.68	0.65	0.73
Quonset Point	0.89	0.94	0.91	1	0.73	0.64
Newport	0.81	0.85	0.89	0.90	1	0.63
Buzzards Bay	0.75	0.82	0.83	0.83	0.90	1

u = the cross-shelf component; v = the along-shelf component.

driven by large-scale weather patterns, sea breezes dominate the daily variation. A correlation analysis of wind direction from measurement sites in Newport, Fall River, Conimicut Point, Providence, and Quonset Point found that wind is highly correlated ($r = 0.75$–0.94) along the coastal direction, but has reduced correlation in the cross-coastal direction (Table 9.1). These findings are consistent with the notion that local sea breeze, generally from the southwest, is a significant component of wind variability over Narragansett Bay. This supports the result from correlation analysis that wind variability in Narragansett Bay–Mount Hope Bay is controlled by short-term fluctuations induced by local sea breeze as well as the episodic passage of atmospheric fronts.

9.3 An Unstructured Grid Narragansett Bay–Mount Hope Bay Model

In the last decade, the basic circulation in Narragansett Bay–Mount Hope Bay has been examined using various oceanographic models (Gordon and Spaulding, 1987; Swanson and Jayko, 1987; Spaulding et al., 1999), all of which were discretized using structured grids. Gordon and Spaulding (1987) applied a traditional finite-difference model to simulate the tidal motion in Narragansett Bay. Forced by the M_2 and M_4 tidal constituents at open boundaries, their model successfully reproduced the M_2- and M_4-induced tidal amplitude and phase in good agreement with tidal gauge observations. A similar effort was made by Spaulding et al. (1999), who included 37 tidal constituents for the purpose of improving the accuracy of tidal simulation. Scientists at Applied Science Associates, Inc. (ASA) applied a curvilinear structured grid coastal ocean model to evaluate the impact of warm water discharge on stratification and circulation in Mount Hope Bay (Swanson and Jayko, 1987). While the curvilinear coordinate model provided improved resolution of the coastline relative to previous studies using Cartesian coordinate models, limitations in their model resolution led to the diffusion of modeled plume. Spaulding and Swanson provide a greater account on the above in Chapter 8.

Most coastal ocean models are based on the same governing equations—the hydrostatic primitive equations (HPE). Driven by the same external forcing, these models should converge toward the same solution as the grid resolution is increased. Ultimately, however, the efficiency by which the models can resolve relevant processes depends on the spatial order of accuracy and the mesh type used for discretization (Chen et al., 2007). The regions at Narragansett Bay–Mount Hope Bay are separated by narrow openings that control the exchange of water among them. The inability to resolve transport through these links will degrade the overall simulation of bay circulation.

A model for resolving the fundamental processes that control circulation in Narragansett Bay–Mount Hope Bay requires: (1) grid flexibility to resolve complex coastline and bathymetry; (2) mass conservation to accurately simulate water, heat, salt, and nutrient transports; (3) proper parameterization of vertical mixing to simulate tidal and wind mixing; and (4) the capability to assimilate observed quantities as real-time atmospheric and coastal ocean measurements become more easily available.

Funded by the Brayton Point Power Plant, the Marine Ecosystem Dynamics Laboratory at the University of Massachusetts-Dartmouth has developed an integrated model for the Narragansett Bay–Mount Hope Bay region. The major components of this system include: (1) a meso-scale atmospheric model (MM5) (Chen et al., 2005), (2) the unstructured-grid Finite-Volume Coastal Ocean circulation Model (FVCOM), and (3) a lower trophic-level food web model. FVCOM is the key component of this integrated system; it solves the hydrostatic primitive equations on unstructured triangular meshes using finite-volume discretization of spatial derivatives (Chen et al., 2003; Chen et al., 2006a,b). Furthermore, it is fully parallelized for efficient multiprocessor execution (Cowles, 2007). Like other coastal models, FVCOM uses the modified Mellor and Yamada level 2.5 (MY-2.5) and Smagorinsky turbulent closure schemes for vertical and horizontal mixing, respectively (Smagorinsky, 1963; Mellor and Yamada, 1982; Galperin et al., 1988), and a sigma coordinate to follow bottom topography. The General Ocean Turbulence Model (GOTM) (Burchard et al., 1999; Burchard, 2002) has been added to FVCOM to provide additional vertical closure schemes. The wet/dry point-treatment method may be incorporated to simulate the flooding/drying process on inter-tidal wetlands. Unlike the existing coastal finite-difference and finite-element models, FVCOM can be solved numerically by calculating fluxes resulting from discretization of the integral form of governing equations on an unstructured triangular grid. This approach combines the best features of finite-element methods (grid flexibility) and finite-difference methods (numerical efficiency and code simplicity), and provides a good numerical representation of momentum, mass, salt, heat, and tracer conservation. An advantage of an unstructured grid model is that grid density can vary spatially, and thus the mesh can be tailored locally to resolve critical processes as well as the complex coastline. This new model is extremely well suited to regions with complex coastlines and bathymetry, meets the requirements described above, and has been successfully applied to simulate circulation in Narragansett Bay–Mount Hope Bay for the last two years (Zhao *et al.*, 2006).

For circulation modeling in Narragansett Bay, a mesh covering the entire Narragansett Bay–Mount Hope Bay region is used (Fig. 9.6). The open boundary of the computational domain forms an arc that runs from Rhode Island Sound to Buzzards Bay. The mesh used here has a variable horizontal grid scale, which ranges from a minimum of 10 m in the northern reaches of Narragansett Bay and Mount Hope Bay to a maximum of 5 km near the open boundary. The vertical coordinate is discretized using 30 equally-spaced layers.

FVCOM has been successfully used to simulate the tidal, river-discharge, and wind-driven currents and the thermal plume in Narragansett Bay–Mount Hope Bay. The model results were validated with direct comparisons to relevant field measurement data. A summary of the model results can be found in Zhao *et al.* (2006).

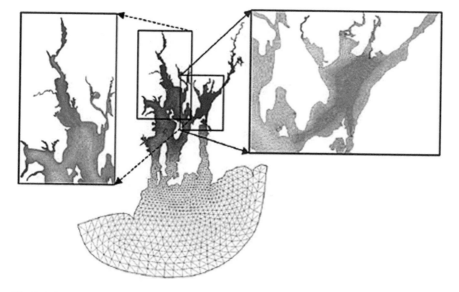

Fig. 9.6 Unstructured grid of the Narragansett Bay–Mount Hope Bay finite-volume coastal ocean model (FVCOM). The horizontal resolution ranges from 10–50 m in upper Narragansett Bay to 100 m over the shelf. This is the new configuration modified from our existing FVCOM to match the dynamic requirements to correctly resolve the tidal flushing and current separation as well as the river plume. 31-sigma levels are used in the vertical.

9.4 Numerical Process Studies in Narragansett Bay–Mount Hope Bay

Process studies were carried out using the FVCOM model to determine the numerical parameters required for resolving principle circulation in Mount Hope Bay, and the extent of the thermal plume generated by the cooling water discharge. These studies included investigations of initial conditions, horizontal and vertical grid resolution, and model time step. Although often disregarded, grid resolution studies are an important component of any model validation effort. The expected resolution is problem dependent, and a careful grid convergence study is necessary to determine proper grid scales.

9.4.1 Case 1: Tidal Flushing in the Narragansett Bay–Mount Hope Bay Channel

Narragansett Bay, Mount Hope Bay, and the Sakonnet River Narrows form an integrated dynamic system. Water exchange between greater Narragansett Bay and Mount Hope Bay occurs principally through the narrow neck separating the two waterbodies. The maximum tidal current through this opening can exceed 100 cm s^{-1}, and current separation on the diverging (downstream) side of the opening generates flow reversal, forming eddies on each side of the current jet around the time of maximum flood (Zhao *et al.*, 2006). These eddies were evident in the measurements made by Kincaid (in press) in the Narragansett Bay–Mount Hope Bay channel. Due to interactions between Narragansett Bay–Mount Hope Bay and Mount Hope Bay–Sakonnet River Narrows water exchanges, water transport in Mount Hope Bay–Sakonnet River Narrows leads the Narragansett Bay–Mount Hope Bay channel by ~90°, even though the transport volume is smaller by a factor of five (Zhao *et al.*, 2006). This phase lead was also observed by Kincaid (2006).

To resolve circulation through these narrow openings, including the development and dissolution of time-dependent lee side eddies, a model must have adequate mesh spacing to resolve the detailed geometries of these narrow necks. The Sakonnet River Narrows are 1,500 m in length and 7 m in mean water depth, and are characterized by two necks with a width of about 70 m at Sakonnet River Bridge and 120 m at Stone Bridge, respectively (Fig. 9.1). The previous models used in this region were discretized using a curvilinear coordinate system (Gordon and Spaulding, 1987; Swanson and Jayko, 1987; Spaulding *et al.*, 1999; Sullivan and Kincaid, 2001). With a horizontal resolution of ~200 m, these models were not able to resolve the geometry of the two narrow necks in the Sakonnet River Narrows, and thus we do not expect that they could simulate a realistic water exchange process between Narragansett Bay and Mount Hope Bay. By using flexible unstructured grids, FVCOM is able to adapt to local scales using variable grid resolution, resulting in a more efficient use of grid cells and enabling the model to resolve tidal-driven water transport processes between Narragansett Bay and Mount Hope Bay.

The current separation observed in our process studies was not reported in previous modeling work. This is likely because the 200-m grid scale used in the previous work was not sufficient to resolve the lateral shear of current jet in the narrow (800 m) Narragansett Bay–Mount Hope Bay channel. After observing the implications of resolution on the computed water exchange, we sought to determine the necessary grid scale for the Narragansett Bay–Mount Hope Bay channel.

Using nominal mesh spacing of 50 and 200 m in the Narragansett Bay–Mount Hope Bay channel, we examined the temporal response of circulation in the channel generated by tidal forcing at the open boundary. In a higher resolution (50 m), the model is able to predict current separation and subsequent eddy shedding. During flood, water that originates from the East Passage flows around Hog Island, merges on the eastern side of the island, and then flushes into Mount Hope Bay through the deep channel (Fig. 9.7a). Current separation on the lee side of the channel generates eddies on each side of the current jet around the time of maximum flood. The eddy on the eastern side enlarges, and its center migrates northeastward after the time of maximum flood, while the eddy on the western side intensifies as a result of an increase of the southward along-coastal tidal flow during the late flood phase. During the ebb period, when tidal currents are reversed, current separation occurs on the Narragansett Bay side of the Narragansett Bay–Mount Hope Bay channel. In this case, the eddy on the northern coast forms around the time of maximum ebb, while the eddy on the southern coast forms several hours after the maximum ebb (Fig. 9.7b).

With a low-resolution (200 m) grid, the model predicts strong tidal currents during the flood and ebb period through the Narragansett Bay–Mount Hope Bay channel. During flood, an eddy clearly forms on the eastern side of the current jet, but there is no clear evidence of flow reversal on the western side (Fig. 9.8a). During the ebb period, no significant eddy shedding was resolved on the lee side of tidal flushing in Narragansett Bay (Fig. 9.8b). The lateral shear of tidal current was much larger in high resolution than in low resolution (Fig. 9.9). Resolving the lateral shear of tidal current jet is critical for the prediction of timing and strength of current separation in a diverging channel (Stommel and Farmer, 1952; Wells and Heijst, 2003).

9.4.2 Case 2: Thermal Plume

One of main features driving the circulation and hydrography in Mount Hope Bay is the thermal plume resulting from cooling water discharge from the Brayton Point Power Plant. The temporal and spatial structure of this plume

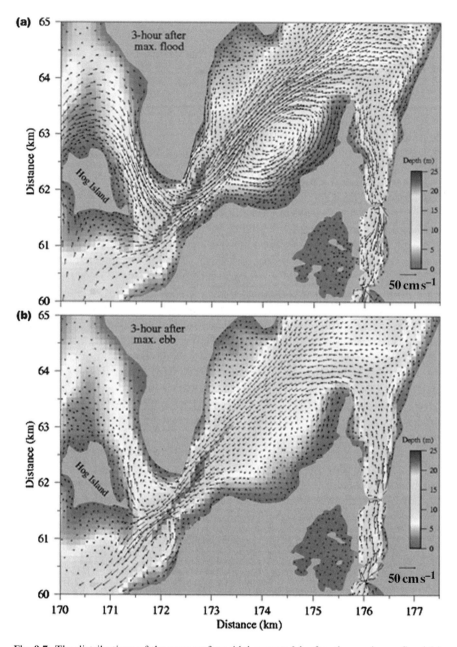

Fig. 9.7 The distributions of the near-surface tidal current 3 h after the maximum flood (a) and 3 h after the maximum ebb (b) in the southern part of the Mount Hope Bay–Narragansett Bay–Sakonnet River regions for the model run with a horizontal resolution of 50 m in the Narragansett Bay–Mount Hope Bay channel. The image shows the bathymetry with depth scales from 0 to 25 m. The number of current points is reduced in the high-resolution region to improve figure clarity.

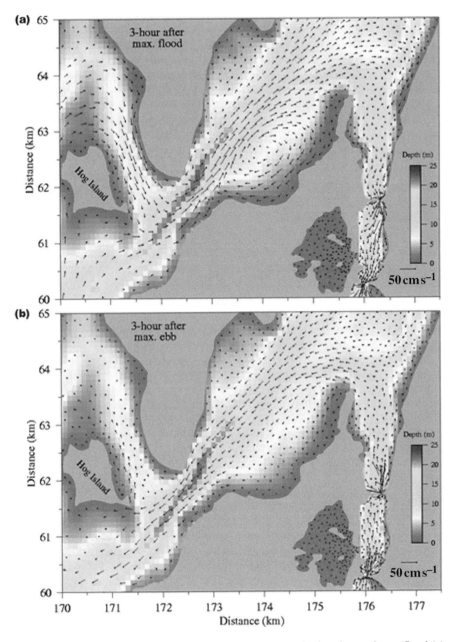

Fig. 9.8 The distributions of the near-surface tidal current 3 h after the maximum flood (a) and 3 h after the maximum ebb (b) in the southern part of the Mount Hope Bay–Narragansett Bay–Sakonnet River regions for the model run with a horizontal resolution of 200 m in the Narragansett Bay–Mount Hope Bay channel. The image shows the bathymetry with depth scales from 0 to 25 m. The number of current points is reduced in the high-resolution region to improve figure clarity.

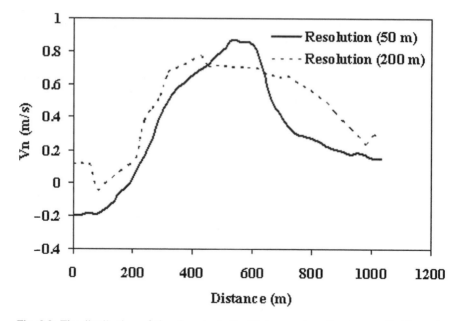

Fig. 9.9 The distribution of the along-isobaths tidal current on the transect L (shown in Fig. 9.1) for the model runs with horizontal resolution of 50 and 200 m in the channel, respectively.

was observed using thermal imaging from an aircraft in September 1998 by scientists at Brown University. The images were taken through the course of one tidal cycle; they clearly show that the basic plume structure is a narrow jet that meanders widely through tidal motion. To map the vertical and lateral structure of the thermal plume, a team of scientists from the University of Massachusetts-Dartmouth (UMASSD) and Woods Hole Oceanographic Institution (WHOI) conducted a high-resolution tow-yo CTD/ADCP survey with repeated tracks across the thermal plume on August 19, 2005, a time of perigee spring tides. Two drifters with portable global positioning system recorders were released at the time of maximum ebb tidal current at the center of the thermal plume to track the trajectory of discharge water. The CTD/ADCP data showed that the plume was a narrow current jet with a width of $\sim$30–40 m, a vertical scale of $\sim$2 m and a maximum velocity exceeding 60 cm s^{-1} during the ebb. A weak return flow was detected on the western side of the plume jet. This evidence was supported by the trajectories of two drifters, which turned clockwise and moved toward the western coast of Mount Hope Bay.

The thermal plume influences the circulation in Mount Hope Bay through a contribution of kinetic energy and buoyancy. Failure to resolve the plume and associated density gradients can lead to unrealistic modeling of buoyancy-driven circulation and stratification. In the previous efforts to study the plume using a curvilinear mesh with inadequate resolution, the plume size was over-predicted, and consequently an exaggeration of the area of influence of warm water (Swanson and Jayko, 1987).

To examine the sensitivity of model-simulated thermal plume to grid spacing, the FVCOM Narragansett Bay–Mount Hope Bay model was executed using horizontal resolutions of 50, 25, 10 and 5 m in northern Mount Hope Bay. The associated impact on water exchange between Mount Hope Bay and Narragansett Bay was also evaluated. For these experiments, the initial conditions of water temperature and salinity were specified using an August climatologic field constructed from historical databases spanning the period 1959–2005. For the same initial conditions (temperature and salinity) and forcing (thermal flux from the power plant and tides), the width of the thermal plume becomes smaller as horizontal resolution increases. Grid convergence is reached when the horizontal resolution is reduced to 10 m. The model comparison results clearly show that insufficiency in horizontal resolution can lead to an over-estimation of lateral diffusion, thus exaggerating the region of influence of heat flux from the power station.

An accurate determination of the contribution of plume heat to the overall temperature in Mount Hope Bay requires an accurate prediction of water exchange between Mount Hope Bay and the adjacent regions. Kincaid (in press) used repeated ADCP transects across the Narragansett Bay–Mount Hope Bay channel to estimate volume transport. He measured the water transport to be $7.5 \times 10^3 \, \text{m}^3 \, \text{s}^{-1}$ at the current peak of spring tide. The FVCOM-modeled volume transport for the same phase in spring tide was $5.2 \times 10^3 \, \text{m}^3 \, \text{s}^{-1}$, $6.2 \times 10^3 \, \text{m}^3 \, \text{s}^{-1}$, and $6.7 \times 10^3 \, \text{m}^3 \, \text{s}^{-1}$ for the cases with horizontal resolutions of 50, 10, and 5 m in the thermal plume area, respectively. Inadequate resolution reduces the exchange of heat between Mount Hope Bay and Narragansett Bay, and can lead to incorrect conclusions on the impact of the thermal plume on Mount Hope Bay.

9.5 Summary

A recent increase in the frequency of hypoxic events in Narragansett Bay reinforces the need to understand the causes of these harmful ecological conditions, as well as the reasons for their observed spatial variability. A well-calibrated model used in concert with an appropriate observation network can elucidate some of the underlying physical mechanisms of ecosystem change in the Narragansett Bay–Mount Hope Bay ecosystem. The bay is divided naturally into subregions separated by narrow channels, and modeling the general circulation of the bay requires resolving the interchange through these narrow channels.

Process-oriented modeling experiments were used to examine the grid resolution required for resolving the exchange of water between Narragansett Bay and Mount Hope Bay, as well as the thermal plume generated by the cooling water discharge of the Brayton Point power station. Results of these experiments indicate that, due to insufficient computing, past modeling efforts lacked the necessary mesh resolution to capture complex circulation generated by tidal forcing through the Narragansett Bay–Mount Hope Bay channel, or the extent of the thermal plume. With an unstructured grid approach, the flexibility in grid resolution allows efficient placement of mesh cells as well as accurate rendering of the detailed coastline. In combination with recent advances in computing, the unstructured grid method offers the potential for grid-converged solutions of the processes controlling circulation in the ecosystem.

It should be pointed out here that the model results for tidal flushing presented in this chapter were based on numerical experiments conducted without river discharge or surface forcing being considered. Because eddies generated by tidal flushing through the Narragansett Bay–Mount Hope Bay channel occur in the shallower area (where water depth is <4 m), they may not be observed when wind is present. The wind in Narragansett Bay and Mount Hope Bay can vary significantly during the day due to local sea breeze. In such shallow areas, wind-driven current is usually in the same direction of surface wind stress (Chen, 2000). Eddies can disappear when the wind-induced current is larger than the tidal-induced eddy current.

Acknowledgment This research is supported by a private research fund awarded by the Fall River Mount Hope Bay Power Plant and Rhode Island Sea Grant under contract number NA040AR4170062. We would like to thank Dr Robert Beardsley at Woods Hole Oceanographic Institution (WHOI) for his encouragement as well as his collaboration in the development of FVCOM, and Jim Churchill (WHOI) for conducting the drifter experiments in the August 2005 Mount Hope Bay survey. We would also like to thank the members of the Marine Ecosystem Dynamics Modeling (MEDM) Laboratory at the School for Marine Science and Technology, University of Massachusetts-Dartmouth for their support in FVCOM development. This paper is #06-1102 in the SMAST Contribution Series, School for Marine Science and Technology, University of Massachusetts-Dartmouth.

References

Bergondo, D.L., Kester, D.R., Stoffel, H.E., and Woods, W. 2005. Time-series observations during the low sub-surface oxygen events in Narragansett Bay during summer 2001. *Marine Chemistry* 97:90–103.

Burchard, H. 2002. Applied Turbulence Modeling in Marine Waters. New York: Springer. 215 pp.

Burchard, H., Bolding, K., and Villarreal, M.R. 1999. GOTM–A general ocean turbulence model. Theory, applications and test cases. Technical Report, EUR 18745 EN, European Commission.

Chen, C. 2000. A modeling study of episodic cross-frontal water transports over the inner shelf of the South Atlantic Bight. *Journal of Physical Oceanography* 30:1722–1742.

Chen, C., Liu, H., and Beardsley, R. 2003. An unstructured grid, finite-volume, three-dimensional, primitive equations ocean model: application to coastal ocean and estuaries. *Journal of Atmospheric and Oceanic Technology* 20(1):159–186.

Chen, C., Beardsley, R.C., Hu, S., Xu, C., and Lin, H. 2005. Using MM5 to hindcast the ocean surface forcing fields over the Gulf of Maine and Georges Bank region. *Journal of Atmospheric and Oceanic Technology* 22(2):131–145.

Chen, C., Beardsley, R.C., and Cowles, G. 2006a. An unstructured grid, finite-volume coastal ocean model (FVCOM) system. Special Issue "Advances in Computational Oceanography", *Oceanography* 19(1):78–89.

Chen, C., Cowles, G., and Beardsley, R.C. 2006b. An unstructured grid, finite-volume coastal ocean model: FVCOM User Manual. Second Edition. SMAST/UMASSD Technical Report-06-0602, pp. 315.

Chen, C., Huang, H., Beardsley, R.C., Liu, H., Xu, Q., and Cowles, G. 2007. A finite-volume numerical approach for coastal ocean circulation studies: comparisons with finite-difference models. *Journal of Geophysical Research* 112.

Cowles, G.W. 2007. Parallelization of the FVCOM Coastal Ocean Model. International Journal of High Performance Computing Applications, in press.

Deacutis, C.F., Murray, D.W., Prell, W.L., Saarman, E., and Korhun, L. 2006. Hypoxia in the Upper Half of Narragansett Bay, RI during August 2001 and 2002, *Northeast Naturalist* 13(Special Issue 4):173–198.

Galperin, B., Kantha, L.H., Hassid, S., and Rosati, A. 1988. A quasi-equilibrium turbulent energy model for geophysical flows. *Journal of Atmospheric Sciences* 45:55–62.

Gordon, R., and Spaulding, M.L. 1987. Numerical simulation of the tidal and wind driven circulation in Narragansett Bay. *Estuarine, Coastal, and Shelf Science* 24:611–636.

Hicks, S.D. 1959. The physical oceanography of Narragansett Bay. *Limnology and Oceanography* 4:316–327.

Jeffries, P. 2002. Rhode Island's ever-change Narragansett Bay. *Maritimes* 41:1–5.

Karentz, D., and Smayda, T.J. 1998. Temporal patterns and variations in phytoplankton community organization and abundance in Narragansett Bay during 1959–1980. *Journal of Plankton Research* 20:145–168.

Keller, A.A., Oviatt, C.A., Walker, H.A., and Hawk, J.D. 1999. Predicted impacts of elevated temperature on the magnitude of the winter-spring phytoplankton bloom in temperate coastal waters: a mesocosm study. *Limnology and Oceanography* 44:344–356.

Keller, A.A., Taylor, C., Oviatt, C., Dorrington, T., Holcombe, G., and Reed, L. 2001. Phytoplankton production patterns in Massachusetts Bay and the absence of the 1998 winter-spring bloom. *Marine Biology* 138:1051–1062.

Kincaid, C. 2006. The exchange of water through multiple entrances to the Mt. Hope Bay estuary. *Northeast Naturalist* 13(Special Issue 4):117–144.

Kremer, J.N., and Nixon, S.W. 1978. A coastal marine ecosystem: Simulation and analysis. New York: Springer Verlag.

Li, Y., and Smayda, T.J. 1998. Temporal variability of chlorophyll in Narragansett Bay, 1973–1990. *ICES Journal of Marine Science* 55:661–667.

Mellor, G.L., and Yamada, T. 1982. Development of a turbulence closure model for geophysical fluid problem. *Reviews of Geophysics and Space Physics* 20:851–875.

Nixon, S.W., Granger, S., Buckley, B.A., Lamont, M., and Rowell, B. 2004. A one hundred and seventeen year coastal water temperature record from Woods Hole, Massachusetts. *Estuaries* 27:397–404.

Oviatt, C.A. 2004. The changing ecology of temperate coastal waters during a warming trend. *Estuaries* 27:895–904.

Pilson, M.E.Q. 1985. On the residence time of water in Narragansett Bay. *Estuaries* 8:2–14.

Smagorinsky, J. 1963. General circulation experiments with the primitive equations, I. The basic experiment. *Monthly Weather Review* 91:99–164.

Spaulding, M.L., and White, F.M. 1990. Circulation dynamics in Mt. Hope Bay and the lower Taunton River, Narragansett Bay. *In* Coastal and Estuarine Studies, Vol. 38. Cheng, R.T. (ed.) New York: Springer-Verlag.

Spaulding, M.L., Mendelsohn, D., and Swanson, J.C. 1999. WQMAP: an integrated three-dimensional hydrodynamic and water quality model system for estuarine and coastal applications. *Marine Technology Society Journal Special issue on State of the Art in Ocean and Coastal Modeling* 33(3):38–54.

Stommel, H., and Farmer, H.G. 1952. On the nature of estuarine circulation. WHOI Technical Report 52–88, 131 pp.

Sullivan, B.K., and Kincaid, C. 2001. Modeling circulation and transport in Narragansett Bay. AGU Spring Meeting Abstract, 21, OS21A05S.

Sullivan, B.K. and D.V. Keuren, 2004. Climate change and zooplankton predator-prey dynamics. Presentation at Science Symposium 2004 "State of Science Knowledge on Nutrients in Narragansett Bay". 17–18 November, Block Island, RI.

Sullivan, B.K., Van Keuren, D., and Clancy, M. 2001. Timing and size of blooms of the ctenophore *Mnemiopsis leidyi* in relation to temperature in Narragansett Bay, RI. *Hydrobiologia* 451:113–120.

Swanson, J.C. and K. Jayko. 1988. A simplified estuarine box model of Narragansett Bay. Final report prepared for the Narrangansett Bay Project and the Environmental Protection Agency, Region I. Applied Science Associates, Inc. No. 85–11. 90 pp.

US Geological Survey (USGS) 2007. *http://ma.water.usgs.gov/basins*.

Weisberg, R.H. 1976. The non-tidal flow in the Providence River of Narragansett Bay: a stochastic approach to estuarine circulation. *Journal of Physical Oceanography* 6:721–734.

Weisberg, R.H., and Sturges, W. 1976. Velocity observations in the west passage of Narragansett Bay: a partially mixed estuary. *Journal of Physical Oceanography* 6:345–354.

Wells, M.G., and Heijst, G. 2003. A model of tidal flushing of an estuary by dipole formation, *Dynamics of Atmosphere and Oceans* 37:223–244.

Zhao, L., Chen, C., and Cowles, G. 2006. Tidal flushing and eddy shedding in Mount Hope Bay and Narragansett Bay: an application of FVCOM. *Journal of Geophysical Research* 111, C10, C10015.

Chapter 10
The Dynamics of Water Exchange Between Narragansett Bay and Rhode Island Sound

Christopher Kincaid, Deanna Bergondo and Kurt Rosenberger

10.1 Introduction

The Narragansett Bay estuary represents an important natural and economic resource. Narragansett Bay has a long history of scientific study, which makes it ideal for understanding the interplay between anthropogenic impacts, estuarine science and management. Over the past 40 years, numerous studies have focused on the biological and chemical processes within Narragansett Bay (e.g., Hicks, 1959; Kremer and Nixon, 1978; Pilson, 1985; Keller, 1988; Bender et al., 1989; Hinga et al., 1989; Nixon, 1997; Granger and Buckley, 1999; Keller et al., 1999; Brush, 2002; Oviatt et al., 2002; Prell et al., 2004; Bergondo et al., 2005). Regions of the bay have been instrumented for long-term monitoring projects (e.g., NB-PORTS sites; Rhode Island Department of Environmental Management buoys) and are sampled by monthly surveys with towed instruments.

Observations suggest that large-scale, climate-induced changes (Hawk, 1998) have occurred and may be linked to modifications in the bay's ecosystem (Hawk, 1998; Oviatt et al., 2002; Sullivan and Van Keuren, 2003). Anthropogenic and natural stresses are also apparent in the increasing number and severity of low oxygen events in more developed regions of the bay (Saarman et al., 2002; Bergondo, 2004; Bergondo et al., 2005; Deacutis et al., 2006; Chapters 11 and 12). Large-scale engineering projects will also have an impact; such as recent dredging of the Providence River shipping channel and a storm-water holding facility planned for Providence to reduce storm-water discharge and ultimately nutrient flux.

Estuarine circulation lies at the heart of multidisciplinary approaches to coastal management. Ecosystem-based models require information on the mixing, flushing and transport of water between sub-regions of the estuary. Moreover, bio-chemical processes, such as those controlling oxygen levels within Narragansett Bay, are also influenced by how efficiently water is

Christopher Kincaid

Graduate School of Oceanography, University of Rhode Island, Narragansett, RI 02882

kincaid@gso.uri.edu

A. Desbonnet, B. A. Costa-Pierce (eds.), *Science for Ecosystem-based Management.*   301
© Springer 2008

exchanged between the estuary and the shelf. The combination of spatially and temporally detailed data is used to characterize circulation within lower Narragansett Bay, and develop models for Narragansett Bay–Rhode Island Sound exchange patterns over multiple time scales and in response to various forcing parameters (tides, wind, and runoff and seasonal density variations).

10.2 Background

The Narragansett Bay estuary is dominated by two distinct north-south oriented branches referred to as the East and West Passages (Fig. 10.1), which are long, linear channels connecting the broad, shallow regions of upper Narragansett Bay and Rhode Island Sound. These serve as conduits both for the flushing of impacted waters of upper Narragansett Bay (Saarman et al., 2002; Bergondo, 2004; Bergondo et al., 2005; Deacutis et al., 2006) and for the re-supply of cooler, saltier, less impacted shelf water of Rhode Island Sound (Shonting and Cook, 1970). The West Passage extends ~55 km from the mouth of the bay northward to a 2-km wide constriction (Warwick Neck to Patience Island), where it connects to upper Narragansett Bay. It has an aspect ratio (length to width) of ~8, with slightly greater widths in the north (~14 km) compared to the southern section (~4 km). The East Passage has a similar length, with an aspect ratio of roughly 10, with widths between 3 and 10 km. The East Passage branches between Mount Hope Bay and the broad shallow Ohio Ledge region of the upper bay. The two channels meet midway up from the mouth, and again through Ohio Ledge.

Bottom topography in the West Passage is, like much of the bay, relatively shallow (6–16 m) (Fig. 10.1). The East Passage is significantly deeper (16–48 m). The deep, southern portion of the East Passage is separated from the shelf by a relatively shallow (30 m) section, similar in some respects to a fjord. The inner shelf of Rhode Island Sound is a broad, gradually deepening region (35–40 m) with multiple connections to lower Narragansett Bay, the Sakonnet River, Buzzards Bay, Block Island Sound and the New England shelf.

Narragansett Bay has been described as a partially to well-mixed estuary (Goodrich, 1988). Recent time series measurements collected in upper Narragansett Bay show that stratification levels vary systematically with the spring–neap cycle, and show complex spatial and temporal patterns with different wind forcing scenarios (Bergondo, 2004). Previous observational and modeling studies have highlighted the importance of wind forcing on Narragansett Bay circulation (Weisberg, 1972; 1976; Weisberg and Sturges, 1976; Gordon and Spaulding, 1987; DeLeo, 2001; Rosenberger, 2001).

Freshwater discharge into Narragansett Bay is low relative to other major east coast estuaries (Goodrich, 1988). Freshwater enters primarily through the Providence and Taunton Rivers with estimates for total discharge ranging from ~20 m^3 s^{-1} in late summer-fall to ~300 m^3 s^{-1} under peak runoff conditions

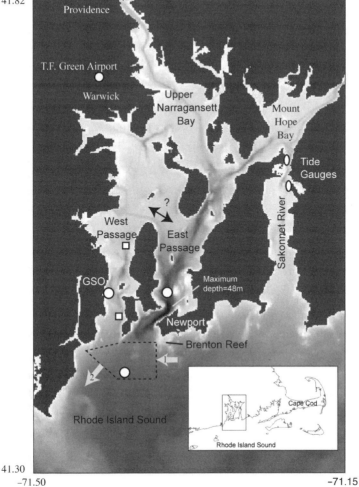

Fig. 10.1 The study area of Narragansett Bay and Rhode Island Sound, with its location in the larger region of the southern New England coast (inset). Bathymetry is presented in gray scale (darker gray = deeper water) with bottom mounted ADCPs (white circles) and survey grid (dashed line; Kincaid *et al.,* 2003) shown. Large arrows (1, 2) show prevailing summertime residual transport; winter flow represented by arrow 1 is not observed. Locations for data from other studies are shown (squares: Weisberg and Sturges, 1976; ovals: DeLeo, 2001).

during the winter–spring months (Pilson, 1985). Underway Acoustic Doppler Current Profiler (ADCP) surveys at the mouth of Narragansett Bay (Fig. 10.1) show that residual transport from the estuary is significantly larger than previously reported runoff values ($\sim$1,000 m^3 s^{-1}) (Kincaid *et al.,* 2003), where higher winter runoffs are largely accommodated through the West Passage. One potential source for higher residual flows from the bay is submarine groundwater discharge (Moore, 1996; 1997; Taniguchi *et al.*, 2002), which remains largely unconstrained for this system.

The water column in central and northern Rhode Island Sound is stratified in summer and vertically mixed in winter (Shonting and Cook, 1970; Rosenberger, 2001; Kincaid *et al.*, 2003). Thermal stratification is large within Rhode Island Sound in summer, with top to bottom differences in temperature and salinity of $\Delta T \sim 6°C$, $\Delta S < 2\%$. Stratification is weaker at the bay–sound interface ($\Delta T \sim 1.5°C$ vs $\Delta S \sim 1\%$) (Kincaid *et al.*, 2003). Strong perturbations in vertical temperature profiles and velocity structures within Rhode Island Sound are correlated with wind events, and show how shelf stratification breaks down in stages during the fall in response to a series of atmospheric cooling and wind events (Rosenberger, 2001). Time series records show that the upper Rhode Island Sound water column exhibits a net west to southwestward flow (Shonting, 1969; Rosenberger, 2001). The deeper non-tidal flow in the sound water is weak and variable, with a strong rotary component (Shonting, 1969).

A number of studies have employed current meter surveys to map spatial and temporal patterns in exchange near the interface between an estuary and the local shelf (Garvine, 1991; Valle-Levinson and Lwiza, 1995; Wong and Münchow, 1995; Valle-Levinson *et al.*, 1996, 1998; Kincaid *et al.*, 2003). Spatially detailed ADCP surveys and hydrographic measurements conducted at the Narragansett Bay–Rhode Island Sound interface reveal characteristics of how Rhode Island Sound water enters Narragansett Bay, and how Narragansett Bay water is advected away from the mouth (Kincaid *et al.*, 2003). These results are discussed in more detail below.

10.3 Methods

10.3.1 Data Collection

Monitoring stations were deployed for collecting velocity time series data to characterize estuarine–shelf exchange processes for lower Narragansett Bay (e.g., East Passage vs West Passage) and Rhode Island Sound for a range of synoptic scale wind forcing and seasonal effects. Velocity data were obtained from self-recording RD Instruments ADCPs deployed within the East Passage (300 kHz) and West Passage (1,200 kHz) of lower Narragansett Bay and outside the entrance to Narragansett Bay (600 kHz) (Fig. 10.1; Table 10.1). Stations within the lower East and West Passages reveal flow within the lower bay, both in terms of flushing of upper Narragansett Bay water and net exchange patterns between the two passages. The Rhode Island Sound site provides information on estuarine–shelf exchange. The East Passage station was located under the Newport Bridge, closer to the western side of the East Passage, in nearly the deepest portion of the channel. The West Passage site was positioned on the western side of the channel, outside of the deepest part of the channel, to

Table 10.1 Location and dates of BM-ADCP deployments.

Station	Location	kHz	Depth (m)	Bin size (m)	Deployment dates
Summer 2000					
Rhode Island Sound	41°24'34.7"N 71°23'19.2"W	600	30	1.0	5/12/2000–7/18/2000
East Passage	41°30'19.8"N 71°21'05.0"W	300	40	2.0	5/13/2000–7/17/2000
West Passage	41°29'50.0"N 71°24'60.0"W	1,200	12	0.5	5/17/2000–7/13/2000
Winter 2000–2001					
Rhode Island Sound	41°24'34.7"N 71°23'19.2"W	600	30	1.0	11/21/2000–1/14/2001
East Passage	41°30'19.8"N 71°21'05.0"W	300	40	2.0	11/15/2000–1/9/2001
Winter 2000					
East Passage	41°30'19.8"N 71°21'05.0"W	300	40	2.0	1/30/2000–3/20/2000
West Passage	41°29'50.0"N 71°24'60.0"W	1,200	12	0.5	2/2/2000–3/18/2000

avoid bottom dragging activity. Within the discussion section we attempt to develop spatial circulation maps by combining these results with prior data sets that were collected within the deeper, eastern portion of the West Passage (Weisberg and Sturges, 1976).

Velocity data were obtained at 6-min intervals for depth bins of 0.5, 1.0 and 2.0 m at West Passage, Rhode Island Sound and East Passage stations, respectively (Table 10.1). Each data ensemble was constructed by averaging information from a burst of ten pings, separated by 1 s. Data quality was generally good within 3 m of the water surface, using a threshold of 95% in the percent good statistics. The first good bin coincided with a height of ~2 m above the bottom.

Values for wind speed and direction were obtained from the National Climate Data Center (NCDC) for the T.F. Green station (PVD) in Warwick, RI (NCDC, Unedited Local Climatological Data at *http://www.ncdc.noaa.gov/oa/ncdc.html*). Daily mean values of freshwater discharge in cubic feet per second were tabulated over the Narragansett basin from the five major United States Geological Survey stream gauges in Rhode Island and Massachusetts (USGS 01112500, 0116500, 01114500, 01114000, and 01108000) (USGS Water Data at *http://ma.water.usgs.gov/basins/*).

10.3.2 Data Analysis

The least-squares harmonic analysis toolbox, T_tide, described in Pawlowicz *et al.,* (2002), was used to extract the tidal signal from the data. Data were decomposed into north–south and east–west components, and data gaps, all less than 1 h, were filled using a linear interpolation between neighboring depth bins.

A number of methods were used to examine long-period, non-tidal flow patterns in Rhode Island Sound for comparison with synoptic-seasonal scale variability in environmental conditions. Near-surface and near-bottom velocities were obtained by averaging the top and bottom five bins of data. Residual east–west and north–south velocity components were calculated by running a MATLAB 5th-order low-pass Butterworth filter to remove variability with periods shorter than 25 h. Hourly values were obtained by interpolating the filtered data for comparison with wind data. Cross-covariance values were determined between processed data and wind data.

10.4 Results

Circulation within Narragansett Bay and Rhode Island Sound is driven by a combination of tides, local and non-local wind forcing, and density differences. In this chapter, we primarily focus on non-tidal circulation within the two primary channels of Narragansett Bay, and between Narragansett Bay and Rhode Island Sound. Local versus non-local winds drive estuarine–shelf exchange in different ways (Wang and Elliott, 1978; Klinck et al., 1982). Winds directed parallel to the estuarine channel tend to produce local gradients in sea level height within the estuary, which drive vertically sheared flow that does not necessarily fill or flush the estuary. Alternatively, coastline parallel winds (normal to the estuarine channel), which drive Ekman transport and result in upwelling or downwelling of shelf water, have been attributed to more substantial estuarine–shelf exchange due to a combination of sea surface slope and baroclinic effects (slope of the pycnocline). For the Narragansett Bay–Rhode Island Sound system, this implies that westward winds drive surface inflow and deep outflow within Narragansett Bay, with an opposite response for eastward winds. Narragansett Bay offers a more complex situation, given the two primary connection points, and a third connection through the Sakonnet River (Fig. 10.1).

Results are summarized for instantaneous tidal velocities before focusing on how seasonal differences in residual flow and wind-induced flows influence estuarine–shelf exchange, given the complex geometry.

10.4.1 Tidally Driven Flow

The tidal forcing at each of the ADCP locations is dominated by the semidiurnal M2 frequency (Table 10.2), with higher M2 tidal amplitudes recorded in the East and West Passages of Narragansett Bay. The M2 amplitude within Rhode

Table 10.2 Tidal harmonics for the three BM-ADCP locations.

Component	Frequency (cph)	Newport Bridge amplitude	GSO Dock amplitude	Rhode Island Sound amplitude
K1	0.0418	25.06	16.53	11.93
N2	0.0790	90.50	39.23	22.49
M2	0.0805	368.57	240.91	88.79
S2	0.0833	88.38	60.51	17.21
K2	0.0836	24.05	16.47	4.68
M4	0.1610	96.07	56.09	20.34

Pawlowicz *et al.* (2002).

Island Sound is only 24% of the amplitude of the East Passage, and 36% of the West Passage value. The next largest tidal constituent in the East Passage is the M4, at 25% of the M2 amplitude, while the S2 and N2 constituents are the next highest amplitude constituents in the West Passage and Rhode Island Sound, respectively.

Near-surface tidal velocities in Narragansett Bay are consistently larger than near-bottom values. As observed in many estuaries (e.g., Valle-Levinson and Lwiza, 1995), near-bottom tidal currents in the East Passage lead near-surface currents by ~1 h. The phase advance of tidal velocities with depth was observed at stations located in the main channel of the West Passage by Weisberg and Sturges (1976), and was represented analytically using a simple model balancing pressure gradients and bottom friction. At our West Passage station, located on the western side of the channel in shallower water, surface and bottom tidal currents are nearly in phase.

Tidal velocities in the West Passage are predominantly axis-parallel (north–south) and vary between $+50/70$ and ±30 cm s^{-1} during spring and neap tides, or nearly 1.5 times higher than values recorded by Weisberg and Sturges (1976) for the deeper, central West Passage channel. East–west velocity fluctuations are <10 cm s^{-1} (<5 cm s^{-1}) in the surface (bottom) water. In the East Passage, peak instantaneous flows vary between $+40/-50$ and ±15 cm s^{-1} for spring and neap periods. East–west rates are slightly lower than north–south values, such that the tidal ellipse in the East Passage is oriented along a north–northeast trend. Tidal circulation in Rhode Island Sound is characterized by NW–SE oriented, elongated ellipses for near-surface water and weaker, more circular ellipses in the near-bottom water (Rosenberger, 2001).

10.4.2 Residual Flow

Non-tidal circulation between estuaries and shelf regions has been shown to be driven by water column density differences, and both local and non-local wind forcing (Elliott, 1978; Kjerve *et al.*, 1978; Klinck *et al.*, 1982; Goodrich, 1988). Seasonal differences in both wind forcing and residual exchange between

Narragansett Bay–Rhode Island Sound are summarized in Fig. 10.2–10.6. During summer, when Rhode Island Sound is stratified, we compare data from May 2000 to July 2000 at all three stations (Fig. 10.2). During the winter when Rhode Island Sound is vertically mixed, results are presented from two periods: November 2000 to January 2001 (Fig. 10.3) (when East Passage and Rhode Island Sound data overlap) and January to March 2000 (Fig. 10.4) (when East Passage and West Passage data overlap). Cross correlations calculated for East Passage and West Passage currents versus winds (Fig. 10.5) show that the East Passage response to winds is nearly identical between the two winter periods. This result, and the goal of deriving basic water column responses to gross seasonal differences and different representative wind conditions, provide support for overlaying results from the two distinct periods.

Winds are predominantly to the northeast during the summer period, but exhibit significant variations in magnitude and direction (Fig. 10.2). Residual

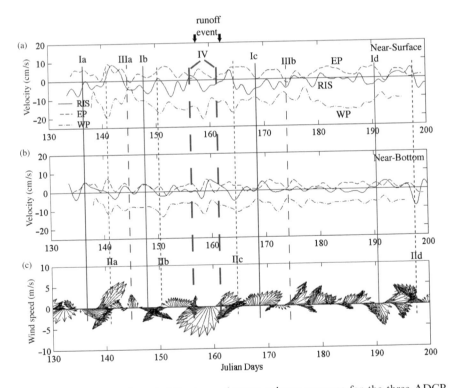

Fig. 10.2 Plots comparing wind forcing and water column response for the three ADCP stations during summer (stratified) conditions, from summer (May–July) 2000 deployment. Plots are: (a) north–south surface velocity, (b) north–south bottom velocity, and (c) wind vectors (showing the direction the wind is blowing) from T.F. Green Airport. Specific events are identified, where prevailing winds have large eastward (I), westward (II) or northward (III) components. The largest wind event is labeled IV.

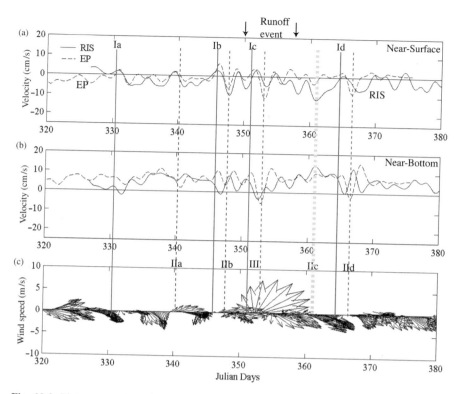

Fig. 10.3 Plots comparing wind forcing and water column response for the three ADCP stations during winter (November, 2000–January, 2001) deployment. Plots are: (a) north-south surface velocity, (b) north–south bottom velocity, and (c) wind vectors from T.F. Green Airport. Specific winds/water column response events are marked, including cases with westward (I), eastward (II) and northward (III) wind components.

flow in lower Narragansett Bay during this period is consistent with a large counterclockwise current, or gyre, running from the East Passage into West Passage (Fig. 10.7a). During a period when winds are consistently northeastward (decimal day 168 to 186) there is a persistent 10–15 cm s^{-1} southward (outflow) past the West Passage station, and a northward inflow past the East Passage station. Mean velocity at the East Passage station is ~5–10 and <5 cm s^{-1} in the near-surface and near-bottom portions of the water column. The mean north–south component of flow in the near-surface water at the Rhode Island Sound station during this time is southward or away from Narragansett Bay. Residual flow of near-bottom Rhode Island Sound water is negligible.

The relationship between residual flow and wind during winter conditions is shown in time series data in Figs. 10.3 and 10.4, and summarized in map view in Fig. 10.8a. Wind directions are predominantly eastward or southeastward during this period, with a few significant northeastward wind events. The

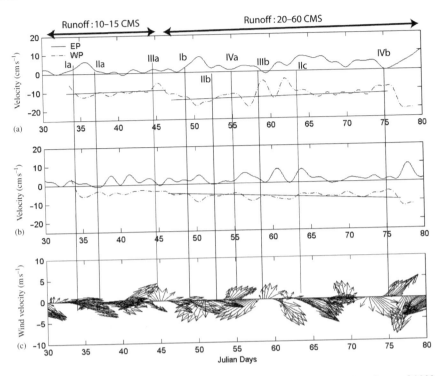

Fig. 10.4 Plots comparing wind forcing and water column response in the winter of 2000, where there is overlap between East Passage and West Passage data sets. Plots include: (a) near-surface low-pass filtered currents, (b) near-bottom low-pass currents and (c) winds. Specific wind events are labeled involving winds with I (westward), II (eastward) and III (northward) components. Events labeled IV highlight periods where winds progresses from northeastward to southwestward, with runoff events. Trends in mean velocity are highlighted for comparison with sharp change in runoff.

early winter data set (Fig. 10.3) shows net outflow from the West Passage and a net inflow to the East Passage, consistent with an overall pattern of counter-clockwise exchange from the East Passage into the West Passage. A similar pattern is apparent in the data from later in winter (Fig. 10.4). An important difference in the winter data is a residual layered exchange between the East Passage and Rhode Island Sound, consistent with estuarine gravitational flow. On average, near-bottom Rhode Island Sound water moves towards the Nar-ragansett Bay–Rhode Island Sound interface. A coincident northward residual flow of near-bottom water is observed in the East Passage. Mean velocities for each deep residual current are 5–10 cm s^{-1}. Near-surface residual currents are also in phase, with a southward flow in both East Passage and Rhode Island Sound of ~5 cm s^{-1}. This southward flow of near-surface water in the East Passage flow becomes negligible towards the end of the deployment.

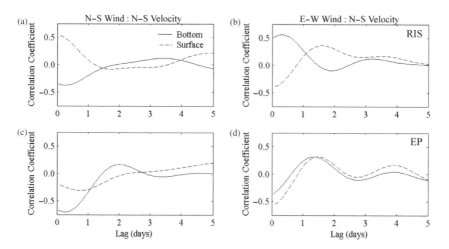

Fig. 10.5 Correlation coefficients for summer deployment (a) Rhode Island Sound north–south velocity versus north–south wind, (b) Rhode Island Sound north–south velocity versus east–west wind, (c) East Passage north–south velocity versus north–south wind, (d) East Passage north–south velocity versus east/west wind, (e) West Passage north–south velocity versus north–south wind, (f) West Passage north–south velocity versus east–west wind.

There is an interesting correlation between residual flow in lower Narragansett Bay and runoff in the second, or late winter (1/31/00–3/21/00) deployment period (Fig. 10.4). Early in this record, mean East Passage flow is summer-like, with a weak inflow of near-surface (<2 cm s^{-1}) and near-bottom (1–5 cm s^{-1}) water, consistent with the reduction in East Passage surface inflow seen at the end of the early winter deployment (Fig. 10.3, early January, 2001). A dramatic change in runoff occurs on decimal day 45 (2/14/00) (Fig. 10.4). Average runoff values before and after this date are 10–15 and 20–60 m^3 s^{-1}, respectively. There are runoff events centered on decimal days 45, 57, and 73 that relate to pulses of near-surface West Passage outflow, which are out of phase by 5–10 days. The most dramatic of these events occurs at the end of the record. A nearly 5-day, 60 m^3 s^{-1} runoff is followed ($\sim$5 days) by a doubling of the near-surface and near-bottom West Passage outflow. The East Passage responds through an increase in prevailing northward flow at both levels, suggesting that runoff events fuel the residual counterclockwise exchange within lower Narragansett Bay. There is a gradual increase in mean outflow of West Passage near-bottom water after 2/14/00 from $\sim$2 to 10 m^3 s^{-1}. A similar pattern of gradually increasing outflow of near-bottom West Passage water is recorded in the summer record (Fig. 10.2; event IIc) which also follows a large runoff event (peak flux of 230 m^3 s^{-1}).

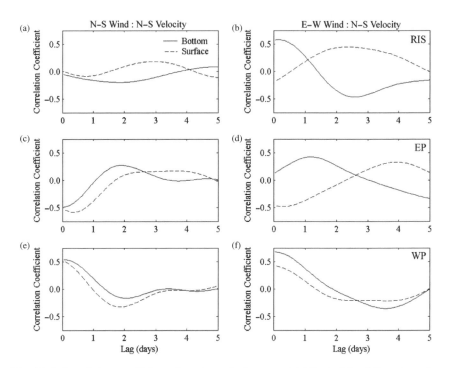

Fig. 10.6 Correlation coefficients for winter deployment (a) Rhode Island Sound north–south velocity versus north–south wind, (b) Rhode Island Sound north–south velocity versus east–west wind, (c) East Passage north–south velocity versus north–south wind, (d) East Passage north–south velocity versus east–west wind.

10.4.3 Wind-Induced Variability in Narragansett Bay and Rhode Island Sound

There is significant variability in both the seasonally averaged wind forcing and residual flows summarized in Figs. 10.7a and 10.8a. Wind events during this experiment involve both sudden shifts in direction (e.g., events Ia–IIa and IIa–IIb in Fig. 10.2; events Id–IId in Fig. 10.3; events IVa, IVb in Fig. 10.4) and more gradual, and progressive fluctuations associated with the passage of high or low pressure systems (event III in Figs. 10.2 and 10.3). During both summer and winter, the variation in winds occurs over time scales from 3 to 10 days, with event magnitudes of 5–7 m s^{-1}. These variations drive significant ± 10 and ± 5 cm s^{-1} perturbations in water column velocity in summer and winter data sets. The most prevalent deviations in summer are to south-westward and southeastward wind directions. During the early winter period, wind directions are more stable, with a single large event (III in Fig. 10.3). The late winter period (Fig. 10.4) includes two large events (IVa, IVb) where

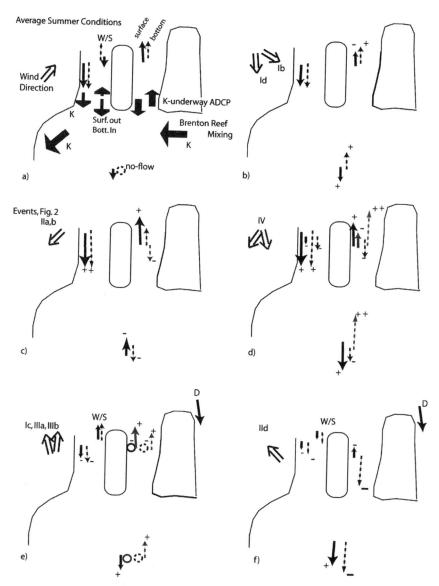

Fig. 10.7 Schematic maps summarizing seasonally averaged flow fields for the summer period (a) and the deviation from these patterns due to specific, representative wind events (b–f). (a) Information is included from underway ADCP surveys (K: Kincaid *et al.*, 2003), Weisberg and Sturges (1976) (W/S) current meter deployments (October–November, 1970) and tide gauge deployments (D: DeLeo, 2001). Corresponding event identifiers from Fig. 10.2 are included. Double-stemmed arrows show prevailing wind direction. Dark (dashed) arrows represent near-surface (near-bottom) subtidal currents. Gray arrows represent later stage, or lagged response of the water column. The timing of Weisberg and Sturges (1976) events in (c), (e) and (f) are 10/25/70, 11/7/70, and 10/23/70. The 10/23 event involves a northwestward wind and weak mid-column outflow. A northwestward wind period around 11/10/70 correlates with enhanced inflow at this site. The events marked D-DeLeo (2001) are from late March, 1997.

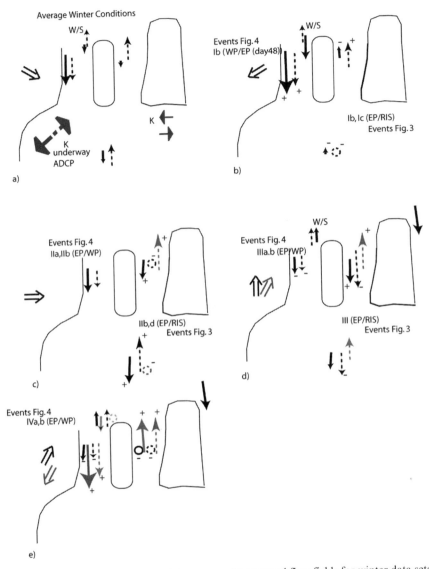

Fig. 10.8 Schematic maps summarizing seasonally averaged flow fields for winter data sets. The seasonally averaged flow patterns are shown in (a), along with information from underway ADCP surveys (K: Kincaid *et al.*, 2003), Weisberg and Sturges (1976) (W/S) current meter deployments (October–November, 1970) and tide gauge deployments (D: DeLeo, 2001). Corresponding event identifiers from Figs. 10.3 and 10.4 are included. Double arrows show prevailing wind direction. Dark (dashed) arrows represent near-surface (near-bottom) subtidal currents. Grey arrows represent later stage, or lagged response of the water column. The timing of Weisberg and Sturges (1976) events in (b), (d) and (e) are 10/23/70, 11/25/70, and 11/7–11/8/70. The timing of the D-DeLeo (2001) event in (d) is 3/31/97 and in (e) they represent pulses seen on 3/13/97 and 3/24/97.

strong winds switched between northeastward and southwestward orientations that also coincided with periods of elevated runoff ($\sim$60 m^3 s^{-1}).

10.4.3.1 Summer Conditions

North–South Winds

Narragansett Bay–Rhode Island Sound exchange patterns clearly differ with east–west versus north–south wind variability. Figure 10.5 shows the cross-correlation coefficients for the north–south response of the water column (near-surface versus near-bottom) relative to north–south (channel parallel) and east–west winds. North–south winds produce the strongest response in the West Passage and East Passage records, with only a weak correlation to non-tidal Rhode Island Sound flow (Fig. 10.5a vs Fig. 10.5c). Channel-parallel winds produce a rapid ($<$12 h) water column response in the West Passage with currents in the direction of the wind (Fig. 10.5e), which is consistent with prior models showing wind-driven Narragansett Bay spin-up time of $\sim$10 h (Gordon and Spaulding, 1987).

Wind forcing in a restricted channel is expected to produce a positive correlation between channel-parallel winds and near-surface currents, and a negative correlation for near-bottom currents in response to the gradient in sea surface height (Weisberg, 1976). However, a comparison of Fig. 10.5c,e shows that the East Passage records are almost mirror images of the West Passage response. Channel-parallel winds tend to produce an opposite and immediate flow in the near-surface and near-bottom water of the East Passage. For example, northward winds perturb West Passage currents in the northward direction and East Passage flow to a southward direction, which is consistent with model results of Gordon and Spaulding (1987). There is little evidence for large-scale East Passage–Rhode Island Sound coupling during this period, such as coherent surface outflow and deep inflow, in response to channel-parallel winds. Deep north–south Rhode Island Sound flow is only weakly correlated with north–south winds with a $\sim$2- to 3-day lag.

Wind events with significant north–south components perturb the prevailing counterclockwise flow within lower Narragansett Bay. Events labeled IIIa (day 145) and IIIb (day 174) in Fig. 10.2 are examples showing coincident stalling of the West Passage outflow and a stalling or modest reversal (to southward) of the prevailing East Passage inflow. Conditions surrounding the stalling of the counterclockwise gyre are represented schematically in Fig. 10.7e. Southwestward wind events produce the opposite effect, or enhancement of the East Passage–West Passage gyre (events IIa, IIb in Fig. 10.2) with stronger West Passage outflow and East Passage inflow. Gyre enhancing conditions are summarized schematically in Fig. 10.7c.

East–West Winds

Figure 10.5d,f shows positive and negative correlation coefficients for channel-parallel flow of West Passage and near-surface East Passage water relative to east–west winds. Such a pattern is consistent with a scenario in which westward winds are enhancing the counterclockwise current with stronger northward flow in the East Passage and southward flow in the West Passage. Figure. 10.2 shows that it is primarily southwestward events that drive such West Passage outflows (events IIa, IIb, represented schematically in Fig. 10.7c). Eastward events have less of an impact on West Passage flow (event Ib).

East–west winds produce a more coherent response between lower Narragansett Bay and Rhode Island Sound during the summer. Whereas the East Passage water column responded more uniformly to north–south wind energy, there is a layered response to east–west winds. Changes in near-bottom flow in first Rhode Island Sound (<1-day lag) and then the East Passage (1-day lag) correlate positively with wind forcing (Fig. 10.5b,d). For example, westward winds correlate with southward perturbations to near-bottom East Passage and Rhode Island Sound flow. East–west winds also tend to trigger longer period oscillations as evidenced by a negatively correlated response in near-bottom records in Rhode Island Sound at a 2- to 3-day lag and a ~4-day rebound in the East Passage near-surface and West Passage near-bottom records (Fig. 10.5d,f).

10.4.3.2 Winter Conditions

Figures 10.3 and 10.4 show how variable wind forcing perturbs basic background winter residual exchange patterns (summarized schematically in Fig. 10.8a). Trends in correlation coefficients for winter show that the pattern of East Passage–West Passage reaction to either north–south or east–west winds is nearly identical to summer. Northward or westward winds correlate with enhanced outflow from the West Passage and inflow to the East Passage. This is evident in the positive correlations between north–south winds and both surface and bottom currents at the West Passage station (identical to Fig. 10.5e) and a negative correlation for the East Passage (Fig. 10.6c). The West Passage response to east–west winds in winter is also identical to summer trends (Fig. 10.5f), where positive correlations throughout the water column contrast with negative values for the East Passage (Fig. 10.6d).

A difference in the winter data is that currents at the Rhode Island Sound station are much more responsive to north–south winds, and the sign of the correlation varies with depth, suggesting that winds are enhancing or stalling the prevailing layered flow. For example, southward winds quickly (<12-h lag) enhance the near-surface outflow (positive correlation) and the near-bottom inflow (negative correlation).

Time series records (Fig. 10.3 and 10.4) show that there is generally less variability in the magnitudes and directions of wind events during the winter period. There are a few relatively weak westward or southwestward wind events (events Ia, Ib, Id in Fig. 10.3) during which onshore Ekman transport and resulting offshore pressure gradients act to offset the prevailing layered East Passage–Rhode Island Sound flow (Fig. 10.8b). Deep northerly flow first stalls in Rhode Island Sound (~1-day lag) and then stalls in the East Passage (1- to 2-day lag). Southward flow of near-surface water also stalls or reverses (e.g., event Ib in Fig. 10.3) at both stations. Eastward winds work in reverse, tending to enhance the southward, near-surface currents in the East Passage and Rhode Island Sound, and the northward flow of near-bottom water at the Rhode Island Sound station (e.g., events IIa, IIc, summarized in Fig. 10.8c). The northward perturbation of near-bottom water in the East Passage under these conditions also lags (~1 day) the signal in near-bottom Rhode Island Sound water (event IIb in Fig. 10.3).

10.5 Discussion

Results of this study show that circulation and exchange within the Narragansett Bay–Rhode Island Sound system is more complex than many other estuarine–shelf systems. The system can respond to winds by either driving lateral exchange between sub-branches of the bay, or through a wind-driven layered flow (surface-with the wind; deep-against the wind). In this section we attempt to further summarize basic modes of Narragansett Bay–Rhode Island Sound exchange in connection with other observational studies. We include in Figs. 10.7 and 10.8 time series data from Weisberg and Sturges (1976), and spatial information from surveys by Kincaid *et al.*, (2003). While the Weisberg and Sturges (1976) data cover the time period of October–November, they document repeatable water column responses to specific wind directions for the deeper portions of the channel, and therefore provide a valuable comparison to our data from the western portion of the channel. We also include results from a tide gauge study on exchange between Mount Hope Bay and the Sakonnet River over a 40-day period in March–April, 1997 (DeLeo, 2001).

10.5.1 Seasonal Differences in Residual Estuarine–Shelf Exchange

Figures 10.7 and 10.8 summarize basic features of this coupled system for seasonally averaged conditions. Residual flow within lower Narragansett Bay is generally similar in winter and summer periods. Time series data presented here show that mean flow is northward in the East Passage and southward on the western side of the West Passage. Such a lateral flow structure is consistent with a conventional two-layer gravitational circulation pattern modified by rotation effects, where a mean outflow of less dense water is expected over the shallow/right portion of the estuary (viewed looking seaward in the northern Hemisphere), and a mean inflow of denser water is expected over the deeper/left portion of the estuary (Pritchard, 1967).

Underway surveys show that residual flow patterns through the mouth of Narragansett Bay are also similar in winter and summer (Kincaid et al., 2003). Inflow to the East Passage occurs through the eastern third of the cross-section, with layered flow (surface-out, bottom-in) through the middle and uniform outflow through the western third of the cross-section (Fig. 10.7a). There is residual southward flow of the whole water column through the western side of the West Passage mouth, and layered flow (surface-out) through the remainder of the cross-section. Time series data from the deeper, central channel of the West Passage (Weisberg and Sturges, 1976) also show a weak (1–3 cm s^{-1}) residual northward flow of water, relative to the persistent outflow recorded at our West Passage station. Interestingly, the combination of results implies the existence of lateral flow structures within each channel, as well as between channels. Estimates for the internal Rossby radius of deformation for the lower bay range from 2 to 4 km (Kincaid et al., 2003), which roughly covers the range in length scales for individual channels and the combined width near the mouth.

A goal of estuarine–shelf interaction studies is defining source regions from which bottom water is drawn into the estuary (Beardsley and Hart, 1978; Masse, 1990). Ship-mounted ADCP results show a summertime residual, cyclonic coastal current with speeds of 5–15 cm s^{-1} confined to within ∼2–3 km of shore (Kincaid et al., 2003), which is consistent with drifter studies (Collins, 1976). This current brings water to the interfacial region from the east-southeast and carries water away from the Narragansett Bay–Rhode Island Sound interface to the southwest. Residual flow at the Rhode Island Sound station is weak to the south-southwest, suggesting that this region of Rhode Island Sound is, on average, isolated from Narragansett Bay by the counterclockwise coastal current in Rhode Island Sound.

The combination of CTD and ADCP measurements along the Narragansett Bay–Rhode Island Sound interface have also been used to suggest a mechanism for pumping deep Rhode Island Sound water into the East Passage in summer (Kincaid et al., 2003). Rhode Island Sound bottom water is mixed upward along Brenton Reef and then efficiently advected into Narragansett Bay by the strong northward residual current along the eastern side of the East Passage. Such a process would be important in summer when Rhode Island Sound bottom water has been shown to contain elevated DIN levels (Nixon et al., 2005).

Residual flow patterns in Rhode Island Sound, and therefore, resupply pathways are different in winter (Fig. 10.8a). A persistent layered flow carries cool (4–5°C), less saline water (29.7‰) outflow from the bay to the southwest along the Rhode Island coast, and supplies warmer, saltier (6°C, 31.3‰) bottom water toward the mouth of Narragansett Bay from the south–southwest

(Kincaid *et al.*, 2003). This persistent deep return flow extends from 2 to 5 km from the northwestern shoreline of Rhode Island Sound.

10.5.2 *Wind-Driven Variability in Estuarine–Shelf Exchange*

One of the pioneering studies on the role of wind forcing in estuaries involved the West Passage (Weisberg and Sturges, 1976). Many of the patterns seen in our data were discussed in this work, involving more spatially limited data. Weisberg and Sturges (1976) used a model of net exchange between the passages to explain observations that wind-induced residual fluxes lasted longer than expected. Exchange between the shallower West Passage and the East Passage is related to gradients in sea level produced by differences in wind setup, which is inversely related to water depth.

Our results are consistent with the ideas put forth by Weisberg and Sturges (1976). Winds strongly influence exchange patterns between the passages and between the estuary and the shelf. Wind forcing has a particularly strong influence on magnitude, timing, and source region for intrusion events to the bay. In summer, southwestward winds enhance the counterclockwise flow within the lower bay (events IIa–IIc, Fig. 10.2). Wind events with stronger southward components do not produce a response at our Rhode Island Sound Station, suggesting that water is drawn into the northward East Passage flow from the east–southeast of the mouth. In summer, north-northwestward winds provide the most favorable conditions for stalling and eventually reversing the counterclockwise gyre through deep residual outflows from the East Passage, and from Mount Hope Bay into the Sakonnet River (DeLeo, 2001; Fig. 10.7e,f).

Results also reveal that prevailing summer flows (Fig. 10.7a) can be overcome to allow shelf water from just outside the mouth to be drawn into the East Passage. Southeastward winds (events Ia–Id, Fig. 10.2) effectively reduce the northward near-surface East Passage flow, drive a southward flow of near-surface Rhode Island Sound water, and trigger a series of deep northward intrusion events from Rhode Island Sound into the East Passage (Fig. 10.7b).

Our findings are also consistent with Weisberg and Sturges (1976) in that the largest exchange events are related to relaxation events, involving a rapid progression in wind direction and magnitude (event IV, Fig. 10.2). Figure 10.7d summarizes schematically the system response to the largest wind event from the summer deployment. An early phase of southwestward winds enhances lateral East Passage to West Passage exchange. A change in wind direction to southward produces a rebound in the system, where the West Passage outflow stalls and strong deep Rhode Island Sound to East Passage intrusions are initiated. The size of the northward response suggests that the preceding period of southwestward wind and associated spin-up of the counterclockwise East Passage–West Passage current lowers East Passage sea level relative to Rhode Island Sound, essentially by draining the East Passage by drawing water off through the West Passage.

Perturbations to both basic Narragansett Bay–Rhode Island Sound interaction modes are also seen in the response of the water column to wind forcing during the winter (Fig. 10.8). Winds with a strong westward component (events Ia–Id Fig. 10.3, events Ia–Ib Fig. 10.4) tend to enhance the East Passage–West Passage lateral exchange flow, as in summer, although it is interesting that the stronger northward flow in the East Passage is in the deeper water during this period when stratification is weaker (Fig. 10.8b).

The influence of Ekman forcing on Rhode Island Sound flow is apparent in the near-bottom Rhode Island Sound response to westward (Fig. 10.8b) and eastward (events IIa–IId in Fig. 10.3, events IIa–IIc in Fig. 10.4, Fig. 10.8c) winds in winter. The prevailing near-bottom northward flow in Rhode Island Sound is stalled (enhanced) during westward (eastward) wind events. Eastward winds tend to drive shallow East Passage outflow that is coherent with southward flow of near-surface Rhode Island Sound water, also consistent with a model of Ekman transport. It is interesting that the near-bottom responses in the East Passage and Rhode Island Sound are out of phase (faster response in Rhode Island Sound). Future observational experiments will require better spatial coverage to distinguish relative roles of baratropic versus baroclinic gradients across the estuarine–shelf interface (e.g., Klinck et al., 1982).

As in summer, the strongest deep intrusions into the East Passage are related to time varying wind events that are common in winter (Fig. 10.8d,e). One example is where wind direction progresses from northward to northeastward (Fig. 10.8d, event III in Fig. 10.3). The northward wind produces a southward flow throughout the water column in the East Passage and Rhode Island Sound, in agreement with numerical model simulations (Gordon and Spaulding, 1987). The change in wind direction to northeastward results in strong northward near-bottom flow at both sites. The rebound is faster in the East Passage.

The East Passage–West Passage winter data set also includes two enhanced exchange events involving a progression from northeastward to southwestward winds, with coincident peaks in runoff (events IVa and IVb in Fig. 10.4). Northeastward winds stall the counterclockwise flow within lower Narragansett Bay, confining water within the system and producing a sea level gradient between the passages (Weisberg and Sturges, 1976). The combination of increased runoff and a switch to southwestward winds produces a very strong return of the counterclockwise flow, where the West Passage outflow and East Passage inflow greatly exceed background levels. In the stronger of these events (IVb), the midpoint of the runoff event is ∼2–3 days prior to the onset of northeastward winds, and 8–10 days prior to the start of the outflow pulse in the West Passage.

10.6 Conclusions

The combination of upward looking ADCPs in the East Passage, West Passage and Rhode Island Sound with spatially detailed ship-mounted ADCP surveys (Kincaid *et al.*, 2003), provides a clearer view of spatial-temporal exchange between sub-regions of Narragansett Bay, and between Narragansett Bay and Rhode Island Sound. Two basic modes of estuarine–shelf interaction are defined, including a large-scale non-tidal gyre and a more typical pattern of layered non-tidal flow (surface flow-out; deep flow-in). The non-tidal currents in the East Passage and West Passage are consistent with a large-scale counterclockwise exchange of water from the East Passage to the West Passage.

The response to most wind directions is to either spin-up or -down the gyre. In both winter and summer data sets, flow is enhanced (stalled) by southwestward (northwestward) winds. In summer, northeastward wind events tend to reverse the prevailing lateral exchange pattern and drive a southward flow (or flushing) of deep East Passage water to Rhode Island Sound. A summertime counterclockwise residual flow brings shelf water to the mouth of the estuary from the southeast and through the primary entry point to Narragansett Bay along the eastern side of the East Passage. Intrusions of Rhode Island Sound water into the East Passage are driven by southeastward winds, with the largest exchanges related to relaxation events, where winds switch between northward and southward directions.

In winter, the counterclockwise current in Rhode Island Sound is replaced by a pattern of layered non-tidal flow that runs parallel to the northwestern Rhode Island Sound coastline [surface (deep) flow away (toward) the estuary mouth]. Southeastward winds in summer episodically produce this exchange pattern, suggesting that the dominant southeastward wind direction in winter is largely responsible for this more persistent seasonal exchange pattern. Eastward wind events and runoff events strengthen the layered exchange flow.

Results of this study show that circulation and exchange within the Narragansett Bay–Rhode Island Sound system is relatively complex. This is because of the multiple connection points between the estuary and the shelf, and within the estuary. Models for exchange between a confined channel-type estuary and inner shelf waters suggest that coastline parallel winds will favor filling of a channel-like estuary due to Ekman transport, or channel-parallel winds will drive layered flow. For the Narragansett Bay–Rhode Island Sound system, the system can respond either by spinning up or reversing the lateral exchange between the West Passage and East Passage, or by layered flow within each channel.

References

Beardsley, R.C., and Hart, J. 1978. A simple theoretical model for the flow of an estuary onto a continental shelf. *Journal of Geophysical Research* 83:873–883.

Bender, M., Kester, D., Cullen, D., Quinn, J., King, W., Phelps, D., and Hunt, C. 1989. Trace metal pollutants in Narragansett Bay waters, sediments and shellfish. Current Report, The Narragansett Bay Project, NBP-89-25.

Bergondo, D.L. 2004. Examining the processes controlling water column variability in Narragansett Bay: time-series data and numerical modeling. Ph.D. Dissertation, University of Rhode Island, Narragansett, RI. 187 pp.

Bergondo, D.L., Kester, D.R., Stoffel, H.E., and Woods, W. 2005. Time-series observations during the low sub-surface oxygen events in Narragansett Bay during summer 2001. *Marine Chemistry* 97:90–103.

Brush, M.J. 2002. Development of a numerical model for shallow marine ecosystems with application to Greenwich Bay, RI. Ph.D. Dissertation, University of Rhode Island, Graduate School of Oceanography, Narragansett, RI. 560 pp.

Collins, B.P. 1976. Suspended Material Transport: Narragansett Bay Area, Rhode Island. *Estuarine and Coastal Marine Science* 4:33–44.

Deacutis, C.F., Murray, D.W., Prell, W.L., Saarman, E., and Korhun, L. 2006. Hypoxia in the Upper Half of Narragansett Bay, RI during August 2001 and 2002. *Northeast Naturalist* 13(Special Issue 4):173–198.

DeLeo, W. 2001. Investigation of the physical mechanisms controlling exchange between Mount Hope Bay and the Sakonnet River. MS Thesis, University of Rhode Island, Narragansett, RI. 168 pp.

Elliott, A.J. 1978. Observations of the meteorologically induced circulation in the Potomac Estuary. *Estuarine and Coastal Marine Science* 6:285–299.

Garvine, R.W. 1991. Subtidal frequency estuary–shelf interaction: Observations near Delaware Bay. *Journal of Geophysical Research* 96:7049–7064.

Goodrich, D.M. 1988. On meteorologically induced flushing in three U.S. East Coast estuaries. *Estuarine, Coastal and Shelf Science* 26:111–121.

Gordon, R., and Spaulding, M.L. 1987. Numerical simulation of the tidal and wind driven circulation in Narragansett Bay. *Estuarine, Coastal, and Shelf Science* 24:611–636.

Granger, S., and Buckley, B. 1999. Nutrient inputs from Rhode Island Sound and the profiling of Narragansett Bay, Progress Reports submitted to Rhode Island Sea Grant, R/EE-982.

Hawk, J.D. 1998. The role of the North Atlantic Oscillation in winter climate variability as it relates to the winter-spring bloom in Narragansett Bay. MS Thesis in Oceanography, University of Rhode Island, Narragansett, RI. pp. 148.

Hicks, S.D. 1959. The physical oceanography of Narragansett Bay. *Limnology and Oceanography* 4:316–327.

Hinga, K.H., Lewis, N.F., Rice, R., Dadey, K., and Keller, A. 1989. A review of Narragansett Bay phytoplankton data: status and trends, Current Report, The Narragansett Bay Project, NBP-89-21.

Keller, A.A. 1988. Estimating phytoplankton productivity from light availability and biomass in MERL mesocosms and Narragansett Bay. *Marine Ecology Progress Series* 45:159–168.

Keller, A.A., Klein-MacPhee, G., and Burns, J. St.Onge. 1999. Abundance and distribution of Ichthyoplankton in Narragansett Bay, Rhode Island, 1989–1990. *Estuaries* 22:149–163.

Kincaid, C., Pockalny, R., and Huzzey, L. 2003. Spatial and temporal variability in flow at the mouth of Narragansett Bay. *Journal of Geophysical Research* 108(C7):3218.

Kjerve, B., Greer, J.E., and Crout, R.L. 1978. Low frequency response of estuarine sea level to non-local forcing. *In* Estaurine interactions. Academic Press, New York, pp. 497–513.

Klinck, J.M., O'Brien, J.J., and Svendsen, H. 1982. A Simple Model of Fjord and Coastal Circulation Interaction. *Journal of Physical Oceanography* 11:1612–1626.

Kremer, J.N., and Nixon, S.W. 1978. A coastal marine ecosystem: Simulation and analysis. New York: Springer Verlag.

Masse, A.K., 1990. Withdrawal of shelf water into an estuary: A barotropic model. *Journal of Geophysical Research* 95:16085–16096.

Moore, W.S. 1996. Large groundwater inputs to coastal waters revealed by ^{226}Ra enrichments. *Nature* 380:612–614.

Moore, W.S. 1997. High fluxes of radium and barium from the mouth of the Ganges-Brahmaputra River during low river discharge suggest a large groundwater source. *Earth and Planetary Science Letters* 150:141–150.

NCDC. Unedited Local Climatological Data. (*http://cdo.ncdc.noaa.gov/ulcd/ULCD*).

Nixon, S.W. 1997. Prehistoric nutrient inputs and productivity in Narragansett Bay. *Estuarine Research Federation* 20(2):253–261.

Nixon, S.W., Buckley, B., Granger, S., Harris, L., Oczkowski, A., Cole, L., and Fulweiler, R. 2005. Anthropogenic nutrient inputs to Narragansett Bay: A twenty five year perspective. A Report to the Narragansett Bay Commission and Rhode Island Sea Grant. Rhode Island Sea Grant, Narragansett, RI. (*www.seagrant.gso.uri.edu/research/bay_commission_report.pdf*).

Oviatt, C.A., Keller, A., and Reed, L. 2002. Annual primary production in Narragansett Bay with no bay-wide winter-spring bloom. *Estuarine, Coastal, and Shelf Science* 54:1013–1026.

Pawlowicz, R., Beardsley, B., and Lentz, S. 2002. Classical tidal harmonic analysis including error estimates in MATLAB using T_TIDE. *Computers and Geosciences* 28:929–937.

Pilson, M.E.Q. 1985. On the residence time of water in Narragansett Bay. *Estuaries* 8:2–14.

Prell, W., Murray, D., Heggie, K., and Saarman, E. 2004. A surface sediment array to monitor how geochemical gradients are related to hypoxic conditions in upper Narragansett Bay, RI, *In* 2004 EMAP Symposium, Integrated Monitoring and Assessment for Effective Water Quality Management, Newport, RI.

Pritchard, D.W. 1967. Observations of circulation in coastal plain estuaries. *In* Estuaries, Lauff, G.H. (ed.) American Association for the Advancement of Science, Washington, DC. pp. 37–44.

Rosenberger, K. 2001. Circulation patterns in Rhode Island Sound: Constraints from a bottom mounted acoustic Doppler current profiler. MS Thesis in Oceanography, University of Rhode Island, Narragansett, RI. pp. 226.

Saarman, E., Prell, W., Murray, D., Deacutis, C., Kester, D., and Oviatt, C. 2002. Hypoxic conditions in Narragansett Bay during the summer of 2001. New England Estuarine Research Society Meeting, Bar Harbor, ME, May 2002.

Shonting, D.H. 1969. Rhode Island Sound square kilometer study 1967: Flow patterns and kinetic energy distribution. *Journal of Geophysical Research* 74:3386–3395.

Shonting, D.H., and Cook, G.S. 1970. On the seasonal distribution of temperature and salinity in Rhode Island Sound. *Limnology and Oceanography* 15:100–112.

Sullivan, B.K., and Van Keuren, D. 2003. Have changes in climate and predation pressure altered timing of seasonal succession of the copepods, *Acartia tonsa* and *A. hudsonica* in a temperate estuary of the northeast Atlantic? PICES 3RD International Zooplankton Production Symposium, Gijon, Spain, May 2003.

Taniguchi, M., Burnett, W.C., Cable, J.E., and Turner, J.V. 2002. Investigations of submarine groundwater discharge. *Hydrological Processes* 16:2115–2129.

USGS Water Data. 2000–2001. (*http://waterdata.usgs.gov/nwis/rt*).

Valle-Levinson, A., and Lwiza, K.M.M. 1995. The effects of channels and shoals on exchange between the Chesapeake Bay and the adjacent ocean. *Journal of Geophysical Research* 100:18551–18563.

Valle-Levinson, A., Klinck, M., and Wheless, G.H. 1996. Inflows/outflows at the transition between a coastal plain estuary and the coastal ocean. *Continental Shelf Research* 16:1819–1847.

Valle-Levinson, A.V., Li, C., Royer, T.C., and Atkinson, L. 1998. Flow patterns at the Chesapeake Bay entrance. *Continental Shelf Research* 18:1157–1177.

Wang, D.P., and Elliott, A.J. 1978. Non-tidal variability in the Chesapeake Bay and Potomac River: Evidence for non-local forcing, *Journal of Physical Oceanography* 8:225–232.

Weisberg, R.H. 1972. Net circulation in Narragansett Bay. MS Thesis, Graduate School of Oceanography, Narragansett, RI. 90 pp.

Weisberg, R.H. 1976. The non-tidal flow in the Providence River of Narragansett Bay: A stochastic approach to estuarine circulation. *Journal of Physical Oceanography* 6:721–734.

Weisberg, R.H., and Sturges, W. 1976. Velocity observations in the west passage of Narragansett Bay: A partially mixed estuary. *Journal of Physical Oceanography* 6:345–354.

Wong, K.C., and Münchow, A. 1995. Buoyancy forced interaction between estuary and inner shelf: Observations. *Continental Shelf Research* 15:59–88.

Chapter 11
Summer Bottom Water Dissolved Oxygen in Upper Narragansett Bay

Emily Saarman, Warren L. Prell, David W. Murray and Christopher F. Deacutis

11.1 Introduction

Narragansett Bay is a medium-sized (328 km^2), relatively shallow (average depth 8.6 m), temperate latitude estuary located in the northeastern United States (Chinman and Nixon, 1985). With relatively low fresh water input (Pilson, 1985; Ries, 1990), the majority of Narragansett Bay is considered to be partially to well mixed (Hicks, 1959; Kremer and Nixon, 1978; Pilson, 1985; Nixon et al., 1995) and only moderately susceptible to adverse effects of nutrient loading, including low oxygen (hypoxic) conditions (Bricker et al., 1999). Despite Narragansett Bay's purportedly low susceptibility to eutrophication, the mid-to-upper bay exhibits many symptoms of excessive nitrogen loading, including hypoxic conditions, macroalgae accumulation, eelgrass loss, and fish kills (RIDEM, 2003; Chapter 12). In response to these symptoms of eutrophication in Narragansett Bay, the Rhode Island Department of Environmental Management (RIDEM) has proposed reducing the summertime (May to October) nitrogen concentrations of wastewater effluent from major treatment plants to 5 mg L^{-1}, which would be about a 50% reduction (Kerr, 1999; RIDEM, 2005). As part of a diverse effort to assess the effectiveness of nitrogen reduction measures, this study seeks to document the current state of summertime hypoxia in Narragansett Bay, and to assess its variability and predictability on the basis of spatial survey and continuous monitoring buoy data. In this chapter, the term "upper bay" refers to all of Narragansett Bay north of the southern end of Prudence Island, including the Providence River, Greenwich Bay, Mount Hope Bay, and the upper East and West Passages, i.e., the upper half of Narragansett Bay (Fig. 11.1). The more formal term "Upper Bay" is reserved for the area south of Conimicut Point and north of Prudence Island.

Emily Saarman

Department of Geological Sciences, Brown University, Providence, RI 02912-1846
esaarman@gmail.com

A. Desbonnet, B. A. Costa-Pierce (eds.), *Science for Ecosystem-based Management.*
© Springer 2008

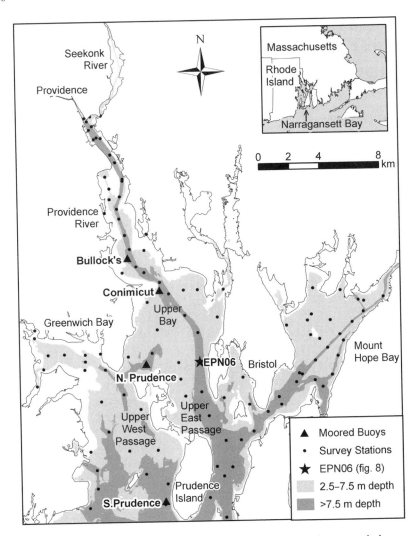

Fig. 11.1 Map of the upper Narragansett Bay dissolved oxygen study area and place names with the location of spatial survey stations (filled circles). Shading indicates water depth, with the deeper (>7.5 m) channels in darker shades. Locations of the four moored buoys (Bullocks, Conimicut, North Prudence, and South Prudence) are shown as filled triangles. The location of the upper East Passage station EPN06 (Fig. 11.8) is indicated by the star.

11.2 Hypoxia in Narragansett Bay: Background and Concerns

Hypoxia—dissolved oxygen concentrations low enough to induce physiological stress—is widely accepted to be a consequence of nitrogen overloading in estuarine waters (NRC, 2000; Diaz, 2001; Rabalais, 2002). In this chapter, hypoxia is defined as less than 2.9 mg L^{-1} dissolved oxygen (Table 11.1).

Table 11.1 Summary of the EPA concentration level and duration criteria for dissolved oxygen impairment, the effects associated with each level, and the descriptive terms used in this paper.

DO level (mg L^{-1})	Effect	Duration criteria	Descriptive term in this paper
2.9–4.8	Larval mortality/ recruitment impairment	Acceptable duration ranges from 1 to 40 days depending on intensity and short term variability of DO	*Suboxic*
2.3–4.8	Juvenile/adult growth impairment	Protection in this range is assured by the larval protection limits (above)	
<2.9	Larval mortality	>24 h duration indicates excessive impairment	*Hypoxic*
<2.3	Juvenile/adult mortality	>24 h duration indicates impairment	
<1.2	Mortality in many species and life stages	Single point-in-time measurement indicates impairment	*Severely hypoxic*

Criteria, effects, and durations are from EPA (2000).

Increased levels of nitrogen in the estuary encourage excessive algal growth including both phytoplankton and macroalgae. The subsequent respiration and decay of this excess organic matter often creates a high oxygen demand. If the oxygen demand is isolated from sources of oxygen at the surface—the atmosphere and actively photosynthesizing phytoplankton—the isolated waters can become hypoxic.

The risk of hypoxia is greatest during the warm summer months. Warmer waters increase respiration rates and tend to stratify the bay, with relatively warm, low salinity surface water (from river inputs) overlying colder, saltier water (from oceanic sources). The density contrast between surface and deeper layers sets up water column stratification that acts to isolate deep water from surface sources of oxygen. In shallow systems like Narragansett Bay, forces that mix and break up water column stratification, such as wind and tidal action, also play a role in the development of hypoxic events (Stanley and Nixon, 1992; Bergondo, 2004). Thus, although nitrogen loading greatly increases the potential for hypoxic conditions, local forces ultimately control the severity and duration of resulting hypoxia.

Freshwater and nutrient inputs are concentrated in the uppermost regions of Narragansett Bay, (Doering *et al.,* 1990a; Ely, 2002) especially the Seekonk and Providence Rivers. An estimated 95% of the total nitrogen entering these areas is exported down bay (Doering *et al.,* 1990a). Water quality in the Providence and Seekonk rivers has long been known to be poor, and most dissolved oxygen studies prior to the late 1990s focused on these areas (US Army Corps of Engineers, 1960; Olsen and Lee, 1979; Oviatt, 1980; Doering *et al.,* 1988a,b,

1990b; Pilson and Hunt, 1989; Granger, 1994; Kester *et al.*, 1996). Weak stratification in the mid and lower bay was assumed to prevent the development of hypoxic conditions. However, in the late 1990s, the occurrence of fish kills, sporadic low oxygen concentrations, and the sediment-based identification of benthic enrichment (Valente *et al.*, 1992) lead scientists to question the validity of this assumption (Deacutis, 1999).

In recognition of the risk to aquatic life posed by hypoxia, the Environmental Protection Agency (EPA) published dissolved oxygen criteria (Table 11.1) for the protection of aquatic life in coastal environments (EPA, 2000). The criteria are based upon laboratory studies that evaluate the effects of low dissolved oxygen on survival and growth of common coastal species that inhabit waters between Cape Cod and Cape Hatteras. The EPA dissolved oxygen criteria consider both oxygen level and duration to assess the level of harm caused to aquatic life. Throughout this chapter, EPA criteria are used to evaluate the risk posed to aquatic life in Narragansett Bay by summertime dissolved oxygen levels.

To better assess the extent and frequency of low summertime dissolved oxygen in the upper bay, this chapter synthesizes 5 years of spatial dissolved oxygen surveys. This approach was partially inspired by the study of Buzzelli *et al.* (2002) who used long-term point estimates of dissolved oxygen from the mid-channel and cross-channel surveys of the Neuse River to estimate the spatial extent of hypoxia. Here, we use both synoptic surveys (point estimates) and time series data to evaluate the status of summertime hypoxia in upper Narragansett Bay. The synoptic surveys provide a spatial view of summertime dissolved oxygen; however, they lack temporal continuity. Time series data from moored buoys provide a temporal view of hypoxia that can be used with the EPA criteria to assess the risk to aquatic life, but their spatial domain is limited (Fig. 11.1). This chapter integrates these data to provide a methodology to better assess the near real-time status of summertime hypoxia and associated risk to marine life.

11.3 Data and Methods

The synoptic mapping of dissolved oxygen in Narragansett Bay was conducted by multi-institutional volunteers who were coordinated by the Narragansett Bay Estuary Program (Deacutis, 1999; Prell *et al.*, 2004; Deacutis *et al.*, 2006). Synoptic surveys were conducted by 4–9 small boats that collectively surveyed 53–87 locations throughout upper Narragansett Bay (Fig. 11.1). Measurements were made over a 6 to 12-h period on 13 separate dates over five summers (1999–2003). Each boat used GPS to occupy 10–12 pre-assigned stations. The survey dates were selected to coincide with periods when physical conditions were most conducive to the onset of hypoxic conditions: warm water, neap tides, night time and early morning hours. The number of stations and the areas sampled in each survey are given in Table 11.2. Station locations and all data, along with profile plots and individual survey results, are available at *http://www.geo.brown.edu/georesearch/insomniacs*. Station location was based on bathymetry to contain a mix of deep (>7.5 m) and shallow stations, historical temperature and salinity gradients, and areas thought most likely to be impacted by low dissolved oxygen. These spatial surveys included the shallower portions of the bay and thus expand on related studies, which were limited to single mooring locations (Bergondo *et al.*, 2005) or NuShuttle data, which are collected along the deeper channels (Berman *et al.*, 2005).

At each station, teams conducted water column profiling of salinity, temperature, and dissolved oxygen (% saturation; concentration) using calibrated electronic sensors (YSI 6000 rapid-pulse dissolved oxygen sensors, and one Hydrolab in 1999 only). High-sensitivity Teflon membranes were used for the dissolved oxygen sensors, and all sondes included nonvented depth (pressure) sensors. All sensors were calibrated according to manufacturers recommended operating procedures prior to surveys, and calibration checks were made during surveys. Oxygen was calibrated to partial pressure in 100% water vapor saturated air. Measurements were made at specified depths, with near-surface (0.5 m) taken first, followed by deepest measurement (within 0.5 m of the bottom), and up-column measurements, at 1.5 m intervals, holding at specific depths to allow for equilibrium of the oxygen sensor. All data are available at *http://www.geo.brown.edu/georesearch/insomniacs*.

Fixed buoys also used YSI sondes to collect continuous time series ($\Delta t = 15$ min) of temperature, salinity, and dissolved oxygen from surface (~1 m) and ~1 m above the bottom throughout the study period at several locations within

Table 11.2 Summary of the geographic areas sampled, the number of sites in each survey, and the % of stations where bottom waters fall below the various EPA criteria for dissolved oxygen (see Table 11.1).

	7/21/ 2000	7/13/ 2001	7/28/ 2003	8/24/ 1999	8/15/ 2001	8/6/ 2002	8/25/ 2003
Total stations	69	82	52	45	76	57	73
% <1.2 mg L^{-1}	3	10	2	0	17	35	10
% <2.3 mg L^{-1}	3	22	8	2	34	58	27
% <2.9 mg L^{-1}	4	37	17	7	43	68	56
% <4.8 mg L^{-1}	30	65	67	60	80	88	81
Area surveyed							
Providence River	INC	Yes	Yes	No	Yes	Yes	Yes
Upper Bay	Yes	Yes	Yes	Yes	Yes	Yes	Yes
Greenwich Bay	Yes	Yes	INC	Yes	Yes	No	Yes
West Passage	Yes	Yes	Yes	No	Yes	Yes	Yes
East Passage	No	Yes	No	No	No	No	No
Mount Hope Bay	Yes	Yes	INC	Yes	Yes	INC	Yes

INC indicates incomplete sampling of the specified geographic area.

upper Narragansett Bay (Fig. 11.1) (Bergondo, 2004; Bergondo *et al.*, 2005; Kester, personal communication; Narragansett Bay Water Quality Monitoring Network). These time series data provide information on tidal-scale variability and temporal evolution of hypoxic events in relation to the timing of the synoptic surveys. Calibration of the moored YSI sondes and quality control of the moored data were conducted independently by Bergondo (2004).

To correct for possible differences between instruments, the synoptic survey sondes were placed in a shared bath of well-mixed seawater immediately prior to, or during, each survey. One sonde (RIDEM-1) was selected as the standard because it was the only common instrument in multiple dip-in comparisons. Most shared measurements had standard deviations for dissolved oxygen, salinity, and temperature within the error range defined in the instrument manual ($\pm 2\%$ sat, $\pm 0.3‰$, and $\pm 0.15°C$, respectively). Salinity and temperature values from the shared bath, as well as values from co-occupied stations, were used to correct selected salinity and temperature data using a best-fit regression equation. If dissolved oxygen, % saturation, salinity, or temperature showed differences exceeding the instrument specifications, the shared-bath correction factor was used to allow comparison of data between sondes. Temperature/salinity (T–S) diagrams of all data collected on a single survey were used to highlight sondes that may have been miscalibrated. Where deviations occurred, they were associated with a particular sonde and were corrected as a constant offset. Approximately 38% of the salinity data and 4% of the temperature data were corrected (Prell *et al.*, 2004).

11.4 Analysis

Once all data were corrected, density was calculated using the IES-80 equation (UNESCO, 1981). Surface, minimum, and bottom dissolved oxygen measurements were extracted and examined for trends. For each survey, statistics were calculated on the number of stations with dissolved oxygen levels indicating suboxic (<4.8 mg L^{-1}) and hypoxic conditions (<2.9 mg L^{-1}).

To examine the spatial patterns of hypoxic waters, maps of bottom and minimum dissolved oxygen concentrations were generated for all surveys using ArcView. Contours of dissolved oxygen distribution were constructed using inverse distance weighting, and interpolation was limited to not more than five neighboring data points within a 3×5.5 km ellipse (13 km^2 oriented along the major axis of the bay; major axis 25° west of north). Experiments using different numbers of neighboring data points revealed that five data points retained the observed gradients without over-smoothing the data. The maps of bottom water dissolved oxygen along with bathymetry were used to estimate the bottom area impacted by hypoxia for each survey. Only areas with depths greater than 2.5 m were included in this estimate because shallower waters were usually within the mixed layer, and very few of our survey stations were located in less than 2.5 m depth, thus we have no reliable estimates of bottom water dissolved oxygen in shallower waters. Area estimates are presented as a percentage of the total survey area (158 km^2) to normalize for the differences in survey coverage. The area estimates provide valuable information about the spatial extent of the risk posed by hypoxic conditions in the upper reaches of Narragansett Bay.

To construct a robust estimate of the magnitude and extent of summertime hypoxia in upper Narragansett Bay, we averaged bottom dissolved oxygen measurements from all available July and August surveys collected between 1999 and 2003. July and August were targeted because this is the interval when hypoxia is most likely to occur. Given survey-to-survey variability in dissolved oxygen concentrations, the mean and standard deviation of bottom water dissolved oxygen should give a better estimate of average conditions than individual surveys. We note that about 18% of the sites are less than 3.5 m of water depth, so their bottom measurements may be within the mixed layer. However, bottom water dissolved oxygen measurement at most sites is from below the pycnocline. The number of sites in each survey and percentage of sites with bottom dissolved oxygen concentrations below the various EPA levels of impairment is given in Table 11.2. To establish areas of spatial coherence within upper Narragansett Bay, the July–August bottom dissolved oxygen data were compared with synchronous data from the moored buoy at Bullock's Reach. Correlation to the Bullock's Reach data was performed because it is proximal to the major nutrient inputs and low dissolved oxygen concentrations, and is the most continuous moored dataset available. Linear regression was used to establish the correlation between continuously recorded Bullock's Reach and synoptic station data. Station-to-station regressions identified natural subsets of stations and the resultant areas of spatial coherence. Buoy-to-station and station-to-station correlations were mapped spatially, but were not interpolated.

11.5 EPA Dissolved Oxygen Criteria

In this chapter, we used the critical dissolved oxygen levels adopted by the EPA (EPA, 2000) to assess the duration and spatial extent of possible impacts on aquatic life. These critical levels, along with the descriptive terms used in this chapter, are presented in Table 11.1. The dissolved oxygen criteria are designed to protect 95% of species against the indicated effect. If dissolved oxygen concentrations ever fall below 1.2 mg L^{-1}, the waters are considered impaired, above that level, however, the duration of the low oxygen event determines the impact on living organisms. For example, dissolved oxygen concentrations between 1.2 mg L^{-1} and 2.3 mg L^{-1} for less than 24 h, should cause less than 5% juvenile/adult mortality for 95% of saltwater species. Early life stages are more sensitive, and the criteria attempt to limit cumulative effects on larval recruitment by adjusting allowable durations of exposure as a function of dissolved oxygen intensity. Thus, exposure to a dissolved oxygen concentration as low as 2.9 mg L^{-1} is limited to 1 day, while exposure to 4.8 mg L^{-1} is allowed for up to 40 days before unacceptable harm to aquatic life. Cumulative growth effects are also expected to be acceptably limited by these criteria.

Since the synoptic surveys provide no temporal resolution, only the moored buoy data can be used to address the risk component. We use the time series data from Bullock's Reach to estimate the percentage of time the bottom waters of the Providence River were below each critical limit because the Bullock's Reach data are most proximal to the major nutrient and waste water inputs, and the data are the most complete. The analysis described above is used to extend these estimates of impairment to upper Narragansett Bay.

11.6 Results

11.6.1 Neap Tide Bias of Synoptic Surveys

Despite the focus on neap tide, comparison of the available Bullock's Reach buoy with adjacent stations (triangles in Fig. 11.2) shows that most surveys were not conducted during the most extreme phase of hypoxic events. The continuous Bullock's Reach data show that hypoxic events generally begin with oxic–suboxic dissolved oxygen values that then decrease over several days to hypoxic levels (<2.9 mg L^{-1}), which typically last between 4 and 9 days and are most often terminated by the occurrence of the spring tides (Fig. 11.2; Bergondo, 2004; Bergondo et al., 2005). Available buoy data shows that, in most cases, dissolved oxygen concentrations continued to decrease after the survey date (Fig. 11.2). Only the August 2002 survey captured the maximum development of hypoxia. Hence, our July–August averages do not represent the most extreme conditions and are, in fact, conservative estimates of the extent of summertime bottom water hypoxia.

11.6.2 Synoptic Survey Statistics

The July–August surveys were characterized by a relatively narrow range of mean temperature (19.9–24.6°C; average 21.9°C) and salinity (27.2–31.1‰; average 29.1‰). The saturation levels of dissolved oxygen dictated by these physical conditions also exhibits rather small variability (6.95–7.68 mg L^{-1}). During July and August, between 30% and 88% of the stations exhibited suboxic (<4.8 mg L^{-1}) conditions (Table 11.2). Hypoxic levels (<2.9 mg L^{-1})

Fig. 11.2 Time series of available bottom water dissolved oxygen concentrations at the Bullock's Reach buoy during July and August surveys. Thin lines are 15-min measurements and thick line is a 24 h smoothed record. The time and dissolved oxygen level of adjacent survey stations are shown as triangles. The EPA suboxic (<4.8 mg L^{-1}) and hypoxic (<2.9 mg L^{-1}) levels are shown for reference.

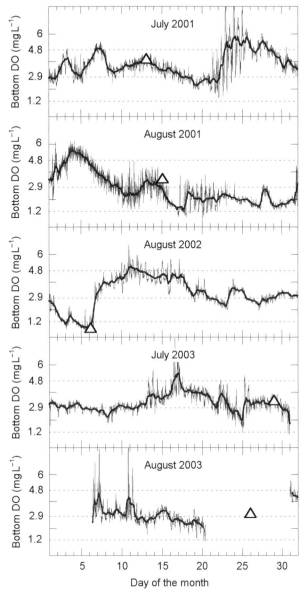

ranged from 4% of the stations on July 21, 2000 to as high as 68% of the stations on August 6, 2002. Table 11.2 gives a summary of the number of stations in each survey, the percentage of stations below each EPA designation, and the geographic areas measured in each survey. Shallower stations (<3.5 m) exhibited increased variability of dissolved oxygen concentrations (average $\sigma = 2.2$, $n = 12$) compared with deeper stations (>5 m, average $\sigma = 1.3$, $n = 50$), probably because they were above the pycnocline and thus more susceptible to wind and tidal mixing forces. Detailed maps of the sample distribution and contoured dissolved oxygen concentrations are available for all surveys at *http://www.geo.brown.edu/georesearch/insomniacs*. The percentage of stations below each EPA criteria reflects both the dissolved oxygen status of the bay and the specific areas included in each survey. For example, in August 1999, the Providence River section, which typically exhibited the lowest dissolved oxygen, was not sampled; hence, the percentage of stations recording hypoxia is lower than other surveys.

11.6.3 Hypoxic Area: Synoptic Surveys

The area and percent area of suboxic (<4.8 mg L^{-1}) waters in the July and August surveys, relative to the entire upper bay, ranges from 19% to 77% (Table 11.3). August 2002 (Fig. 11.3a) was the most severe hypoxic event captured by the synoptic surveys, with 16% below 1.2 mg L^{-1} (severely hypoxic), 45% below 2.9 mg L^{-1} (hypoxic), and 67% below 4.8 mg L^{-1} (suboxic). In contrast, the August 2003 survey (Fig. 11.3b) captured milder, but more widespread hypoxic conditions, with 77% of upper Narragansett Bay below 4.8 mg L^{-1}, and 35% below 2.9 mg L^{-1}, but only 1% below 1.2 mg L^{-1}. During August 2003, hypoxic waters were concentrated in the West Passage while the Providence River was merely suboxic. On August 23, 2003, a large fish kill in western Greenwich Bay was attributed to low dissolved oxygen (RIDEM, 2003). However, the August 25 survey could not verify these conditions because strong winds mixed much of the water column just prior to and during the survey.

None of the June or September surveys captured severe hypoxic events. The four June surveys indicated that 16–28% of the upper bay was suboxic (<4.8 mg L^{-1}), with hypoxic (<2.9 mg L^{-1}) areas less than 8%. Only two surveys were conducted in September—one showed no hypoxic waters, although no data were available for the Providence River; the other showed widespread (73%) suboxic (<4.8 mg L^{-1}) conditions but only a limited area (3%) of hypoxia (<2.9 mg L^{-1}).

11.6.4 Summertime Bottom Dissolved Oxygen Averages

The multi-survey mean and standard deviation of bottom water dissolved oxygen should give a more representative estimate of the distribution of summertime hypoxic waters than individual surveys. A plot of mean and standard deviation of July–August bottom water dissolved oxygen (sorted by increasing values) (Fig 11.4) reveals that 79% of upper bay stations had average dissolved oxygen below 4.8 mg L^{-1} and, within the uncertainty envelope, all stations were below 4.8 mg L^{-1}. Only a few stations had average bottom dissolved oxygen concentrations below 2.9 mg L^{-1}. However, within one standard deviation of

Table 11.3 Summary of the area (km²) and the % of total area (158 km²) for each survey and the area (km²) and area % of bottom waters that fall below EPA dissolved oxygen criteria for the summertime average and for individual surveys.

	July–August	7/21/2000	7/13/2001	7/28/2003	8/24/1999	8/15/2001	8/6/2002	8/25/2003
Area km²	137 (127)	139	151	130	76	142	116	136
% 158 km²	87% (81%)	88%	96%	83%	48%	90%	74%	86%
<1.2 mg L⁻¹	0% (3%)	0	1 km²	0	0	4 km²	26 km²	1 km²
		0%	1%	0%	0%	2%	16%	1%
<2.3 mg L⁻¹	1% (10%)	0	5 km²	2 km²	0	15 km²	59 km²	31 km²
		0%	3%	1%	0%	10%	37%	20%
<2.9 mg L⁻¹	6% (16%)	0	16 km²	5 km²	1 km²	29 km²	71 km²	55 km²
		0%	10%	3%	1%	18%	45%	35%
<4.8 mg L⁻¹	70% (49%)	29 km²	70 km²	71 km²	61 km²	88 km²	105 km²	122 km²
		19%	44%	45%	38%	56%	67%	77%

The spatial distribution of the July–August average values are shown in Fig. 11.5. Average values in ( ) are from averaging the individual survey levels rather than the mean values of the surveys.

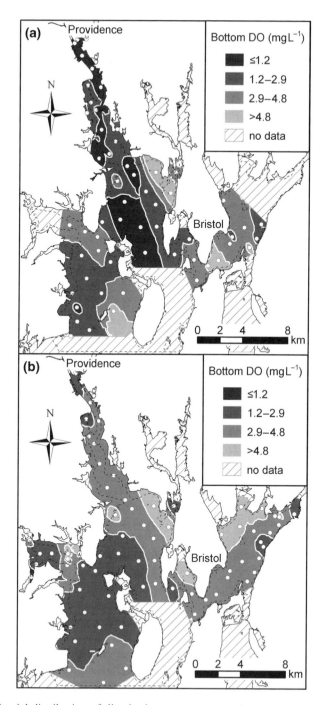

Fig. 11.3 Spatial distribution of dissolved oxygen concentrations observed in the bottom waters of upper Narragansett Bay on (a) August 6, 2002 and (b) August 25, 2003. Dashed line is the 2.5 m contour. Stations are a subset of those shown in Fig. 11.1.

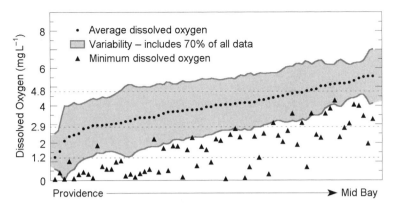

Fig. 11.4 Average July–August bottom water dissolved oxygen concentrations from 69 survey stations. Stations are ordered by increasing average dissolved oxygen concentration, which roughly reflects a down-bay gradient. The smoothed envelope surrounding the averages (± 1 sd) contains roughly 70% of the observed data. Minimum dissolved oxygen measurements (triangles) from each station are shown along with the critical EPA criteria for suboxic (<4.8 mg L^{-1}), hypoxic (<2.9 mg L^{-1}), and severely hypoxic (<1.2 mg L^{-1}) waters.

the mean, 68% of the stations surveyed were below 2.9 mg L^{-1}. The minimum observed dissolved oxygen at each station (shown as triangles) presents a worst case scenario and reveals that 82% of stations had at least one measurement below 2.9 mg L^{-1} (hypoxic) and 46% of stations had a measurement below 1.2 mg L^{-1} (severely hypoxic).

The spatial distribution of the average bottom water dissolved oxygen (Fig. 11.5, Table 11.3) reveals that about 110 km^2, or about 70% of upper Narragansett Bay (>2.5 m), was covered by suboxic waters during the summer. Areas consistently exhibiting hypoxic waters (averages <2.9 mg L^{-1}) include the upper Providence River, the western side of the upper bay, and westernmost Greenwich Bay. The only areas with average summertime dissolved oxygen above 4.8 mg L^{-1} were the southern part of Mount Hope Bay and the southeastern portion of the West Passage. The East Passage was not systematically surveyed, but in comparison to the July 2001 survey, the East Passage is likely to average above the 4.8 mg L^{-1} level. We note that the average suboxic (<4.8 mg L^{-1}) area of the bay is 70%, whereas the average of the individual surveys is about 49% (Table 11.3). However, the individual surveys have greater areas below 2.9, 2.3, and 1.9 mg L^{-1}. This difference reflects the influence of extreme low dissolved oxygen values in depressing the mean value.

11.6.5 Comparison of Synoptic Survey Averages and Buoy Data

The summertime (July–August) average bottom water dissolved oxygen at the Bullock's Reach, Conimicut, and North Prudence buoys (Fig. 11.1) is nearly

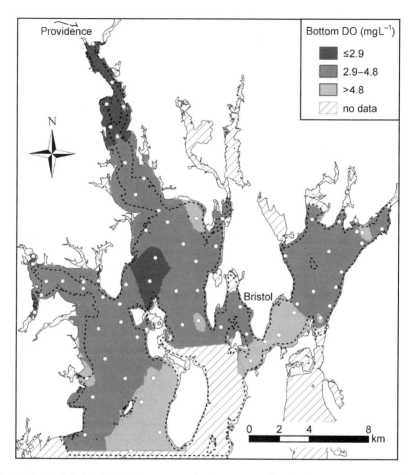

Fig. 11.5 Spatial distribution of average bottom water dissolved oxygen concentrations for summertime (July and August) surveys (averages shown in Fig. 11.4). Dashed line is the 2.5 m contour. Dissolved oxygen divisions are based on the EPA criteria (Table 11.1).

identical to the average bottom water dissolved oxygen of adjacent stations (Table 11.4). Both the buoy and station averages exhibit a down bay (Bullock's Reach to southern Prudence Island) increase in bottom water dissolved oxygen. The buoy data, approximately 6,000 measurements, should provide the most reliable estimate of mean bottom water dissolved oxygen, and the close correspondence to the synoptic survey data confirms the reliability of the survey station averages and provides a link between spatial and temporal data. We note, however, that the averages of the buoy data do not capture short-term hypoxic conditions that may be associated with tidal-scale or weather-related variability.

Table 11.4 Summary of the July–August seasonal average bottom water dissolved oxygen from the 1999–2003 buoy time series data and adjacent spatial survey stations.

	Bullock's (mg L^{-1})	Conimicut (mg L^{-1})	North Prudence (mg L^{-1})	South Prudence (mg L^{-1})
1999	–	–	4.51	5.32
2001	3.28	–	3.73	5.62
2002	3.48	–	4.20	5.48
2003	3.20	3.59	3.00	–
Average from buoy data	3.31	3.59	3.85	5.47
Average from adjacent stations	3.36	3.58	3.67	5.46

Buoy averages are from 15-min measurements averaged over July and August. Buoy data are ordered from Providence River to mid Narragansett Bay. Adjacent data are from the closest stations with at least five surveys.

11.6.6 Spatial Projection of Time Series Data

Given the correspondence between coincident buoy and adjacent survey station data, the question arises, "Can continuous buoy and station data be combined to estimate the extent of hypoxia throughout the upper bay?" The Bullock's Reach buoy data was selected as a reference because the site is proximal to nutrient and wastewater inputs into the bay, is proximal to low dissolved oxygen in the Providence River, and has the most complete time series. If buoy-to-station correlations are high, the Bullock's Reach time series could be used to estimate the near real-time aerial extent of hypoxic events throughout upper Narragansett Bay. High correlation does not imply that the buoy and station dissolved oxygen are identical, but merely that the station dissolved oxygen concentrations can be reliably estimated from buoy data. Regression analysis of station data with synchronous buoy data reveals that large areas of upper Narragansett Bay are highly correlated, and may act as a spatially coherent region (Fig. 11.6). Most of the deeper stations in upper Narragansett Bay and the Providence River are highly correlated to the Bullock's Reach buoy ($r = 0.75$). However, correlation is low in the shoal stations, such as Greenwich Bay, and at stations where dissolved oxygen is consistently low. Intriguingly, some stations in the West Passage and Mount Hope Bay also exhibit high correlations to the Bullock's Reach buoy data.

Station-to-station regressions revealed four distinct regions of spatial coherence within upper Narragansett Bay (Fig. 11.7): (1) the Providence River, the Upper Bay, and the Shipping Channel, (2) Greenwich Bay, (3) upper West Passage, and (4) Mount Hope Bay. These areas have the highest correlations to adjacent stations and can be expected to respond in a spatially coherent manner. In the Providence River, the Upper Bay, and the shipping channel stations are highly correlated (Fig. 11.7), similar to the Bullock's Reach analysis

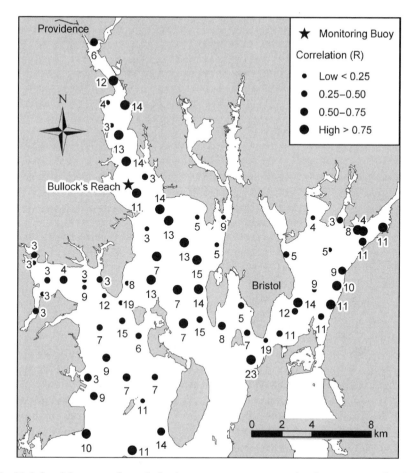

Fig. 11.6 Spatial pattern of correlation between average summertime bottom water dissolved oxygen concentrations from the spatial surveys, and the observed bottom water dissolved oxygen concentration at the Bullock's Reach monitoring buoy (indicated by a star) at the time of each survey. The degree of linear correlation is indicated by the by the size of the symbol; water depths are shown adjacent to each station location.

(Fig. 11.6). In general, deeper waters below the mixed layer are highly correlated to adjacent stations. Mount Hope Bay, Greenwich Bay, and the upper West Passage also reveal high internal coherence. Stations in shallow waters or close to tributaries are quite variable and tend to exhibit low station-to-station correlation. Dissolved oxygen values at these stations are highly variable and strongly dependent on the depth of the mixed layer. However, we note that several stations in Greenwich Bay do form a coherent group. The high degree of inter correlation within these four regions of upper Narragansett Bay implies that strategically located buoys could be expected to collect data representative of large areas of upper Narragansett Bay.

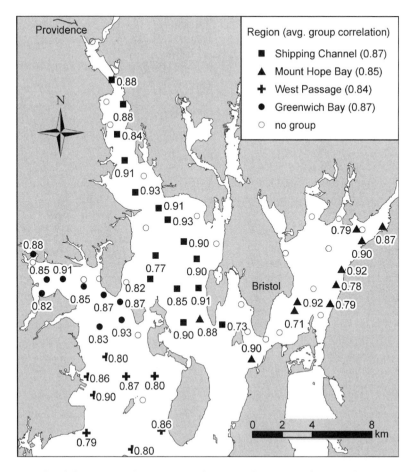

Fig. 11.7 Spatial pattern of average station-to-station correlation within geographic groupings of stations. Symbols define the geographic groups, and values indicate the average correlation between that station and all members of a geographic group. Open circles were not highly correlated to adjacent stations and are mostly shallow (<3.5 m) sites. Simple linear correlation among stations indicates that large areas of upper Narragansett Bay are spatially coherent and could be expected to respond in a similar manner.

11.6.7 Application of Buoy and Spatial Data to Estimate the Impacts of Hypoxia

If dissolved oxygen time series data are available, EPA criteria for the level and duration of critical dissolved oxygen conditions can be used to make specific estimates of the degree of impairment. Given high correlations, continuous buoy data can be used with the spatial regressions to provide time series estimates of near real-time extent and duration of low dissolved oxygen events over much of upper Narragansett Bay, as follows.

Application of EPA dissolved oxygen criteria to the Bullock's Reach time series data for the summer of 2001 is shown in Fig.11.8, and the number of days recorded and cumulative days below each dissolved oxygen level are summarized for the Bullock's Reach and North Prudence buoys in Table 11.5. The results show that for the summers (July–August) of 2001 and 2002, the Bullock's Reach site was below 4.8 mg L^{-1} for 53.9 days (87%) and 48.9 days (91%) and below 2.9 mg L^{-1} for 25.5 days (41%) and 15.5 days (29%), respectively (Fig. 11.8; Table 11.5). The North Prudence buoy data illustrate that the durations decrease down bay, but still exhibit 15 days (29%) and 12 days (19%) below 2.9 mg L^{-1} in 2001 and 2002, respectively (Table 11.5). Comparing the full season (May–October), summer values constitute the bulk of the low dissolved oxygen duration, confirming that July–August is the most likely time for hypoxia. The duration data also display more variability in the Providence River (Bullock's Reach) than in the more central North Prudence site, especially in the number of days below 2.9 mg L^{-1}. Almost all of the observations below 2.9 mg L^{-1} are observed during July and August.

The Bullock's Reach buoy-to-station regression was used to estimate the dissolved oxygen levels for an upper East Passage station (EPN06) (Fig. 11.1). The EPN06 linear regression:

$$DO_{EPN06} = 0.99 + 1.18 DO_{Bullock's Reach}$$

has a correlation coefficient of 0.94. Hence, the estimates for the upper East Passage site EPN06 are a constant offset ($\sim$1 mg L^{-1}) from the Providence River site, and they maintain the temporal pattern observed at Bullock's Reach. This approach offers the potential of evaluating the dissolved oxygen status of Narragansett Bay for any real-time or retrospective buoy data. For July–August of 2001, we estimate that dissolved oxygen in the upper East Passage was below 4.8 mg L^{-1} for 31 days, or about 50% of the interval, and below 2.9 mg L^{-1} for only 5 days, about 8% (Table 11.5). Comparison of observed dissolved oxygen from the spatial surveys (Prell et al., 2004) and the Nu-Shuttle surveys (Berman et al., 2005) with estimated dissolved oxygen concentrations at EPN06 show reasonable fit, although no potential tidal excursion lags were taken into account (Fig. 11.8). These estimates illustrate that dissolved oxygen generally increases down-bay, but also indicate that even the upper East Passage is most likely subjected to suboxic to hypoxic waters during the summer. Similar analysis could be done for all stations that are highly correlated to the buoy data to develop spatial estimates of dissolved oxygen patterns and durations throughout the summer.

Table 11.5 Summary of the number of days recorded at Bullock's Reach and North Prudence monitoring buoys and the cumulative time (July–August and May–October) and % time that bottom water dissolved oxygen was below EPA criteria during 2001 and 2002. Duration times are the summation of 15-min intervals below a specific level and the % duration is relative to the days recorded. Estimates of dissolved oxygen and the durations below specific levels for the upper East Passage station EPN06 were calculated from the linear regression between the Bullock's Reach buoy data and the spatial survey station (EPN06).

		Bullock's Reach buoy				North Prudence buoy				Upper East Passage (EPN06)		
		# days	<4.8		<2.9	# days	<4.8		<2.9	# days	<4.8		<2.9
2001	July–August	62	53.9	87%	25.5	41%	51	38.2	75%	14.7	29%	62	31.0	50%	5.0	8%
2001	May–October	173	94.7	55%	36.8	21%	81	45.2	56%	14.9	18%	173	47.9	28%	5.8	3%
2002	July–August	54	48.9	91%	15.5	29%	62	37.7	61%	11.7	19%	54	21.8	40%	4.1	8%
2002	May–October	131	77.7	59%	15.7	12%	170	43.5	26%	11.8	7%	131	22.9	17%	4.1	3%

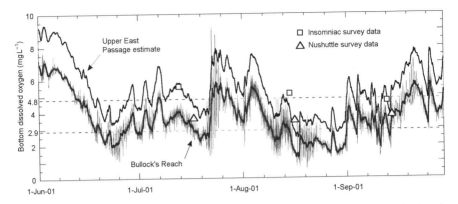

Fig.11.8 Time-series of bottom water dissolved oxygen concentrations from Bullock's Reach buoy (gray line is 15-min. data; thick solid line is 24 h smoothed data) and the regression-based estimate of bottom water dissolved oxygen for the upper East Passage station EPN06 (thin solid line). The East Passage estimates are daily smoothed to ease comparison. Observed values of dissolved oxygen measured at EPN06 during the spatial surveys (squares) and the Nu-Shuttle surveys (triangles) are shown for comparison. This plot illustrates the strategy for projecting buoy data to develop spatial estimates of dissolved oxygen concentrations for upper Narragansett Bay.

11.7 Conclusions

This synthesis of the spatial survey data from 1999 to 2003 and its integration with fixed buoy time series data reveals that much of upper Narragansett Bay is suboxic (dissolved oxygen <4.8 mg L^{-1}) during the summer. Comparison with the Bullock's Reach time series data reveals that most of the synoptic surveys were not conducted during the most hypoxic conditions, and that station bottom water dissolved oxygen concentrations averaged over five summers are almost identical to the continuous time series data collected at fixed locations.

The most hypoxic conditions were sampled in August 2002 when 35% of the stations and 16% of the area of upper Narragansett Bay was severely hypoxic (<1.2 mg L^{-1}). In upper Narragansett Bay, 46% of the stations exhibited at least one dissolved oxygen measurement below 1.2 mg L^{-1}. During summer (July–August), 70% (by area) of upper Narragansett Bay bottom water was suboxic (dissolved oxygen <4.8 mg L^{-1}). All upper Narragansett Bay stations surveyed experienced suboxic conditions during at least one survey. The hypoxic area of upper Narragansett Bay ranged from 0% to 45% among the surveys, but averaged 6%.

Simple buoy-to-station and station-to-station regressions of summer bottom water dissolved oxygen concentrations reveal high correlations over much of upper Narragansett Bay, and identify four areas of high internal correlation—Providence River, Upper Bay, shipping channel, Greenwich Bay, upper

West Passage, and Mount Hope Bay. These areas of internal spatial coherence indicate that much of upper Narragansett Bay can be monitored with a few critically placed buoys, and that buoy results can be projected to other parts of upper Narragansett Bay to estimate the areas impacted by low dissolved oxygen events.

This initial study was limited by the small amount of data in the station-to-buoy regression analysis, and could be greatly strengthened by additional spatial station data to better constrain the regressions, and by collection of buoy data within spatially coherent areas of upper Narragansett Bay. This study suggests that coherent spatial structure exists in upper Narragansett Bay and could be more formally defined by better data and additional analysis techniques.

Acknowledgments We thank all the institutions and individuals that participated in the "Insomniac" dissolved oxygen surveys. A complete listing of organizations and individuals can be found at *http://www.geo.brown.edu/georesearch/insomniacs/*. We also thank Don Pryor for his critical analysis, continuous encouragement, and stimulating discussions; Philip Howell for assistance in the data handling and analysis, Lynn Carlson for GIS assistance, and reviewers of the manuscript for many useful suggestions. Partial support for this research was provided by Brown University, RIDEM, and the Bay Windows project.

References

Bergondo, D.L. 2004. Examining the processes controlling water column variability in Narragansett Bay: time-series data and numerical modeling. Ph.D. Dissertation, University of Rhode Island, Narragansett, RI. 187 pp.

Bergondo, D.L., Kester, D.R., Stoffel, H.E., and Woods, W. 2005. Time-series observations during the low sub-surface oxygen events in Narragansett Bay during summer 2001. *Marine Chemistry* 97:90–103.

Berman, M., Melrose, C., and Oviatt, C. 2005. *http://www.narrbay.org/d_projects/nushuttle/shuttletree.htm*.

Bricker, S.B., Clement, C.G., Pirhalla, D.E., Orland, S.P., and Farrow, D.G.G. 1999. National Estuarine Eutrophication Assessment: A Summary of Conditions, Historical Trends, and Future Outlook. National Ocean Service, National Oceanic and Atmospheric Administration, Silver Springs, MD.

Buzzelli, C.P., Luettich, R.A., Jr, Powers, S.P., Peterson, C.H., McNinch, J.E., Pinckney, J.L., and Paerl, H.W. 2002. Estimating the spatial extent of bottom-water hypoxia and habitat degradation in a shallow estuary. *Marine Ecology Progress Series* 230:103–112.

Chinman, R.A., and Nixon, S.W. 1985. Depth–area–volume relationships in Narragansett Bay. NOAA/Sea Grant Marine Technical Report 87. Graduate School of Oceanography, University of Rhode Island, Narragansett, RI.

Deacutis, C.F. 1999. Nutrient impacts and signs of problems in Narragansett Bay. *In* Proceedings of a Workshop on Nutrient Removal from Wastewater Treatment Facilities. Nutrients and Narragansett Bay, pp. 7–23. Kerr, M. (ed.) Rhode Island Sea Grant, Narragansett, RI.

Deacutis, C.F., Murray, D.W., Prell, W.L., Saarman E., and Korhun, L. 2006. Hypoxia in the Upper Half of Narragansett Bay, RI during August 2001 and 2002. *Northeast Naturalist* 13(Special Issue 4):173–198.

Diaz, R.J. 2001. Overview of Hypoxia around the World. *Journal of Environmental Quality* 30(2):275–281.

Doering, P.H., Weber, L., Warren, W., Hoffman, G., Schweitzer, K., Pilson, M.E.Q., and Oviatt, C.A. 1988a. SPRAY Cruise Dissolved Oxygen and Chlorophyll, #Narragansett BayP-89-24 (GC512.R4 D637 × 1988 Brown).

Doering, P.H., Weber, L., Warren, W., Hoffman, G., Schweitzer, K., Pilson, M.E.Q., and Oviatt, C.A. 1988b. Monitoring of the Providence and Seekonk Rivers for Trace Metals and Associated Parameters, Vol. 2. Data Report for Spray Cruises I—VI. Marine Ecosystems Research Laboratory, Graduate School of Oceanography, University of Rhode Island, Narragansett, RI.

Doering, P.H., Oviatt, C.A., and Pilson, M.E.Q. 1990a. Control of nutrient concentrations in the Seekonk-Providence River Region of Narragansett Bay, Rhode Island. *Estuaries* 13:418–430.

Doering, P.H., Oviatt, C.A., and Pilson, M.E.Q. 1990b. Characterizing Late Summer Water Quality in the Seekonk River, Providence River and Upper Narragansett Bay, Final Report, NBP 90–49. The Narragansett Bay Project. 58 pp.

Ely, E. 2002. An overview of Narragansett Bay, Rhode Island. RI Sea Grant, Narragansett, RI. 8 pp.

Granger, S.H. 1994. The basic hydrography and mass transport of dissolved oxygen in the Providence and Seekonk Rivers estuaries, MS Thesis, University of Rhode Island, Narragansett, RI. 230 pp.

Hicks, S.D. 1959. The physical oceanography of Narragansett Bay. *Limnology and Oceanography* 4:316–327.

Kerr, M. (ed). 1999. Nutrients and Narragansett Bay: Proceedings of a Workshop on Nutrient Removal for Wastewater Treatment Facilities. Rhode Island Sea Grant, Narragansett, RI.

Kester, D.R., Fox, M.F., and Magnuson, A. 1996. Modeling, measurements and satellite remote sensing of biologically active constituents in coastal waters. *Marine Chemistry* 53:131–145.

Kremer, J.N., and Nixon, S.W. 1978. A Coastal Marine Ecosystem: Simulation and Analysis. Springer Verlag, New York. 217 pp.

Narragansett Bay Water Quality Monitoring Network, 2005. Data and references available at (*http://www.narrbay.org/d_projects/buoy/buoydata.htm*).

National Research Council. 2000. Clean Coastal Waters: Understanding and Reducing the Effects of Nutrient Pollution. National Academies Press, Washington, DC. 405 pp.

Nixon, S.W., Granger, S.L., and Nowicki, B.L. 1995. An assessment of the annual mass balance of carbon, nitrogen, and phosphorus in Narragansett Bay. *Biogeochemistry* 31:15–61.

Olsen, S., and Lee., V. 1979. A Summary and Preliminary Evaluation of Data Pertaining to the Water Quality of Upper Narragansett Bay. Coastal Resource Center Report to EPA Region 1 (TD224.35 R4 O78 1979 Narragansett Bay). Coastal Resources Center.

Oviatt, C. 1980. Some Aspects of Water Quality in and Pollution Sources to the Providence River, prepared by Graduate School of Oceanography, University of Rhode Island, Narragansett, prepared for EPA Region I, Boston, MA.

Pilson, M.E.Q. 1985. On the residence time of water in Narragansett Bay. *Estuaries* 8:2–14.

Pilson, M.E.Q., and Hunt, C.D. 1989. Water Quality Survey of Narragansett Bay, A Summary of Results from the SINBADD Cruises: 1985–86. Marine Ecosystem Research Laboratory, Graduate School of Oceanography, University of Rhode Island, Narragansett, RI.

Prell, W., Saarman, E., Murray, D., and Deacutis, C. 2004. Summer-Season, Nighttime Surveys of Dissolved Oxygen in Upper Narragansett Bay (1999–2003). (*http://www.geo.brown.edu/georesearch/insomniacs*).

Rabalais, N.N. 2002. Nitrogen in aquatic systems. *Ambio* 31:102–112.

Rhode Island Department of Environmental Management (RIDEM). 2003. The Greenwich Bay fish kill—August 2003 Causes, impacts and responses. Providence, RI. (*www.dem.ri.gov/pubs/fishkill.pdf*).

Rhode Island Department of Environmental Management (RIDEM). 2005. Plan for Managing Nutrient Loadings to Rhode Island Waters. Report to the RI General Assembly Pursuant to RIGL 46-12-3(25). Rhode Island Department of Environmental Management.

Ries, K.G., III. 1990. Estimating Surface-Water Runoff to Narragansett Bay, Rhode Island and Massachusetts. US Geological Survey, Water Resources Investigations. Report 89-4164. 44 pp.

Stanley, D.W., and Nixon, S.W. 1992. Stratification and bottom-water hypoxia in the Pamlico River Estuary. *Estuaries* 15(3):270–281.

UNESCO. 1981. Tenth Report of the Joint Panel on Oceanographic Panels and Standards. UNESCO Technical Papers in Marine Science No. 35, pp. 24. UNESCO.

United States Environmental Protection Agency. 2000. Ambient Aquatic Life Water Quality Criteria for Dissolved Oxygen (Saltwater): Cape Cod to Cape Hatteras, US EPA, Office of Water, November 2000: EPA 822-R-00-012. *http://www.epa.gov/waterscience/criteria/dissolved/index.html.*

US Army Corps of Engineers. 1960. Effects of Proposed Hurricane Barriers on Water Quality of Narragansett Bay (TC224.R4 U56 Narr Bay). US Army Corps of Engineers.

Valente, R.M., Rhoads, D.C., Germano, J.D., and Cabelli, V.J. 1992. Mapping of benthic enrichment patterns in Narragansett Bay, RI. *Estuaries* 15:1–17.

Chapter 12
Evidence of Ecological Impacts from Excess Nutrients in Upper Narragansett Bay

Christopher F. Deacutis

12.1 Introduction

A review of global responses to anthropogenic-driven nutrient loads in marine systems discloses a lengthy list of estuaries, with the severity of impacts appearing to increase over the last several decades (Bricker *et al.*, 1999; Diaz, 2001; Diaz and Rosenberg, 2001). Compared with other estuaries of similar size, Narragansett Bay (370 km^2, Ries, 1990) is considered to have significantly large anthropogenic loads due to a high population density within its 4,714 km^2 urbanized watershed. A major source of nutrients comes from the large volume of secondary-treated wastewater released into bay waters (Table 12.1) (Nixon and Pilson, 1983; Nixon *et al.*, 1995; Bricker *et al.*, 1999; Pryor *et al.*, 2006). Narragansett Bay's susceptibility to negative impacts of such large nutrient loads, however, has been considered low due to low freshwater inflow (Ries, 1990) and high average mixing energies from tides and prevailing winds (Weisberg, 1976; Weisberg and Sturges, 1976; Kremer and Nixon, 1978; Olsen and Lee, 1979; Granger, 1994; Bricker *et al.*, 1999). These factors are known to minimize stratification and maximize dilution capacity, minimizing eutrophic impacts (Kremer and Nixon, 1978; Nixon and Pilson, 1983).

This chapter presents evidence that Narragansett Bay has exhibited negative responses to excess nutrients, including dense macroalgal biomass, tidally linked summertime hypoxia and benthic community changes in the upper, urbanized area of the bay for at least the past two decades. The chapter closes with a discussion of some potential responses of the ecosystem to a projected 50% decrease in nitrogen loading from permitted point sources (State of Rhode Island, 2004; RIDEM, 2005b), and suggestions of areas where monitoring could be focused to capture the strongest ecosystem response signals to this change.

Christopher F. Deacutis
Narragansett Bay Estuary Program, University of Rhode Island, Graduate School of Oceanography, Box 27 Coastal Institute Building, Narragansett, RI 02882
deacutis@gso.uri.edu

A. Desbonnet, B. A. Costa-Pierce (eds.), *Science for Ecosystem-based Management.*
© Springer 2008

Table 12.1 Major wastewater treatment facilities discharging into Narragansett Bay (USEPA, 2001).

Facility (State)	Design flow ($m^3 \times 10^3\ d^{-1}$)
Attleboro (MA)	32.7
No. Attleboro (MA)	17.5
Fall River (MA)	117.4
UBWPAD (Worcester, MA)	212.8
Bristol (RI)	14.4
Cranston (RI)	76.8
East Greenwich (RI)	4.7
East Providence (RI)	39.5
Bucklin Point (East Providence, RI)	117.8
Field's Point (Providence, RI)	246
Newport (RI)	40.5
Quonset Point (RI)	6.7
Warren (RI)	7.6
Warwick (RI)	29.3
West Warwick (RI)	39.9
Woonsocket (RI)	60.8

12.2 Nutrient Gradient in Narragansett Bay

Narragansett Bay experiences large anthropogenic loads of nutrients, especially dissolved inorganic nitrogen (DIN) (Nixon *et al.*, 1995). Field measurements have shown nutrients to exhibit a clear north–south gradient within the bay (Oviatt *et al.*, 2002), with 62–73% of the nutrient load arising from wastewater treatment facility (WWTF) discharges (State of Rhode Island, 2004). Large WWTFs discharge directly into the tidal Seekonk and Providence Rivers, or just upstream within the watershed (Fig. 12.1, Table 12.1; Nixon *et al.*, 1995; Nixon *et al.*, 2005; Pryor *et al.*, 2006). The Providence River receives 66–68% of its total nitrogen load from WWTFs (State of Rhode Island, 2004; Pryor *et al.*, 2006).

There is limited mixing capability in the restricted areas of the Providence and Seekonk Rivers under normal estuarine flow ($\sim$89 $\times$ 10^6 m^3, Chinman and Nixon, 1985). In this region, lower salinity surface water is often separated from higher salinity waters moving up bay in the deeper ship channel (Weisberg, 1976; Granger, 1994), further limiting dilution of near-surface freshwater discharges. The largest WWTF in the watershed (Field's Point—246×10^3 m^3 d^{-1}) releases its effluent from a bulkhead at Field's Point in the upper Providence River. Another significant WWTF discharge, Bucklin Point (117×10^3 m^3 d^{-1}), discharges to a shallow mud flat in the Seekonk River. These same areas of the bay also exhibit the most severe eutrophication impacts (hypoxia and anoxia), although impacts continue to a lesser extent beyond these areas.

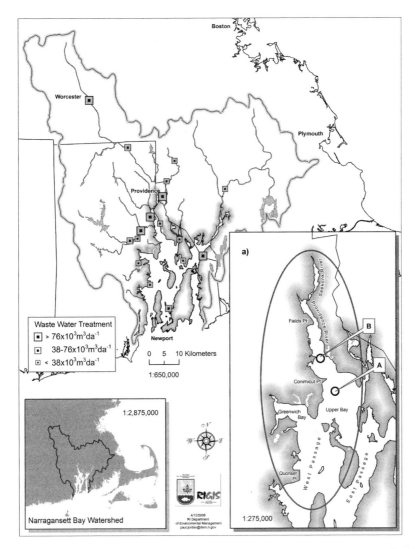

Fig. 12.1 Narragansett Bay watershed showing approximate discharge points of major wastewater sewage treatment facilities. Largest wastewater facilities discharge to the Providence-Seekonk Rivers. (a) Area of greatest hypothesized ecological change following nutrient reductions. Circle A = present area of maximum abundance of small opportunistic individuals. Circle B = hypothesized northward migration of this point.

12.3 Indications of Nutrient Impacts

Various environmental indicators have been proposed to assess eco-system response to anthropogenic nutrient stress. Bricker *et al.* (2003) and Cloern (2001) have provided recent assessment frameworks to categorize eutrophication of estuaries using various primary and secondary responses. Others have provided evidence of specific repetitive patterns in benthic community structure and biodiversity in response to organic enrichment (Pearson and Rosenberg, 1978; Rosenberg, 2001; Llanso *et al.*, 2003). Changes in the ratio of demersal to pelagic fish species have also been associated with eutrophication (Caddy, 2000; de Leiva Moreno *et al.*, 2000). These multiple indicators are used here to provide a framework that strongly implicates moderate to severe impacts to the upper half of Narragansett Bay from excess external nutrient loads (Table 12.2).

12.4 Benthic Primary Producers

12.4.1 Submerged Aquatic Vegetation

High-quality benthic submerged aquatic vegetation (SAV) such as eelgrass (*Zostera marina*) is considered critical habitat within shallow areas of Narragansett Bay (Raposa and Oviatt, 2000). Decline of eelgrass meadows have been linked to increased nutrient loading associated with land use changes and increasing population density (Short *et al.*, 1993; 1996; Nixon *et al.*, Chapter 5). Increasing water temperature may further exacerbate the negative response to nutrient additions (Moore *et al.*, 1997; Bintz *et al.*, 2003; Lee *et al.*, 2004).

Loss of eelgrass is considered a secondary impact from anthropogenic nutrient loads (Cloern, 2001; Bricker *et al.*, 2003); though such losses are often one of the first observed negative ecological impacts to shallow estuarine systems because of the visible change in habitat structure (Havens *et al.*, 2001). Chlorophyll levels likely increase first (primary response, Bricker *et al.*, 2003), causing declines in light levels within the water column. Light availability is considered a principle factor governing growth of seagrass in moderately enriched areas (Moore and Wetzel, 2000), and Dennison *et al.* (1993) have linked SAV declines with increased chlorophyll at or above a threshold range of 15 µg L^{-1}. Seasonal severe light limitation from rapid overgrowth by drift macroalgae can also cause eelgrass decline in shallow estuaries (Hauxwell *et al.*, 2001; 2003).

12.4.2 SAV Losses in Narragansett Bay

Loss of eelgrass meadows throughout the northwest Atlantic occurred in the 1930s, with most of this loss attributed to a slime mold. A recent review of coastal biological changes points out that eelgrass loss also coincided with a warming trend, and suggests water temperature may have played a role in the

Table 12.2 Eutrophic indicators for upper Narragansett Bay.

Indicator	Threshold	Bay area experiencing threshold	First observed
Submerged (Vascular) Aquatic Vegetation (SAV) losses	Observed impacts on SAV; losses of SAV from low transparency, epiphytes, etc.	Providence River + upper bay + Greenwich Bay	1930s + 2nd losses in late 1940s–early1950s
Submerged Aquatic Vegetation (present coverage of *Zostera marina*)	High: $\geq 50\%$, $\leq 100\%$ coverage of suitable habitat		
	Med: $\geq 25\%$, $<50\%$	Lower half of Narragansett Bay south of Prudence Island	
	Low: $\geq 10\%$, $<25\%$		
	Very Low:$=0$, $<10\%$	Upper half of Narragansett Bay including shallow subembayments + coves	1930s + 2nd losses in late 1940s–early1950s
Nuisance and toxic blooms (incl. benthic macroalgae)	Impact observed to biological resources (H_2S, hypoxia)	Seekonk River, mid and lower Providence R. + Sakonnet R. + Greenwich Bay shallows <3 m	Interference w/ trawl gear 1980s; Increased complaints 1980s–1990s
Water Column Algal Blooms Chl *a* 90th percentile concentration in growing season ($\mu g\ L^{-1}$)	Hypereutrophic: $>60\ \mu g\ L^{-1}$	Seekonk River	Unknown
	High: >20, ≤ 60 $\mu g\ L^{-1}$	Providence.R. + upper bay + Greenwich Bay	Unknown
	Med: >5, $\leq 20\ \mu g$ L^{-1}	Upper + mid East + West Passages	Unknown
	Low: >0, $\leq 5\ \mu g$ L^{-1}	South of the Jamestown and Newport bridges	Unknown
Dissolved Oxygen 10th percentile value	Anoxic to severely hypoxic ≥ 0 to ≤ 2 mg L^{-1}	Seekonk + Providence River ship channel + western Greenwich Bay cove areas	Unknown–assumed early to mid 1900s for Prov. R., more recent (1980s) for Greenwich Bay
	Hypoxic ≤ 3 mg L^{-1}	Seekonk + Providence River shallows + upper bay + central Greenwich Bay +	Unknown–assumed early to mid 1900s for Prov. R., more recent (mid 70s to ea. 80s) for upper

Table 12.2 (continued)

Indicator	Threshold	Bay area experiencing threshold	First observed
		upper West + East Passages	Passages + Greenwich Bay
	Stressed >3, ≤4.8 mg L^{-1} (new EPA chronic criteria)	Mid-bay East + West Passages	Unknown–perhaps 1980s
BenthicCommunity Change ◆	Shift to r-strategy opportunists and low sensitivity sp. and/or kills (fish + invert.)	Seekonk River + Providence River + upper bay + western Greenwich Bay	Unknown–assumed ea. to mid 1900s for Prov. R., more recent (mid 70s to ea. 80s) for upper bay + upper Passages + Greenwich Bay
Demersal–Pelagic Fish Complex ◆	Shift from demersal to pelagic species	All of Narragansett Bay	Early 1980s

Modified from Bricker *et al.* (2003); ◆ indicates non-Bricker indicator.

decline. Recovery of many of these eelgrass beds along the northwest Atlantic occurred during a cooling period in the 1940s and 1950s (Oviatt, 2004).

Evidence of the historical existence of extensive eelgrass meadows in Narragansett Bay is found in documented oral interviews of retired scallopers who worked bay waters, and reviews of old geodetic survey maps and herbarium collection sources from Rhode Island waters. These sources indicate that eelgrass meadows and linear beds covered large areas of the shallow embayments in the upper half of the bay and along parts of the Providence River during the period from the late 1800s to the early 1950s (Doherty, 1995; 1997; Kopp *et al.*, 1997; Nixon *et al.*, Chapter 5). The scallopers noted visible declines in scallop grounds (eelgrass meadows) after the Great Hurricane of 1938 and into the late 1940s, with final large-scale losses associated with Atlantic hurricane Carol (Aug. 31, 1954; Doherty, 1995). Based on this work, qualitative maps of approximate locations and extent of major beds in Narragansett Bay from 1848 to 1996 have been produced (Kopp *et al.*, 1997). Extensive meadows occurred in Greenwich Bay, the Palmer and Warren Rivers, and the mouths of the rivers on the north side of Mount Hope Bay (Fig. 12.2 based on Kopp *et al.*, 1997; RICRMC, 2005). Today, no significant eelgrass beds exist north of the southern end of Prudence Island, approximately halfway up the bay, although a small number of isolated beds can be found along the west side of Prudence Island (Short *et al.*, 1993; Cottrell, 2001; Lee *et al.*, 2004; RICRMC, 2005).

Although the cause of initial losses up to the 1950s is unclear, light limitation impacts from phytoplankton and macroalgal blooms in the northern half of

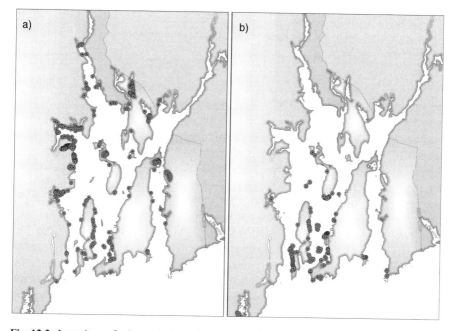

Fig. 12.2 Locations of eelgrass beds in Narragansett Bay. (a) approximately 1848–1954, and (b) contemporary eelgrass beds (1996). Maps from Rhode Island Coastal Resources Management Council eelgrass website: *http://www.edc.uri.edu/Eelgrass/mapshome.html* based on Kopp *et al.* (1997) and Cottrell (2001).

Narragansett Bay (Oviatt *et al.*, 2002; Nixon *et al.*, 2005) is likely limiting the current northern extent of *Z. marina*. Studies in a number of estuaries, including Narragansett Bay, have shown strong negative response of eelgrass to high nutrient levels such as those found in upper Narragansett Bay (Lee *et al.*, 2004). Nitrogen loads to the Providence River area, and to Greenwich Bay, are >750 kg ha^{-1} yr^{-1} and >130 kg ha^{-1} yr^{-1}, respectively (based on loadings from Granger *et al.*, 2000). This is well over the 60 kg ha^{-1} yr^{-1} threshold for loss of SAV (Hauxwell *et al.*, 2003). Based on data from the Chesapeake Bay system, changes in turbidity due to suspended solids linked to runoff during late spring may also be a factor, at least near river mouths (Moore *et al.*, 1996; 1997).

Warm water temperatures may further exacerbate negative impacts to SAVs from high nutrient loadings (Bintz *et al.*, 2003; Lee *et al.*, 2004). Annual average water temperatures at the mouth of the bay (Newport, RI) increased 1–3° C in the 1980s–1990s compared with a 1977–2002 baseline (Oviatt, 2004). An annual water temperature increase in the 1990s of at least 1.2° C has been found for Southern New England waters compared with the mean for 1870–1970 (Nixon *et al.*, 2004). Warmest temperatures are found in the upper half of Narragansett Bay, especially the shallow embayments like

Greenwich Bay (Fischer and Mustard, 2004; Prell *et al.*, 2004). With warm water and high nutrient loads, such areas are no longer viable environments for eelgrass.

12.4.3 Benthic Microalgae

Intertidal and shallow subtidal benthic microalgae can be an important component of primary production for macrozoobenthic consumers (Malin *et al.*, 1992; Cahoon, 1999; Wainright *et al.*, 2000; Kang *et al.*, 2003; Tobias *et al.*, 2003). Changes in microphytobenthos can be an early indicator of changes in nutrient loads and other environmental variables (Hillebrand and Sommer, 2000; Lever and Valiela, 2005). Where pelagic primary production has responded strongly to high nutrient loadings, subtidal (1 m) microphytobenthic productivity can be impacted negatively due to shading at the sediment–water interface (Meyercordt and Meyer-Reil, 1999).

Little is known about the level or importance of primary production from benthic microalgae in Narragansett Bay, and therefore the lack of data precludes any discussion of changes within the system. It may be prudent to monitor these primary producers along the nutrient gradient to understand their present role in the ecosystem, and how they respond to changing nutrient loads.

12.4.4 Benthic Macroalgae

Macroalgae are natural components of shallow estuarine environments, including those with low nutrient loadings (Meng and Powell, 1999), and vegetated areas often exhibit increased species diversity for benthic fish and macroinvertebrates compared with unvegetated areas (Meng *et al.*, 2004). Excessive macroalgal growth, however, is considered an early response of shallow water estuarine systems to introduced nutrient loads (Hauxwell *et al.*, 2001; 2003). In areas with high biomass of ephemeral rapid-growth species like *Ulva*, an annual dystrophic "crisis" may occur during the hottest months (Flindt *et al.*, 1997). In such a "crisis" algal mats senesce and decompose rapidly, bringing localized but severe hypoxia or anoxia, the release of sulphides from sediments, and significant loss of benthic community diversity. Even prior to senescence, benthic diversity beneath mats is low (Reise and Siebert, 1994; Viaroli *et al.*, 1996; Flindt *et al.*, 1997; Siebert *et al.*, 1997; Tagliapietra *et al.*, 1998; Havens *et al.*, 2001; Sfriso *et al.*, 2001). High seasonal biomass of such "green tide" species has permanently shifted some shallow systems from areas characterized by stable, grazing-controlled perennial vascular rooted plants (eelgrass), to unstable detritus/mineralization systems with temporary macroalgal complexity. Nutrient fluxes become much more dynamic in these conditions, and oxygen concentrations often wildly oscillate between aerobic and anaerobic states (Flindt *et al.*, 1997).

The extent of macroalgal growth in Narragansett Bay was synoptically examined in a single, limited, bay-wide survey in June 1990. In that survey, macroalgae species were found along all shorelines of Narragansett Bay. Mapped results suggested slower growth species such as fucoids and *Chondrus crispus* were more common in the mid and lower half of the bay, while high densities of rapid-growth "nuisance" species like *Ulva lactuca,* appeared to be common in shallow areas of the northern half and mid bay. *Ascophyllum nodosum* and *Laminaria spp.* were found mainly in the lower bay (French *et al.*, 1992). Recent pilot studies in the Providence River and Greenwich Bay have verified that the densest *ulvoid* macroalgal beds seem to occur in shallow subtidal (<2.5–3 m) zones, although ephemeral decaying biomass may build up in deeper, low energy areas following storms (Deacutis 2005, unpublished video surveys).

A more recent study attempted to quantify macroalgae during the summer of 1997 in Greenwich Bay (Granger *et al.*, 2000). These researchers found dense patches of *Ulva lactuca* and *Gracilaria tikvahiae* (to 1,000 g dry weight m^{-2}) with peak biomass in late June and late August. Distribution and biomass were highly variable and extremely patchy, though most macroalgal biomass was located at depths <2.5 m. The majority of annual primary production biomass, however, was from the phytoplankton (Granger *et al.*, 2000).

Because of the lack of a regular monitoring program to track biological changes in Narragansett Bay, there are no data to discern trends in macro-algal biomass over the last few decades. However, some indirect evidence does suggest changes in biomass occurring. RIDEM Fish and Wildlife monitor fisheries stocks with 20-min standardized bottom trawls, and have done so seasonally since 1979, increasing to monthly sampling in 1990. In the mid 1980s, all stations in less than ~6 m of water depth from the upper half of the bay were removed from the survey due to interference from excessive macroalgal biomass. Captured drift macroalgal biomass was so large that the otter trawl closed within 2 min of deployment, precluding the ability to sample, and sometimes even causing endangerment of the boat due to excessive weight in the net (Lynch, personal communication). In 2004, macroalgal interference forced further abandonment of stations at all depths within the entire upper half of Narragansett Bay. The state fisheries biologist, who has conducted these surveys since the early 1980s, indicated that these changes have been especially noticeable over the last 10–15 years (Lynch, personal communication). In addition, increased complaints of foul odor from hydrogen sulfide at low tide have occurred in the mid and lower Providence River and Greenwich Bay between 2002 and 2004. The odors have been tracked to decomposing mats of *Ulva* trapped in shallow coves, with parts of the biomass exposed at extreme low tides. The state has now developed a seaweed removal program to eliminate rotting vegetation in these shallow coves during the mid and late summer months (Mulhare, personal communication).

12.5 Water Column Primary Productivity

12.5.1 Phytoplankton Blooms

The major annual primary production biomass response to nutrients occurs in the phytoplankton within most of Narragansett Bay proper (Nixon, 1997). Phytoplankton are considered a critical source of the primary production that fuels secondary production (Nixon and Buckley, 2002), and is also the source for organic matter that can be shunted into a microbial loop (Turner, 2001). Nitrogen is considered the limiting nutrient for phytoplankton growth throughout most of Narragansett Bay, although phosphorous may play a role in parts of the Providence River under certain conditions (Kremer and Nixon, 1978; Granger, 1994; Oviatt et al., 1995). Smayda and Borkman, in Chapter 15, provide detailed analysis of plankton dynamics in Narragansett Bay.

12.5.2 Chlorophyll Levels in Narragansett Bay and Eutrophication State

An assessment of chlorophyll in Narragansett Bay is provided here using the 90th percentile chlorophyll a water column concentration thresholds from Bricker et al. (2003). Hypereutrophic, high, medium, and low eutrophic expressions are associated with 90th percentile values >60 μg L^{-1}, >20 μg L^{-1}, >5 μg L^{-1}, and ≤5 μg L^{-1}, respectively. An embayment/regional assessment resolution is provided because the state environmental agencies make regulatory decisions at this level (RIDEM, 2005b).

Assessments are based on chlorophyll data from 15 stations for 1997–1998 (Oviatt et al., 2002), 7 stations from 1985 to 1986 (Li and Smayda, 2001), 16 stations from 1995 to 1996 (Turner, 1997), and 12 stations from 1996 to 1997 (Granger et al., 2000) spread along the N–S nutrient gradient (Table 12.3). Starting at the northernmost end of the bay, 90th percentile values for the Seekonk River appear to exceed the hypereutrophic threshold (>60 μg L^{-1} chlorophyll a). Providence River values appear to fall well into the "high" threshold, sometimes entering the hypereutrophic threshold, with a range between 30 and 64 μg L^{-1} depending on which dataset, stations, and years are considered. The upper bay appears to fall just above the lower limit of the "high" threshold (∼22–25 μg L^{-1}) for one published dataset (2 stations, Oviatt et al., 2002), while Li and Smayda (2001) and Turner (1997) are in the range of ∼40 to 50 μg L^{-1}. These surveys represent single measurements taken on a weekly (Li and Smayda, 2001), bi-weekly (Oviatt *et al.*, 2002) or monthly basis (Turner, 1997). Nine continuous monitoring stations have recently been put into operation, and a better understanding of intra- and interannual variability will develop from these data, and help bracket the range of values under different climatic conditions (RIDEM, 2005a).

Greenwich Bay ranges from 20 to 22 μg L^{-1} chlorophyll *a* at the mouth to 18–20 μg L^{-1} mid-Greenwich Bay (Granger *et al.*, 2000; Li and Smayda, 2001; Oviatt *et al.*, 2002), while the coves are slightly higher at 25–30 μg L^{-1} (Granger *et al.*, 2000). Unpublished data from a continuous monitoring site at the mouth of a western-shore cove (Apponaug Cove) suggest this area has a summer mean around 15–25 μg L^{-1}, and may see occasional levels of 30–50 μg L^{-1} during significant blooms, with the area highly variable on a temporal basis (Stoffel, unpublished data). Thus, Greenwich Bay seems to be on the upper level of the "medium" or low end of the "high" category, with the western coves extending into the "high" level.

Data for the East and West Passages are sparse. A single station for upper West Passage is available (Li and Smayda, 2001) along with a station merged with other West Passage stations for Oviatt *et al.* (2002). The 90th percentile for the former dataset falls in the range of ∼25–28 μg L^{-1} chlorophyll *a*, while the latter is lower at ∼9–10 μg L^{-1}. The station used by Li and Smayda (2001) is in a more northerly location in the upper West Passage, located much closer to higher nutrient levels of the upper bay. The Oviatt *et al.* (2002) data include a station much closer to the mouth of the bay that would be expected to produce lower values. A station near Fox Island (mid Bay in the West Passage) appears to have a 90th percentile around 15–17 μg L^{-1} for 1985–1987 (Li and Smayda, 2001), but the 1997–1998 value was lower (∼7 μg L^{-1}; Oviatt *et al.*, 2002), suggesting this area of the West Passage may fall into the "medium" threshold some years, and below that threshold in other years. The estimated 90th percentile for the upper East Passage was ∼16–18 μg L^{-1} for the single station available (Oviatt *et al.*, 2002); suggesting this area also is in the "medium" threshold category. Both East and West Passage lower bay stations exhibited low chlorophyll *a* levels of 5–7 μg L^{-1} (Oviatt *et al.*, 2002), falling just above or within the "low" threshold category.

12.5.3 Changes in Phytoplankton Patterns in Narragansett Bay

Phytoplankton primary production has been tracked for over 40 years in the mid-to-lower bay at two long-term monitoring stations (Fox Island and the mouth of the bay). Less long-term data are available for the upper half of Narragansett Bay (Pratt, 1959; Karentz and Smayda, 1998).

Upper bay primary production from phytoplankton is thought linked to intermittent hypoxia formation (Bergondo *et al.*, 2005). If hypoxia is a phenomenon of recent origin (last two to three decades), then a change in

Table 12.3 Present conditions in Narragansett Bay relative to indicator thresholds for chlorophyll a (based on Bricker et al., 2003).

Trophic State	Range ~90th Percentile value
Hypereutrophic	>60 µg L^{-1}
High	>20 µg L^{-1} but ≤60 µg L^{-1}
Medium	>5 µg L^{-1} but ≤20 µg L^{-1}
Low	>0 µg L^{-1} but ≤5 µg L^{-1}

Data Source / Bay Area	Li and Smayda (2001) (~weekly) 1985–1987	Oviatt et al. (2002) (Biweekly) 1997–1998	Turner (1997) (monthly) May–October 1995–1996	Granger et al. (2000) (~biweekly May 1996–May 1997)
Seekonk River	NA	NA	72; 81; 62 µg L^{-1} (upper; mid-river; lower river)	NA
Providence River	~40 µg L^{-1} (Field's Point) ~54 µg L^{-1} (Gaspee Point)	~30–40 µg L^{-1} (4 stations Fox Point to Conimicut Pt.)	57–69 µg L^{-1} (Fox Pt. to Field's Pt.) 44–46 µg L^{-1} (east of Pawtuxet Cove; Gaspee Pt.) 47–64 µg L^{-1} (just No. of Conimicut Pt.)	NA
Upper Bay	~40–42 µg L^{-1} (Just South of Conimicut Point and just Northwest of Patience Is.)	~22–25 µg L^{-1} (just So. of Conimicut Pt. and No. of Patience Is.)	28–53 µg L^{-1} (just So. of Conimicut Pt.)	NA
Upper West Passage	~28 µg L^{-1} (just North of Quonset Point)	~9 µg L^{-1} (Hope Is.)	NA	NA
Upper East Passage	NA	~16–18 µg L^{-1} (just So. of Poppasquash Pt.)	NA	
Greenwich Bay	~22 µg L^{-1} (mouth of Greenwich Bay)	20 µg L^{-1} (mid Bay)	NA NA	~18–20 µg L^{-1} inner-mid Greenwich Bay
Lower Bay	~17 µg L^{-1} (Fox Is.)	4–7 µg L^{-1} (Fox Is; GSO; Gould Is.)	NA	~25–30 µg L^{-1} western Coves

Note: 90th Percentile estimate from published graphs or actual data.

some aspect of primary production must have fueled the hypoxia. It is difficult to track past changes in primary production through sediment cores or other historical reconstruction due to bioturbation and diagenesis, but there is some suggestion of increased organic carbon production reaching the sediments in the upper bay over recent decades. King *et al.* (Chapter 7) provide core data analysis for three sites with the bay ecosystem. Sediment core studies of dinoflagellate cysts may also help provide a timeline for changes due to anthropogenic impacts (Chmura *et al.,* 2004; Pospelova *et al.,* 2005).

One hypothesis of potential causal change is the historical shift to secondary sewage treatment over the last 30 years. All major WWTFs were required to expand from primary treatment (settling and chlorination) to secondary treatment to remove greater amounts of particulate organics in the effluent. In Rhode Island, this shift began in the 1970s at all the municipal sewage treatment facilities, and was completed by the early 1990s, with the majority upgraded by the mid to late 1980s (Desbonnet and Lee, 1991).

When secondary treatment was introduced in the New York area, a decrease in total nitrogen ($\sim$20%, mainly particulate organic nitrogen) was observed (O'Shea and Brosnan, 2000). For Rhode Island waters, total nitrogen concentrations did not change significantly, but the concentrations of various forms of nitrogen in the effluents were altered. Nixon *et al.* (2005) found a large fraction (60%) of total N in effluents to occur as organic (particulate) N in the 1970s (along with high total suspended solids—TSS). They repeated these measurements in the 1980s and 1990s, finding similar levels of total N, but TSS decreased and organic N now accounted for only 13–24% of total N in these same effluents, while DIN concentrations had increased (Nixon *et al.,* 2005). Such shifts to more bioavailable forms may have altered the nutrient availability for phytoplankton, and increased the aerial extent of microalgal blooms (Parker and O'Reilly, 1991; Nixon *et al.,* 2005).

12.6 Dissolved Oxygen

Dissolved oxygen is a critical, highly dynamic parameter of estuarine environments, controlled by both physical and biological factors. Increased salinity and increased temperature both decrease the maximum (100% saturation) concentration of oxygen, so seasonal and weather-related variability is inherent in this parameter. This parameter, however, is used as an indicator of negative impacts due to excess nutrients (Ritter and Montagna, 1999; Breitburg, 2002; Bricker *et al.,* 2003; Bergondo *et al.,* 2005).

When bacteria decompose high concentrations of organic matter, both in the water column and at the sediment–water interface, a significant decrease in oxygen concentrations to biologically stressful levels (hypoxia) often occurs. In order for this to happen, several physical factors must co-occur, including both warm water temperatures and thermohaline stratification of the water column to minimize mixing of oxygenated surface waters with waters below the pycnocline (Bergondo *et al.,* 2005).

Significant physiological stress for many estuarine organisms is typically considered to occur when oxygen levels drop below 2.0 mg L^{-1}. However, sensitive species are considered to experience acute impacts at 2.9–3.0 mg L^{-1} (Ritter and Montagna, 1999; USEPA, 2000). For this chapter, hypoxia is defined as oxygen concentrations of $\leq$3.0 mg L^{-1}.

Hypoxia has a wide range of negative impacts on estuarine organisms, including fish kills and mass mortality of benthic marine invertebrates (Baden *et al.,* 1990; Diaz and Rosenberg, 1995; Lu and Wu, 2000; Diaz, 2001; Breitburg, 2002; Wu, 2002). Hypoxia can significantly alter benthic communities, and cause loss of biogenic structural species such as oysters and mussels in depth-specific zones (Llanso, 1992; Rosenberg *et al.,* 1992; Gray *et al.,* 2002; Altieri and Witman, 2006; Deacutis *et al.,* 2006). In addition, increased levels of hydrogen sulfide associated with anoxic events may play a role in fish kills and lobster mortality (Luther *et al.,* 2004; Valente and Cuomo, 2005). Decreased immune system response and surface lesions have been associated with severe hypoxia (Burnett, 1997; Burnett and Stickle, 2001), while sublethal hypoxia has been shown to decrease growth in some fish species (Bejda *et al.,* 1992; Thetmeyer *et al.,* 1999; Meng *et al.,* 2001). Moderate hypoxia can reduce growth rates of marine organisms, cause shifts in benthic and pelagic community structure, control distribution of mobile pelagic species, and alter predatorprey interactions (Pihl, 1994; Diaz and Rosenberg, 1995; Ritter and Montagna, 1999; Rabalais *et al.,* 2001; Turner, 2001; Breitburg *et al.,* 2003; Eby and Crowder, 2004; Altieri and Witman, 2006).

12.6.1 Patterns of Hypoxia in Narragansett Bay

Hypoxia in Narragansett Bay is not the seasonal, summer-long hypoxia experienced in the deep areas of Chesapeake Bay and western Long Island Sound. Rather, it is similar to the tidally linked stratification and hypoxia of the James and York Rivers, VA (Haas, 1977; Ruzecki and Evans, 1986; Stanley and Nixon, 1992; Bergondo *et al.,* 2005). Low oxygen events occur in summer months during the mid to latter part of weak neap tides, with oxygen being replenished during high-energy mixing periods such as strong spring tides and/ or strong steady winds (Bergondo *et al.,* 2005; Deacutis *et al.,* 2006). These intermittent hypoxic events can persist for hours to approximately a week, with rarer durations extending over 2–3 weeks in the lower Providence River, upper bay (Bergondo, 2004) and Greenwich Bay under certain stratification-linked conditions (RIDEM, 2003; Bergondo, 2004). Severity and duration are also linked to freshwater flow, which changes surface density, flushing rates, and water column stratification (Pilson, 1985; Bergondo *et al.,* 2005). Continuous time series data suggest that the upper West Passage may have shorter hypoxic bouts (hours to days; Stoffel 2005, unpublished data).

Bricker *et al.* (2003) recommend using the lower 10th percentile for characterization of the minimum oxygen concentration as an expression of eutrophication state. Observed 10th percentile values for the tidal Seekonk and upper Providence Rivers exhibit extreme hypoxia and even anoxia (0.0–1.3 mg L^{-1}) on a seasonal intermittent basis (Turner, 1997). The mid Providence River can fall well below 3.0 mg L^{-1} (1.4–2.2 mg L^{-1}) while the lower Providence River and upper bay fall into or just above the hypoxic range (2.5–3.8 mg L^{-1}; Turner, 1997), depending on the tidal stage (Table 12.2).

Recent continuous field measurements from water quality monitoring sites (Bergondo *et al.,* 2005; RIDEM, 2005a) and temporally targeted oxygen surveys conducted in Narragansett Bay (Prell *et al.,* 2004) have shown oxygen levels often lower than those found by Turner (1997), with the lower Providence River and areas of the upper bay sometimes dipping below 2.0 mg L^{-1} (Prell *et al.,* 2004; Deacutis *et al.,* 2006; Stoffel, unpublished data). Thus, significant areas in the upper half of the bay experience hypoxia (≤ 3.0 mg L^{-1}) and even severe hypoxia (<2.0 mg L^{-1}) on an intermittent basis associated with neap tides (Bergondo, 2004; Bergondo *et al.,* 2005; Deacutis *et al.,* 2006). Smaller zones in western Greenwich Bay and the mid and upper Providence River reach <1.0 mg L^{-1}, and even experience occasional short-lived periods of anoxia (RIDEM, 2003; Deacutis *et al.,* 2006). Mount Hope Bay experiences severe hypoxia at several of the tidal river mouths (Lee, Cole, and Taunton Rivers) and in the vicinity of the ship channel near the Fall River WWTF, but much of the central area of this shallow embayment seems to remain above 3.0 mg L^{-1} (Howes and Schlezinger, 2003; Deacutis *et al.,* 2006). Saarman *et al.,* in Chapter 11, provide an in-depth assessment of dissolved oxygen conditions in Mount Hope Bay waters based on broad-scale survey data.

There may be a relationship between the lesser response of Mount Hope Bay and the extent of the lengthy (~19 km) meso- and oligohaline tidal reaches of the Taunton River, the longest undammed coastal river in New England (Napolitano, 2005). The extended oligohaline mixing zone of the Taunton River may process riverine nutrient loads farther upstream prior to reaching the polyhaline zone. A nutrient loading analysis is presently underway for Mount Hope Bay and should provide clearer answers (Howes, 2005).

12.6.2 Historic or Recent Response?

Previous reviews of dissolved oxygen all concluded that hypoxic events were either rare or do not occur below the Providence River (Olsen and Lee, 1979; Doering *et al.,* 1990; Desbonnet and Lee, 1991; Granger, 1994). More recent data indicate that it clearly does (Prell *et al.,* 2004; Deacutis *et al.,* 2006; Saarman *et al.,* Chapter 11), but what is unclear is how long this more extensive tidally linked hypoxia has been occurring in Narragansett Bay (Bergondo *et al.,* 2005; Deacutis *et al.,* 2006). A comparison of tidal state against historical available oxygen datasets for Narragansett Bay finds that hypoxia did occur in the Providence River, but was not a regular summer occurrence on neap tides farther down the bay (Bergondo *et al.,* 2005). However, high commercial landings data for leased oyster sites in the Providence River and upper bay from 1898 to 1931 (Oviatt *et al.,* 2003) provide strong circumstantial evidence that low oxygen events were not particularly severe or of extended duration while these landings levels were sustained.

Temperature can play a synergistic role with such nutrient additions (Bintz *et al.,* 2003), and plays a clear role in the risk of developing hypoxia by increasing metabolic rates while simultaneously decreasing oxygen maximum saturation levels (RIDEM, 2004). Mean annual water temperatures for the southern New England coast, including Narragansett Bay, have increased by at least 1.2° C for the 1990s compared with conditions for 1890–1970 (Nixon *et al.,* 2004; Oviatt, 2004). Greenwich Bay in particular shows a very high maximum temperature, and an earlier date of maximum temperature attainment compared with other areas of the bay based on Landsat data (Fischer and Mustard, 2004).

As noted in the chlorophyll discussion, another possible factor in increased hypoxic events is the shift to secondary treatment. The shift to a larger fraction of total nitrogen in dissolved inorganic forms could have played a role by increasing phytoplankton response, thereby increasing organic productivity in the upper bay (Nixon *et al.,* 2005). A measurable increase in water clarity in the lower bay was observed between 1984 and 1994, correlated to decreased suspended solids from the sewage plant upgrades (Borkman and Smayda, 1998; Smayda and Borkman, Chapter 15). Perhaps this increased clarity was a potential signal of removal of a larger component of the sewage nutrient load farther up the bay due to its more readily available form following secondary treatment.

An additional factor to consider in the phytoplankton increase is the decrease in toxic releases into the Providence River. Heavy metal loads decreased by over fivefold between 1981 and 1988 through both economic shifts in local industrial discharges (jewelry and metal plating), and increased regulation of toxics, especially metals through pretreatment regulations applied during the 1980s (Narragansett Bay Commission, 2005). Toxics such as copper, with strong inhibitory influence on phytoplankton and zooplankton growth, have significantly decreased over the last two decades in the Providence River (NBC, 2005) and could be a factor in a positive response of phytoplankton growth (O'Shea and Brosnan, 2000).

Today, maximum chlorophyll levels and minimum oxygen zones are found just south of the major wastewater treatment discharge at Field's Point, sometimes extending into the upper bay (Turner, 1997; Oviatt *et al.,* 2002; Prell *et al.,* 2004; Deacutis *et al.,* 2006). Findings by King *et al.* in Chapter 7 suggest increased nutrient impact and possible hypoxia/anoxia in this region, further corroborating that the Providence River area is highly impacted by nutrients.

12.7 Benthic Community Change: Responses to Low Oxygen

Intermittent low oxygen events, if severe enough, can potentially alter benthic community structure through attrition of sensitive species, followed by dominance of shallow-dwelling opportunistic species (Dauer *et al.*, 1992; Llanso, 1992; Ritter and Montagna, 1999). Benthic communities follow a predictive pattern of change associated with hypoxic stress (Pearson and Rosenberg, 1978; Diaz and Rosenberg, 1995; Rosenberg, 2001). For gradients of organic enrichment, benthic communities usually show an increase in biomass and diversity up to a point along the gradient, and then a decline in these community aspects. The density of small, shallow-dwelling opportunistic species increases to a maximum inflection point, followed by a rapid decline along the enrichment gradient, reaching a minimum-life azoic zone where the most severe ecological impact (anoxia) occurs (Pearson and Rosenberg, 1978; Dauer *et al.*, 1992; Turner, 2001).

Frequency, severity, and duration of stress (including impacts from organic load) are considered to strongly influence both the level of change in marine benthic communities, and the length of time to return to an equilibrium state (Rosenberg *et al.*, 1992; Rosenberg, 2001). Longer lived sensitive species are thought capable of achieving equilibrium state (Type III) within their approximate lifespan, usually 4–5 years (Zajac and Whitlatch, 1988). Thus, if rare events occur at a frequency below 4–5 years, such areas should be able to return to stable equilibrium (Type III) communities, while areas subject to severe stress (e.g., hypoxic events <2 mg L^{-1}) at frequencies <2–4 years are likely to experience permanent loss of equilibrium species (Diaz and Rosenberg, 1995).

A classic Pearson–Rosenberg benthic pattern (Pearson and Rosenberg, 1978; Diaz and Rosenberg, 1995; Rosenberg, 2001) appears to be occurring in the upper reaches of Narragansett Bay. Maximum macrobenthic biomass is found in the mid and lower reaches of the bay, followed by a steep increase in the density of small opportunistic organisms up to a maximum found at Conimicut Point in the lower Providence River. When data from environmental impact studies associated with a recent Providence River dredging project are considered along with the Oviatt data (Calabretta and Oviatt, 2004; Oviatt, Chapter 18), the maximum impact (near-azoic) zone appears in the vicinity between the Hurricane Barrier and Field's Point, close to where the major sewage treatment effluents enter (US Army Corps of Engineers (USACE) 2001). Examination of other benthic data found in the EIS literature for the area (RICRMC, 1999) shows the zone of maximum total density just south of Conimicut Point to be shared by the opportunistic species *Streblospio benedicti* and *Mediomastus ambiseta*. These species are known to have high tolerance to hypoxia (Llanso, 1992; Ritter and Montagna, 1999).

Depth-specific stress may produce unusual benthic community structure where shallow benthic populations are altered by lethal but intermittent hypoxic levels (Llanso, 1992; Rosenberg *et al.,* 1992; Ritter and Montagna, 1999). In upper Narragansett Bay, the most severe hypoxia is often found just below the pycnocline. This condition may be influencing benthic communities and even ecosystem functions within this depth stratum (3–5 m in the lower Providence River, and 6–9 m in mid bay areas, Deacutis *et al.,* 2006). Severe loss of biogenic structure and filtering capacity have recently been linked to an order-of-magnitude loss of blue mussel (*Mytilus* edulis) reefs at 5–6 m depth in upper Narragansett Bay by hypoxia. Losses represented a >75% decrease in filtration capacity provided by this species. Such a large functional loss may cause an increase in phytoplankton biomass, increasing the likelihood of severe hypoxia (Altieri and Witman, 2006).

Seasonally intermittent but severe hypoxic stress is likely to have influenced the benthic community in Narragansett Bay over the last two to three decades (Deacutis, 1999; Deacutis *et al.,* 2006). Frithsen (1990) noted a shift in the late 1970s from a *Nephtys-Nucula* community to a *Mediomastus-Nucula* community around the mid-bay region. Grassle *et al.* (1985) compared mid bay communities from 1976 benthic collections with prior benthic studies, and found a complete absence of *Mediomastus ambiseta* in mid bay prior to 1975, while later surveys found *M. ambiseta* to be a dominant species. Recent data from a Final Environmental Impact Statement Report confirm continued absence of equilibrium-type species such as *Nephtys incise*, while *M. ambiseta* is now common in the mid-bay area (USACE, 2001). Sediment profile camera work completed in Narragansett Bay during August 1988 (Valente *et al.,* 1992) reinforces the evidence for organic enrichment impacts to the mid bay and Greenwich Bay areas—the shallow apparent RPD (Redox Potential Discontinuity Depth) recorded for stations in the upper half of Narragansett Bay showed a remarkably similar shape to the hypoxic patterns found by Deacutis *et al.* (2006).

12.8 Benthic–Pelagic Fish Community Shifts

The RIDEM Fish & Wildlife Division has recorded species counts from seasonal benthic trawls across Narragansett Bay since 1979 using mesh nets small enough to capture juvenile-sized fish. Frequency of trawls was increased to monthly at 12 fixed stations in 1990. In addition, the University of Rhode Island has maintained a weekly benthic trawl station at mid bay near Fox Island since 1959.

Shifts from demersal to pelagic fish species have been observed in systems experiencing hypoxia (Caddy, 2000; 2001; Turner, 2001). The RIDEM fisheries data for Narragansett Bay show a drop in demersal species and biomass around the 1980s, together with a shift to increased biomass of pelagic species

(RIDEM, 2001; Oviatt *et al.*, 2003). Because the benthic trawls are not well suited to pelagic sampling, the pelagic results are most likely an underestimate.

It is difficult to separate fishing pressure impacts from water quality issues like hypoxia because many demersal species are highly prized commercial targets (e.g., winter flounder, *Pseudopleuronectes americanus*, and summer flounder, *Paralichthys dentatus*). However, other demersal species not commercially fished, such as the oyster toadfish, *Opsanus tau*, and the hogchoker, *Trinectes maculates*, are also showing decreases in biomass (RIDEM, 2001). The latter species is a smaller flounder that would pass through the minimum commercial mesh size and therefore would not be impacted by bycatch stress. The toadfish, being a summer benthic egglayer (Collette and Klein-Macphee, 2002), may be highly susceptible to stress (hypoxia) at the sediment–water interface.

The pelagic:demersal fish ratio is thought to be a possible indicator of differential impacts of nutrients on pelagic and benthic ecosystem components (Caddy, 2000; de Leiva Moreno *et al.*, 2000). Pelagic:Demersal ratios based on RIDEM Fish & Wildlife trawl data indicate a radical shift from <1.0 in 1979 to 31 around 1988, up to a maximum for 1992 and 1993 (100 and 227, respectively), followed by a drop to ~15 in 2000 (Fig. 12.3; Gibson, unpublished trawl data 1979–2004). This drop is due to decreases in pelagic species rather than increases in demersal fish. The ratio swings back up to ~70 in 2002 and 2003, then drops in 2003–2004 (48 and 34, respectively) (Fig. 12.3). The pelagic:demersal fish ratio cannot be assigned to specific areas of the bay, but applies to Narragansett Bay as a system-wide indicator. Similar shifts from demersal to pelagic fish species have been seen in the URI trawl dataset. Oviatt, in Chapter 18, however, suggests that shifts in fish species abundance and composition are more related to fishing pressures and climate change, particularly temperature, than to nutrient availability or nutrient gradients.

Interestingly though, a recent data comparison between trawl stations as well as on-line recreational survey data concluded that maximum fish abundance and species richness occurred in the lower mid-bay region (the vicinity of the two major bridges), while the mouth of the bay displayed moderate abundance but high species richness. The upper bay (north of Prudence Island) was considered of least recreational value due to lower abundance and low species richness (RIDEM, 2006). This observation shows possible links to nutrient gradients, but to thermal change as well.

12.9 Present Level of Eutrophication Impacts for Narragansett Bay

Eutrophic status and benthic community state (Type I, II, and III) are described here along the N–S nutrient gradient for Narragansett Bay, based on present eutrophication impacts as expressed by the various eutrophication indictors discussed earlier, as well as benthic community concepts from Rosenberg (Rosenberg *et al.*, 1992; Rosenberg, 2001) (Fig. 12.4 and Table 12.2).

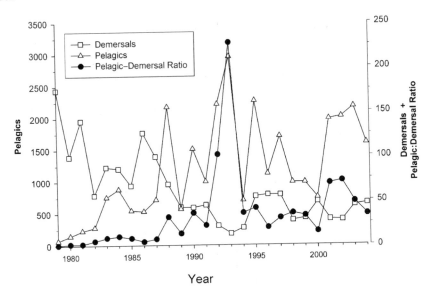

Fig. 12.3 Changes in pelagic and demersal fish populations in Narragansett Bay, and in the Pelagic:Demersal Ratio for 1979–2004. Based on RIDEM 20-min fish trawls (RIDEM Fish and Wildlife unpublished fisheries trawl data 1979–2004).

The area from the mouth of the bay northward to the two major bridges (∼38–30 km from Field's Point) is normoxic and produces no obvious expression of eutrophic condition. This area is characterized by low chlorophyll a bloom levels (90th percentile ≤5 µg L^{-1}), presence of SAVs, and equilibrium benthic (Type III) communities. The mid-bay area just north of the bridges to southern Prudence Island and Quonset Point exhibits no clear eutrophic expression, presence of SAVs, moderate blooms of chlorophyll a (90th percentile ≤20 µg L^{-1}), and low risk of hypoxia. However, this appears to be a transition zone, and rare, severe events can have an impact. A survey of the bay with an undulating sampler during a severe August 2003 fish kill event found hypoxic waters (≤3 mg L^{-1}) extending from the Providence River down to the Jamestown Bridge in the West Passage (Berman, unpublished data; RIDEM, 2003).

Upper West and East Passages (Quonset on the West Passage and Poppasquash Point on the East Passage) to the upper bay, and mid and eastern Greenwich Bay are areas experiencing low but clear eutrophic expression. Moderately high chlorophyll a (90th percentile ∼20–30 µg L^{-1}), lack of coherent eelgrass beds, Type II and Type III benthic communities, and sporadic but at least biennial hypoxia (≤3 mg L^{-1}) associated with weak neap summer tides occur in these areas (Bergondo et al., 2005; Deacutis et al., 2006). Hypoxic events last hours to days, and may on occasion fall below 2 mg L^{-1} (Bergondo, 2004; Bergondo et al., 2005; Stoffel, unpublished data). The upper bay and lower Providence River exhibit moderate eutrophic expression, with high

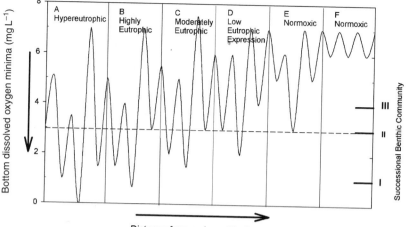

Fig. 12.4 Eutrophic status along an N–S nutrient gradient in Narragansett Bay, based on threshold indicators and hypothesized relationships between hypoxia severity and summer benthic successional stage loosely based on concepts in Diaz and Rosenberg (1995). The cyclic nature of bottom water dissolved oxygen concentrations is portrayed for each geographic zone for June–August, based on Bergondo et al. (2005). See (A) Ship channel, Seekonk and Providence River, and the western coves in Greenwich Bay. (B) Shallows, Seekonk and Providence River, and the western portion of Greenwich Bay. (C) Southern Providence River and upper Narragansett Bay, the mouths of tidal rivers and parts of the ship channel in Mount Hope Bay. (D) Upper West and East Passages to Quonset Point and Poppasquash Point, and the central and eastern portion of Greenwich Bay. (E) Mid Narragansett Bay from southern Prudence Island to the Jamestown and Newport bridges. (F) Jamestown and Newport bridges south to the mouth of Narragansett Bay.

chlorophyll a (90th percentile ∼25–50 µg L^{-1}), complete lack of SAV, Type I and Type II benthic communities below the pycnocline (5–7 m), and bouts of seasonal hypoxia to severe hypoxia ($\leq$3 to $\leq$2 mg L^{-1}) commencing on neap tides, with duration of hours to ∼1 week (Bergondo, 2004; Stoffel, unpublished data). Rarer events in the Providence River may extend for longer durations under strongly stratified conditions (Bergondo, 2004). Extreme hypoxia ($<$1 mg L^{-1}) or even near-anoxia, can occur in this region, but with lesser frequency and duration (hours to 1–3 days based on Bergondo, 2004).

The mid and northern Providence River, Seekonk River, and the western third of Greenwich Bay, are characterized as highly eutrophic, with benthic habitat below a shallow pycnocline (3–4 m) dominated by opportunistic species (Type I communities). Shallower areas experience less severe impacts and often have a mix of Type I and Type II communities. The deep ship channel in the Providence and Seekonk Rivers, and the western Greenwich Bay coves, all exhibit the most severe symptoms of excess organic loading, with high chlorophyll a levels—sometimes approaching hypereutrophic levels (90th percentile ∼30–70 µg L^{-1})—extremely severe hypoxic events at approximately biennial periodicity, and possible biennial anoxic events, although duration varies between years. Interannual variability in severity, frequency, and duration of negative impacts is considered linked to freshwater flow, summer water temperatures, wind energies, and other mixing forces that affect stratification and the depth of the pycnocline (Bergondo, 2004). The shallow coves in these zones are often plagued by thick mats of nuisance macroalgae (usually *Ulva sp.*) as well. The ship channel in the Seekonk and the northern Providence Rivers exhibits azoic zones in those areas experiencing anoxia (USACE, 2001).

12.9.1 Areas of Increased Susceptibility to Nutrient Additions: Greenwich Bay

Local hydrodynamics may enhance sustainability for local benthic populations that possess a planktonic larval life-history phase (e.g., scallops, *Aequipectin irradians*). Part of Greenwich Bay, for instance, once known as "Scalloptown," was covered with extensive eelgrass meadows from the late 1800s to the late 1940s (Doherty, 1995; Oviatt *et al.,* 2003). Abdelrhman (2005) has suggested counterclockwise residual circulation in central and western Greenwich Bay using a simplified hydrodynamic model. Such local hydrodynamics may have been ideal for retaining sensitive planktonic larvae of scallops until settlement, ensuring large populations within this embayment. This same retention factor may be magnifying present effects of nutrient additions, causing Greenwich Bay to be extremely sensitive to increased nutrient loads. This area is now devoid of eelgrass and scallops, and is plagued by heavy phytoplankton blooms and rapid-growth macroalgal mats. The area is subject to some of the most severe bouts of hypoxia and even anoxia found in Narragansett Bay (Purcell *et al.,* 2001a,b; RIDEM, 2003; Deacutis *et al.,* 2006; Saarman *et al.,* Chapter 11). Interestingly, Greenwich Bay has high concentrations of gelatinous zooplankton (B. Sullivan, personal communication; Sullivan *et al.,* Chapter 16), which tend to concentrate in poorly flushed areas, and are known to be fairly insensitive to low oxygen, having the ability to survive anoxia for short periods (Keister *et al.,* 2000; Purcell *et al.,* 2001a,b; Rutherford *et al.,* 2002; Breitburg *et al.,* 2003; Decker *et al.,* 2004).

12.10 Can Nutrient Management Go Too Far?

One aspect of concern is the possibility that excessive efforts to decrease nitrogen loads might have negative impacts on secondary production, including commercially important macrobenthic species such as hard shell clams (*Mercenaria mercenaria*) and various commercially unimportant benthic fish species (Nixon and Buckley, 2002). Examination of secondary productivity has shown some positive correlations between phytoplankton concentrations and fish yields, especially pelagic planktivores like menhaden (Nixon and Buckley, 2002; Brandt and Mason, 2003). Some bivalves are also known to respond positively with rapid growth to increased chlorophyll levels (Weiss *et al.,* 2002; Carmichael, 2004), but most such studies have used shallow stations (1 m depth) that are unlikely to suffer any significant periods of hypoxia. In fact, lower survival rates (<50%) were found in the above study for protected bivalve transplants at sites where dawn oxygen levels were <4 mg L^{-1} (Carmichael, 2004).

There is an apparent lack of association between secondary productivity levels and nutrients along the present nutrient gradient in Narragansett Bay (see Oviatt, Chapter 18), though good relationships occur for primary production along the gradients, confounding the possibility of a crystal clear answer to nutrient reduction. However, maximal oyster landings in the historical record for leased oyster beds in the upper bay occurred in the late 1890s (Oviatt *et al.,* 2003), a period when nutrient loads were far below today's levels. Primary production in this earlier era was clearly capable of supporting a significantly larger commercial filter feeding shellfish population when compared with today's lower shellfish catch (Oviatt *et al.,* 2003). It is doubtful that we could push nitrogen loadings below levels found at that time, so inadequate primary production from projected nutrient changes to sustain present shellfish stocks in the upper half of Narragansett Bay (maximum yield zone) is an unlikely consequence of present management plans. Future filter feeding populations may change, but it likely will be due to climate-related changes rather than responses to management efforts to control nutrients. Any improvements in benthic habitat quality in terms of oxygen may allow some macroinvertebrate biomass improvement, and if water clarity is significantly improved, SAVs may respond positively in the lower mid bay, but confounding climate change impacts make such predictions difficult, if not impossible.

12.11 What Is Next: After the Change

Nitrogen loadings from the major Rhode Island WWTFs are projected to decrease ~50% by 2008–2010, once required tertiary treatment is fully operational. This represents a change of ~30–35% of Total N load to Narragansett Bay (RIDEM, 2005b; Pryor *et al.,* 2006). In addition, an estimated 40% of the combined sewer-stormwater overflows (CSO) annual volumes will be captured (~3.3 × 10^6 m^3 yr^{-1}) by a huge storage-tunnel system just completed beneath Providence (NBC, 2005). This water will be diverted to the NBC Field's Point WWTF for treatment after storms.

Although system responses to these decreases have not been successfully modeled for Narragansett Bay (RIDEM, 2004), the likely first response of the Providence River area would be a decrease in phytoplankton density (blooms) and increase in water clarity. Based on nutrient enrichment experiments (Oviatt *et al.,* 1986) and estimates of decreased loading rates following tertiary treatment, the amount of organic material (phytoplankton) decomposing would be expected to be lower, perhaps leading to a decrease in minimum dissolved oxygen concentrations experienced, and possibly the spatial extent of hypoxic events as well (RIDEM, 2004). It could be possible that some nuisance species of macroalgae may find a period of competitive advantage as nutrient levels begin to decrease, perhaps showing an increase in biomass. Based on experiences in Tampa Bay in Florida (Johansson, 2002) and Mumford Cove in Connecticut (Vaudrey *et al.,* 2002), there should follow a decrease in nuisance rapid-growth macroalgae such as *Ulva* as well. In Mumford Cove, Long Island Sound, extensive *Ulva* mats disappeared within 2 years after removal of the Groton sewage treatment wastewater discharge in 1987. Vascular rooted species (*Ruppia maritime* and *Z. marina*) returned within 10 years, and eelgrass meadows returned to a theorized initial stable high-quality trophic state in 15 years (Vaudrey *et al.,* 2002). Submerged aquatic vascular rooted grass beds (*Thalassia testudinum*) increased in bottom coverage in both Tampa and Sarasota Bays within 8–10 years following significant decreases in nitrogen loads (Johansson, 2002).

It should be noted that the total N load following a 35% decrease would still far exceed the 60 kg ha^{-1} yr^{-1} threshold for SAV (Hauxwell *et al.,* 2003) in the upper bay and Providence River area. Perhaps the southernmost area of the "improved area" ellipse (Fig. 12.1) will see an increase from present low density, patchy spatial SAV distribution (Fig. 12.2). Small transplant or seeded test plots of eelgrass (*Z. marina*) could be used within the lower elliptical area outlined in Fig. 12.1 to signal improvements in water clarity and other conditions. One variant test design might be to include a depth-stratified design ranging from 1 m to the maximum edge depth found for eelgrass farther down bay (~4 m). Moore *et al.* (2003) note that maximum depth of growth for seagrass populations is positively correlated with water clarity and water quality. An increase in maximum depth in successful test plots would indicate water quality improvement. In addition, a research project assessing the microphytobenthic communities within the elliptical zone (Fig. 12.1) may be useful as a means to follow changes in the nutrient gradient following nutrient management reductions (Hillebrand and Sommer, 2000; Lever and Valiela, 2005).

A hypothesis is put forward here that benthic community changes will follow a logical extension of the Pearson–Rosenberg benthic gradient response following a projected decrease in nitrogen load. The density maxima of small, shallow dwelling, opportunistic species may be easier to distinguish than the ecotone point (where number of individuals equals the total number of species). It is suggested that the most easily measurable change is the maximum density inflection zone, the area experiencing the maximum total benthic organism density of small opportunistic species, which would be expected to move northward from the upper bay into the mid-to-lower Providence River (Fig. 12.1, points A and B, respectively).

It is critical that monitoring of chlorophyll, dissolved oxygen, and other important parameters be maintained for the upper half of Narragansett Bay, at a minimum to track actual response of the system to lowered nutrient loads. In addition, biomass and distribution of macroalgae, and pelagic:demersal fish ratios and fish and invertebrate biomass, should be tracked, all at adequate resolutions to note changes if they occur in response to management efforts. It is strongly recommended that a full estuary baseline survey be developed, and regular surveys be maintained in order to track changes in these indicator measures.

Acknowledgements I would like to thank Don Pryor, Warren Prell, and Candace Oviatt for stimulating discussions concerning the material, and I am especially indebted to the late Dana Kester, who provided much advice on the temporal issues involved in understanding hypoxia and stratification in Narragansett Bay, a critical component of this issue. Warren Prell, David Murray, and Emily Saarman of Brown have helped immensely through their final processing and mapping of low dissolved oxygen data for Narragansett Bay. Christian Turner and Mark Gibson of RIDEM both generously provided unpublished state agency data. Suzanne Bricker, Stephen V. Smith, and three anonymous reviewers provided helpful advice and suggestions. I am grateful to Paul Jordan, RIDEM for the GIS maps (Figs 12.1 and 12.2). I also gratefully acknowledge the huge volunteer effort from all the members of the "Insomniacs," who generously donated so much of their time and efforts to help map oxygen concentrations in Narragansett Bay.

References

Abdelrhman, M.A. 2005. Simplified modeling of flushing and residence times in 42 embayments in New England, USA, with special attention to Greenwich Bay, Rhode Island. *Estuarine Coastal and Shelf Science* 62:339–351.

Altieri, A., and Witman, J. 2006. Local extinction of a foundation species in a hypoxic estuary: integrating individuals to ecosystem. *Ecology* 87(3):717–730.

Baden, S.P., Loo, L.O., Pihl, L., and Roseberg, R. 1990. Effects of eutrophication on benthic communities including fish: Swedish west coast. *Ambio* 19:113–122.

Bejda, A.J., Phelan, B., and Studholme, A. 1992. The effect of dissolved oxygen on the growth of young-of-the-year winter flounder, *Pseudopleuronectes americanus*. *Environmental Biology of Fishes* 34(3):321–327.

Bergondo, D.L. 2004. Examining the processes controlling water column variability in Narragansett Bay: time-series data and numerical modeling. Ph.D. Dissertation, University of Rhode Island, Narragansett, RI. 187 pp.

Bergondo, D.L., Kester, D.R., Stoffel, H.E., and Woods, W. 2005. Time-series observations during the low sub-surface oxygen events in Narragansett Bay during summer 2001. *Marine Chemistry* 97:90–103.

Bintz, J.C., Nixon, S.W., Buckley, B.A., and Granger, S.L. 2003. Impacts of temperature and nutrients on coastal lagoon plant communities. *Estuaries* 26(3):765–776.

Borkman, D.G., and Smayda, T.J. 1998. Long-term trends in water clarity revealed by Secchi-disk measurements in lower Narragansett Bay. *ICES Journal of Marine Science* 55:668–679.

Brandt, S.B., and Mason, D.M. 2003. Effect of nutrient loading on Atlantic Menhaden (*Brevoortia tyrannus*) growth rate potential in the Patuxent River. *Estuaries* 26(2A):298–309.

Breitburg, D.L. 2002. Effects of hypoxia, and the balance between hypoxia and enrichment, on coastal fishes and fisheries. *Estuaries* 25(4b):767–781.

Breitburg, D.L., Adamack, A., Rose, K.A., Kolesar, S.E., Decker, M.B., Purcell, J.E., Keister, J.E., and Cowan, J.H., Jr. 2003. The pattern and influence of low dissolved oxygen in the Patuxent River, a seasonally hypoxic estuary. *Estuaries* 26:280–297.

Bricker, S.B., Clement, C.G., Pirhalla, D.E., Orland, S.P., and Farrow, D.G.G. 1999. National Estuarine Eutrophication Assessment: A Summary of Conditions, Historical Trends, and Future Outlook. Silver Springs, MD: National Ocean Service, National Oceanic and Atmospheric Administration.

Bricker, S.B., Ferreira, J.G., and Simas, T. 2003. An integrated methodology for assessment of Estuarine Trophic Status. *Ecological Modelling* 169:39–60.

Burnett, L.E. 1997. The challenges of living in hypoxic and hypercapnic aquatic environments. *American Zoologist* 37:633–640.

Burnett, L.E., and Stickle, W.B. 2001. Physiological responses to hypoxia. *In* Coastal Hypoxia: Consequences for Living Resources and Ecosystems, pp. 101–114. Rabalais, N.N., and Turner, R.E. (eds) *Coastal and Estuarine Studies No. 58,* Washington, DC: Publisher AGU. 463 pp.

Caddy, J.F. 2000. Marine catchment basin effects versus impacts of fisheries on semi-enclosed seas. *ICES Journal of Marine Science* 57:628–640.

Caddy, J.F. 2001. A brief overview of catchment basin effects on marine fisheries. *In* Coastal Hypoxia: Consequences for Living Resources and Ecosystems, pp. 129–146. Rabalais, N.N., and Turner, R.E. (eds) *Coastal and Estuarine Studies No. 58,* Washington, DC: Publisher AGU. 463 pp.

Cahoon, L.B. 1999. The role of benthic microalgae in neritic ecosystems. *Oceanography and Marine Biology: Annual Review* 37:47–86.

Calabretta, C.J., and Oviatt, C.A. 2004. Benthic macrofauna in Narragansett Bay, Rhode Island: an evaluation of different approaches for describing community assemblages. Oral Presentation and published abstract, Spring 2004 New England Estuarine Research Society, March 7–9, 2004. Burlington, VT.

Carmichael, R.H. 2004. The effects of eutrophication on *Mya arenaria* and *Mercenaria mercenaria*: growth, survival, and physiological responses to changes in food supply and habitat across estuaries receiving different nitrogen (N) loads. Ph.D. Dissertation, Boston University Graduate School of Arts and Sciences. Boston, MA, USA. 209 pp.

Chinman, R.A., and Nixon, S.W. 1985. Depth–area–volume relationships in Narragansett Bay. NOAA/Sea Grant Marine Technical Report 87. Graduate School of Oceanography, University of Rhode Island, Narragansett, RI.

Chmura, G.L., Santos, A., Pospelova, V., Spasojevic, Z., Lam, R., and Latimer, J.S. 2004. Response of three paleo-primary production proxy measures to development of an urban estuary. *Science of the Total Environment* 320:225–243.

Cloern, J.E. 2001. Our evolving conceptual model of the coastal eutrophication problem. *Marine Ecology Progress Series* 210:223–253.

Collette, B.B., and Klein-MacPhee, G. (eds) 2002. Bigelow and Schroeder's Fishes of the Gulf of Maine, 3rd edn. Washington DC: Smithsonian Press. 748 pp.

Cottrell, H. 2001. Atlas of Narragansett Bay Coastal Habitats. Narragansett Bay Estuary Program Report # 01–118. October 2001, 17pp. Univ. of RI Coastal Institute, Narragansett, RI, USA.

Dauer, D.M., Rodi, A., Jr. and Ranasinghe, J. 1992. Effects of low dissolved oxygen events on the macrobenthos of the lower Chesapeake Bay. *Estuaries* 15(3):384–391.

Deacutis, C.F. 1999. Nutrient impacts and signs of problems in Narragansett Bay. *In* Proceedings of a Workshop on Nutrient Removal from Wastewater treatment Facilities. Nutrients and Narragansett Bay, pp. 7–23. Kerr, M. (ed.) Narragansett, RI: Rhode Island Sea Grant.

Deacutis, C.F., Murray, D.W., Prell, W.L., Saarman E., and Korhun, L. 2006. Hypoxia in the upper half of Narragansett Bay, RI during August 2001 and 2002. *Northeast Naturalist* 13(Special Issue 4):173–198.

Decker, M.B., Breitburg, D., and Purcell, J. 2004. Effects of low dissolved oxygen on zooplankton predation by the ctenophore *Mnemiopsis leidyi*. *Marine Ecology Progress Series* 280:163–172.

de Leiva Moreno, J.I., Agostini, V., Caddy, J.F., and Carocci, F. 2000. Is the pelagic-demersal ratio from fishery landings a useful proxy for nutrient availability? A preliminary data exploration for the semi-enclosed seas around Europe. *ICES Journal of Marine Science* 57(4):1091–1102.

Dennison, W., Orth, R., Moore, K., Stevenson, J., Carter, V., Kollar, S., Bergstrom, P., and Batiuk, R. 1993. Assessing water quality with submersed aquatic vegetation. *Bioscience* 43:86–94.

Desbonnet, A., and Lee, V. 1991. Water quality and fisheries- Narragansett Bay. A report to National Ocean Pollution Program Office, National Oceanic and Atmospheric Administration, RI Sea Grant, Narragansett, RI.

Diaz, R.J. 2001. Overview of Hypoxia around the World. *Journal of Environmental Quality* 30(2):275–281.

Diaz, R.J., and Rosenberg, R. 1995. Marine benthic hypoxia: a review of its ecological effects and the behavioral responses of benthic macrofauna. *Oceanography and Marine Biology Annual Review* 33:245–303.

Diaz, R.J., and Rosenberg, R. 2001. Overview of anthropogenically-induced hypoxic effects on Marine Benthic Fauna. *In* Coastal Hypoxia: Consequences for Living Resources and Ecosystems, pp. 129–146. Rabalais, N.N., and Turner, R.E. (eds) *Coastal and estuarine Studies No. 58*. Washington, DC: Publisher AGU. 463 pp.

Doering, P.H., Oviatt, C.A., and Pilson, M.E.Q. 1990. Characterizing Late Summer Water Quality in the Seekonk River, Providence River and Upper Narragansett Bay, Final Report, #Narragansett BayP-90-49. Univ. of RI Coastal Institute, Narragansett, RI, USA.

Doherty, A. 1995. Historical distributions of eelgrass (Zostera marina L.) in Narragansett Bay, Rhode Island, 1850–1995. Senior Independent Project for Geology-Biology BSc. Advisors: S. Hamburg, T. Webb, III, J. Witman, Brown University, Providence RI.

Doherty, A. 1997. Historical Distribution of eelgrass in Narragansett Bay, RI. *In* Proceedings of 14th Biennial Estuarine Research Federation International Conference. "The State of Our Estuaries" October 12–16, 1997. Providence, RI.

Eby, L., and Crowder, L.B. 2004. Effects of Hypoxic Disturbances on an Estuarine Nekton Assemblage across Multiple Scales. *Estuaries* 27(2):342–351.

Fischer, J., and Mustard, J.F. 2004. High spatial resolution sea surface climatology from Landsat thermal infrared data. *Remote Sensing of Environment* 90:293–307.

Flindt, M., Salomonsen, J., Carrer, M., Brocci, M., and Kamp-Nielsen, L. 1997. Loss, growth and transport dynamics of *Chaetomorpha aerea* and *Ulva rigida* in the Lagoon of Venice during an early summer field campaign. *Ecological Modelling* 102:133–141.

French, D., Rines, H., Boothroyd, J., Galagan, C., Harlin, M., Keller, A., Klein-McPhee, G., Pratt, S., Gould, M., Villalard-Bohnsack, M., Gould, L., Steere, L., and Porter, S. 1992. Atlas and habitat inventory/resource mapping for Narragansett Bay and associated coastlines, Rhode Island and Massachusetts: Final Report for the Narragansett Bay Project, Providence, RI.

Frithsen, J. 1990. The Benthic Communities within Narragansett Bay.NBP Report #NBP-90-28. 90 pp. and App A-C. Univ. of RI Coastal Institute, Narragansett, RI, USA.

Granger, S.H. 1994. The basic hydrography and mass transport of dissolved oxygen in the Providence and Seekonk Rivers estuaries, MS Thesis, University of Rhode Island, Narragansett, RI. 230 pp.

Granger, S., Brush, M., Buckley, B., Traber, M., Richardson, M., and Nixon, S.W. 2000. An assessment of eutrophication in Greenwich Bay. Paper No. 1, Restoring water quality in Greenwich Bay: a whitepaper series. Rhode Island Sea Grant, Narragansett, RI. 20 pp.

Grassle, J.F., Grassle, J.P., Brown-Leger, L.S., Petrecca, R.F., and Copley, N.J. 1985. Subtidal macrobenthos of Narragansett Bay. Field and mesocosm studies of the effects of eutrophication and organic input on benthic populations. *In* Marine Biology of Polar Regions and Effects of Stress on Marine Organisms, pp. 421–434. Gray, J.S., and Christiansen, M.E. (eds) New York: Wiley.

Gray, J.S., Shiu-sun Wu, R., and Or, Y.Y. 2002. Effects of hypoxia and organic enrichment on the coastal marine environment. *Marine Ecology Progress Series* 238:249–279.

Haas, L. 1977. The effect of the spring-neap tidal cycle on the vertical salinity structure of the James, York, and Rappahannock Rivers, Virginia, USA. *Estuarine and Coastal Marine Science* 5:485–496.

Hauxwell, J., Cebrian, J., Furlong, C., and Valiela, I. 2001. Macroalgal canopies contribute to eelgrass (*Zostera marina*) decline in temperate estuarine ecosystems. *Ecology* 82(4):1007–1022.

Hauxwell, J., Cebrian, J., and Valiela, I. 2003. Eelgrass *Zostera marina* loss in temperate estuaries: relationship to land-derived nitrogen loads and effect of light limitation imposed by algae. *Marine Ecology Progress Series* 247:59–73.

Havens, K.E., Hauxwell, J., Tyler, A.C., Thomas, S., McGlathery, K., Cebrian, J., Valiela, I., Steinman, A., and Hwang, S.J. 2001. Complex interactions between autotrophs in shallow marine and freshwater ecosystems: implications for community responses to nutrient stress. *Environmental Pollution* 113(1):95–107.

Hillebrand, H., and Sommer, U. 2000. Diversity of benthic microalgae in response to colonization time and eutrophication. *Aquatic Botany* 67(3):221–236.

Howes, B.L. 2005. Sampling and Analysis Plan. Mount Hope Bay Estuarine Monitoring. Mt. Hope Bay & Taunton River Estuary. University of Massachusetts Dartmouth, School for Marine Science and Technology (SMAST). EPA RFA # 05188 MA DEP # 2005-04/604.

Howes, B.L., and Schlezinger, D.R. 2003. Nutrient related habitat quality of Mount Hope Bay. Published abstract. New England Estuarine Research Society (NEERS) Spring Symposium, May 8–10, 2003. Fairhaven, MA, State of MA, USA.

Johansson, J.O.R. 2002. Historical and current observations on macroalgae in the Hillsborough Bay Estuary (Tampa Bay), Florida. *In* Understanding the Role of Macroalgae in Shallow Estuaries, pp. 26–28. McGinty, M., and Wazniac, C. (eds) Maritime Institute, Linthicum, MD. January 10–11, 2002. Annapolis, MD: Maryland Department of Natural Resources.

Kang, C-K., Jeong, B-K., Lee, K-S., Jong, B-K., Lee, P-Y., and Hong, J-S. 2003. Trophic importance of benthic microalgae to macrozoobenthos in coastal bay systems in Korea: dual stable C and N isotope analyses. *Marine Ecology Progress Series* 259:79–92.

Karentz, D., and Smayda, T.J. 1998. Temporal patterns and variations in phytoplankton community organization and abundance in Narragansett Bay during 1959–1980. *Journal of Plankton Research* 20:145–168.

Keister, J., Houde, E., and Breitburg, D. 2000. Effects of bottom-layer hypoxia on abundances and depth distributions of organisms in Patuxent River, Chesapeake Bay. *Marine Ecology Progress Series* 205:43–59.

Kopp, B., Doherty, A., and Nixon, S. 1997. A guide to the site-selection for eelgrass restoration projects in Narragansett Bay, RI. Final Report to the RI Aqua Fund Council. 22 pp. and App. Univ. of RI Coastal Institute, Narragansett, RI, USA.

Kremer, J.N., and Nixon, S.W. 1978. A Coastal Marine Ecosystem: Simulation and Analysis. New York: Springer Verlag. 217 pp.

Lee, K.S., Short, F.T., and Burdick, D.M. 2004. Development of a nutrient pollution indicator using the seagrass, *Zostera marina*, along nutrient gradients in three New England estuaries. *Aquatic Botany* 78:197–216.

Lever, M.A., and Valiela, I. 2005. Response of microphytobenthic biomass to experimental nutrient enrichment and grazer exclusion at different land-derived nitrogen loads. *Marine Ecology Progress Series* 294:117–129.

Li, Y., and Smayda, T.J. 2001. A chlorophyll time series for Narragansett Bay: assessment of the potential effect of tidal phase on measurement. *Estuaries* 24(3):328–336.

Llanso, R.J. 1992. Effects of hypoxia on estuarine benthos: the lower Rappahannock River (Chesapeake Bay), a case study. *Estuarine, Coastal and Shelf Science* 35(5):491–515.

Llanso, R.J., Dauer, D.M., Volstad, J.H., and Scott, L.C. 2003. Application of the benthic index of biotic integrity to environmental monitoring in Chesapeake Bay. *Environmental Monitoring and Assessment* 81:163–174.

Lu, L., and Wu, R.S.S. 2000. An experimental study on recolonization and succession of marine macrobenthos in defaunated sediment. *Marine Biology* 136:291–302.

Luther, G.W., III, Shufen, M., Trouwborst, R., Glazer, B., Blickley, M., Scarborough, R.W., and Mensinger, M.G. 2004. The Roles of Anoxia, H sub(2)S, and Storm Events in Fish Kills of Dead-end Canals of Delaware Inland Bays. *Estuaries* 27(3):551–560.

Malin, M.A., Burkholder, J.M., and Sullivan, M.J. 1992. Contributions of benthic microalgae to coastal fishery yield. *Transactions of American Fisheries Society* 121(5):691–693.

Meng, L., and Powell, J.C. 1999. Linking juvenile fish and their habitats: an example from Narragansett Bay, Rhode Island. *Estuaries* 22(4):905–916.

Meng, L., Powell, J.C., and Taplin, B. 2001. Using winter flounder growth rates to assess habitat quality across an anthropogenic gradient in Narragansett Bay, Rhode Island. *Estuaries* 24(4):576–584.

Meng, L., Cicchetti, G., and Chintala, M. 2004. Nekton habitat quality at shallow water sites in two Rhode Island coastal systems. *Estuaries* 27(4):740–751.

Meyercordt, J., and Meyer-Reil, L.A. 1999. Primary production of benthic microalgae in two shallow lagoons of different trophic status in the southern Baltic Sea. *Marine Ecology Progress Series* 178:179–191.

Moore, K.A., and Wetzel, R.L. 2000. Seasonal variations in eelgrass (*Zostera marina* L.) responses to nutrient enrichment and reduced light availability in experimental ecosystems. *Journal of Experimental Biology and Ecology* 244:1–28.

Moore, K.A., Neckles, H.A., and Orth, R.J. 1996. *Zostera marina* L. (eelgrass) growth and survival along a gradient of nutrients and turbidity in the lower Chesapeake Bay. *Marine Ecology Progress Series* 142:247–259.

Moore, K.A., Wetzel, R.L., and Orth, R.J. 1997. Seasonal pulses of turbidity and their relations to eelgrass (*Zostera marina* L.) survival in an estuary. *Journal of Experimental Marine Biology and Ecology* 215:115–134.

Moore, K.A., Anderson, B.A., Wilcox, D.J., Orth, R.J., and Naylor, M. 2003. Changes in seagrass distribution as evidence of historical water quality conditions. *Gulf of Mexico Science* 21(1):142–143.

Napolitano, W. 2005. Advocates seek "Wild & Scenic" designation for Taunton River. The Narragansett Bay Journal # 10. Summer 2005, pp. 1–10.

Narragansett Bay Commission. 2005. Combined Sewer Overflow (CSO) Comprehensive Plan. Retrieved June 20, 2005. (*http://www.narrbay.com/CSO.asp*).

Nixon, S.W. 1997. Prehistoric nutrient inputs and productivity in Narragansett Bay. *Estuarine Research Federation* 20(2):253–261.

Nixon, S.W., and Buckley, B. 2002. "A strikingly rich zone"-nutrient enrichment and secondary production in coastal marine ecosystems. *Estuaries* 25(4b):782–796.

Nixon, S.W., and Pilson, M. 1983. Nitrogen in estuarine and coastal marine ecosystems. *In* Nitrogen in the Marine Environment, pp. 565–648. Carpenter, E., and Capone, D. (eds) Academic Press, New York.

Nixon, S.W., Granger, S.L., and Nowicki, B.L. 1995. An assessment of the annual mass balance of carbon, nitrogen, and phosphorus in Narragansett Bay. *Biogeochemistry* 31:15–61.

Nixon, S.W., Granger, S., Buckley, B.A., Lamont, M., and Rowell, B. 2004. A one hundred and seventeen year coastal water temperature record from Woods Hole, Massachusetts. *Estuaries* 27:397–404.

Nixon, S.W., Buckley, B., Granger, S., Harris, L., Oczkowski, A., Cole, L., and Fulweiler, R. 2005. Anthropogenic nutrient inputs to Narragansett Bay: A twenty five year perspective. A Report to the Narragansett Bay Commission and Rhode Island Sea Grant. Rhode Island Sea Grant, Narragansett, RI. www.seagrant.gso.uri.edu/research/bay_commission_report.pdf.).

Olsen, S., and Lee, V. 1979. A Summary and Preliminary Evaluation of Data Pertaining to the Water Quality of Upper Narragansett Bay. Coastal Resource Center Report to EPA Region 1 (TD224.35 R4 O78 1979 Narragansett Bay), Univ. of RI, Narragansett, RI, USA.

O'Shea, M.L., and Brosnan, T.M. 2000. Trends in indicators of eutrophication in western Long Island Sound and the Hudson-Raritan estuary. Estuaries 23(6):877–901.

Oviatt, C.A. 2004. The changing ecology of temperate coastal waters during a warming trend. Estuaries 27:895–904.

Oviatt, C.A., Doering, P., Nowicki, B., Reed, L., Cole, J.J., and Frithsen, J. 1995. An ecosystem level experiment on nutrient limitation in temperate coastal marine environments. Marine Ecology Progress Series 116:171–179.

Oviatt, C.A., Keller, A., and Reed, L. 2002. Annual primary production in Narragansett Bay with no bay-wide winter-spring bloom. Estuarine, Coastal, and Shelf Science 54: 1013–1026.

Oviatt, C., Olsen, S., Andrews, M., Collie, J., Lynch, T., and Raposa, K. 2003. A century of fishing and fish fluctuations in Narragansett Bay. Reviews in Fisheries Science 11:221–242.

Parker, C., and O'Reilly, J. 1991. Oxygen depletion in Long Island Sound: a historical perspective. Estuaries 14:248–264.

Pearson, T.H., and Rosenberg, R. 1978. Macrobenthic succession in relation to organic enrichment and pollution of the marine environment. Oceanography and Marine Biology: An Annual Review 16:229–311.

Pihl, L. 1994. Changes in the diet of demersal fish due to eutrophication-induced hypoxia in Kattegat, Sweden. Canadian Journal of Fisheries and Aquatic Sciences 51:321–336.

Pilson, M.E.Q. 1985. On the residence time of water in Narragansett Bay. Estuaries 8:2–14.

Pospelova, V., Chmura, G.L., Boothman, W.S., and Latimer, J.S. 2005. Spatial distribution of modern dinoflagellate cysts in polluted estuarine sediments from Buzzards Bay (Massachusetts, USA) embayments. Marine Ecology Progress Series 292:23–40.

Pratt, D.M. 1959. The phytoplankton of Narragansett Bay. Limnology and Oceanography 4:425–440.

Prell, W., Saarman, E., Murray, D., and Deacutis, C. 2004. Summer-Season, Nighttime Surveys of Dissolved Oxygen in Upper Narragansett Bay (1999–2003). (http://www.geo.brown.edu/georesearch/insomniacs).

Pryor, D. Saarman, E., Murray, D., and Prell, W. 2006. Nitrogen Loading from Wastewater Treatment Plants to Upper Narragansett Bay. Narragansett Bay Estuary Program Report NBEP-2007-122. 22 pp. + App. Univ. of RI Coastal Institute, Narragansett, RI, USA.

Purcell, J., Breitburg, D.L., Decker, M., Graham, W., Youngbluth, M.J., and Raskoff, K.A. 2001a. Pelagic cnidarians and ctenophores in low dissolved oxygen environments: a review. In Coastal Hypoxia: Consequences for Living Resources and Ecosystems, pp. 77–100. Rabalais, N.N., and Turner, R.E. (eds) Coastal and Estuarine Studies No. 58. Washington, DC: Publisher AGU. 463 pp.

Purcell, J., Graham, W., and Dumont, H. 2001b. Jellyfish blooms: ecological and societal importance. International Conference on Jellyfish Blooms, Proceedings, Gulf Shores, Alabama, January 12–14, 2000. Hydrobiologia 451:333.

Rabalais, N.N., Harper, D., and Turner, R.E. 2001. Responses of nekton and demersal and benthic fauna to decreasing oxygen concentrations. In Coastal Hypoxia: Consequences for Living Resources and Ecosystems, pp. 115–128. Rabalais, N.N., and Turner, R.E. (eds) Coastal and Estuarine Studies No. 58. Washington, DC: Publisher AGU. 463 pp.

Raposa, R., and Oviatt, C. 2000. The influence of contiguous shoreline type, distance from shore, and vegetation biomass on nekton community structure in eelgrass beds. *Estuaries* 23(1):46–55.

Reise, K., and Siebert, I. 1994. Mass occurrence of green algae in the German Wadden Sea. *Deutsche Hydrographische Zeitschrift Supplement* 1:171–180.

Rhode Island Coastal Resources Management Council (RICRMC). 2005. Present and historic eelgrass location maps. Site visited January 2005. (*htttp://www.edc.uri.edu/Eelgrass/mapshome.html*).

Rhode Island Coastal Resources Management Council Final Report (RICRMC). 1999. Study and Analysis of Potential In-Water Dredged Material Disposal Sites in Narragansett Bay. August 12, 1999. 72 pp. + App + Tables & Figures. State of RI, USA.

Rhode Island Department of Environmental Management (RIDEM). 2001. Report on the Status of Marine Fisheries Stocks and Fisheries Management Issues in Rhode Island. RIDEM Fish and Wildlife, Jamestown, RI. 64 pp.

Rhode Island Department of Environmental Management (RIDEM). 2003. The Greenwich Bay fish kill—August 2003 Causes, impacts and responses. Providence, RI. (*www.dem.ri.gov/pubs/fishkill.pdf*).

Rhode Island Department of Environmental Management (RIDEM). 2004. Evaluation of Nitrogen Targets and WWTF Load Reductions for the Providence and Seekonk Rivers. Providence, RI: RIDEM Office of Water Resources. 31 pp.

Rhode Island Department of Environmental Management (RIDEM). 2005a. Fixed Monitoring Stations. Available at the RIDEM "BART." (*http://www.dem.ri.gov/bart*).

Rhode Island Department of Environmental Management (RIDEM). 2005b. Plan for Managing Nutrient Loadings to Rhode Island Waters. Report to the RI General Assembly pursuant to RIGL 46-12-3(25). State of Rhode Island, Providence, RI, USA.

Rhode Island Department of Environmental Management. (RIDEM). 2006. Public Access to Shoreline Recreational Fishing in Narragansett Bay. 49pp. + App. A–F. App. C. Fish Abundance and Diversity Assessment. 24 pp. Prepared by J.H. McKenna for G.R. Archibald, Inc. February 2006. (*http://www.dem.ri.gov*).

Ries, K.G., III. 1990. Estimating surface-water runoff to Narragansett Bay, Rhode Island and Massachusetts. US Geological Survey, Water Resources Investigations. Report (WRIR) 89-4164. 44 pp.

Ritter, C., and Montagna, P. 1999. Seasonal hypoxia and models of benthic response in a Texas bay. *Estuaries* 22(1):7–20.

Rosenberg, R. 2001. Marine benthic faunal successional stages and related sedimentary activity. *Scientia Marina* 65(2):107–119.

Rosenberg, R.B., Loo, L-O., and Moller, P. 1992. Hypoxia, salinity and temperature as structuring factors for marine benthic communities in a eutrophic area. *Netherlands Journal of Sea Research* 30:121–129.

Rutherford, L., Jr, Brommer, P., Winet, J., and Thuesen, E. 2002. An ecophysiological study of the effects of hypoxia and anoxia on gelatinous zooplankton of southern Puget Sound. *In* Proceedings of the 2001 Puget Sound Research Conference, February 12–14, 2001, Puget Sound Action Team Publication. Evergreen State College.

Ruzecki, E.P., and Evans, D.A. 1986. Temporal and spatial sequencing of destratification in a coastal plain estuary. *In* Tidal mixing and planktonic dynamics, Lecture Notes on Coastal and Estuarine Studies, Vol. 17. pp. 368–389. Bowman, J., Yentsch, M., and Peterson, W.T. (eds) New York: Springer-Verlag.

Sfriso, A., Birkemeyer, T., and Ghetti, P. 2001. Benthic macrofauna changes in areas of Venice lagoon populated by seagrasses or seaweeds. *Marine Environmental Research* 52(4):323–349.

Short, F.T., Burdick, D.M., Wolf, J., and Jones, G.E. 1993. Eelgrass in Estuarine Research Reserves along the East Coast, USA, Part I: Declines from pollution and disease; Part II:

Management of eelgrass meadows. NOAA, Coastal Ocean Program through the Sea Grant College Program. 107 pp.

Short, F.T., Burdick, D.M., Granger, S., and Nixon, S.W. 1996. Long-term decline in eelgrass, *Zostera marina* L., linked to increased housing development. *In* Seagrass Biology: Proceedings of an International Workshop, pp. 291–298. Kuo, J., Phillips, R.C., Walker, D.I., Kirkman, H. (eds.) Rottnest Island, Western Australia, 25–29 January 1996. Nederlands, Western Australia: Sciences UWA.

Siebert, I., Reise, K., Buhs, F., Herre, E., Metzmacher, K., Parusel, E., Schories, D., and Wilhelmsen, U. 1997. *Distribution of green algae in the Wadden Sea.* (English Summary) Texte. no. 21, May 1997. Umweltbundesamt, Berlin (FRG) Publisher. 180 pp.

Stanley, D.W., and Nixon, S.W. 1992. Stratification and bottom-water hypoxia in the Pamlico River Estuary. *Estuaries* 15(3):270–281.

State of Rhode Island. 2004. Governor's Narragansett Bay and Watershed Planning Commission. Nutrient and Bacteria Pollution Panel Report, March 2004. 33 p. Retrieved July 20, 2004 from URI Coastal Institute website: (*http://www.ci.uri.edu/GovComm/Documents/Phase1Rpt/Docs/Nutrient-Bacteria.pdf*).

Tagliapietra, D., Pavan, M., and Wagner, C. 1998. Macrobenthic community changes related to eutrophication in Palude della Rosa (Venetian Lagoon, Italy). *Estuarine, Coastal and Shelf Science* 47(2):217–226.

Thetmeyer, H., Waller, U., Black, K.D., Inselmann, S., and Rosenthal, H. 1999. Growth of European sea bass (*Dicentrarchus labrax* L.) under hypoxic and oscillating oxygen conditions. *Aquaculture* 174(3–4):355–367.

Tobias, C.R., Cieri, M., Peterson, B.J., Deegan, L.A., Vallino, J., and Hughes, J. 2003. Processing watershed-derived nitrogen in a well-flushed New England estuary. *Limnology and Oceanography* 48(5):1766–1778.

Turner, C. 1997. Development of a Total Maximum Daily Loading for nutrients: Results in the Providence and Seekonk Rivers. RIDEM Report, August 20, 1997. 51pp + App. Rhode Island Dept. of Environmental Management, State of RI, Providence, RI.

Turner R.E. 2001. Some effects of eutrophication on pelagic and demersal marine food webs. *In* Coastal Hypoxia: Consequences for Living Resources and Ecosystems, pp. 371–398. Rabalais, N.N., and Turner, R.E. (eds) *Coastal and Estuarine Studies No. 58.* Washington, DC: Publisher AGU. 463 pp.

United States Environmental Protection Agency. 2000. Ambient Aquatic Life Water Quality Criteria for Dissolved Oxygen (Saltwater): Cape Cod to Cape Hatteras, US EPA, Office of Water, November 2000: EPA 822-R-00-012 *http://www.epa.gov/waterscience/criteria/dissolved/index.html*.

United States Environmental Protection Agency. 2001. Municipally Owned Wastewater Treatment facilities in New England. Compiled by Municipal Assistance Unit, Office of Ecosystem Protection, US EPA New England. October 2001. 116 p.

US ACE. 2001. Providence River and Harbor maintenance dredging project, Final Environmental Impact Statement. US Army Corps of Engineers, New England District, Concord, MA, August.

Valente, R.M., and Cuomo, C. 2005. Did multiple sediment-associated stressors contribute to the 1999 lobster mass mortality event in western Long Island Sound, USA? *Estuaries* 28(4):529–540.

Valente, R.M., Rhoads, D.C., Germano, J.D., and Cabelli, V.J. 1992. Mapping of benthic enrichment patterns in Narragansett Bay, Rhode Island. *Estuaries* 15:1–17.

Vaudrey, J., Branco, B., and Kremer, J. 2002. Mumford Cove: a system in Rebound. Abstract for Poster Presentation. Fall 2002 New England Estuarine Research Society. 24–26, October 2002. University of Connecticut Avery Point Campus, Groton, CT.

Viaroli, P., Bartoli, M., Bondavalli, C., Christian, R.R., Giordani, G., and Naldi, M. 1996. Macrophyte communities and their impact on benthic fluxes of oxygen, sulphides and nutrients in shallow eutrophic environments. *Hydrobiologia* 239(1–3):105–119.

Wainright, S.C., Weinstein, M.P., Able, K.W., and Currin, C.A. 2000. Relative importance of benthic microalgae, phytoplankton and the detritus of smooth cordgrass Spartina alterniflora and the common reed Phragmites australis to brackish-marsh food webs. *Marine Ecology Progress Series* 200:77–91.

Weisberg, R.H. 1976. The non-tidal flow in the Providence River of Narragansett Bay: a stochastic approach to estuarine circulation. *Journal of Physical Oceanography* 6:721–734.

Weisberg, R.H., and Sturges, W. 1976. Velocity observations in the west passage of Narragansett Bay: a partially mixed estuary. *Journal of Physical Oceanography* 6:345–354.

Weiss, E.T., Carmichael, R., Shriver, A.C., and Valiela, I. 2002. The effect of nitrogen loading on the growth rates of quahogs (*Mercenaria mercenaria*) and soft-shell clams (*Mya arenaria*) through changes in food supply. *Aquaculture* 211(1–4):275–289.

Wu, R.S. 2002. Hypoxia: from molecular responses to ecosystem responses. *Marine Pollution Bulletin Article* 45:35–45.

Zajac, R.N., and Whitlatch, R.B. 1988. Population ecology of the polychaete *Nephtys incisa* in Long Island Sound and the effects of disturbance. *Estuaries* 11(2):117–133.

Chapter 13
An Ecosystem-based Perspective of Mount Hope Bay

Christian Krahforst and Marc Carullo

13.1 Introduction

Water-column characteristics of Mount Hope Bay, a large shallow embayment located in the northeastern portion of Narragansett Bay (Fig. 13.1), are described here for selected periods during 1999–2003. These observations provide new information that adds to our knowledge base about the functioning of shallow estuarine systems, Mount Hope Bay's relevance to the functioning of the greater Narragansett Bay ecosystem, and Mount Hope Bay's relevance to management issues of the region. In this chapter, we explore the potential links among the structure, function, and composition of the Mount Hope Bay ecosystem, and possible ecosystem-based management approaches that may be used for restoring lost functions of the bay ecosystem. Of particular importance to this discussion is the fact that the coastline of Mount Hope Bay is shared by two states: Rhode Island and Massachusetts. Hence, the Mount Hope Bay ecosystem provides unique challenges to those management strategies that adopt guiding principles of ecosystem-based management.

13.2 Physical Setting and Land Use

Mount Hope Bay is a large shallow embayment in the uppermost, northeastern corner of Narragansett Bay. The bay connects with Narragansett Bay proper at the East Passage, and stretches approximately 11 km north and east to the mouth of the Taunton River estuary. Surface waters of Mount Hope Bay cover an area of approximately 35 km^2 (Kauffman and Adams, 1981) with a mean water column depth and volume of 5.7 m and $2.0 \times 10^8 \ m^3$ at mean low water, respectively (Chinman and Nixon, 1985). Bathymetrically, Mount Hope Bay

Christian Krahforst
Massachusetts Bays National Estuary Program, 251 Causeway Street, Suite 800, Boston, MA 02114
christian.krahforst@state.ma.us

A. Desbonnet, B. A. Costa-Pierce (eds.), *Science for Ecosystem-based Management.*
© Springer 2008

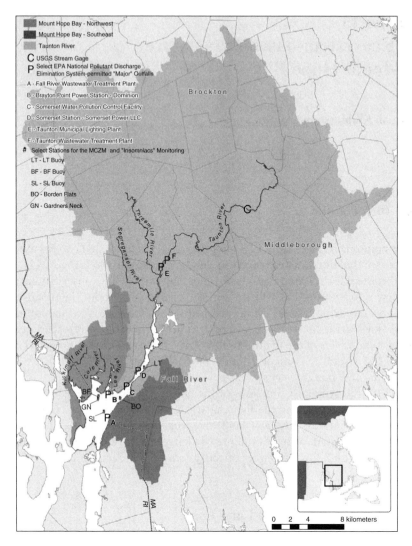

Fig. 13.1 Mount Hope Bay region, including three watersheds (Taunton River, Northwest Mount Hope Bay, and Southeast Mount Hope Bay), and the locations of autonomous monitoring buoys (#), WWTFs (A–F), major NPDES sites (P), and the USGS Taunton River gauge (C).

consists of two regions: the shallow flat area that occupies the north and western area, and the deeper waters of the main shipping channel that runs south–north along the eastern boundary. The Taunton River is the second largest river in Massachusetts and flows approximately 64 km before draining into Mount Hope Bay.

Approximately 70% of Mount Hope Bay's surface water lies within the boundary of the state of Rhode Island, yet nearly the entire watershed is located within the state of Massachusetts. The Taunton River watershed is the largest of the three sub-watersheds (Fig. 13.1), representing 89% of the total drainage area for this system. The US Geological Survey (USGS) maintains three gauges within the Taunton River watershed; one each located in the Taunton, Three Mile, and Segreganset Rivers. These gauges account for approximately 60% of the total drainage area for the entire Mount Hope Bay watershed. The annual mean discharge of the gauged portion of the Taunton River derived from monthly averages for the period of record (USGS gauge, Bridgewater, MA; 1927–2003, not inclusive) is $1.08 \times 10^7 \, \text{m}^3 \, \text{d}^{-1}$. The monthly averaged flow from the second largest tributary below the USGS gauge (Three Mile River) was found to significantly co-vary with Taunton River flow ($r^2 = 0.99$; Fig. 13.2) for comparable years (1967–2003). Using USGS reported annual flows for the Taunton River near Bridgewater, MA, and linearly extrapolating to the total drainage area of the Mount Hope Bay watersheds (Pilson, 1985), the estimated total annual riverine input into Mount Hope Bay is $2.5 \times 10^7 \, \text{m}^3 \, \text{d}^{-1}$. When compared to mean daily flows of major rivers (e.g., Blackstone and Pawtuxet) discharging into Narragansett Bay (Nixon $et\ al.$, 2005), the Taunton is the

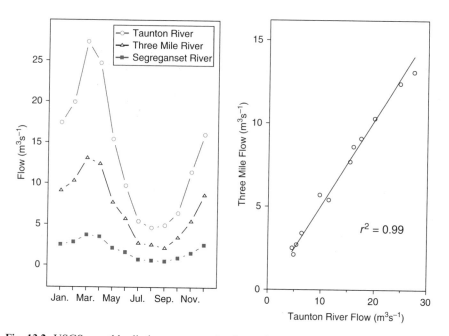

Fig. 13.2 USGS monthly discharge summaries for each period of record for the Taunton, Three Mile, and Segreganset Rivers in the Taunton River watershed and simple linear regression plot between the Taunton and Three Mile Rivers. Mean monthly averages from the period 1967–2003 were derived from USGS gauged flows *(http://ma.water.usgs.gov/ basins/tauntonstw.htm)*.

Table 13.1 Estimates of freshwater inputs to Mount Hope Bay (USGS gauge data and MAGIS data layers).

Surface Area (km^2)	Taunton (gauged)	676
	Three Mile (gauged)	217
	Segreganset (gauged)	27
	Total	1380
Annual mean flows ($\times 10^6$ m^3 d^{-1}) adjusted	Taunton	2.48
mL, for non-gauged portion of watershed	Wastewater[a]	0.12
	CSOs[b]	0.01
	Direct Rainfall	0.01
Estimated total freshwater flows	2.6×10^7 m^3 d^{-1}	

[a] Below the USGS Taunton River gauge near Brockton.
[b] Prior to the Fall River CSO abatement.

largest riverine source of fresh water to the greater Narragansett Bay ecosystem. A summary of freshwater discharge to Mount Hope Bay from the watersheds is provided in Table 13.1.

13.2.1 Land Use Classification

There are three major sub-watersheds in the Mount Hope Bay system: the Taunton River, Mount Hope Bay–Northwest, and Mount Hope Bay–Southeast (Fig. 13.1). The 1:24,000 scale land use data for 1985 and 1999 (MA) and for 1988 and 1995 (RI) were rectified to coincide for analytical purposes, but are herein referred to as 1985 and 1999, since 97% of the watershed is within Massachusetts. With a focus on water quality, 29 MacConnell classes for Massachusetts and 31 Anderson classes for Rhode Island were aggregated to form eight land use bins (Table 13.2).

13.2.1.1 Mount Hope Bay Watershed Land Use Results

Between 1985 and 1999, rates of low-density residential development were the greatest component of the developed land category for each of the three Mount Hope Bay sub-watersheds (Table 13.3). Medium-density residential development (i.e., greater than or equal to 0.25 acre and less than one acre) also increased significantly, with a total net gain of 27 km^2 (21%). From an N loading perspective, forest loss and increases in areas classified as urban and residential improve rapid N transport through watersheds (Alexander et al., 2002).

13.2.1.2 Sub-watershed Results

Between 1985 and 1999, rates of low-density residential development were the greatest component of the developed land category for each of the three sub-

Table 13.2 Land use class descriptions for Mount Hope Bay watersheds.

Land Use Class[a]	Land Use Description
Agricultural	Cropland, pasture, orchard, cranberry bog, nursery
Disturbed Open	Strip mine, quarry, gravel pit, landfill, junkyard
Maintained Open	Parks, playfield, playground, marina, golf course, tennis court, swimming pool, power line, pipeline, cemetery, vacant undeveloped land (urban)
Natural Open	Forest, salt marsh, forested and non-forested freshwater wetland, abandoned agricultural field, meadow
Residential Low	Lots greater than or equal to 1 acre
Residential Medium	Lots greater than or equal to 0.25 and less than 1 acre
Residential High	Lots less than 0.25 acre
Urban (i.e., Commercial, Industrial, Transportation)	Shopping center (i.e., primary sale of products and services), manufacturing, industrial parks, airport, divided highway, railroad, freight storage, dock, pier, storage tank

[a] Land use classes were derived from the aggregation of MacConnell (1973) and Anderson et al. (1976) land use classification systems.

watersheds, though medium and high residential land use categories also featured gains. Table 13.4 shows the Taunton River sub-watershed with an increase in residential low density land use of 45 km² (51%). The northwest and southeast sub-watersheds show respective increases in the residential low category by 32 and 25%, but these sub-watersheds make up just 13% of the total land area in the Mount Hope Bay watershed. Half of the 25 inclusive cities and towns in the watershed had greater than 60% of their buildable land developed by 2001 (MA EOEA, 2003).

The largest urban centers are Fall River (located in the southeast sub-watershed) with a human population of 92,000 and Brockton (located in the northern reaches of the Taunton River sub-watershed) with 94,000 (US Census

Table 13.3 Mount Hope Bay watershed land use trends.

	Land Use Class[a]	1985 Area (km²)	1999 Area (km²)	Change (km²)	Change (%)
Mount Hope Bay (1,550 km²)	Agricultural	138	118	−20	−15
	Disturbed open	18	14	−4	−24
	Maintained open	48	46	−2	−4
	Natural open	994	927	−67	−7
	Residential high	48	53	5	10
	Residential medium	96	143	47	49
	Residential low	130	156	27	21
	Urban	77	91	15	19

[a] Land use classes were derived from the aggregation of MacConnell (1973) and Anderson et al. (1976) land use classification systems.

Table 13.4 Mount Hope Bay sub-watershed land use trends.

	Land use class[a]	1985 Area (km^2)	1999 Area (km^2)	Change(km^2)	Change (%)
Taunton River 1,370 km^2	Agricultural	122	104	−18	−14
	Disturbed open	17	13	−4	−24
	Maintained open	42	40	−2	−4
	Natural open	891	826	−65	−7
	Residential high	32	36	4	13
	Residential medium	89	135	45	51
	Residential low	116	140	24	21
	Urban	63	77	14	23
Southeast (SE) 93 km^2	Agricultural	4	3	0	−10
	Disturbed open	0	0	0	4
	Maintained open	4	4	0	−2
	Natural open	56	54	−1	−3
	Residential high	10	10	0	2
	Residential medium	2	2	0	25
	Residential low	7	8	1	16
	Urban	10	10	0	0
Northwest (NW) 86 km^2	Agricultural	13	11	−2	−16
	Disturbed open	1	1	0	−40
	Maintained open	3	3	0	3
	Natural open	48	47	−1	−2
	Residential high	6	7	0	6
	Residential medium	5	6	2	32
	Residential low	7	8	1	17
	Urban	4	4	0	12

[a] Land use classes were derived from the aggregation of MacConnell (1973) and Anderson *et al.* (1976) land use classification systems.

Bureau, 2000). These municipalities are served by the two largest wastewater treatment facilities (WWTFs) in the Mount Hope Bay watershed.

13.3 Human Uses and Values

Mount Hope Bay has been used historically for fishing, navigation, and shipping, though today fisheries resources remain largely unavailable for human use because of pollution, habitat loss, and changes in biological components similar to those observed for the larger Narragansett Bay ecosystem (MRI, 1983; Gibson, 1996).

The upper reaches of Mount Hope Bay, including the lower Taunton River, were used for the harvest of oysters (*Crassostrea virginica*) and became a viable commercial fishery in the late 1800s (Belding, 1921). By 1907, however, oysters from portions of the Taunton River estuary were considered enough of a threat to human health because of pollution that commercial interests in the fishery ended.

River herring (*Alosa pseudoharengus* and *A. aestivalis*) supported another important early fishery, and early local news accounts report of Native Americans using the Taunton River each spring for harvesting river herring (Belding, 1921). By the turn of the 20th century, the Taunton River and its tributaries were considered to be substantially polluted with sewage and industrial waste. Because of declining herring stocks and antiquated uses (fish oils were used in the manufacturing of paint and cosmetics and medicinal applications), little demand existed for the river herring fishery, and commercial interests in Taunton River herring ceased altogether by the 1960s. Today, the Mount Hope Bay system is recognized by the Commonwealth of Massachusetts as an important habitat for alewives and other herring species. The Nemasket River for instance, located within the Taunton River watershed, contains one of the most prolific herring runs in Massachusetts. Efforts by state and federal agencies, as well as other nongovernmental organizations, are being conducted to remove impediments to anadromous fish and return the Taunton River and its tributaries to a more pre-industrial state. Recent concerns about dramatic declines in the number of herring returning for spawning have resulted in a moratorium in Massachusetts on the harvesting, possession, and sale of river herring through 2008.

At present, human use continues primarily as commercial shipping (coal supply and cargo shipping), waste management (municipal wastewater disposal, stormwater runoff, combined sewer overflows and industrial cooling), and secondary recreation (boating and fishing). The Port of Fall River, which is located in upper Mount Hope Bay, is the second largest cargo shipping port in Massachusetts. There are seven Massachusetts-based wastewater treatment facilities within the watersheds of Mount Hope Bay. From monthly mean data reported to the US Environmental Protection Agency (US EPA) for 2002, the two large WWTFs serving the municipalities of Fall River and Taunton averaged an annual discharged of treated effluent of 32 and 8.4 $\times$ 10^6 m^3, respectively, directly into Mount Hope Bay. Based on effluent reporting data to EPA for 2004, the Brockton WWTF, located in the upper portion of the Taunton River (above the USGS gauge in Bridgewater) adds an additional 23 $\times$ 10^6 m^3 y^{-1}. Prior to recent improvements by the City of Fall River, municipal combined sewer overflow (CSO) outfalls discharged approximately 4.9 $\times$ 10^6 m^3 of rainwater runoff and partially treated sewage to Mount Hope Bay each year. The freshwater volumes from these sources below the USGS gauging station on the Taunton River are minor (about 8.6 $\times$ 10^7 m^3 y^{-1}) when compared to the Taunton's discharge, which is estimated to be below 8 $\times$ 10^9 m^3 y^{-1}, but may be important with respect to contaminant loadings to the bay.

Because of the shallow nature of Mount Hope Bay and its relatively large surface area, water temperature responds readily to changes in heat flux

(Mustard *et al.*, 1999). Late summer temperature averages (using infrared satellite imagery from 1984 to 1995) found Mount Hope Bay to be 0.8°C warmer than other shallow embayments in the upper Narragansett Bay region. Industrial cooling to the bay is dominated by 1.4×10^{10} W of heat discharged annually from the Brayton Point Electric Power Station (US EPA, 2003; Fan and Brown, 2006). The Brayton Point facility is a 1600-MW electric power facility located on the northern shore of Mount Hope Bay. Spaulding and Swanson (Chapter 8) and Chen *et al.* (Chapter 9) detail circulation impacts of this thermal input to Mount Hope Bay. Fan and Brown (2006) produced a heat budget for Mount Hope Bay using simple box models and found that the Taunton River contributed negligible heat load (~4%) when compared with the heat load from the Brayton Point facility. The US EPA identified the Brayton Point facility as the major contributor to significant detrimental changes observed in the aquatic and biological conditions of Mount Hope Bay (US EPA, 2002). By 2002, the Brayton Point facility was withdrawing bay water at a rate of approximately 4×10^6 m^3 d^{-1}. In 2003, EPA issued a new National Pollution Discharge Elimination System (NPDES) permit that identified near total reductions in both heat loading and water withdrawals (96% and 94%, respectively) as necessary in order to achieve compliance with state and federal regulations as outlined in the NPDES section of the Clean Water Act (33 U.S. Code §§ 1342).

13.4 Environmental Management

13.4.1 Designated Uses

Management strategies for aquatic ecosystems typically rely on water quality standards as a starting point for implementing measures that mitigate anthropogenic impacts. These impacts are often evaluated under the umbrella of established "designated uses" (e.g., aquatic life habitat, fish consumption, shellfish harvesting, and swimming) and have traditionally focused on water quality. Currently, water quality standards that guide local management contain two important elements: designated beneficial use or uses (e.g., recreation, water supply, fishing and others) and numerical or narrative targets for specific contaminant levels above which contaminants have been shown to interfere with the ecosystem's healthy, natural biological community.

Mount Hope Bay shares borders with Rhode Island and Massachusetts and use designations are distinct to each jurisdiction. The differences among each state's class-specific criteria are highlighted in Table 13.5. A move towards ecosystem-based management strategies may require states that share ecosystems to redefine designated uses and align supporting criteria to be more sensitive to ecosystem functioning.

Table 13.5 Comparison of class-specific water quality criteria for marine surface water classifications among Massachusetts and Rhode Island states. Compiled from the current Code of Massachusetts Regulation, Massachusetts Division of Water Pollution Control (314CMR 4.00) and RIDEM (2000).

State	Rhode Island			Massachusetts		
	SA	SB/SB1	SC	SA	SB	SC
DO (mg L^{-1})	≥6 except as naturally occurs	≥5 except as naturally occurs.		≥6 unless background conditions are lower; ≥75% saturation due to a discharge	>5	≥5 for at least 16 h in a 24-h period
Temp (°C)	28.3 max. Raised no more than 0.9 (16 June–September) Raised no more than 2.2 (October–16 June) No increase above the recommended limit on most-sensitive water use.			29.4 max. 26.7 daily, temperature rise due to discharge <0.8	29.4 max. 26.7 daily	29.4 max. rise in temperature due to discharge <2.8
Fecal C (100 mL^{-1})	14 and 49	50 and 500	None in such concentrations that would impair specified use	14 and 43[a]	88 and 260[a]	1,000 and 2,000
Solids	Discharges of sludge, solid refuse, floating solids, oil, grease, scum are not allowed		None in such concentrations that would impair specified use	Surface waters shall be free from floating, suspended, and settleable solids in concentrations or combinations that would impair any use assigned, would cause aesthetically objectionable conditions, or impair benthic biota or degrade the chemical composition of the bottom		

Table 13.5 (continued)

State	Rhode Island			Massachusetts		
	SA	SB/SB1	SC	SA	SB	SC
Nutrients			Concentrations shall not exceed levels that would impair any usages or cause undesirable or nuisance aquatic species associated with cultural eutrophication, be preventive or minimize accelerated or cultural eutrophication. Total phosphorus, nitrates and ammonia may be assigned site-specific limits based on reasonable best available technologies. Where waters have low tidal flushing rates, applicable treatment to prevent or minimize accelerated or cultural eutrophication may be required for regulated non-point source activities.	Shall not exceed the site-specific limits necessary to control accelerated or cultural eutrophication		

[a] Geometric mean or median of Most Probable Number method, second number reflects the value at which 10% of samples can not exceed. Different standards (higher) apply for those areas not designated for shellfishing.

13.4.2 Environmental Monitoring

State environmental managers continue to design monitoring programs that conform to their jurisdictional boundaries, probably due in part to the traditional use of jurisdictional boundaries in state land management practices. Aquatic systems that transcend these boundaries are often not evaluated or monitored in a manner that allows for the application of sound ecosystem-based principles, as evidenced in Curley *et al.* (1974).

Since the early 1970s, water quality monitoring undertaken in Mount Hope Bay has been conducted to assess impacts caused by water usage for condenser cooling by electric power generation and from wastewater discharges from local municipalities. By the mid-1990s, dramatic changes were observed in the fish populations of Mount Hope Bay (Gibson, 1996). Trawl survey abundance trends for a number of species connected the collapse of winter flounder stocks in Mount Hope Bay to coincidental increases in power generation at the Brayton Point station (Gibson, 2002).

Near continuous monitoring of water quality parameters from efforts sponsored by the Brayton Point facility in 1997 revealed dramatic changes of water quality in both the surface and bottom water of upper Mount Hope Bay. Episodic anoxic and low dissolved oxygen levels were observed in surface and bottom waters for two fixed monitoring stations during a 6-week, late-summer monitoring period near the facility. Figure 13.3 shows the details of dissolved oxygen and salinity concentrations during the last week in August, 1997. On several occasions, dissolved oxygen levels at both stations fell below 3 mg L^{-1}, a level below which aquatic biologists believe to be detrimental to supporting healthy ecosystems (Baden *et al.*, 1990; Johansson, 1997). Of particular interest in the 1997 dissolved oxygen data is surface concentrations that were lower than the corresponding dissolved oxygen concentrations in the bottom waters at the Gardners Neck site, near where the Taunton River enters into Mount Hope Bay. Historically, dissolved oxygen values below 3 mg L^{-1} have been reported for areas in Mount Hope Bay specifically in bottom waters of the lower and upper reaches of the tidally influence Taunton River (Curley *et al.*, 1974; MRI, 1983).

In Mount Hope Bay, impairment of use is often linked with excess contamination by fecal coliform, heat, suspended matter, nutrients, or chemicals of environmental concern, and manifested by changes in the biological community, related losses of essential aquatic habitat, low dissolved oxygen, or fish kills. Excessive nutrient loading to Mount Hope Bay has been identified as a major stressor by resource managers since the late 1960s (Curley *et al.*, 1974). Because of increases in the urbanization of coastal areas, many marine embayments today may be receiving nutrient loads in excess of their capacities to adequately assimilate these pollutants (Bricker *et al.*, 1999; Howarth *et al.*, 2000; Bowen and Valiela, 2001; Howarth *et al.*, 2002). The resultant consequence, referred to as eutrophication, is frequently becoming the source of some

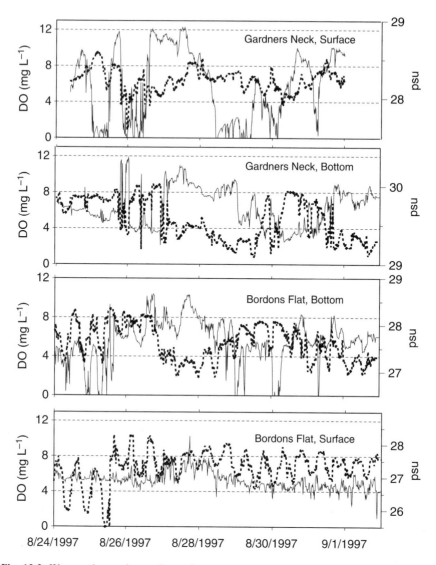

Fig. 13.3 Water column observations of salinity (dashed line) and dissolved oxygen (solid line) at two stations in upper Mount Hope Bay. Gardners Neck (41.7013N, 71.2130W) and Bordons Flats (41.7022N, 71.1755W).

of the most widespread and serious impacts occurring in coastal waters (Diaz and Rosenberg, 1995; NRC, 2000; Bricker *et al.*, 2006).

For upper Narragansett Bay, the ability of Rhode Island WWTFs to meet end-of-the-pipe water quality standards (3–8 mg L^{-1}) are currently being evaluated as a means to reduce nitrogen loading and achieve desired water quality goals. Furthermore, efforts are underway by the Massachusetts Department of

Environmental Protection's (MA DEP) Estuaries Project to develop a nitrogen TMDL for Mount Hope Bay. The Massachusetts process attempts to identify acceptable water quality characteristics, namely site-specific water column N concentrations known to be supportive of key aquatic habitats, and link to an understanding of the system's hydrodynamics in order to determine allowable levels of nitrogen loading. Monitoring of water column nutrient species and freshwater flow from most of the streams and rivers by MA DEP is in progress during the preparation of this chapter.

13.5 Selected Water Quality Observations, 1999–2003

The Massachusetts Office of Coastal Zone Management (MCZM) began a pilot project designed to document some of the occurrences of hypoxia and/or anoxia in Mount Hope Bay waters, and to assess the potential of cumulative effects from multiple stressors on the biological integrity of the bay ecosystem. MCZM targeted their efforts to discerning the role of the Taunton River and its watershed on the frequency and magnitude of hypoxic events in upper Mount Hope Bay, and to begin evaluating the importance of water quality on observed changes in the biology of the ecosystem. Monitoring was designed to answer questions such as: What are the sources and/or mechanisms controlling low dissolved oxygen in surface and bottom waters? What are the timing, duration, and frequency of these events, and are they ecologically significant? Are the Taunton River and its watershed significant contributors to the low dissolved oxygen observed in Mount Hope Bay?

Observations from water quality surveys in Mount Hope Bay are from near-continuous monitoring via autonomous modules deployed by MCZM in the Taunton River estuary as a pilot project in 1999. This project was later expanded to include stations located in upper Mount Hope Bay near the Lee and Cole Rivers, and in waters straddling the Massachusetts–Rhode Island border near the center of Mount Hope Bay (Fig. 13.1). The pilot project was further expanded in 2000 to include a survey of contemporary levels of nutrients, chlorophyll, particulate organic carbon, and total suspend matter (TSM) in the water column. As part of this analysis, data from bay-wide dissolved oxygen surveys conducted in the Mount Hope Bay system during the MCZM pilot are presented.

13.5.1 Procedure and Methods

Commercially available autonomous environmental monitoring buoys (YSI/ Endeco) were deployed in Mount Hope Bay by MCZM beginning in 1999 in the lower Taunton River, and in the central and upper portion of Mount Hope Bay the following year. These systems were fitted with sondes programmed to monitor salinity, temperature, dissolved oxygen, pH, and on occasion, turbidity, chlorophyll fluorescence, and photosynthetically active radiation (PAR: 390–710 nm) every 15 min. PAR light attenuation data were collected using LICOR® 2π sensors. All data were reviewed for adherence to initial and continuing calibration. Monitoring buoys were typically deployed in late spring/ early summer and sonde performance checked every two weeks, or weekly when bio-fouling significantly increased near the end of July and into August. Sondes were swapped out with pre-calibrated replacements, or removed for a few days and serviced prior to re-deployment.

For dissolved oxygen, sensors were checked for in-air 100% saturation by wrapping the sensor housing with a wet towel and obtaining a reading after minimum equilibration time of 3 min. Periodic discrete water samples were obtained during field surveys at sensor depth and compared with the determination of dissolved oxygen concentrations via classic Winkler procedure (Strickland and Parsons, 1972). Salinity calibration and periodic performance checks were conducted using conductivity standards (Myron L Co.) for brackish (16.6 mS) and coastal marine waters (30.1 mS). Periodic performance checks were conducted at the beginning of each field excursion using a secondary standard of Massachusetts Bay water (30.44‰) that had been collected and stored in a large carboy (20 L) and periodically re-analyzed and compared to simultaneous analyses with the University of MA/Boston Seabird® CTD system.

Water quality surveys conducted during the late summer through early fall in 2000 included analysis of dissolved inorganic nutrients (DIN) from ten stations in Mount Hope Bay. Each survey included hydrocasts for temperature and salinity, and the collection of discrete water samples for chlorophyll a (Chl a), DIN ($NO_2 + NO_3$, NH_3, $O\text{-}PO_4$, SiO_4, etc.), particulate organic carbon and nitrogen (POC/N), and TSM at the surface, middle (depth permitting), and bottom of the water column. Discrete samples were collected using Niskin water samplers (General Oceanics) and transferred to acid-cleaned (HCl) 2-L polycarbonate bottles. Subsamples for nutrients were syringe-filtered in the field, the filtrate quickly frozen using dry ice and stored frozen for later analysis. The remaining sample was filtered (Whatman GF/F glass microfiber media, nominal pore size 0.7 μm) in the field and the filters returned to the lab for processing of chlorophyll, POC/N, and TSM. Analyses for DIN and Chl a were performed at the University of MA/Boston. Analytical methods, desired method accuracy, and detection thresholds for monitoring parameters are listed in Table 13.6.

Data from Prell *et al.* (2004), as well as MCZM's DIN surveys, are used here to describe water column conditions during the late summer periods of 1999–2003. The resultant dissolved oxygen and surface water nutrient contour plots present nutrient concentration gradients generated using the kriging option in Surfer® (Golden Software, Inc.). Kriged concentration gradients are derived from a regression technique used in geostatistics to approximate or interpolate data for representation in two- and three-dimensional space. The kriging method was used to interpolate the sampling grid because it is a less sensitive method to the non-random sampling design (Cressie, 1991). Because the spatial interpolation is a minimum variance-based estimation technique,

Table 13.6 Target accuracy and detection limits for laboratory measurements.

Parameter	Method	Reference	Accuracy[d]	Lower Detection ($\mu g\ L^{-1}$)
Ammonia	Phenol/hypochlorite (autoanalyzer)	Guffy et al. (1988)	5%	1
Nitrite + nitrate	Cd–Cu reduction/ sulfanilamide/N-ED HCl$_2$	Guffy et al. (1988)	5%	1
Total particulate carbon/ nitrogen	Elemental analyzer	Hedges and Stern (1983)[a]	5%[e]	6
Reactive silicate	Molybdatium blue autoanalyzer	Guffy et al. (1988)	5%	3
Orthophosphate	Molybdenum Blue (autoanalyzer)	Guffy et al. (1988)	5%	3
Chlorophyll a	Acetone extraction fluorometric	Strickland and Parsons (1972)[b]	10%	0.01
Total suspended solids (TSS)	Desiccator, gravimetric microbalance	UMass Boston protocol[c]	5%	0.5 mg L^{-1}

[a] Modified for the Perkin–Elmer Model 2400 CHN Elemental Analyzer.
[b] pp. 201–203.
[c] Using 0.4 μm polycarbonate (Poretics) membrane filtration.
[d] Accuracy based on results of laboratory control standards and spiked samples.
[e] Precision based on relative percent difference of sub-sample analysis. No spikes are available for POC/PON analysis.

these concentration maps ultimately contain less variability than the actual sampling data, which should be noted when drawing conclusions from these plots. The uncertainties associated with these estimates can be large, especially for interpolations with lower sample density. Interpretation must be sensitive to two important points: (1) a small portion of the water column was sampled and (2) these descriptions generalize processes that occur over a period of many hours (typically 8) during which significant tidal oscillation has occurred.

13.5.2 Autonomous Water Quality Monitoring

Results from autonomous monitoring of salinity, temperature, and dissolved oxygen from 2000 to 2003 in upper Mount Hope Bay are summarized from three stations: culminating in 2003 with monitoring buoy systems in the lower Taunton River, the bottom waters off of Brayton Point in an area labeled as Brayton Flats, and adjacent to the shipping channel near the Fall River WWTF straddling the state border, termed the State Line station. In 2003, these three stations were monitored as a supplemental to EPA's winter flounder habitat characterization efforts in Mount Hope Bay (CoastalVision, 2004).

For most of the summer, Mount Hope Bay surface waters (~20–27‰) remain somewhat isolated from the deeper, more saline water (~30‰) of greater Narragansett Bay. In 2003, bay waters near the State Line station remained stratified with respect to salinity and temperature from the beginning of monitoring (late July) through mid-September, except for a brief period near

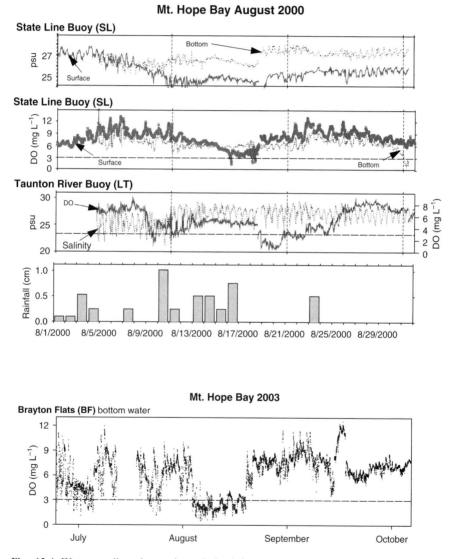

Fig. 13.4 Water quality observations derived from autonomous monitoring buoys for selected periods in 2000 and 2003.

the end of June. Similar observations for continued salinity-induced stratification during summer periods were reported at the State Line station for 2001 (Howes and Sundermeyer, 2003). Mixing periods may persist for several days as evidenced by the convergence of salinity concentrations in surface and deep waters for August 2000 (Fig. 13.4). Surface water temperatures at all stations typically exceeded 25°C during late summer.

For most of the monitoring period, Mount Hope Bay showed signs of relatively healthy dissolved oxygen levels (>4 mg L^{-1}). However, low dissolved oxygen concentrations were observed at all autonomous monitoring stations, typically falling below critical levels in late summer (see Fig. 13.4). Though the autonomous data for Mount Hope Bay is not inclusive for 1999–2003, the most critical period with respect to dissolved oxygen appears to occur during August when freshwater runoff is low (see Fig. 13.2), heat flux is near maximum (Fan and Brown, 2006), and water column stratification is most pronounced. In several instances during the late summer, episodic dissolved oxygen concentrations at or below the critical levels of 2–3 mg L^{-1} were observed in the upper portion of the water column (Fig. 13.4). The role of the Taunton River on summer dissolved oxygen concentrations in Mount Hope Bay may be important, and is implied by the data from the lower Taunton River station (Fig. 13.4, top panel, center). Here, water column dissolved oxygen concentrations <3 mg L^{-1} persisted for nearly 40 h in mid-August, 2000. Recovery of dissolved oxygen concentrations followed a series of step-like increases over a period of 5–6 days. However, this transient low dissolved oxygen event was not evident at the mid-bay State Line station where dissolved oxygen concentrations were observed to be ≥ 6 mg L^{-1}.

Dissolved oxygen concentrations reported for the Brayton Flats station in 2003 (CoastalVision, 2004) show episodic behavior in the shallow region near the Lee and Cole Rivers in upper Mount Hope Bay (Fig. 13.4, lower panel), with two pronounced periods of depressed dissolved oxygen concentrations (early July and August). During the more severe August 2003 event, dissolved oxygen levels in the bottom waters at Brayton Flats were typically less than 3 mg L^{-1}, lasting for a period of 2 weeks, with the lowest concentrations near 0.6 mg L^{-1}. Comparable water quality data for the lower Taunton River is not as complete during this period, but show concentration levels above 4 mg L^{-1} at least until August 15, 2003, well after the onset of low dissolved oxygen at Brayton Flats. Dissolved oxygen concentrations in bottom waters at the State Line station during this period were also near or above 4 mg L^{-1} (CoastalVision, 2004). A clearer understanding of seasonal circulation in upper Mount Hope Bay, and an additional monitoring station near the entrance to the Taunton River estuary, is needed in order to better understand the nature of these surface water low dissolved oxygen observations and their significance to the ecology of the system. Chen et al. in Chapter 9 of this volume provide initial model results for Mount Hope Bay, and provide insight into requirements for future progress in this direction.

13.5.3 Summer 1999–2003 Mount Hope Bay Dissolved Oxygen Surveys

Synoptic surveys designed to assess spatial dissolved oxygen levels (along with standard water quality parameters such as salinity and temperature) at over 100 stations in Narragansett Bay were performed during summer periods from 1999 to 2003 (Prell et al., 2004). Further details regarding these data are presented by Saarman et al. and Deacutis in Chapters 11 and 12, respectively.

Water column observations of salinity and dissolved oxygen collected in Mount Hope Bay during this period were used to estimate synoptic water quality concentration fields and represent late summer conditions for Mount Hope Bay. Typical transects extend from Narragansett Bay stations beyond the Mount Hope Bridge in the south, to just north of the Braga Bridge in the lower Taunton River estuary along the axis of the main shipping channel (Fig. 13.5). Figure 13.6 shows summer cross-sections from 1999 to 2003. Each transect plot contains station locations (Prell et al., 2004) and an overlay of sampling depths (white open circles) to illustrate the sample density used for generating contour plots. Five surveys, typically conducted during the month of August, are summarized here to provide a cursory view of summer dissolved oxygen dynamics in Mount Hope Bay, and to identify potentially sensitive

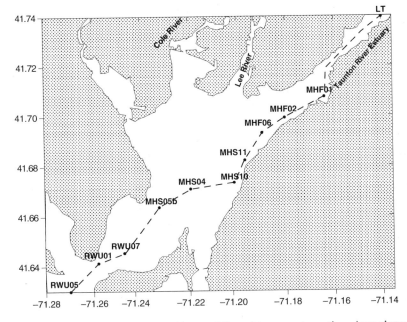

Fig. 13.5 Stations [Prell et al. (2004); Chapter 11] used to generate section views along the long axis of Mount Hope Bay. Station LT represents the MCZM autonomous monitoring site located in the Taunton River estuary during 1999–2003.

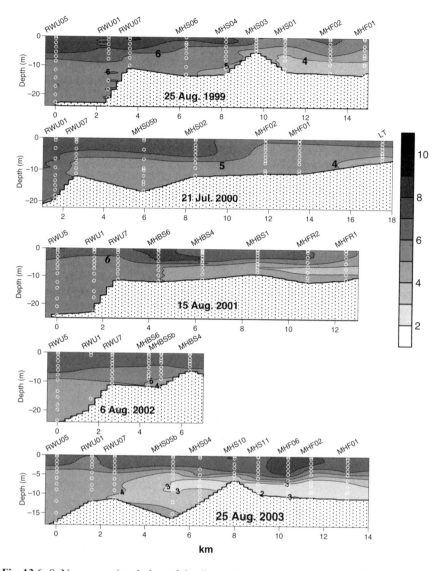

Fig. 13.6 S–N cross-sectional plots of the dissolved oxygen concentration field (milligram of O_2 per liter) in Mount Hope Bay for mid–late summer seasons, 1999–2003. Stations (see Chapter 11, and Fig. 13.5) used in each plot are shown. Open circles represent the samples taken at depths that were used to in interpolating each concentration field.

areas. No August data were collected for upper Mount Hope Bay during 2000, so late July data for 2000 are shown instead.

During these surveys, most of Mount Hope Bay exhibited relatively healthy levels of dissolved oxygen—greater than the designated water quality standards for these waters—and were generally not considered as impaired with respect to dissolved oxygen when compared to the greater Narragansett Bay system. The water column for these bay-wide surveys was typically stratified, with respect to salinity, temperature, and dissolved oxygen (CoastalVision, 2004). In general, higher dissolved oxygen concentrations were observed in surface waters relative to deeper waters. Surface waters with the highest dissolved oxygen concentrations ($\sim$7 mg L^{-1}) typically occurred in the lower portion of Mount Hope Bay, near the East Passage of Narragansett Bay. However, higher dissolved oxygen levels (>8 mg L^{-1}) associated with freshwater input in the upper portion of Mount Hope Bay was observed during the August 2003 survey. Because semi-diurnal tidal mixing dominates the circulation of the bay (Spaulding *et al.*, 1999), it is important to note the described surveys were conducted over the period of 6–8 h during which significant tidal mixing occurred and may have resulted in some artifacts in our portrayal of the position of maxima and minima dissolved oxygen layers. Tidal excursions of heated effluent from Brayton Point from satellite imagery (Mustard *et al.*, 1999) indicate excursion distances of near one-half the length of the bay (5–6 km), and may be of importance to these interpretations.

The dissolved oxygen concentration fields generated from survey data across the central portion of Mount Hope Bay along west–east transects offer little information for generalizations, partly because of the poor spatial coverage in our sampling design, and are not shown. However, these analyses hint of higher surface dissolved oxygen concentrations in the western portion of the bay, especially near station MHN05, and could be indicative of processes such as longer water residence times and/or higher primary productivity, and subsequent *in situ* production of oxygen from photosynthesis.

Lower dissolved oxygen concentrations were usually observed in the upper reaches of Mount Hope Bay, associated with the deeper waters of the main shipping channel in the vicinity where the lower Taunton River exchanges with Mount Hope Bay (Braga Bridge, US Interstate 195). The August 2003 survey yielded lowest dissolved oxygen concentrations ($\sim$3 mg L^{-1}) in the shipping channel, extending from the lower Taunton River to approximately 9 km into the central portion of Mount Hope Bay. The sub-surface minimum was observed at 10 m depth at Station MHS05b approximately 1 h prior to low tide, and appears to be associated with the Taunton River outflow (Fig. 13.6, bottom). Interestingly, similar distribution patterns were observed for salinity, and simple linear regression was used to further explore the relationship between salinity and dissolved oxygen.

A least squares fit to the August 2003 Mount Hope Bay hydrographic data was highly significant ($r^2 = 0.69$). However, an examination of the residuals (Fig. 13.7, top panel) reveals a cluster of points with relatively higher dissolved oxygen values in the mid-salinity range near 28–29‰. All 14 of these points represent sample observations from the lower portion of Mount Hope Bay that is associated with a portion of cooler, deeper Narragansett Bay water that exchanges across the open boundary near the Mount Hope Bridge. When these points are treated as outliers, the salinity-dissolved oxygen relationship

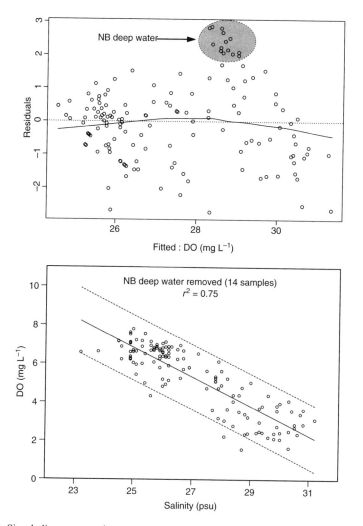

Fig. 13.7 Simple linear regression analyses of the August 2003 data (Prell *et al.*, 2004). Top panel shows a cluster of residuals that represent samples of Narragansett Bay "deep water".

improved ($r^2 = 0.75$, Fig. 13.7, bottom panel) and point to the importance of the higher salinity surface water from Narragansett Bay for establishing the conditions for water quality in Mount Hope Bay, at least for dissolved oxygen during the late summer period of August 2003.

13.5.4 2000 Summer–Fall Nutrient Surveys

The distributions of DIN and Chl *a* were monitored by MCZM at 10 stations in Mount Hope Bay from discrete water samples collected during the summer and early fall in 2000. Summaries of surface water concentration fields for sampled nutrients and Chl *a* at 2-week intervals beginning in July 2000 (7/21, 8/3, 8/18, and 8/31), and a late October survey (10/27), are illustrated in Fig. 13.8a–d. Surface contour plots for sampled nutrients and Chl *a* were generated by kriged interpolation of 10 stations using Surfer®, and at best show generalities in spatial trends.

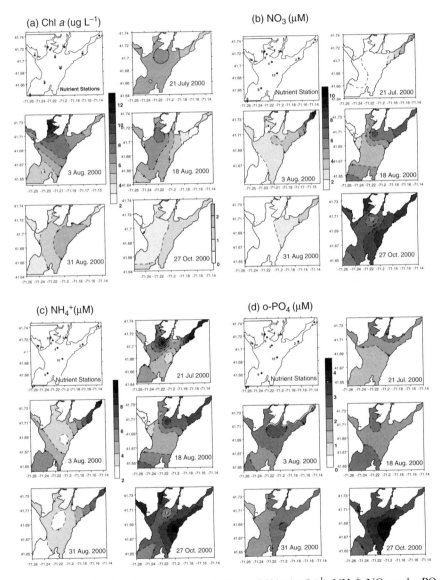

Fig. 13.8 Generalized surface water concentrations of Chl *a* (μg L^{-1}), NH$_4^+$, NO$_3$, and o-PO$_4$ (μM) derived from water quality monitoring by MCZM during the summer, 2000.

Overall, Chl *a* concentrations were generally higher in the central shallow portions of Mount Hope Bay during most of the surveys (Fig. 13.8a). Much of the bay area was typically at or above 5 µg L^{-1}. By the end of October, all surface water Chl *a* concentrations were closer to 1 µg L^{-1}. In early August, the highest concentrations of Chl *a* were observed in central and northwestern portions of the system. Stations in the central portion of the Mount Hope Bay were near 7 µg L^{-1} and the highest Chl *a* level (>12 µg L^{-1}) was observed at a single station in the Lee River (Station 5). It is important to note that Station 6, which is adjacent to Station 5, was not sampled during the August 3, 2000 survey, and therefore cannot be used to further substantiate the observations. Mean surface concentrations of NO_3 and NH_4^+ during this period are summarized in Fig. 13.8b,c, respectively. Mean values for NO_3 and NH_4^+ in the lower Taunton River Estuary (Stations 1 and 2) were computed for each survey during 2000 and found to be significantly higher ($p < 0.02$, repeated measures general linear model with Mauchly's *post hoc* test of sphericity; SPSS 13.0 Chicago, IL) than mean values computed for the remaining stations (i.e., Mount Hope Bay). These differences signify the importance of the Taunton River with respect to nutrient loading to Mount Hope Bay.

The general spatial trend in dissolved inorganic phosphate (o-PO_4) is shown in Fig. 13.8d. For most of the sampling events, o-PO_4 seldom exceeded 2.5 µM in the bay proper. Our kriging analyses show evidence of a strong o-PO_4 concentration gradient in surface waters near the outfall of the Fall River WWTF, possibly during the late August survey but clearly by late October. The o-PO_4 concentrations suggest that N is limiting to phytoplankton growth, based on the principle that organisms use nutrients like N and P roughly in the same proportion as chemical composition of their organic soft tissue (Broecker and Peng, 1982). This N:P ratio—also known as part of the Redfield ratio—is often expressed as 16:1. In all cases, the ratios of N:P were <10, averaging around 4.0 (± 2.0, $n = 51$).

Dissolved nutrient concentrations observed in Mount Hope Bay and the lower Taunton River during 2000 were comparable to the 1982–1983 observations made by MRI (1983). The relatively similar DIN concentrations observed for Mount Hope Bay over the past 20 or so years support the conclusion by Nixon *et al.* (2005) that total nitrogen entering Narragansett Bay by way of sewage discharges has not changed significantly since the mid-1980s (and see Chapter 5, Nixon *et al.*). It is important to note that both MCZM and MRI did not monitor for dissolved organic nitrogen, which can comprise a significant portion of the dissolved nitrogen pool for these waters (Pilson and Hunt, 1989; Nixon *et al.*, 2005).

Because of the paucity of recent nutrient data for Mount Hope Bay, we rely on the MRI data to illustrate the highly variable nature of dissolved inorganic nitrogen and the seasonal response of the Mount Hope Bay system to nitrogen loading. Figure 13.9 shows NO_3 and NH_4^+ concentrations for three MRI stations: near the confluence of the Segreganset River in the upper most, tidally influenced portion of the Taunton River Estuary; mid-estuary, near the

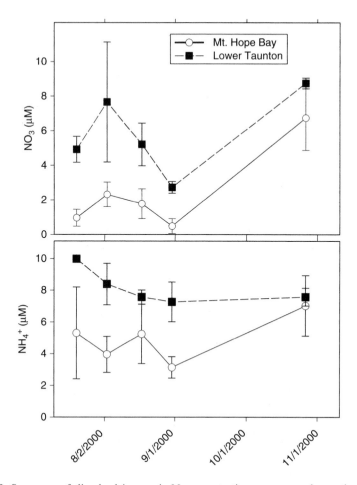

Fig. 13.9 Summary of dissolved inorganic N concentrations as mean observations from stations in Mount Hope Bay (Stations 3–10) and the lower Taunton River estuary (Stations 1 and 2) during the later part of 2000. Error bars reflect 1 SD about the mean, except for the Lower Taunton river means, where they represent the range of the two observations.

Montaup electric power station in Somerset, MA; and mid-bay, adjacent to the main shipping channel. Mean concentration values for NO_3 and NH_4^+ increased from 9 and 7 µM, respectively, in the surface waters of the lower and central bay, to 27 and 22 µM, respectively, in the upper reaches of the Taunton Estuary near the Segreganset River. Figure 13.10 indicates that a significant portion of the inorganic N load to the estuary appears to be readily diluted with higher salinity Mount Hope Bay water before exiting into Mount Hope Bay from the lower Taunton River.

The Taunton River receives significant amounts of nutrients from direct discharges of wastewater from municipal WWTFs (Save the Bay, 1998). From monthly mean data reported by the Taunton facility to EPA for 2002

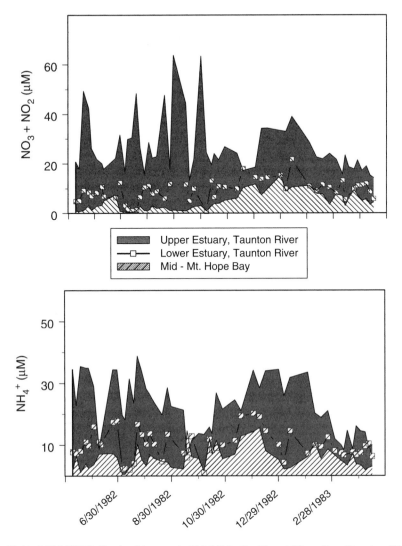

Fig. 13.10 MRI (1983) dissolved inorganic N (µM) in the Mount Hope Bay–Taunton River system from May, 1982 to April, 1983.

and the Brockton facility based on 2004 effluent data, it is estimated that these WWTFs add approximately 2.7×10^7 mol N annually directly to the lower portion of the Taunton River (Table 13.7). The Somerset and Mansfield WWTFs discharge minor flows beyond the Taunton River USGS gauge in Brockton, and contribute comparatively negligible amounts of N to the estuary. Taunton river nutrient data collected by investigators from Bridgewater State College and the Taunton River Watershed Alliance in 2000 *(www.glooskapandthefrog.org/phosphorus.htm)* provide the basis for

Table 13.7 Estimated N fluxes to Mount Hope Bay based on characteristics of WWTF derived from monthly averages reported to EPA[a,b], Taunton River discharges to the Taunton River estuary near the USGS gauge in Bridgewater, MA, and from atmospheric loadings using Howarth *et al.* (2002). Relative standard deviations are shown in parenthesis (RSD%) where monthly data from WWTF were available.

Source	Mean Flow (m^3 d^{-1}) $\times$ 10^4	Total Estimated N flux (mol N d^{-1}) $\times$ 10^3	%N as NO_3 + NO_2	%N as NH_3	Save the Bay (1998) N flux (mol N d^{-1}) $\times$ 10^3
Taunton River (Bridgewater)	108	59.6[d]	Unknown		
Fall River WWTF[a]	8.7 (16%)	84.5 (19%)	2	78	99.0
Taunton WWTF[a]	2.3 (17%)	15.1 (33%)	63	18	17.8
Somerset WWTF[c]	1.3 (5%)	0.4 (25%)	21	68	NA
Mansfield WWTF[c]	0.94 (6%)	NA	NA	NA	0.3
Brockton WWTF[b]	6.2	59.6 (32%)	2	74	5.9
Atmospheric Deposition	Indirect Direct	30–60 6.8			

NA: not available.

[a] NPDES data reported to EPA for 2002.

[b] NPDES data reported to EPA for 2004. Note significant performance changes at Brockton WWTF for 2004. Total N flux for 2002 and 2003 were calculated as 86 and 81 mol N d^{-1}($\times 10^3$) respectively.

[c] NPDES data reported to EPA for 2005. Mansfield WWTF report only N species concentration minima.

[d] Upper Taunton River (Bridgewater) nutrient data from Bridgewater State College 2000 Study (20.7 $\times$ 10^3 mol N d^{-1} as TKN, 38.9 $\times$ 10^3 mol N d^{-1} as NO_3) from *www.glooskapandthefrog.org/phosphorus.htm*, includes N from Brockton and Bridgewater WWTF.

estimating nutrient loads to the upper Taunton, and they are remarkably similar to the N loading from the Brockton WWTF shown in Table 13.7.

We lack sufficient understanding to ascertain what portion of N is retained within the Mount Hope Bay system, either through deposition as particles to the sediment, uptake by fringing marshes, sequestering by or remineralization from sediments, exportation to Narragansett Bay as organic forms or other important N species, or the amount lost to the atmosphere as N_2 via denitrification.

Also important, and beyond the scope of this chapter, is an assessment of the role of atmosphere N deposition, either directly to the surface waters of the Mount Hope Bay system or to its watershed. Howarth (Chapter 3) constructs estimates of atmospheric N deposition to the Narragansett Bay ecosystem. The rate of atmospheric N deposition to the northeast is considered among the highest in North America, contributing approximately 7.1 $\times$ 10^4 mol N km^{-2} y^{-1} (Howarth *et al.*, 2002). However, in many forested watersheds, only 10–20% of atmospherically derived N enters the receiving waters (Jaworski *et al.*, 1992; Howarth *et al.*, 2002). To the surface waters of Mount Hope Bay, this then translates to a loading between 1 and 2 $\times$ 10^7 mol N y^{-1}, comparable in terms of magnitude to loadings of N associated with the larger WWTFs in the Mount Hope Bay region.

To further complicate atmospheric N deposition estimates is the potential of local atmospheric loading of N from the Brayton Point power station. Power station N emissions deposited on land and water are considered an important contributor to declining water quality (Schlesinger, 1997). Not considering the potential of a localized atmospheric contribution of N from Brayton Point to the immediate watersheds of Mount Hope Bay, we estimate between 7.2 and 8.3×10^7 mol N y^{-1} are added to Mount Hope Bay annually from wastewater discharges and atmospheric deposition (Table 13.7). If we consider the upper loading estimate of 2×10^7 mol N y^{-1} derived from atmospheric deposition, approximately 70% of the N entering Mount Hope Bay is the result of WWTF discharges. However, this estimate also does not take into account exchanges of N between Mount Hope Bay and greater Narragansett Bay (i.e., the potential of sewage-derived N from Rhode Island WWTFs to Mount Hope Bay).

13.5.5 *Photosynthetically Active Radiation*

The characteristics of light attenuation, specifically in the range considered as photosynthetically active (390–710 nm), were determined by MCZM on a limited set of water quality observations during the 2003 monitoring period. Habitat structure and quality are intrinsically linked to the amount and quality of solar radiation received, and therefore may serve as one of the attributes (e.g., key species growth and mortality, food) to help quantify or qualify habitat quality or suitability for resource management analyses. In fact, light attenuation is one of the key indicators used for assessing aquatic ecosystem condition (Bricker et al., 1999; Batiuk et al., 2000; Biber et al., 2005) and is thought to be one of the major abiotic factors affecting the survivorship and health of submerged aquatic vegetation (SAV) (Dennison et al., 1993; Batiuk et al., 2000; Moore and Wetzel, 2000; Kemp et al., 2004). Light penetration in the water column is typically described as a function of the concentrations of suspended particles and colored dissolved organic matter (CDOM) in the water column.

The PAR penetration in the water column of Mount Hope Bay, specifically during the 2003 season, is considered here to provide some initial measure of the changes of water column light penetration; in the form of light extinction coefficients (K_d), and the amount of light available at depth (I_Z) for use by primary producers. The depth at which critical light levels meet the minimum physiological requirements of seagrasses is often reported to be between 10 and 30% of the incident surface light (I_0), or around 5–14 mol quanta m^{-2} d^{-1} (Kemp et al., 2004). We use the average of 46.1 mol quanta m^{-2} d^{-1} for average solar irradiance (based on field observations typical for June and July in the northern hemisphere in Bugbee, 1994).

In Mount Hope Bay, the only significant eelgrass beds were historically limited to the western shores of the Kickamuit and Lee Rivers in the northwest regions of the Bay (RI Coastal Resource Management Council, 2001). Villalard-Bohnsack *et al.* (1988) reported a complete absence of eelgrass beds in Mount Hope Bay and an extensive coverage of macroalgae beds in the Lee and Cole Rivers region during the mid-1980s. Abundant sea lettuce (*Ulva lactuca*) has been observed in many of the shallow areas of Mount Hope Bay during this study period, and may be symptomatic of increased loading of bio-available nitrogen (Short and Wyllie-Echeverria, 1996).

Water column PAR attenuation characteristics are reported here as a measure for assessing the suitability for SAV survivorship in Mount Hope Bay, and to illustrate some important limits and related controls on primary production in the Mount Hope Bay system. The extinction coefficient (K_d) was calculated from hydrocasts conducted during summer cruises on Mount Hope Bay in 2003 by performing regression analysis on the PAR versus depth data. Typical light attenuation data follow the relationship:

$$I_Z = I_0 e^{-K_d z}$$

where z is the depth (from surface), I_Z is the light intensity at depth z, and I_0 is the light intensity at the surface.

A typical Mount Hope Bay water column profile of PAR is shown in Fig. 13.11 (top panel). The model fit of the PAR versus depth data generally showed good agreement (typically, $r^2 > 0.9$) and yielded K_d values that ranged from 0.5 to $2.4\,\mathrm{m^{-1}}$ (Fig. 13.11, bottom panel). Higher values of K_d, representing greater light attenuation (i.e., more turbid water), occurred later in the year during fall observations. Though PAR data were collected at different times during daylight (09:00–17:00), and thus under variable I_0, this does not affect the estimate for K_d since light extinction is only a function of the scattering properties of the medium, dependent mainly on the concentration and types of material suspended (particles) or dissolved (i.e., CDOM) in the water column (Kemp *et al.*, 2004).

13.6 Summary and Conclusions

Ecosystem-based management requires an integrated approach to environmental management that incorporates human elements as part of the ecosystem (McLeod *et al.*, 2005), and requires integrated knowledge of the functions of the ecosystem. Much of the information needed to answer the monitoring questions poised by MCZM in 1999 remains unavailable. However, the recent water quality information presented earlier provides valuable insight into the

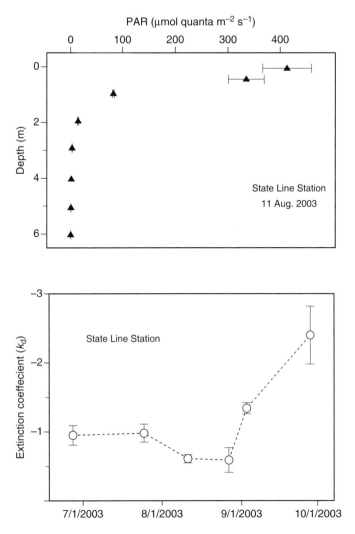

Fig. 13.11 Typical profile of PAR concentrations as a function of depth in the Mount Hope Bay water column (top panel) and PAR extinction coefficients (K_d) observed for surveys conducted between July–October, 2003 (bottom panel).

function of the Mount Hope Bay ecosystem and a direction for environmental management.

Mount Hope Bay is characterized by episodic low dissolved oxygen events in the upper central and western reaches of the bay. These low dissolved oxygen events are unique in the sense that some occur in surface waters, occasionally dropping below dissolved oxygen levels simultaneously observed in corresponding bottom waters. The link between these surface events and freshwater input from the Taunton River, by far the largest source of fresh water to the ecosystem, is thought to be important, but has not been firmly established to date. Modeling results for August, 1997 by Spaulding *et al.* (1999), however, predict that portions of the surface plume of the Taunton River discharge to be transported to the northwest toward the Lee and Cole Rivers. More detailed monitoring from July, and extending through September for several years, and an additional autonomous monitoring site near the entrance of the Taunton River, could aid in evaluating the importance of Taunton River inputs to the late summer dissolved oxygen dynamics and its influence on the biological character of Mount Hope Bay.

Much of the water column in Mount Hope Bay remained well-stratified during summer periods, mostly a result of salinity and temperature conditions, and mixing with bottom water was often limited during these periods. If surface low dissolved oxygen events are "biologically" significant, their impacts on the benthic community may be minimized by water column stratification. We have also shown that surface waters from greater Narragansett Bay can, at times, control the concentration of dissolved oxygen in Mount Hope Bay waters. This reflects, in part, the importance of exchange between Mount Hope Bay and Narragansett Bay both in magnitude (volume), and over relatively rapid (on the order of days) time scales of mixing. The critical period for low dissolved oxygen concentrations appears to be during the late summer months, when freshwater input is low and bay water is at its warmest.

Mount Hope Bay is relatively turbid, and potentially light-limited, based on limited PAR data. Details about suspended particle concentration and composition are still needed in order to evaluate the roles of phytoplankton productivity and the input of terrigenous material on water column light quality. Because light requirements for typical SAV survivorship appear to be seldom met at depths below 1 m in the water column, seagrass beds and other SAVs are probably not appropriate biological targets for habitat restoration in Mount Hope Bay. Biological criteria developed for ecosystem-based management and/or TMDLs need to be sensitive to the recent values placed on the nursery and spawning habitats of the Mount Hope Bay ecosystem, especially those related to anadromous fish habitats and bottom-dwelling finfish populations.

Mount Hope Bay is considered to be experiencing over-enrichment and appears to mirror the symptoms of eutrophication—loss of critical habitats, hypoxic events, abundance of *Ulva* sp., etc.,—in adjacent Narragansett Bay. Wastewater discharges from Massachusetts could account for as much 70% of the nitrogen flux to the bay. However, consideration should also be given to the role of exchange of material between Narragansett Bay and Mount Hope Bay as the extent of this exchange, and its significance on water quality, remains unknown. Improved modeling of inter-bay circulation, as noted by Chen *et al.* in Chapter 9, should be considered a priority research area.

Atmospheric deposition of N to Mount Hope Bay, assuming N depositional rates recently reported for the Northeastern US (Howarth *et al.*, 2002; Howarth in Chapter 3 of this volume) and N attenuation of about 80% in the Mount Hope Bay watershed, is also thought to be significant, being on the scale of loading from the two largest WWTFs discharging to Mount Hope Bay, but further research effort is needed here as well. Further evaluation of N loading from the Brayton Point power station, in the form of atmospheric emissions, could elevate the importance of atmospheric N deposition to the overall loading of N to Mount Hope Bay.

Bathymetrically, Mount Hope Bay consists of two regions: the shallow flat region of the central and northwestern and the deeper waters of the main shipping channel along its eastern boundary. Hydrodynamic models point to reduced circulation in the shallow portion of the bay (Spaulding and White, 1990; Spaulding *et al.*, 1999; Swanson *et al.*, 2006) though these models are limited when evaluating biologically important differences in these areas with respect to water circulation, water quality, water residence times, and contaminant loading. The biological/ecological component noted by Chen *et al.* in Chapter 9 should be further pursued as its application could greatly expand our understanding of ecosystem function and response in both Mount Hope and Narragansett Bay. The shallow region of the bay may contain important areas of productivity (indicated by higher values of Chl *a*) and potentially an important source of organic matter produced *in situ* for exportation to greater Narragansett Bay. Numerical models with the potential for sub-system analyses and incorporation of ecological parameters will highlight the importance of smaller scale processes such as eddy-mixing and local wind forcing. These processes may be biologically significant and key to understanding the ecological function and interaction between Mount Hope Bay and Narragansett Bay ecosystems.

Changing land use in the watershed of the Mount Hope Bay ecosystem demonstrates certain human use trends that may have important ramifications on future water quality and, subsequently, the ecological integrity of Mount Hope Bay. Land uses that experienced the greatest gains since 1985—residential and urban—reflect increases in human population to the area and pose concerns about the potential for increases in impervious cover and greater volumes of wastewater, and hence N delivery to the system. The loss of river herring from tributaries feeding into Mount Hope Bay has been linked to excessive anthropogenic pollution since the beginning of the 20th century, most notably from wastewater pollution. However, changes in land use within the Mount Hope Bay watershed that accompanies continued development contribute important factors, such as increases in the percentage of impervious cover, that affect the efficiency of N transport to bay waters. With respect to eutrophication of Mount Hope Bay, therefore, an ecosystem-based management strategy should consider improvements to wastewater treatment foremost, but should also consider concurrent efforts to better manage stormwater and on-site wastewater systems because of the impending rapid growth and development projected for the watershed (SRPEDD, 2005).

Clearly, for effective management that seeks to restore the ecological integrity of Mount Hope Bay, a departure from traditional management practices at the state-level is required. New mechanisms that allow for inter-state dialog, cooperation, and jurisdiction have yet to be defined. State-level management of the Mount Hope Bay ecosystem and greater Narragansett Bay is often hindered by jurisdictional boundaries, often times reflected in the scope of efforts to monitor and manage the environmental impacts on the ecosystem. The drafting of the 2004 NPDES permit for Brayton Point electric power station is one recent example where agencies from Rhode Island and Massachusetts, as well as federal agencies, were able to transgress these boundaries and work collectively to improve on the management of the bay ecosystem.

The management of ecosystems that extend among multiple jurisdictional boundaries is challenging. For some of these systems, regional entities have formed that, while limited with respect to the regulatory aspects of environmental management, are developing frameworks that incorporate the principles of ecosystem-based management. Examples within the US include the Gulf of Maine Council on the Marine Environment—which includes three US States as well as two Canadian Provinces—and the Council of Great Lakes Governors, which has helped to develop a regional information exchange network, providing not only environmental data but also information on economy, tourism, and education. While Mount Hope Bay is of much smaller scale than these larger regions, environmental management of the ecosystem may be better served through an environmental management council that allows for meaningful input from Massachusetts, Rhode Island, and regional and national organizations. An ecosystem-based management framework can then be applied to the greater ecosystem, capturing those management issues that often elude the more traditional state-level approach to management of coastal surface waters.

Acknowledgements The authors are grateful for the statistical assistance provided by Siobhan McGurk and comments from Dr. Todd Callaghan at MA Office of Coastal Zone Management, as well as those of an anonymous reviewer.

References

Alexander, R.B., Johnes, P.J., Boyer, E.W., and Smith, R.A. 2002. A comparison of models for estimating the riverine export of nitrogen from large watersheds. *Biogeochemistry* 57/58:295–339.

Anderson, J.R., Hardy, E.E., Roach, J.T., and Witmer, R.E. 1976. A land use and land cover classification system for use with remote sensor data. Washington, DC: United States Geological Survey.

Baden, S.P., Loo, L.O., Pihl, L., and Roseberg, R. 1990. Effects of eutrophication on benthic communities including fish: Swedish west coast. *Ambio* 19:113–122.

Batiuk, R.A., Bergstrom, P., Kemp, M., Koch, E., Murray, L., Stevenson, J.C., Bartleson, R., Carter, V., Rybicki, N.B., Landwehr, J.M., Gallegos, C., Karrh, L., Naylor, M., Wilcox, D., Moore, K.A., Ailstock, S., and Teichberg, M. 2000. Chesapeake Bay Submerged Aquatic Vegetation Water Quality and Habitat-based Requirements and Restoration Targets: A Second Technical Synthesis. Chesapeake Bay Program, US EPA, Washington, DC. 231 pp.

Belding, D.L. 1921. A report upon the quahog and oyster fisheries of Massachusetts. Commonwealth of Massachusetts, Department of Conservation, Division of Marine Fish and Game. 135 pp.

Biber, P.D., Paerl, H.W., Gallegos, C.L., and Kenworthy, W.J. 2005. Evaluating indicators of seagrass stress to light. *In* Estuarine Indicators, pp. 193–209. Borton, S.A. (ed.) Boca Raton, FL: CRC Press.

Bowen, J.L., and Valiela, I. 2001. The ecological effects of urbanization of coastal watersheds: historical increases in nitrogen loads and eutrophication of Waquoit Bay estuaries. *Canadian Journal of Fisheries and Aquatic Sciences* 58(6):1489–1500.

Bricker, S.B., Clement, C.G., Pirhalla, D.E., Orland, S.P., and Farrow, D.G.G. 1999. National Estuarine Eutrophication Assessment: A Summary of Conditions, Historical Trends, and Future Outlook. Silver Springs, MD: National Ocean Service, National Oceanic and Atmospheric Administration.

Bricker, S., Lipton, D., Mason, A., Dionne, M., Keeley, D., Krahforst, C., Latimer, J., and Pennock, J. 2006. Improving methods and indicators for evaluating coastal water eutrophication: A pilot study in the Gulf of Maine. Final Report to the Cooperative Institute for Coastal and Estuarine Environmental Technology (CICEET). University New Hampshire, Durham, NH. 88 pp.

Broecker, W.S., and Peng, T.H. 1982. Tracers in the Sea. Eldigio Press Lamont Doherty Geological Observatory, 1982, 690 pp.

Bugbee, B. 1994. Effects of radiation quality, intensity, and duration on photosynthesis and growth. International Lighting in Controlled Environments Workshop, T.W. Tibbitts (ed.) NASA-CP-95-3309. *http://ncr101.montana.edu/Light1994Conf/index.htm.*

Chinman, R.A., and Nixon, S.W. 1985. Depth–area–Volume Relationships in Narragansett Bay. NOAA/Sea Grant Marine Technical Report 87. Graduate School of Oceanography, University of Rhode Island, Narragansett, RI.

CoastalVision. 2004. Water quality monitoring in Mt. Hope Bay, Massachusetts and Rhode Island, 2003; Supplement to the habitat suitability assessment in Mt. Hope Bay. Newport, RI. 36 pp., plus appendices.

Cressie, N. 1991. Statistics for Spatial Data. New York: John Wiley and Sons, Inc. 900 pp.

Curley, J.R., Lawton, R.P., Chadwick, D.L., Reback, K., and Hickey, J.M. 1974. Study of the marine resources of the Taunton River and Mount Hope Bay. Massachusetts Division of Marine Fisheries Monograph Series No. 15. 37 pp. OCEANS '88. 'A Partnership of Marine Interests'. Proceedings:1173–1177.

Dennison, W., Orth, R., Moore, K., Stevenson, J., Carter, V., Kollar, S., Bergstrom, P., and Batiuk, R. 1993. Assessing water quality with submersed aquatic vegetation. *Bioscience* 43:86–94.

Diaz, R.J., and Rosenberg, R. 1995. Marine benthic hypoxia: a review of its ecological effects and the behavioral responses of benthic macrofauna. *Oceanography and Marine Biology Annual Review* 33:245–303.

Fan, Y., and Brown, W. 2006. On the heat budget for Mt. Hope Bay. *Northeastern Naturalist* 13(Special Issue 4):47–70.

Gibson, M.R. 1996. Comparison of trends in the finfish assemblage of Mt. Hope Bay and Narragansett Bay in relation to operations at the New England Power Brayton Point Station. A report to the Brayton Point Technical Advisory Committee. Rhode Island Division Fish and Wildlife Research Reference Document 95/1. RIDFW, Wickford, RI.

Gibson, M.R. 2002. Winter Flounder Abundance near Brayton Point Station, Mt. Hope Bay Revisited: Separating Local from Regional Impacts using Long Term Abundance Data. Rhode Island Division Fish and Wildlife Research Reference Document 02/1. RIDFW, Wickford, RI. 27 pp.

Guffy, J.D., Spears, M.A., and Biggs, D.C. 1988. Automated analyses of nutrients in seawater with the Technicon TRAACS-800 Autoanalyzer System.

Hedges, J.I., and J.H. Stern. 1984. Carbon and Nitrogen Determinations of Carbonate-Containing Solids. *Limnology and Oceanography*, 29, 657–663.

Howarth, R., Anderson, D., Cloern, J., Elfring, C., Hopkinson, C., Lapointe, B., Malone, T., Marcus, N., McGlathery, K., Sharpley, A., and Walker, D. 2000 Nutrient pollution of coastal rivers, bays, and seas. *Issues in Ecology* 7:1–15.

Howarth, R., Walker, D., and Sharpley, A. 2002. Sources of nitrogen pollution to coastal waters of the United States. *Estuaries* 25:656–676.

Howes, B., and Sundermeyer, M. 2003. Chapter 3. Habitats and habitat quality. *In* Framework for Formulating the Mt. Hope Bay Natural Laboratory: A synthesis and Summary, pp. 57–97. Roundtree, R.A., Borkman, D., Brown, W., Fan, Y., Goodman, L., Howes, B., Rothschild, B., Sundermeyer, M., and Turner, J. School for Marine Science and Technology Technical Report No. SMAST-03-0501. University of Massachusetts, Dartmouth, MA. 306 pp.

Jaworski, N.A., Groffman, P.M., Keller, A.A., and Prager, J.C. 1992. A watershed nitrogen and phosphorus balance: the upper Potomac River basin. *Estuaries* 15:83–95.

Johansson, B. 1997. Behavioral response to gradually declining oxygen concentration by Baltic Sea macrobenthic crustaceans. *Marine Biology* 129(1):71–78.

Kauffman, J.T., and Adams, E.E. 1981. Coupled near- and far-field thermal plume analysis using finite element techniques. Massachusetts Institute of Technology. Energy Laboratory Report: MIT-EL: 81-036. Cambridge, MA.

Kemp, M., Batiuk, R., Bartleson, R., Bergstrom, P., Carter, V., Gallegos, C.L., Hunley, W., Karrh, L., Koch, E.W., Landwehr, J.M., Moore, K.A., Murray, L., Naylor, M., Rybicki, N.B., Stevenson, J.C., and Wilcox, J.C. 2004. Habitat requirements for submerged aquatic vegetation in Chesapeake Bay: water quality, light regime, and physical--chemical factors. *Estuaries* 27:63–377.

MacConnell, W.P. 1973. Massachusetts Map Down: Land-use and Vegetation Cover Mapping Classification Manual. Cooperative Extension Service. Amherst, MA: University of Massachusetts.

Marine Research, Inc. (MRI). 1983. Final Report: Taunton River Estuary Study, May 1982–1983. Vol. 1&2. Falmouth.

Massachusetts Department of Environmental Protection (MA DEP). (2006) Watershed analyst tools coefficients. *http://mass.gov/mgis/wa-coeff.pdf*.

Massachusetts Executive Office of Environmental Affairs (MA EOEA). 2003. Community preservation initiative. Regional buildout analysis. *(http://commpres.env.state.ma.us/content/buildout.asp)*.

Massachusetts Office of Geographic and Environmental Information (MassGIS). (2006) Geographic Information System Database. *(http://mass.gov/mgis/database.htm)*.

McLeod, K.L., Lubchenco, J., Palumbi, S.R., and Rosenberg, A.A. 2005. Scientific consensus statement on marine ecosystem-based management. Signatures from 219 academic scientists and policy experts. Communication Partnership for Science and the Sea. *(http://compassonline.org/?q EBM.)*.

Moore, K.A., and Wetzel, R.L. 2000. Seasonal variations in eelgrass (*Zostera marina* L.) responses to nutrient enrichment and reduced light availability in experimental ecosystems. *Journal of Experimental Biology and Ecology* 244:1–28.

Mustard, J.F., Carney, M., and Sen, A. 1999. The use of satellite data to quantify thermal effluent impacts. *Estuarine, Coastal, and Shelf Science* 49:509–524.

National Research Council (NRC). 2000. Clean Coastal Waters: Understanding and Reducing the Effects of Nutrient Pollution. Washington, DC: National Academies Press. 405 pp.

Nixon, S.W., Buckley, B., Granger, S., Harris, L., Oczkowski, A., Cole, L., and Fulweiler, R. 2005. Anthropogenic nutrient inputs to Narragansett Bay: A twenty five year perspective. A Report to the Narragansett Bay Commission and Rhode Island Sea Grant. Rhode Island Sea Grant, Narragansett, RI. *(www.seagrant.gso.uri.edu/research/bay_commission_report.pdf.)*.

Pilson, M.E.Q. 1985. On the residence time of water in Narragansett Bay. *Estuaries* 8:2–14.

Pilson, M.E.Q., and Hunt, C.D. 1989. Water quality survey of Narragansett Bay, A summary of results from the SINBADD cruises: 1985–86. Marine Ecosystem Research Laboratory, Graduate School of Oceanography, Narragansett, RI: University of Rhode Island.

Prell, W., Saarman, E., Murray, D., and Deacutis, C. 2004. Summer-Season, Nighttime Surveys of Dissolved Oxygen in Upper Narragansett Bay (1999–2003). *(http:// www.geo.brown.edu/georesearch/insomniacs)*.

Rhode Island Coastal Resource Management Council. 2001. 1848–1994 Historical Eelgrass Beds of Narragansett Bay. Environmental Data Center, University of Rhode Island. *(www.edc.uri.edu/eelgrass)*.

Rhode Island Department of Environmental Protection (RIDEM). 2000. Water Quality Regulations, Rule 8. Providence, RI. EVM 112-88.97-1. pp. 10–19.

Save the Bay. 1998. The Good, The Bad, The Ugly: A Special Report 1996–1997 Narragansett Bay Watershed Wastewater Treatment Plants Performance Survey published by Save the Bay, Providence, RI. 49 pp.

Schlesinger, W.H. 1997. Biogeochemistry: An Analysis of Global Change. San Diego, CA: Academic Press.

Short, F.T., and Wyllie-Echeverria, W. 1996. Natural and human-induced disturbance of seagrasses. *Environmental Conservation* 23:17–27.

Southeastern Regional Planning amp; Economic Development District (SRPEDD). 2005. Toward a More Competitive Southeastern Massachusetts. Comprehensive Economic Development Strategy (CEDS). 88 Broadway, Taunton, MA. 81 pp.

Spaulding, M.L., and White, F.M. 1990. Circulation dynamics in Mt. Hope Bay and the lower Taunton River, Narragansett Bay. *In* Coastal and Estuarine Studies, Vol. 38. Cheng, R.T. (ed.) New York: Springer-Verlag.

Spaulding, M.L., Mendelsohn, D., and Swanson, J.C. 1999. WQMAP: an integrated three-dimensional hydrodynamic and water quality model system for estuarine and coastal applications. *Marine Technology Society Journal Special Issue on State of the Art in Ocean and Coastal Modeling* 33(3):38–54.

Strickland, J.H., and Parsons, T.R. 1972. A Practical Handbook of Seawater Analysis. Bulletin 167 (2nd ed.) *Journal of Fisheries Research Board of Canada*. Ottawa, Ontario, CA. 310 pp.

Swanson, C., Kim, H.S., and Sankaranarayanan, S. 2006. Modeling of temperature distributions in Mount Hope Bay due to thermal discharges from the Brayton Point Station. *Northeast Naturalist* 13(Special Issue 4):145–172.

US Census Bureau. Census 2000. *(http://www.census.gov/main/www/cen2000.html)*.

US Environmental Protection Agency (US EPA). 2002. Brayton Point Station: Final NPDES Permit Fact Sheet. *(www.epa.gov/NE/braytonpoint/pdfs/finalpermit/ braytonpointfactsht2003.pdf)*.

US Environmental Protection Agency (US EPA). 2003. Brayton Point Station: Final NPDES Permit. *(http://www.epa.gov/NE/braytonpoint/index.html)*.

Villalard-Bohnsack, M., Peckol, P., and Harlin, M.M. 1988. Marine macroalgae of Narragansett Bay and adjacent sounds. *In* Freshwater and Marine Plants of Rhode Island, pp. 101–118. Sheath, R.G., and Harlin, M.M. (eds.) Dubuque, IA: Kendall/ Hunt Publishing Co.

Chapter 14
Natural Viral Communities in the Narragansett Bay Ecosystem

Marcia F. Marston

14.1 Introduction

Viruses in marine environments are extremely abundant and diverse. In productive coastal waters, 10^7 to 10^8 viral particles per milliliter are commonly observed (Fuhrman, 1999; Wommack and Colwell, 2000; Weinbauer and Rassoulzadegan, 2004). In a metagenomic study of near-shore marine waters, an estimated 5,000 genotypes of viruses were detected in 200 l of seawater (Edwards and Rohwer, 2005). Many viruses that infect marine heterotrophic bacteria and cyanobacteria have been isolated and characterized (Fuhrman, 1999; Suttle, 2000; Wommack and Colwell, 2000; Mann, 2003; Weinbauer, 2004). Marine viruses that infect eukaryotic phytoplankton have also been isolated (Brussaard, 2004). These include viruses that infect bloom-forming algae, such as *Phaeocystis globosa* (Brussaard *et al.,* 2004; Baudoux and Brussaard, 2005), *Aureococcus anophagefferens* (Garry *et al.,* 1998), *Heterosigma akashiwo* (Tarutani *et al.,* 2000), and *Emiliania huxleyi* (Wilson *et al.,* 2002). Recently, viruses that infect diatoms belonging to the genera *Chaetoceros* (Bettarel *et al.,* 2005; Nagasaki *et al.,* 2005) and *Rhizosolenia* (Nagasaki *et al.,* 2004) have been described. It is likely that there is at least one virus type to infect each microbial species, making the diversity of viruses as high as or higher than the diversity of microbes (Weinbauer and Rassoulzadegan, 2004).

While it has been difficult to obtain accurate *in situ* estimates of viral-mediated host mortality (Suttle, 2005), many studies have found that viruses are responsible for 5 to 70% of daily mortality of marine bacteria and phytoplankton (Wommack and Colwell, 2000; Brussaard, 2004). In some cases, these rates are comparable to mortality rates due to zooplankton grazing (Suttle, 2005). Many viruses are host-specific, and therefore may affect the abundance, composition, and diversity of marine microbial species. A number of studies have shown that adding viruses to natural bacterial or phytoplankton communities can result in changes in the community compositions (Weinbauer and

Marcia F. Marston
Department of Biology, Roger Williams University, Bristol, RI 02809
mmarston@rwu.edu

A. Desbonnet, B. A. Costa-Pierce (eds.), *Science for Ecosystem-based Management.*
© Springer 2008

Rassoulzadegan, 2004). In addition, viral-mediated lysis (death) of bacteria and phytoplankton can influence nutrient cycling in marine environments by increasing the pool of dissolved and particulate organic matter and decreasing the nutrients and carbon available to higher trophic levels (Fuhrman, 1999; Wilhelm and Suttle, 1999; Wommack and Colwell, 2000; Weinbauer, 2004; Suttle, 2005). It has been estimated that viral lysis of algal cells may recycle 6 to 26% of photosynthetically fixed organic carbon back to DOM (Wilhelm and Suttle, 1999).

Although viruses are one of the most diverse and dynamic biological entities in marine waters with potentially far-reaching impacts, few studies have examined the effects of natural viral communities in the Narragansett Bay ecosystem. In 1999, a research program was initiated to examine the abundance, diversity, and potential impact of viruses (i.e., cyanophages) on cyanobacteria belonging to the genus *Synechococcus* in Rhode Island waters. *Synechococcus* spp. are important contributors to primary production, are ubiquitous in coastal waters, and can be cultured in the lab, which is an important consideration in the isolation and characterization of viruses. Many isolates of *Synechococcus* spp. have been well characterized (Waterbury *et al.*, 1986; Scanlan and West, 2002), and genetic markers are available to characterize at least some types of viruses that infect *Synechococcus* (Fuller *et al.*, 1998; Zhong *et al.*, 2002). Thus, this system could serve as a model to examine the dynamics and impact of viral communities in an estuarine system. The approach used to study cyanophages in Narragansett Bay could also be applied to study viruses that affect other important species of bacteria and/or phytoplankton in this ecosystem.

14.2 Materials and Methods

Cyanophages were enumerated and isolated from surface seawater collected from three different locations in Rhode Island waters. The samples were collected at least once a month from 2001 to 2004 at the Roger Williams University dock located at the southern end of Mount Hope Bay (Mount Hope Bay site), and from June 2003 to 2004 at two additional locations: Colt State Park in Narragansett Bay (Narragansett Bay site) and Brenton Point State Park on Rhode Island Sound (Rhode Island Sound site; see map in Chapter 1 for approximate sample site locations). The abundance of viruses capable of infecting *Synechococcus* sp. strain WH7803 was estimated from seawater samples using the most-probable-number (MPN) technique as previously described by Marston and Sallee (2003). Several studies had previously shown that *Synechococcus* sp. WH7803 is particularly susceptible to a broad range of cyanophages, and this strain has been commonly used to enumerate and isolate cyanophages from natural marine communities (Waterbury and Valois, 1993; Lu *et al.*, 2001; Mann, 2003). To estimate the abundance of

Synechococcus spp. in seawater samples, 2–5 ml of seawater was filtered onto 0.22 μm black polycarbonate membrane filters, and the cells in 5 to 10 random fields were counted using epifluorescence microscopy (Waterbury *et al.*, 1986).

Cyanophages were isolated from seawater samples by diluting it to viral extinction before enrichment (i.e., extinction-dilution enrichment) and by plaque purification (Ward *et al.*, 1998; Marston and Sallee, 2003). Most cyanophages belong to one of three viral families (e.g., *Myoviridae*, *Podoviridae*, or *Siphoviridae*) (Mann, 2003), but several studies suggest that a majority of cyanophages in coastal ecosystems are myoviruses (Lu *et al.*, 2001; Zhong *et al.*, 2002; Mann, 2003). Thus, in this study, myoviruses are the focus. Once viral isolates were obtained, the polymerase chain reaction (PCR) was used to amplify regions of the viral g20 gene (capsid assembly protein gene) from myoviruses as previously described (Marston and Sallee, 2003). Cyanophage genotypes were differentiated by digesting the g20 PCR products with a series of restriction endonucleases (PCR-RFLP), and the g20 genes of cyanophages with different restriction profiles were sequenced (Marston and Sallee, 2003). To double check genotyping, viral isolates with identical PCR-RFLP profiles isolated months or even years apart were sequenced. Seven pairs of isolates were tested in this manner. In all cases but one, the sequences were identical, or differed by no more than two nucleotides from each other, suggesting that PCR-RFLP analysis can be used to distinguish cyanophage g20 genotypes.

The relative abundance of different cyanophage genotypes was estimated using extinction-dilution enrichment followed by PCR-RFLP analysis of g20 genes (Marston and Sallee, 2003). Aliquots of diluted seawater (e.g., 10^{-2}, 10^{-3}, or 10^{-4}) were incubated with *Synechococcus* sp. strain WH7803 in 48-well microtiter plates. The dilution in which approximately 50% of the wells lysed was used for further analysis. Lysate from each lysed well was analyzed separately in PCR-RFLP assays in order to identify the genotype of the virus in that well. In 2002, samples were analyzed from the Mount Hope and Narragansett Bay sites, and in 2004, samples from the Rhode Island Sound site were also included. Forty to fifty viral isolates from each site were analyzed each month.

In an effort to understand the possible impact of viruses on host populations, a continuous culture system utilizing small culture vessels (chemostats) was established in the laboratory. *Synechococcus* strain WH7803 was grown in 35 ml of artificial media (AN) (Waterbury *et al.*, 1986) supplied at a rate of 1.4 ml hr^{-1}. Four chemostats were inoculated with marine cyanomyophage S-RIM8 (an isolate from Mount Hope Bay), and an additional chemostat without phage served as the control. The chemostats were maintained for 24 weeks after the addition of phage. The concentrations of *Synechococcus* cells and phages were measured weekly using epifluorescent microscopy and plaque assays, respectively. In addition, aliquots of cells and phage lysates were periodically saved, and individual *Synechococcus* and cyanophage genotypes were isolated to determine if either the cells or phage had evolved.

14.3 Results and Discussion

Our studies have shown that the Narragansett Bay cyanophage community is genetically diverse, very dynamic, and may potentially influence the abundance and composition of the *Synechococcus* community in the ecosystem. Annual seasonal variations in the abundance of cyanophages were observed with cyanophage titers ranging from over 10^5 viruses per milliliter of seawater during summer months to less than 10^2 viruses per milliliter of seawater during winter months (Fig. 14.1). In summer, *Synechococcus* abundance ranged from 10^3 to 10^5 cells per milliliter, while in winter *Synechococcus* abundance decreased, and accurate counts were difficult to obtain, but were less than 10^2 cells per milliliter. Both the bacterial (Staroscik and Smith, 2004) and phytoplankton (Karentz and Smayda, 1998) communities in Narragansett Bay display strong seasonal variations. Thus, it is not surprising to observe annual or seasonal variability in viruses since viral abundance is usually correlated with host abundance (Wang and Chen, 2004; Waterbury and Valois, 1993). Annual cycles of *Synechococcus* and cyanophage abundance with comparable population sizes have been reported in Woods Hole Harbor (Waterbury *et al.*, 1986; Waterbury and Valois, 1993) and Baltimore's Inner Harbor (Wang and Chen, 2004).

The abundances of cyanophages in seawater collected from three different locations (i.e., Mount Hope Bay, Narragansett Bay, and Rhode Island Sound) during fall, winter, and spring months were similar; however, the peak viral

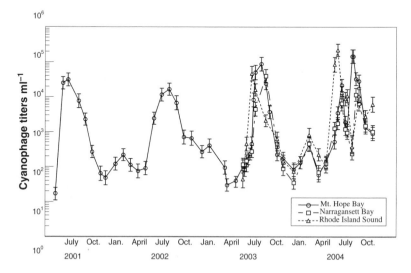

Fig. 14.1 Abundance of cyanophages in surface seawater samples collected from Mount Hope Bay, Narragansett Bay, and Rhode Island Sound. Estimates of viral titers were determined using the MPN technique with marine *Synechococcus* sp. strain WH 7803.

abundance during summer months varied by one to three weeks between sites (Fig. 14.1). In all the years of monitoring, the maxima occurred first at the Rhode Island Sound site and then in the two inner bay locations. Continued monitoring will be necessary to see if this trend persists. The effect of localized nutrient dynamics or other environmental factors on the timing of peak abundance is not known, and could not be discerned given the current analysis.

From September 1999 to August 2004, over 600 cyanophage clones were isolated and characterized from Narragansett Bay. From 1999 to 2002, 36 unique g20 cyanomyoviral genotypes were identified in seawater collected from the Mount Hope Bay site (Marston and Sallee, 2003). These viruses are phylogenetically diverse and fall into three different clades with nucleotide sequence similarities ranging from 47.3 to 98.6% (Marston and Sallee, 2003). In 2004, another 20 unique g20 viral genotypes were isolated from Rhode Island waters. Even after five years of monthly sampling, cyanophage genotypic diversity present in this ecosystem has not been completely characterized as new genotypes are frequently isolated at each new sampling date. Interestingly, many of these g20 viral genotypes are not unique to Rhode Island waters; viruses with g20 sequences identical to those found in Rhode Island have been isolated from Woods Hole Harbor in Massachusetts, in the Gulf of Mexico, and in a river estuary in Georgia (Marston and Sallee, 2003).

The diversity of cyanophages in a single water sample was much lower than the total cyanophage diversity observed over the 5-year study period. A total of 56 different cyanophage g20 genotypes have been identified in Rhode Island waters, but only four to ten different g20 genotypes were detected at any sampling date. Some viral genotypes could be detected in almost every sample, regardless of time of the year, while other genotypes appeared more sporadically. Although temporal variation was observed in both the relative abundance of specific g20 genotypes and the overall composition of cyanophage communities in 2002 (Fig. 14.2; Marston and Sallee, 2003), and again in 2004, the genotypic composition of cyanophage communities at three different sample locations was similar on any given sampling day. Preliminary data suggest that there may be cyclical, possibly seasonal patterns, in the genotypic composition of the viral community. For example, many of the viral genotypes observed during June 2002 in Mount Hope Bay and Narragansett Bay were also observed in June 2004 at these two sites, and at the Rhode Island Sound site (Fig. 14.3). More sampling will need to be conducted to determine if patterns really exist.

The g20 genotypic analysis of cyanophages was used to characterize viruses in the *Myoviridae* family. On some sampling dates, the g20 gene from up to 50% of the viral isolates could not be amplified using PCR primers specific for myoviruses. This suggests that other viral types, perhaps belonging to the *Podoviridae* or *Siphoviridae* viral families, may be an important component of the Rhode Island cyanophage community. Moreover, there appears to be temporal variation in the relative abundance of these other cyanophage types. Clearly, the viral community infecting *Synechococcus* in this ecosystem is genetically diverse and

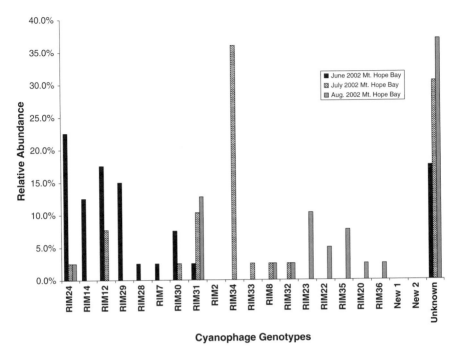

Fig. 14.2 Relative abundance of cyanophage g20 genotypes in Mount Hope Bay seawater samples collected in June, July, and August of 2002 (modified from Marston and Sallee, 2003). Sequences of "RIM" g20 genes are available from GenBank. "New 1" and "New 2" refer to unique viral genotypes isolated in 2004. Viruses were classified as "Unknown" if the g20 gene could not be amplified via PCR using the available PCR primers.

very dynamic. Although it is not known how temporal changes in viral community composition affect host populations in this ecosystem, studies elsewhere have suggested that cyanophage infection can influence seasonal changes in *Synechococcus* communities (Wang and Chen, 2004; Muhling *et al.,* 2005).

A continuous culture system in the laboratory is being used to examine the potential impacts of cyanophages on *Synechococcus* populations (Pierciey *et al.,* 2005). When a cyanophage isolated from Mount Hope Bay (S-RIM8) was added to *Synechococcus* sp. WH 7803 growing in chemostats, *Synechococcus* concentrations dropped by three orders of magnitude within the first week. However, after six to eight weeks, the *Synechococcus* populations recovered to the same levels as populations in the control chemostats with no virus. Cells isolated from cyanophage-infected chemostats were found to be resistant to the original cyanophage isolate S-RIM8. Subsequently, mutants of S-RIM8 capable of infecting the resistant hosts were isolated. Our chemostat results suggest that population studies of bacteria and viruses will need to incorporate evolutionary processes, as these may influence communities over relatively small time scales. Indeed, resistant host strains have been reported for most isolated viruses

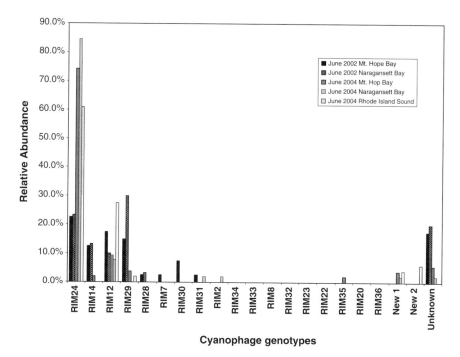

Fig. 14.3 Relative abundance of cyanophage g20 genotypes in surface seawater samples collected in June of 2002 and 2004 from Mount Hope Bay, Narragansett Bay, and Rhode Island Sound. Sequences of "RIM" g20 genes are available from GenBank. "New 1" and "New 2" refer to unique viral genotypes isolated in 2004. Viruses were classified as "Unknown" if the g20 gene could not be amplified via PCR using the available PCR primers.

of eukaryotic phytoplankton (Brussaard, 2004) and bacteria (Waterbury and Valois, 1993; Weinbauer, 2004). If viral mutants capable of overcoming resistant host genotypes are common in natural populations, the Rhode Island cyanophage community is almost certainly more diverse and complex than our analyses of g20 genotypes have revealed thus far.

While our studies have focused on viruses infecting cyanobacteria, it is likely that there are viruses infecting most, if not all, heterotrophic bacteria and eukaryotic phytoplankton that reside in Narragansett Bay. In fact, one of the first indications that viruses may affect phytoplankton in Narragansett Bay came from a study of an algal bloom. In 1985, a bloom of *A. anophagefferens* was first observed in Narragansett Bay (Sieburth *et al.,* 1988), and "brown tide" blooms of this species have been a recurring problem in the coastal bays of New Jersey and Long Island ever since. Seawater collected from Narragansett Bay during the 1985 bloom contained virus-like particles (Sieburth *et al.,* 1988), and subsequently, viruses capable of infecting *A. anophagefferens* were isolated and characterized (Milligan and Cosper, 1994; Garry *et al.,* 1998; Gastrich *et al.,* 1998; 2002). The impact of viruses on *A. anophagefferens* population dynamics in Narragansett Bay has not been further examined; however, a recent study suggests that viruses may be a significant source of mortality of natural populations of *A. anophagefferens* in the coastal bays of New York and New Jersey (Gastrich *et al.,* 2004), and thus it is possible that they have a similar effect in Narragansett Bay. Viruses are also important in regulating the blooms of other algal species, including *P. globosa* (Brussaard *et al.,* 2005), *E. huxleyi* (Castberg *et al.,* 2001), and *H. akashiwo* (Tarutani *et al.,* 2000).

By increasing the mortality rates of bacteria and phytoplankton, viruses are known to impact the cycling of organic compounds and nutrients in marine ecosystems (Fuhrman, 1999; Scanlan and Wilson, 1999; Wilhelm and Suttle, 1999). However, changes in nutrient availability can also influence the dynamics of viral–host interactions (Wommack and Colwell, 2000; Brussaard, 2004). Nutrient availability may influence *Synechococcus*-cyanophage interactions by affecting the mode of phage replication. Depending on the viral genotype, physiological conditions of the host, and environmental conditions, viral infection of *Synechococcus* can result in several possible outcomes, including lysis and lysogeny. In a lytic infection, the virus immediately replicates in the host cell and releases progeny by lysing the host cell. Most of the cyanophages isolated thus far, including the ones from Narragansett Bay, can reproduce via the lytic infection cycle (Suttle and Chan, 1993; Waterbury and Valois, 1993; Lu *et al.,* 2001; Marston and Sallee, 2003). Nevertheless, nutrient limitation may lead to lysogenic infections observed in some natural populations of *Synechococcus* and cyanophages (Scanlan and Wilson, 1999; Mcdaniel *et al.,* 2002; Mcdaniel and Paul, 2005). During lysogenic infections, viral genomes become integrated into the bacterial genome, delaying the release of progeny virus. In particular, it is thought that phosphate concentrations may be responsible for determining whether a cyanophage establishes a lytic or lysogenic infection, with lysogeny occurring in phosphate-limited conditions (Wilson *et al.,* 1996; Scanlan and Wilson, 1999). Other studies report that manipulating phosphorus concentrations in natural seawater samples resulted in significant changes in viral activity (Scanlan and Wilson, 1999).

While studies in Narragansett Bay have shown that grazing can modify phytoplankton species abundance and composition (e.g., Martin, 1970), similar studies examining the effects of viruses have not been conducted. The extent to which viruses influence the observed interannual and seasonal variations in phytoplankton and bacterial assemblages is unknown. As nutrient levels change in Narragansett Bay, the impact of these changes on the abundance and diversity of natural viral communities, the dynamics of viral–host interactions, and the levels of viral-induced mortality of bacteria, cyanobacteria, and phytoplankton should be evaluated.

Acknowledgements This work was supported by NSF Award #0314523 and the Roger Williams University research foundation. I would like to thank Jennifer Hughes for help analyzing the community data and John Pierciey, James Torbett, Lauren Stoddard, and Alicia Shepard for their help in the lab.

References

Baudoux, A.C., and Brussaard, C.P.D. 2005. Characterization of different viruses infecting the marine harmful algal bloom species *Phaeocystis globosa*. *Virology* 341:80–90.

Bettarel, Y., Kan, J., Wang, K., Williamson, K.E., Cooney, S., Ribblett, S., Chen, F., Wommack, K.E., and Coats, D.W. 2005. Isolation and preliminary characterization of a small nuclear inclusion virus infecting the diatom *Chaetoceros* cf. *gracilis*. *Aquatic Microbial Ecology* 40:103–114.

Brussaard, C.P.D. 2004. Viral control of phytoplankton populations—a review. *Journal of Eukaryotic Microbiology* 51:125–138.

Brussaard, C.P.D., Short, S.M., Frederickson, C.M., and Suttle, C.A. 2004. Isolation and phylogenetic analysis of novel viruses infecting the phytoplankton *Phaeocystis globosa* (Prymnesiophyceae). *Applied and Environmental Microbiology* 70:3700–3705.

Brussaard, C.P.D., Kuipers, B., and Veldhuis, M.J.W. 2005. A mesocosm study of *Phaeocystis globosa* population dynamics—1. Regulatory role of viruses in bloom. *Harmful Algae* 4:859–874.

Castberg, T., Larsen, A., Sandaa, R.A., Brussaard, C.P.D., Egge, J.K., Heldal, M., Thyrhaug, R., Van Hannen, E.J., and Bratbak, G. 2001. Microbial population dynamics and diversity during a bloom of the marine coccolithophorid *Emiliania huxleyi* (Haptophyta). *Marine Ecology-Progress Series* 221:39–46.

Edwards, R.A., and Rohwer, F. 2005. Viral metagenomics. *Nature Reviews Microbiology* 3:504–510.

Fuhrman, J.A. 1999. Marine viruses and their biogeochemical and ecological effects. *Nature* 399:541–548.

Fuller, N.J., Wilson, W.H., Joint, I.R., and Mann, N.H. 1998. Occurrence of a sequence in marine cyanophages similar to that of T4 g20 and its application to PCR-based detection and quantification techniques. *Applied and Environmental Microbiology* 64:2051–2060.

Garry, R.T., Hearing, P., and Cosper, E.M. 1998. Characterization of a lytic virus infectious to the bloom-forming microalga *Aureococcus anophagefferens* (Pelagophyceae). *Journal of Phycology* 34:616–621.

Gastrich, M.D., Anderson, O.R., Benmayor, S.S., and Cosper, E.M. 1998. Ultrastructural analysis of viral infection in the brown-tide alga, *Aureococcus anophagefferens* (Pelagophyceae). *Phycologia* 37:300–306.

Gastrich, M.D., Anderson, O.R., and Cosper, E.M. 2002. Viral-like particles (VLPS) in the alga, *Aureococcus anophagefferens* (Pelagophyceae), during 1999–2000 brown tide blooms in Little Egg Harbor, New Jersey. *Estuaries* 25:938–943.

Gastrich, M.D., Leigh-Bell, J.A., Gobler, C.J., Anderson, O.R., Wilhelm, S.W., and Bryan, M. 2004. Viruses as potential regulators of regional brown tide blooms caused by the alga, *Aureococcus anophagefferens*. *Estuaries* 27:112–119.

Karentz, D., and Smayda, T.J. 1998. Temporal patterns and variations in phytoplankton community organization and abundance in Narragansett Bay during 1959–1980. *Journal of Plankton Research* 20:145–168.

Lu, J., Chen, F., and Hodson, R.E. 2001. Distribution, isolation, host specificity, and diversity of cyanophages infecting marine Synechococcus spp. in river estuaries. *Applied and Environmental Microbiology* 67:3285–3290.

Mann, N.H. 2003. Phages of the marine cyanobacterial picophytoplankton. *FEMS Microbiology Reviews* 27:17–34.

Marston, M.F., and Sallee, J.L. 2003. Genetic diversity and temporal variation in the cyanophage community infecting marine *Synechococcus* species in Rhode Island's coastal waters. *Applied and Environmental Microbiology* 69:4639–4647.

Martin, J.H. 1970. Phytoplankton-zooplankton relationships in Narragansett Bay: IV. The seasonal importance of grazing. *Limnology and Oceanography* 10:185–191.

Mcdaniel, L., and Paul, J.H. 2005. Effect of nutrient addition and environmental factors on prophage induction in natural populations of marine *Synechococcus* species. *Applied and Environmental Microbiology* 71:842–850.

Mcdaniel, L., Houchin, L.A., Williamson, S.J., and Paul, J.H. 2002. Lysogeny in marine *Synechococcus*. *Nature* 415:496.

Milligan, K.L.D., and Cosper, E.M. 1994. Isolation of virus capable of lysing the brown tide microalga, *Aureococcus Anophagefferens*. *Science* 266:805–807.

Muhling, M., Fuller, N.J., Millard, A., Somerfield, P.J., Marie, D., Wilson, W.H., Scanlan, D.J., Post, A.F., Joint, I., and Mann, N.H. 2005. Genetic diversity of marine *Synechococcus* and co-occurring cyanophage communities: evidence for viral control of phytoplankton. *Environmental Microbiology* 7:499–508.

Nagasaki, K., Tomaru, Y., Katanozaka, N., Shirai, Y., Nishida, K., Itakura, S., and Yamaguchi, M. 2004. Isolation and characterization of a novel single-stranded RNA virus infecting the bloom-forming diatom *Rhizosolenia setigera*. *Applied and Environmental Microbiology* 70:704–711.

Nagasaki, K., Tomaru, Y., Takao, Y., Nishida, K., Shirai, Y., Suzuki, H., and Nagumo, T. 2005. Previously unknown virus infects marine diatom. *Applied and Environmental Microbiology* 71:3528–3535.

Pierciey, F.J., Hughes, J.B., and Marston, M.F. 2005. Coevolution of Marine *Synechococcus* and Cyanophages in Chemostats. *American Society for Microbiology General Meeting.* Abstract, Atlanta, GA.

Scanlan, D.J., and West, N.J. 2002. Molecular ecology of the marine cyanobacterial genera *Prochlorococcus* and *Synechococcus*. *Fems Microbiology Ecology* 40:1–12.

Scanlan, D.J., and Wilson, W.H. 1999. Application of molecular techniques to addressing the role of P as a key effector in marine ecosystems. *Hydrobiologia* 401:149–175.

Sieburth, J.M., Johnson, P.W., and Hargraves, P.E. 1988. Ultrastructure and ecology of *Aureococcus anophagefferens* gen. et sp. nov. (Chrysophyceae): the dominant picoplankter during a bloom in Narragansett Bay, Rhode Island, summer 1985. *Journal of Phycology* 24:416–425.

Staroscik, A.M., and Smith, D.C. 2004. Seasonal patterns in bacterioplankton abundance and production in Narragansett Bay, Rhode Island, USA. *Aquatic Microbial Ecology* 35:275–282.

Suttle, C.A. 2000. Cyanophages and their role in the ecology of cyanobacteria. *In* The ecology of cyanobacteria, pp. 563–589. Whitton, B.A., and Potts, M. (eds.) Netherlands: Kluwer Academic Publishers.

Suttle, C.A. 2005. Viruses in the sea. *Nature* 437:356–361.

Suttle, C.A., and Chan, A.M. 1993. Marine cyanophages infecting oceanic and coastal strains of *Synechococcus*: abundance, morphology, cross-infectivity and growth characteristics. *Marine Ecology Progress Series* 92:99–109.

Tarutani, K., Nagasaki, K., and Yamaguchi, M. 2000. Viral impacts on total abundance and clonal composition of the harmful bloom-forming phytoplankton *Heterosigma akashiwo*. *Applied and Environmental Microbiology* 66:4916–4920.

Wang, K., and Chen, F. 2004. Genetic diversity and population dynamics of cyanophage communities in the Chesapeake Bay. *Aquatic Microbial Ecology* 34:105–116.

Ward, D.M., Ferris, M.J., Nold, S.C., and Bateson, M.M. 1998. A natural view of microbial biodiversity within hot spring cyanobacterial mat communities. *Microbiology and Molecular Biology Reviews* 62:1353–1370.

Waterbury, J.B., and Valois, F.W. 1993. Resistance to co-occurring phages enables marine *Synechococcus* communities to coexist with cyanophages abundant in seawater. *Applied and Environmental Microbiology* 59:3393–3399.

Waterbury, J.B., Watson, S.W., Valois, F.W., and Franks, D.G. 1986. Biological and ecological characterization of the marine unicellular cyanobacterium *Synechococcus*. *Canadian Bulletin of Fisheries and Aquatic Sciences* 214:71–120.

Weinbauer, M.G. 2004. Ecology of prokaryotic viruses. *FEMS Microbiology Reviews* 28:127–181.

Weinbauer, M.G., and Rassoulzadegan, F. 2004. Are viruses driving microbial diversification and diversity? *Environmental Microbiology* 6:1–11.

Wilhelm, S.W., and Suttle, C.A. 1999. Viruses and nutrient cycles in the sea. *Bioscience* 49:781–788.

Wilson, W.H., Carr, N.G., and Mann, N.H. 1996. The effect of phosphate status on the kinetics of cyanophage infection in the oceanic cyanobacterium *Synechococcus* sp. WH7803. *Journal of Phycology* 32:506–516.

Wilson, W.H., Tarran, G.A., Schroeder, D., Cox, M., Oke, J., and Malin, G. 2002. Isolation of viruses responsible for the demise of an *Emiliania huxleyi* bloom in the English Channel. *Journal of the Marine Biological Association of the United Kingdom* 82:369–377.

Wommack, K.E., and Colwell, R.R. 2000. Virioplankton: viruses in aquatic ecosystems. *Microbiology and Molecular Biology Reviews* 64:69–114.

Zhong, Y., Chen, F., Wilhelm, S.W., Poorvin, L., and Hodson, R.E. 2002. Phylogenetic diversity of marine cyanophage isolates and natural virus communities as revealed by sequences of viral capsid assembly protein gene g20. *Applied and Environmental Microbiology* 68:1576–1584.

Chapter 15
Nutrient and Plankton Dynamics in Narragansett Bay

Theodore J. Smayda and David G. Borkman

15.1 Introduction

This chapter has been prepared with the aim of contributing toward the development of an ecosystem-based management strategy for Narragansett Bay. Of specific concern is the unresolved question: What would be the ecological impacts of a sustained increase or decrease in anthropogenic nutrient delivery to Narragansett Bay? Two recent ecological disturbances in Narragansett Bay, interpreted by some as responses to increasing nutrification, propel this concern: the apparent regional spreading of hypoxic bottom water in upper Narragansett Bay (Chapters 11 and 12; Altieri and Witman, 2006), and two extensive fish and benthic community die offs in Greenwich Bay attributed to nutrient-stimulated blooms of harmful algal species or related ecological disruptions. In the summer of 2003, a major fish kill in Greenwich Bay was followed by an equally devastating die-off of the clam *Mya arenaria* (Rhode Island Department of Environmental Management, 2003).

This mortality has become a local *cause célebre*, resulting in unusual and sustained attention by the news media, and political concern, over the health of the Narragansett Bay ecosystem. Hypoxia has been implicated as the cause of the Greenwich Bay die offs, the oxygen deficit widely believed to have been caused by the decomposition of ungrazed phytoplankton blooms stimulated by eutrophication. This eutrophication theory has been bolstered by well-documented evidence that Greenwich Bay is a relatively poorly flushed *cul de sac* vulnerable to nutrient inputs from the surrounding watershed and, perhaps, extensive marina development (Granger *et al.*, 2000; Kennedy and Lee, 2003). The Greenwich Bay disruptions have also heightened concern over the seasonal hypoxia that occurs in upper Narragansett Bay, and its causes,

Theodore J. Smayda
Graduate School of Oceanography, University of Rhode Island, Kingston, RI 02881
tsmayda@gso.uri.edu

A. Desbonnet, B. A. Costa-Pierce (eds.), *Science for Ecosystem-based Management.*
© Springer 2008

possible spreading, and potential biotic consequences. Remarkably, investigation of the phytoplankton bloom species composition, dynamics, and accompanying nutrient levels and cycles during the Greenwich Bay die offs, was not undertaken to provide the data needed to determine the cause(s) of the mortalities. As a result, and at best, the generally accepted conclusion that elevated nutrients induced the Greenwich Bay hypoxia, and not the physical hydrographic conditions or other non-nutrient-related conditions, is largely anecdotal.

The Greenwich Bay episodes have been extrapolated in a general sense to Narragansett Bay, resulting in the widely held contention that Narragansett Bay is undergoing a "creeping eutrophication" that is jeopardizing ecosystem health. This disputed claim has sparked a lively, continuing debate. However, the debate over the ecosystem health of Greenwich Bay specifically, and Narragansett Bay generally, suffers from two knowledge gaps beyond the failure to carry out field studies during the 2003 mortalities: (1) uncertainty over historical patterns, trends, cycles, and interannual and bay-wide variability in plankton, nutrients, and other habitat conditions, and (2) the unresolved impacts of the confluence of climate and anthropogenically altered ecology in driving and altering Narragansett Bay ecosystem behavior.

Mean annual sea surface and winter temperatures in Narragansett Bay have progressively increased and altered plankton dynamics since the early 1960s (Borkman, 2002; Smayda *et al.*, 2004). This climatic effect blurs the distinction between plankton behavior that may have been altered because of climate warming versus changes induced by anthropogenically altered nutrient concentrations. Long-term changes in grazer communities, both fish and benthic invertebrates, have also occurred in Narragansett Bay (Jeffries and Terceiro, 1985; Oviatt *et al.*, 2003), which further confound prediction of the impacts that either an increase or decrease in nutrients would have on the Narragansett Bay ecosystem.

As a step toward resolution of these uncertainties, and to facilitate development of a nutrient-based ecosystem management strategy for Narragansett Bay, this chapter describes the surface distribution or the integrated water column concentrations of nutrients, phytoplankton chlorophyll, primary production, and zooplankton and jellyfish (ctenophore) abundance at seven stations located along the salinity gradient in Narragansett Bay extending from the Providence River estuary to lower Narragansett Bay near the entrance into the West Passage (Fig. 15.1). The study extended over a 2-year period, from late July 1985 to late June 1987, with the 7-station transects sampled at a very high frequency, corresponding to an average sampling interval of 12 days between survey cruises. This study was triggered by a prodigious and prolonged (5 months) brown tide event during 1985 by a species and genus previously unknown to science, the pelagophyte *Aureococcus anophagefferens*. This bloom caused massive mortality of bivalves and macroalgae, impaired zooplankton growth and spawning of bay anchovy, among other negative consequences (Durbin and Durbin,

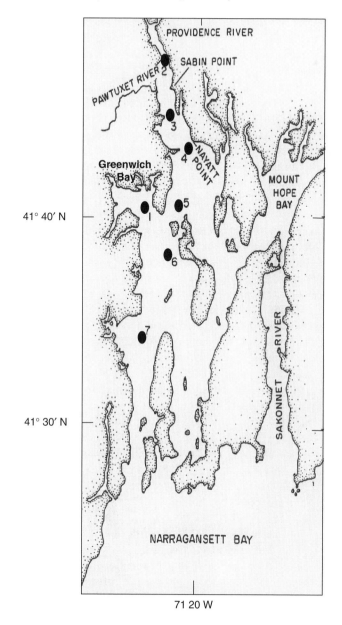

Fig. 15.1 Station locations sampled during the surveys.

1989; Sieburth *et al.,* 1988; Smayda and Fofonoff, 1989; Smayda and Villareal, 1989a,b; Tracey *et al.,* 1988). This novel bloom developed synchronously in Narragansett Bay, in Long Island coastal embayments and in Barnegat Bay (Cosper *et al.,* 1989). This indicates that Narragansett Bay is responsive to regional scale events in addition to local, watershed-driven and *in situ* processes. This regional aspect cannot be neglected in developing an ecosystem management strategy for Narragansett Bay.

The trophic effects of the brown tide dissipated during 1986 and through mid-1987 when the surveys ended. The plankton cycles that developed post-brown tide conformed to the behavioral patterns and variations described both previously and subsequently for Narragansett Bay (Pratt, 1959; 1965; Smayda, unpublished). Thus, the ecological patterns described in this chapter include both anomalous and "normal" bloom-year behavior, which provides insight into the range and variability of the interlinked processes of nutrients–phytoplankton–zooplankton in Narragansett Bay. This chapter will not focus on the brown tide and its consequences since these aspects have been treated elsewhere, as referenced earlier. The physical, chemical, and plankton measurements made during the bay-wide surveys are emphasized since these data have not previously been evaluated, but are highly relevant to the development of an ecosystem management strategy for Narragansett Bay. For this reason, the average physical, chemical, and plankton conditions along the axis of Narragansett Bay will be the focus rather than detailed examination of the individual transects comprising the surveys. The average bay-wide conditions are the important element in developing an ecosystem-based management strategy, not the station-to-station or week-to-week variations in habitat conditions and plankton behavior. The data are presented as average responses during each of the two continuous survey years, 1985–1986 and 1986–1987, with each survey year commencing in July and extending to the following June. This manipulation is made to facilitate identification of average bay-wide habitat conditions and plankton ecology, and to achieve insight into the interannual variations in these properties.

The material presented in this chapter will not resolve the specific question of the most probable, quantitative impacts that either a reduction or increase in nutrient loading into Narragansett Bay would have on plankton dynamics and overall ecosystem behavior. However, it provides perspectives into the type of quantification, nutrient delivery manipulations, and experimentation required to develop an ecosystem-based management protocol to achieve desired goals. A further caveat: this chapter follows a mass balance approach—plankton biomass, primary production, and nutrient levels are considered without regard to species composition, functional groups, or their temporal and spatial dynamics; space limitations preclude description and analysis of the seasonal nutrient and plankton cycles. Yet, it is increasingly evident that while nutrient and biomass concentrations are fundamental to ecosystem functioning, the species making up different trophic groups are equally important in their influence on the flow of energy, nutrient recycling, growth and fecundity of the grazers, and whether the blooms will be harmful or ecologically advantageous (Borkman and Smayda, in review). This aspect should also not be ignored in developing an ecosystem management strategy.

15.2 Methods

Water samples were collected irrespective of tidal phase from three depths (top = surface, mid and bottom) at seven stations located along a 30-km transect in Narragansett Bay extending from the Providence River estuary to Fox Island in lower Narragansett Bay (Fig. 15.1). In all, 62 transects were made: 34 transects from 25 July 1985 to 18 June 1986, and 28 transects from 2 July 1986 to 29 June 1987, with a sampling interval averaging 12 days. Water column depth at each station is given in Table 15.1. This chapter summarizes measurements made on temperature, salinity, Secchi Disc depth, NH_4, NO_3, PO_4, $Si(OH)_4$, chlorophyll, primary production, zooplankton biomass, and ctenophore numerical abundance, with correlation analyses of copepod and benthic larvae numerical abundance as well.

Samples were collected with PVC Niskin® bottles and transported under ice to the laboratory for chemical and biological analyses. The concentrations of NH_4, NO_3, PO_4, and $Si(OH)_4$ were measured, in duplicate, using standardized methods for macronutrients (Strickland and Parsons, 1972) and carried out on a Technicon Autoanalyzer® using slight modifications of these methods. The analytical manifolds and reagents for ammonia, nitrate + nitrite and silicate analysis were as given by Friederich and Whitledge (1972), and the phosphate manifold and reagents as given by Grasshof (1966). Chlorophyll a was measured using the fluorescent technique (Yentsch and Menzel, 1963) emended by Lorenzen (1966). Productivity measurements were made using the ^{14}C method

Table 15.1 Mean water column depth (m), surface and bottom water temperature and salinity, Secchi Disc depth (m), extinction coefficient (k, m^{-1}) and euphotic zone (= 1% Isolume) depth (m) at the seven transect stations during the1985–1986 and 1986–1987 surveys.

Station	1	2	3	4	5	6	7
Depth (m)	8.1	12.2	13.4	13.5	10.8	6.8	7.0
1985–1986							
Surface °C	12.4	12.8	13.0	12.2	12.7	12.2	12.0
Bottom °C	11.8	11.9	12.1	11.6	12.7	11.9	11.7
Surface Salinity	28.5	20.7	23.7	26.6	28.6	29.8	30.3
Bottom Salinity	29.5	29.9	30.2	30.5	29.7	30.2	30.7
Secchi Depth (m)	1.90	1.71	1.80	1.88	2.12	2.20	2.42
k (m^{-1})	0.76	0.84	0.80	0.77	0.68	0.65	0.60
1% Isolume Depth (m)	6.1	5.5	5.8	6.0	6.8	7.1	7.7
1986–1987							
Surface °C	11.8	12.6	12.6	11.9	12.8	12.9	12.4
Bottom °C	11.5	12.5	12.0	11.9	12.9	12.8	11.5
Surface Salinity	28.1	21.2	23.1	25.3	28.0	29.0	29.4
Bottom Salinity	29.1	29.7	29.7	29.9	29.2	29.5	30.1
Secchi Depth (m)	2.35	2.06	2.10	2.24	2.44	2.55	2.73
k (m^{-1})	0.61	0.70	0.69	0.64	0.59	0.56	0.53
1% Isolume Depth (m)	7.6	6.6	6.7	7.2	7.8	8.2	8.7

(Steemann Nielsen, 1952). A pooled sample for each station containing equal proportions from top, mid-, and bottom-depth samples was incubated in 50 mI glass bottles, inoculated with $2\mu Ci$ $H^{14}CO_3$ and incubated under ambient temperature and light conditions in an outdoor flow-through incubator located on the laboratory dock located on the University of Rhode Island Graduate School of Oceanography campus, and through which Narragansett Bay water flowed. Flushing time of the incubator was 30 min. The productivity samples, in duplicate, were exposed to 100%, 60%, 25%, 10%, and 3% natural irradiance for 24 h, and ^{14}C uptake measured in a scintillation counter. Light intensity was monitored continuously during the incubation by an Eppley pyrheliometer located near the incubation platform.

Non-gelatinous zooplankton were sampled in two ways: with a 153-μm mesh net fitted with a TSK flow meter, and a 64-μm mesh net fitted with a General Oceanics flow meter, each net having a mouth diameter of 0.305 m. At Station 7, a 20-μm net was towed in place of the 64 μm mesh net to collect phytoplankton species <64 μm. A double oblique tow was made, during which the net was slowly lowered to within 1 m of the bottom, and then raised at a towing speed of 1–2 knots. Each tow filtered from 1 to 4 m^3 of water. In the laboratory, zooplankton collected in the 153-μm net tows were split into two halves using a sediment splitter. Half of the sample was sieved, rinsed with deionized water, and dried for 4 weeks at 60°C in aluminum weighing pans for determination of dry weight using a Mettler H-16 balance. The other half of the sample was preserved in 5% formalin for estimation of the numerical abundance of copepodite and adult copepod stages and decapod larvae. The 64-μm net samples were preserved without splitting to estimate numbers of copepod nauplii and smaller meroplankton. Ctenophores and large medusae were sampled with a 1-m^2 net with 1 mm mesh that was lowered to within 1 m of the bottom and then hauled vertically with the ship at rest. Replicate tows ($n = 2$) were made. The ctenophores were removed from the net by spoon, sorted and counted by size classes (>1 cm, 1–2 cm, 2–4 cm, and <4 cm) using a gridded dish.

15.2.1 Statistical Treatment and Curve Fitting

The relationships between nutrient or plankton variables and salinity along the estuarine gradient and the relationships between nutrient and plankton variables were quantified using regression analysis applied to mean values. Three regression models—linear, exponential, and quadratic (second-order polynomial)—were evaluated using SAS (Version 9.1) software. For each regression, the model having the best fit was chosen. Only regression results (regression equation, r^2, and plotted regression line) explaining significant portions of the variance ($p \leq 0.05$) are presented here.

15.3 Results

15.3.1 Temperature

Mean surface and bottom water temperature, salinity, and water column transparency at the seven stations for the 62 transects are given in Table 15.1. The 1986–1987 winter was relatively mild, with minimum temperature approximately 1°C occurring briefly in February. During winter 1985–1986, in January, temperature decreased to −0.5°C and moderate icing occurred. Summer surface temperatures during both the survey years were 24–25°C. The annual temperature range, seasonal pattern, and weekly levels at the transect stations were similar, unlike salinity.

15.3.2 Salinity, Mixing Characteristics, Circulation

A pronounced, year-round gradient in surface salinity occurs in Narragansett Bay. The difference in mean annual salinity between end-member Stations 2 and 7 (Fig. 15.1) was about 10‰ and 8‰ during the 1985–1986 and 1986–1987 surveys, respectively. In absolute terms, mean surface salinity during the 1985–1986 survey period progressively increased down bay, from 20.7‰ in the Providence River region near Field's Point (Station 2) to 30.3‰ in lower Narragansett Bay near Fox Island (Station 7). The corresponding 1986–1987 salinity was 21.2‰ and 29.4‰, respectively. Since marine phytoplankton generally are euryhaline, salinity-induced changes in their cellular rates of photosynthesis, nutrient uptake, and growth along the osmotic gradient in Narragansett Bay are not expected to be important. The differences in population dynamics along the salinity gradient are expected to be related more to associated water mass optical properties, nutrient concentrations, and grazer abundance than to cellular stress and morbidity because of lowered salinity.

Despite the pattern in the salt gradient, considerable week-to-week variations in salinity occurred at all stations, particularly Stations 2, 3, and 4 in upper Narragansett Bay which are directly exposed to river runoff (Table 15.1). The difference between minimum and maximum salinity recorded at these stations during the 1986–1987 surveys was 19.7‰ at Station 2; 16.6‰ at Station 3, and 18.0‰ at Station 4. River runoff has a reduced influence down bay from this region, as reflected in the narrower range in week-to-week salinity fluctuations recorded at Stations 5, 6, and 7 (Table 15.2). Nonetheless, riverine dilution periodically can have a significant down bay impact, as revealed by the salinity minimum of 23.9‰ recorded at Station 7 during the 1986–1987 survey, which equaled the survey mean for Station 3 (Table 15.1). The converse is also true— reduced river flow into the upper Narragansett Bay region of Stations 2, 3, and 4, in combination with other physical processes, can elevate surface salinity levels

Table 15.2 Maximum and minimum surface salinity (%) recorded at the survey stations.

Station	1985–1986			1986–1987		
	Minimum	Maximum	ΔSalinity	Minimum	Maximum	ΔSalinity
1	17.1	31.7	14.6	20.2	30.1	9.9
2	10.9	28.6	17.7	6.8	26.5	19.7
3	11.8	28.6	16.8	11.4	28.0	16.6
4	16.5	30.7	14.2	11.6	29.6	18.0
5	24.4	30.7	6.3	23.3	31.7	8.4
6	27.5	31.7	4.2	25.4	30.7	5.3
7	28.6	31.7	3.1	23.9	31.7	7.8

that approach the mean levels at down bay Stations 6 and 7, i.e., >28.0‰. Clearly, the salinity gradient in Narragansett Bay, and salinity levels within a given region along the gradient, are dynamic features rather than invariant, i.e., fixed into a specific pattern, or remaining at a constant degree of dilution.

A well-defined vertical salinity gradient accompanies the surface salinity gradient, particularly in upper Narragansett Bay (Stations 2, 3, and 4) where the mean bottom water salinity was 4.0–9.0‰ higher than at the surface (Table 15.1). The regional variation in the vertical salinity gradient influences the seasonal and regional water mass stratification patterns and the intensity of vertical mixing along the Narragansett Bay axis. This physical dynamic, through its influence on the light-photosynthesis relationship, is an important regulator of phytoplankton growth. Calculations of water column density (σ_t) reveal (data not presented) that water-column mixing characteristics along the salinity gradient change, from an upper bay region of intense, and virtually continuous year-round water mass stratification to a down bay condition of year-round mixing. The transitional area between the vertically mixed and stratified segments in Narragansett Bay lies between Stations 4 and 5 (Fig. 15.1). Upper bay Stations 2, 3, and 4 are stratified year-round, with a distinct halocline present. The down bay region (Stations 1, 5, 6, and 7) is vertically mixed down to the bottom year-round. The relative shallowness of Narragansett Bay contributes to the prevalence of vertical mixing down bay (Table 15.1; Hitchcock and Smayda, 1977). The regional stratification-mixing pattern can be temporarily obliterated during periods of particularly heavy freshwater runoff when, given suitable wind conditions, the entire bay can briefly stratify, as during the 2 April 1986 transect. However, stratification of down bay waters appears to be infrequent, just as vertical mixing of the entire water column is infrequent at upper bay Stations 2, 3, and 4. The mixing-stratification characteristics not only influence phytoplankton photosynthesis, but also the vertical admixture of nutrients from bottom waters and oxygen ventilation of the water column (Bergondo et al., 2005).

The higher bottom-water salinity in Narragansett Bay (Table 15.1) is consistent with a two-layer estuarine circulation pattern needed to maintain hydrostatic balance. The down bay surface flow of less salty water, diluted by river inflow that exits Narragansett Bay is compensated by the inflow of saltier, bottom water from offshore (Hicks, 1959; Kincaid and Pockalny, 2003; Kincaid *et al.,* Chapter 10). Salinity profiles for Stations 1, 5, 6, and 7 reveal bottom water flows into the West Passage that proceeds up bay, countering the offshore flow of less saline, near-surface water. A consequence of this hydrology is that both river runoff and bottom water inflow inject nutrients along the Narragansett Bay axis and influence their concentrations, as shown for nitrogen (Nixon *et al.,* 1995; Culver-Rymsza, 1988). The relative importance of these two nutrient sources at a given segment along the gradient is a function of its distance from the upper Narragansett Bay enrichment zone, and from Rhode Island Sound (Fig. 15.1).

The transition from a stratified to a vertically mixed water column along the salinity gradient in the regions of Stations 4 and 5 has been pointed out. The source of the bottom water in that region appears to vary since the vertical temperature, salinity, and density (σ_t) profiles at Station 4 are often distinct from those at Station 5. The mean bottom salinity (29.9‰) at Station 4 during the 1986–1987 survey was 0.7‰ more saline than at Station 5 (29.2‰), and the mean bottom water temperature −1.0 °C colder (Table 15.1). The corresponding differences at Station 4 were 0.8‰ and −1.1°C. On average, Station 4 bottom water is therefore more saline and colder than at Station 5, features inconsistent with an influx of bottom water solely from the region of Station 5. Another conspicuous local divergence—the mean surface temperature at Station 4 during both the surveys (11.6° and 11.9°C) was −0.6 to −0.8°C colder than at Stations 3 and 2 (Table 15.1). The frequent influx of colder, more saline bottom water at Station 4 and its periodic upwelling to the surface, unlike at Stations 2, 3, and 5 is evident in the profiles of the individual transect data during 1985–1986. Upwelling was less evident during the 1986–1987 surveys. There was a periodic influx of warmer, more saline water at Stations 3 and 4 during winter. Hence, upwelling of bottom water, possibly flowing from the East Passage, may be another source of nutrient influx into upper Narragansett Bay, an incursion which may also ventilate the water mass. Bergondo *et al.* (2005) have shown that oxygenation of the subsurface waters in this region, which influences the potential for hypoxia development, is influenced by the extent to which there is lateral advection of bottom waters and vertical mixing.

These physical–chemical features point to Stations 4 and 5 as being located in a particular unique segment hydrographically along the salinity gradient (Fig. 15.1). This region appears to be an important ecotone and major node in the habitat gradient in Narragansett Bay with regard to water column mixing, bottom water movement, nutrient sources, and delivery of nutrients and oxygen. Transitioning of vertical mixing characteristics along the Narragansett Bay axis from primarily an annually stratified to a well-mixed habitat also occurs in this region. In addition, it is the region where bottom water incursions from the East Passage merge with inflow through the West Passage. The resultant circulation pattern suggested by the data is more complex than a simple, two-layer estuarine flow. Kincaid *et al.,* in Chapter 10, further explore the circulation patterns exhibited at the mouth of the bay. Aperiodic upwelling of bottom water, and probably gyre flow, occur as a consequence of the horizontal and vertical density gradients. The physical–chemical features at Station 4, in functioning as a buffer zone and transitional region between upper and lower Narragansett Bay with regard to freshwater input, water quality and mixing characteristics, are expected to influence plankton dynamics in several key ways, as described later in this chapter.

15.3.3 Water Column Transparency

A turbidity (i.e., light extinction) gradient occurs along the salinity gradient in Narragansett Bay characterized by a progressive down bay increase in water clarity. This gradient in light transmission influences the amount of light available for photosynthesis by phytoplankton, benthic macroalgae, and eelgrass (*Zostera marina*). Based on Secchi Disk measurements (i.e., water clarity), this gradient develops because of light absorbance by suspended particles, dissolved organic matter, and phytoplankton chlorophyll whose collective abundance decreases down bay along the salinity gradient. Mean Secchi depth increased down bay by 42% during the 1985–1986 survey, from 1.7 m at Station 2 to 2.4 m at Station 7, and by 33% during 1986–1987, from 2.1 to 2.7 m at Stations 2 and 7, respectively.

The seven stations sampled during the 1985–1986 transects segregated into two optical water mass types, each exhibiting a strong positive correlation between mean Secchi Disk Depth and salinity. Station 1 (Greenwich Bay) and lower bay Stations 2, 3, and 4 comprised one optical group ($r = 0.99$), and lower bay Stations 5, 6, and 7 the other ($r = 0.96$; the slopes of the regression lines differed significantly). In contrast, the 1986–1987 survey stations clustered into a single group, with a strong positive correlation ($r = 0.92$) found between Secchi Disc depth and surface salinity.

The mean extinction coefficient, $k(m^{-1})$, was calculated from the Holmes (1970) equation:

$$k = \frac{1.44}{D} \tag{15.1}$$

where D is Secchi Disk depth in meters. The increase in water column turbidity with decreasing salinity is shown in Table 15.1. A strong inverse correlation occurred between mean salinity and the mean light attenuation coefficient (k) for both the survey periods: $r = -0.89$ and -0.95, respectively. This inverse relationship is not unexpected given that (1) *in situ* light transmission is reduced by the absorbance of suspended particles and dissolved organic matter, (2) the suspended solids loading in river runoff should be related to salinity because of dilution of recipient waters and their subsequent down bay flow, and (3) there is an increase in chlorophyll in response to riverine nutrient delivery. Each contributes to the inverse relationship found between mean salinity and the mean light attenuation coefficient (k). The "bell-shaped" distribution of chlorophyll along the salinity gradient (discussed later) suggests riverine delivery of terrigenous matter may be more important in governing turbidity at Stations 2 and 3 in the Providence River estuary than at contiguous Stations 4 and 5. At those stations, higher chlorophyll levels may be more important. Down bay from this region along the gradient there is a combined and progressive reduction of particulate and dissolved organic matter concentrations.

The 1% isolume depth is commonly used as the lower depth of the euphotic zone (i.e., depth of the primary production layer), and was calculated from:

$$I_z = I_0 e^{-kz} \tag{15.2}$$

where I_z is the irradiance at a given isolume depth (z), I_0 is the incident irradiance, and k the extinction coefficient. Mean euphotic zone depth (I% isolume depth) deepened approximately 2.0 m along the salinity gradient—it ranged from 6.6 to 8.7 m during 1986–1987, and was approximately 1 m deeper at all stations compared with the 1985–1986 surveys (Table 15.1). Relative to total water column depth, the euphotic zone at down bay Stations 6 and 7 extended to the bottom sediments. At Station 5 and in Greenwich Bay, the euphotic zone comprised 70–90% of the water column. At upper bay Stations 2, 3, and 4 located in the low salinity, highly turbid segment of Narragansett Bay, the euphotic zone was restricted to the upper half of the water column. The progressive deepening of the euphotic zone down bay along the salinity gradient is expected to favor development of an autotrophic bottom community, i.e., epibenthic microalgae, macroalgae, and eelgrass. This autotrophy, together with the abundant shellfish community, is expected to lead to regional differences in the strength and level of benthic–pelagic coupling, including nutrient fluxes from bottom sediments and benthic filter feeding.

In summary, the salinity gradient in Narragansett Bay influences light transmission and euphotic zone depth, with turbidity progressively increasing up bay. The proportion of the total water-column depth in which photosynthesis occurs decreases with salinity. During the 1985–1986 survey, the euphotic zone at Stations 2, 3, and 4 was approximately 45% of the total water column; it increased to 63% at Station 5, and extended to bottom sediments at Stations 6 and 7, allowing photosynthesis and growth of microscopic epibenthic algae, macroalgae, and *Zostera marina*.

15.3.4 Nutrient Gradient

The annual mean, maximum, and minimum surface concentrations of $NH_4 + NO_3$, PO_4, and $Si(OH)_4$ during the surveys are given in Table 15.3. All nutrients exhibited a pronounced down bay decrease in mean concentration that was

Table 15.3 Mean, maximum, and minimum nutrient concentrations (mg-at m^{-3}) at surface during the 1985–1986 and 1986–1987 surveys.

Station	1	2	3	4	5	6	7
Mean concentrations							
NH$_4$							
1985–1986	2.66	33.34	24.22	11.07	4.26	2.7	1.94
1986–1987	3.56	29.63	17.37	10.62	4.20	3.47	2.39
Δ	+0.90	−3.71	−6.85	−0.45	−0.06	+0.77	+0.45
NO$_3$							
1985–1986	1.64	20.70	13.00	7.80	3.70	1.90	1.30
1986–1987	4.31	21.23	14.41	9.07	5.34	4.52	3.47
Δ	+2.67	+0.53	+0.71	+1.27	+1.64	+2.62	+2.17
NH$_4$ + NO$_3$							
1985–1986	4.30	54.04	37.92	18.87	7.96	4.60	3.24
1986–1987	7.87	50.86	31.78	20.32	9.54	7.99	5.86
Δ	+3.57	−3.18	−6.14	+1.45	+1.58	+3.39	+2.62
PO$_4$							
1985–1986	1.36	5.43	4.14	3.56	1.49	1.55	1.02
1986–1987	1.67	4.08	3.01	2.27	1.54	1.34	1.20
Δ	+0.31	−1.35	−1.13	−1.29	+0.05	−0.21	+0.18
Si(OH)$_4$							
1985–1986	16.51	34.76	28.39	17.59	13.07	12.23	10.69
1986–1987	20.94	49.38	37.65	24.72	20.75	18.97	17.43
Δ	+4.43	+14.62	+9.26	+7.13	+7.68	+6.74	+6.74
Maximum and minimum concentrations							
1985–1986 NH$_4$							
Max	11.0	78.5	71.7	45.0	14.6	10.4	8.4
Min	0.4	2.5	1.6	0.6	0.6	0.0	0.5
1986–1987 NH$_4$							
Max	11.5	56.3	40.2	45.3	16.3	13.5	8.9
Min	0.2	8.3	2.8	0.8	0.3	0.1	0.4
1985–1986 NO$_3$							
Max	11.9	43.5	27.8	19.7	13.9	13.5	12.1
Min	0.1	5.4	0.1	0.1	0.2	0.1	0.1
1986–1987 NO$_3$							
Max	18.5	37.9	27.8	19.7	13.9	13.5	12.1
Min	0.2	8.7	1.3	0.0	0.1	0.1	0.0
1985–1986 PO$_4$							
Max	9.9	30.0	19.0	19.4	4.5	4.8	3.0
Min	0.1	1.5	0.3	0.2	0.1	0.2	0.2
1986–1987 PO$_4$							
Max	6.1	8.2	6.6	5.4	3.2	10.8	2.5
Min	0.0	0.6	0.1	0.1	0.1	0.0	0.0
1985–1986 Si(OH)$_4$							
Max	83.1	74.0	70.1	44.7	40.8	42.2	42.2
Min	0.0	11.0	2.4	0.0	0.0	0.0	0.0
1986–1987 Si(OH)$_4$							
Max	> 60.0	164.7	84.5	49.0	52.1	50.5	45.9
Min	1.2	25.1	6.8	5.7	1.1	1.0	1.0

0 entries indicate below minimum detection level.

strongly coupled to the salinity gradient (Fig. 15.2). The inverse correlations found between nutrient concentrations and salinity were highly significant statistically, whether grouping the data by nutrient and by survey year ($r \geq -0.96$), or combining the data for both the survey periods ($r = -0.91$ to -0.98). The nutrient gradient is driven primarily by the copious discharge of riverine and sewage effluent nutrients into the Providence River estuary, with the zone of initial dilution found in the region of Stations 2, 3, and 4. The region between Stations 4 and 5 transitions to lower nutrient conditions found down bay, the gradients set up by progressive dilution and the uptake of nutrients by phytoplankton as the enrichment plume moves down bay.

The magnitude of the nutrient differences along the salinity–nutrient gradient between Stations 2 and 7 varied significantly among nutrients and by year. The mean $NH_4 + NO_3$ concentration at Station 2 during 1985–1986 was approximately 17-fold greater than at lower bay Station 7, and ninefold greater during the 1986–1987 survey (Table 15.3). For PO_4, the corresponding differences between the survey periods were about five- and threefold, respectively, and for $Si(OH)_4$ approximately threefold during both the surveys. Thus, on a relative basis, the down bay decrease in inorganic N introduced into the low salinity segment of the gradient in the upper bay exceeds that for P and Si, whose down bay decline along the gradient is more gradual. This greater diminution rate reflects the relatively greater dilution and uptake of nitrogen evident at Stations 3 and 4 (Table 15.3). The sharp nutrient-transition zone in the region of Stations 4 and 5 subdivides Narragansett Bay into an upper, nutrient-enriched region and a lower, nutrient-poorer region. This regional separation is consistent with the subdivision evident in the salinity data (Table 15.1).

In reality, two different, non-biological nutrient sources (i.e., mechanisms) occur along the salinity–nutrient gradient in Narragansett Bay. The Providence River estuary (Stations 2 and 3) functions as a major anthropogenic nutrient pump, delivering N, P, and Si that are then transported down bay. At the entrance into Narragansett Bay, "new" nutrient—primarily NO_3—is advected through inflow of enriched bottom water and transported up bay from Station 7 (Nixon et al., 1995; Culver-Rymsza, 1988). The salinity distribution along the horizontal and vertical axes of Narragansett Bay proxies this dual system of nutrient input. Nutrients recycled by food web dynamics along the salinity gradient supplement these two major input mechanisms (Vargo, 1976; Verity, 1985). The relative importance of the physically and biologically regulated nutrient fluxes along the gradient changes regionally and seasonally. In lower Narragansett Bay, in situ biological recycling and offshore input of nutrients become progressively more important along the gradient than the down bay nutrient flux from the Providence River estuary. The very high nutrient concentrations that persist at Stations 2 and 3, and the low concentrations at Station 7, with intermediate concentrations at Stations 1, 4, 5, and 6, reflect these differing nutrient accretion mechanisms, which Nixon et al. (1995) considered in their assessment of nutrient mass balance in Narragansett Bay.

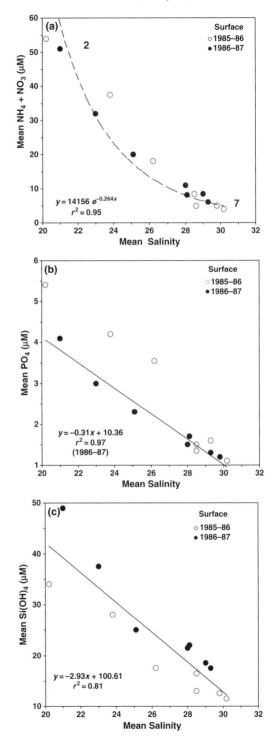

Fig. 15.2 Mean concentrations of (a) $NH_4 + NO_3$ (b) PO_4, and (c) $Si(OH)_4$ along the surface salinity gradient during the surveys.

15.3.5 Interannual Differences in Nutrients along the Salinity–Nutrient Gradient

Mean nutrient concentrations along the salinity gradient differed interannually, i.e., between the two survey periods (Table 15.3). The specific annual behavior varied among nutrients and stations, but did not prevent development of the nutrient gradients. The differences in $NH_4 + NO_3$, PO_4, and $Si(OH)_4$ between the survey periods were primarily in concentration, and not in the pattern of their down bay decline (Fig. 15.2). Mean PO_4 and NH_4 concentrations at Stations 2, 3, 4, and $NH_4 + NO_3$ concentrations at Stations 2 and 3 in upper Narragansett Bay during 1985–1986 exceeded 1986–1987 levels, whereas NO_3 and $Si(OH)_4$ concentrations were substantially lower (Fig. 15.2; Table 15.3). The lower NH_4 concentrations in 1986–1987 at Stations 2 and 3 corresponded to reduced loadings of 13% and 39%, respectively. Unlike NH_4, mean NO_3 concentrations during 1986–1987 generally exceeded 1985–1986 levels by 2.5-fold in Greenwich Bay (Station 1) and lower Narragansett Bay (Stations 6 and 7); by 15–44% in the transitional region at Stations 4 and 5, but were approximately equal at upper bay Stations 2 and 3. Mean $NH_4 + NO_3$ concentrations at Stations 2 and 3 during 1985–1986 exceeded 1986–1987 levels by only 6% and 19%, respectively, a modest increase given the high $NH_4 + NO_3$ concentrations at those stations (Table 15.3). At the other stations, $NH_4 + NO_3$ concentrations during 1985–1986 were 1.45–3.57 mg-at m^{-3} lower than in 1986–1987, but about 80% higher during 1986–1987 at Stations 1, 6, and 7.

Mean annual PO_4 concentrations were similar during both the surveys, excluding upper bay Stations 2, 3, and 4 where concentrations were 33–57% higher during 1986–1987. $Si(OH)_4$ concentrations were elevated bay-wide during the 1986–1987 survey, similar to NO_3, exceeding 1985–1986 levels by 4.4–14.6 mg-at m^{-3} (Table 15.3). Mean $Si(OH)_4$ concentrations, in contrast to the reduced PO_4 and NH_4 concentrations at Stations 2, 3, and 4, exceeded 1985–1986 levels by 33–42%, and were 55–63% higher at Stations 5, 6, and 7. The most modest increase (27%) occurred in Greenwich Bay (Station 1).

The interannual variations in nutrients reflect their variable supply from climate-influenced riverine runoff, anthropogenic enrichment, biological remineralization, assimilation by phytoplankton, and other processes. These interactions are further influenced by various physical, chemical, and biological processes. We emphasize that the concentrations of nutrients measured and reported here represent the residual of nutrient supply and uptake, and give no direct indication of the magnitude of the various processes contributing to the nutrient levels measured. But the survey results indicate that the amount of nutrient delivered into Narragansett Bay, both annually and seasonally, and available for assimilation is not fixed, but varies and (as will be shown) leads to significant interannual differences in productivity.

15.3.6 Annual Nutrient Maxima and Variations along the Salinity Gradient

A feature of the annual nutrient cycles along the salinity gradient that is obscured by averaging the data is the temporal variability of the annual maximum. The annual maxima of the four macronutrients, collectively and individually, were neither spatially coincident nor temporally synchronous along the gradient. Each nutrient had its own pattern. The time of the annual maximum for each nutrient also varied between the two survey periods. Some general examples of this variability are given *in lieu* of a detailed discussion of the annual nutrient cycles, which is beyond the scope of this chapter.

There was a conspicuous down bay temporal progression in the annual NH_4 maximum during 1986–1987: mid-September at Stations 2, 3, and 4; late October in Greenwich Bay (Station 1); early November at Station 5, and late-November at Stations 6 and 7 (Table 15.3). NO_3 did not exhibit this progression, although the timing of the NH_4 and NO_3 maxima was generally similar regionally: mid-September at Stations 2, 3, and 4, and late-November at Stations 5, 6, and 7. The annual NO_3 maximum in Greenwich Bay occurred in mid-December. The NH_4 and NO_3 maxima during 1985–1986 were asynchronous in Greenwich Bay and at lower bay Stations 5, 6, and 7, a NO_3 maximum in August and NH_4 maximum in December. NH_4 and NO_3 concentrations were high throughout the year at upper bay Stations 2 and 3, and during summer in the transitional region (Station 4).

The regional and temporal patterns in the annual PO_4 and $Si(OH)_4$ maxima also varied. For the 1986–1987 period, the annual PO_4 maximum at the seven stations ranged from July to December—in late July at Station 2; in August at Station 5; mid-September at Stations 1, 3, and 6, and in December at Stations 4 and 7. During the 1985–1986 survey period, the annual PO_4 maximum at Stations 1, 5, 6, and 7 occurred during the summer brown tide bloom, and at Stations 2, 3, and 4 in mid-September. The annual $Si(OH)_4$ maximum during 1986–1987 occurred in July at Stations 1, 4, 5, 6, and 7, and in January at Stations 2 and 3. The surface $Si(OH)_4$ maximum at Station 5 (52.1 mg-at m^{-3}), in early December, slightly exceeded maximal July levels (48.2 mg-at m^{-3}). A summer $Si(OH)_4$ maximum (in August) also occurred during 1985–1986 at Stations 1, 4, 5, 6, and 7; at Stations 2 and 3, $Si(OH)_4$ concentrations were high year round.

Each nutrient exhibited a conspicuous range in concentration, another feature embedded within the down bay decrease in mean nutrient concentrations along the salinity gradient (Fig. 15.2). This variance is illustrated using Station 2 as an example, and where nutrient accretion from riverine discharge and sewage effluent is pronounced. Station 2 lies within, or adjoins the zone of initial dilution of nutrients discharged from the Narragansett Bay Commission sewage treatment facility located at Field's Point in the Providence River (Fig. 15.1). Stations 3 and 4 also lie within this region, directly exposed to sewage discharge inputs during downstream flow of the enrichment plume.

The maximal vs. minimal surface concentrations at Station 2 for NH_4, NO_3, PO_4, and $Si(OH)_4$, combining all transect data were—78.5 vs. 2.5; 43.5 vs. 5.4; 30.0 vs. 0.6; 164.7 vs. 11.0 mg-at m^{-3}, respectively (Table 15.3). The markedly lower minimal concentrations reflect the combined effect of phytoplankton uptake, water mass dilution, and reduced accretion processes whose relative importance undoubtedly varied between sampling dates. Down bay from Stations 2, 3, and 4 the minimal concentrations of NH_4, NO_3, PO_4, and $Si(OH)_4$ recorded were usually <1 mg-at m^{-3}, and near analytical detection limits. These very low concentrations suggest phytoplankton growth can become nutrient limited along the entire axis of the nutrient–salinity gradient in Narragansett Bay, both in yield (i.e., biomass) and growth rate (i.e., physiologically), with this limitation progressively becoming more intense proceeding down bay from Station 2.

15.3.7 Nutrient Surges along the Salinity Gradient

Sampling protocol influences detection of the actual maximal and minimal nutrient concentrations, the time of their occurrence, spatial coherence and temporal variations. The sampling frequency of the 62 transects over the 2-year period corresponds to a sampling interval of 12 days between surveys. The high sampling frequency and the seven-station sampling grid make it probable that the major physical, chemical, and biological features in Narragansett Bay along the salinity gradient during the survey years were captured. Nonetheless, a habitat feature important to phytoplankton growth, but most likely inadequately surveyed, is the aperiodic surge in nutrient delivery that results when storm-driven, nutrient pulses enter into the Providence River estuary and spread down bay. Nutrient surveys carried out at the seven transect stations at 3–4 day intervals, from July to October 1982, revealed high nutrient loadings pulsing into upper Narragansett Bay during storm and runoff events, with the nutrient enrichment plumes then moving down bay, often reaching Station 7 near the entrance into Narragansett Bay (Smayda unpublished). The elevated, downstream nutrient levels persist only 1–2 days, being rapidly assimilated by the phytoplankton. During the 1985–1987 surveys, storm and runoff-driven nutrient surges periodically occurred between sampling dates, this influx evident as transect-to-transect differences at the upper bay stations directly influenced by riverine and sewage nutrient discharges. For example, surface NH_4 concentrations at Station 3 sampled on 3, 10, and 17 September 1986 were 18.1, 9.4, and 40.2 mg-at m^{-3}, respectively. The weekly total rainfall during this period, from weeks 1 through 3, was 2.06, 0.15, and 2.67 cm, respectively. During such aperiodic and rapid, short-term pulsing of nutrients into down bay regions, Narragansett Bay functions as a chemostat, with the flux in nutrients boosting phytoplankton growth. The significance of this cryptic nutrient–phytoplankton relationship in the overall nutrient regulation of phytoplankton dynamics in Narragansett Bay remains to be determined, but cannot be neglected in the development of ecosystem-based management strategies, or in ongoing mitigation projects to reduce runoff surges.

This summary of the spatial and temporal differences in macronutrient concentrations and behavior should not be interpreted as an indication that regional nutrient dynamics in Narragansett Bay, particularly changes resulting from phytoplankton uptake, are uncoupled, despite the well-defined nutrient–salinity gradient (Fig. 15.2). This is not the case—all nutrients show more or less synchronous regional decreases resulting from phytoplankton assimilation and biomass growth as reported by Pratt (1959; 1965).

15.3.8 Nutrient Ratios

Nutrient ratios are of interest because they influence functional group selection. Functional groups and their shifts are of interest because of significant differences in their physiology and ecological impacts. Cyanobacterial and harmful algal blooms and their potential regulation by the N:P ratios, for example, are events of great contemporary interest (Smayda, 2004). Cyanobacterial blooms are relatively rare in Narragansett Bay, while harmful algal blooms are not unusual (Smayda and Villareal, 1989a,b; Li and Smayda, 2000). The diatom: flagellate abundance ratio as a potential indicator of eutrophication is of special interest since it has been hypothesized that this ratio should decrease with increasing nutrient enrichment and, consequently, might serve as an indicator of eutrophication status (Smayda, 1990). An evaluation of long-term blooms and nutrient conditions in various regions led Smayda (1990) to suggest that anthropogenic enrichment of N and P has led to long-term declines in the ratios of Si:N and Si:P, and these altered nutrient conditions potentially favor flagellate blooms in regions so impacted. Silica is the primary nutrient expected to regulate the shift in functional groups from diatoms to flagellates, since diatoms, unlike other microalgal groups, require Si (Officer and Ryther, 1980; Smayda, 1990). Silica is assimilated by diatoms stoichiometrically in the Redfield Ratio in the atomic proportions of 1:1 with N, and 16:1 with P. At Si:N supply ratios of <1:1, diatoms will be Si-limited, but N-limited at Si:N supply ratios >1:1 [see Smayda (2004) for a detailed discussion of Redfield Ratio kinetics and eutrophication]. Diatoms are the major phytoplankton component driving productivity in Narragansett Bay (Pratt, 1959; 1965), with evidence that long-term shifts in their specific abundance (Borkman, 2002) and in the diatom:flagellate abundance ratios (Smayda and Borkman, unpublished) have occurred. This changing behavior is of great interest given the potential involvement of Si in influencing these changes.

The mean annual ratios of N:P, N:Si, and Si:P (by atoms) during the surveys (Table 15.4) have been plotted against the salinity gradient (Fig. 15.3). The ratios varied along the gradient, reflecting the differing sources and magnitude

Table 15.4 Mean nutrient ratios (by atoms) during the surveys.

Station	1	2	3	4	5	6	7
N:P							
1985–1986	3.2	10.0	9.2	5.3	5.3	3.6	3.2
1986–1987	4.7	12.5	10.6	9.0	6.2	6.0	4.9
Δ	+1.5	+2.5	+1.4	+3.7	+0.9	+2.4	+1.7
N:Si							
1985–1986	0.26	1.56	1.33	1.08	0.61	0.38	0.30
1986–1987	0.38	1.03	0.84	0.82	0.46	0.42	0.34
Δ	+0.12	+0.53	−0.49	−0.26	−0.15	+0.04	+0.04
Si:P							
1985–1986	12.1	6.4	6.9	4.9	8.8	7.9	10.5
1986–1987	12.5	12.1	12.5	10.9	13.5	14.2	14.5
Δ	+0.4	+5.7	+5.6	+6.0	+4.7	+6.3	+4.0

of nutrient input along its axis. The N:P and N:Si ratios were strongly and inversely correlated with mean salinity (Fig. 15.3; Table 15.5). The mean annual N:P ratio (using $NH_4 + NO_3$) during both the survey periods progressively decreased down bay, from 10:1 (Station 2) to 3.2:1 (Station 7) during 1985–1986, and from 12.5:1 to 4.9:1 during 1986–1987 (Fig. 15.3a). The increase in mean N:P ratio at the stations during 1986–1987 (0.9–3.7) corresponded to a 53% (Station 7) to 70% (Station 6) increase in the amount of nitrogen per unit phosphorus in lower Narragansett Bay; 47% in Greenwich Bay (Station 1), and 15–25% in upper Narragansett Bay. Plotting the mean N:P ratios against mean salinity, either for a given survey year or combining all data, yielded highly significant inverse correlations that exceeded $r = -0.92$ (Table 15.5).

The down bay distribution in N:P ratio along the salinity–nutrient gradient suggests three distinct regional zones in this stoichiometry occur, and are indicative of a variable regional sensitivity to the supply of N and P in Narragansett Bay. The mean N:P ratios at inner bay Stations 2 and 3 varied from about 9.0–12.5:1; at mid-bay Stations 4 and 5, the ratio was about 5:1 (i.e., 50% less N available per unit P). In Greenwich Bay and at lower bay Stations 1, 6, and 7, the N:P ratio was about 3:1. Based on Redfield Ratio kinetics, the distribution of the mean N:P ratio indicates N becomes progressively more available relative to P up bay along the salinity gradient. The corollary is that upper Narragansett Bay, on average, is more sensitive to the amount of phosphorus available relative to nitrogen. Conversely, lower Narragansett Bay, particularly the region down bay from Station 4 and extending toward the bay entrance, becomes progressively more sensitive to nitrogen availability, i.e., N progressively becomes more limiting to productivity (Fig. 15.3a). Nitrate reductase measurements during the 1985–1986 surveys provide experimental support for this expectation (Culver-Rymsza, 1988). Based on the primary production measurements and Redfield stoichiometry, the annual mean contribution of NO_3 advected into Narragansett Bay from

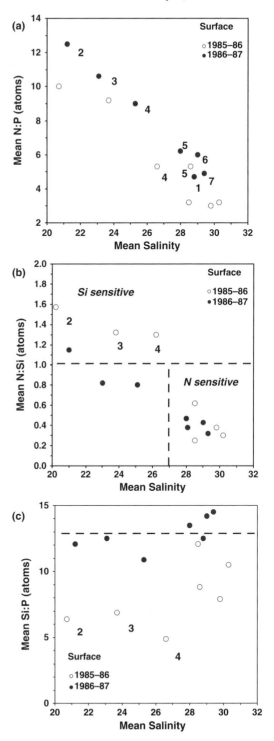

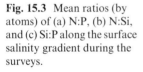

Fig. 15.3 Mean ratios (by atoms) of (a) N:P, (b) N:Si, and (c) Si:P along the surface salinity gradient during the surveys.

Table 15.5 Statistically significant ($p \leq 0.0001$) Pearson correlation coefficients (r) between mean nutrient concentrations and mean salinity, and between mean nutrient ratios (by atoms) and mean salinity along the salinity gradient during the surveys.

Variables	1985–1986	1986–1987	1985–1987
PO_4 vs Salinity	−0.98	−0.97	−0.93
NH_4 vs Salinity	−0.98	−0.97	−0.97
NO_3 vs Salinity	−0.99	−0.97	−0.98
$NH_4 + NO_3$ vs Salinity	−0.99	−0.97	−0.98
$Si(OH)_4$ vs Salinity	−0.98	−0.96	−0.91
N:P vs Salinity	−0.95	−0.97	−0.92
N:Si vs Salinity	−0.95	−0.97	−0.89
Si:P vs Salinity	+0.58	+0.67	–

Rhode Island Sound, relative to the total N demand by the phytoplankton, was greatest at Station 7 (24%). The contribution of this exogenous NO_3 to the total N demand progressively decreased up bay as the supply and residual levels of NH_4 progressively increased along the gradient. The contribution of NO_3 was elevated (25%) in semi-enclosed Greenwich Bay. Lower Narragansett Bay (Station 7) has been demonstrated to be N-limited by a variety of experimental approaches: C-compound labeling (Hitchcock, 1978); cellular C and N content (Sakshaug, 1977); estimated N-turnover time (Furnas *et al.*, 1976), and nutrient enrichment bioassay experiments (Smayda, 1974).

The mean annual N:Si ratio, pooling the $NH_4 + NO_3$ concentrations, decreased along the salinity gradient exhibiting a strong, inverse correlation with mean salinity (Fig. 15.3b; Table 15.5). During the 1985–1986 survey, the N:Si ratio progressively decreased, from 1.6:1 at Station 2 to 0.3:1 at Station 7 (Table 15.4). During the 1986–1987 surveys, $Si(OH)_4$ concentrations were elevated relative to N, resulting in a 25–37% reduction in the N:Si ratio at upper bay Stations 2, 3, 4, and 5 (Table 15.4). Nonetheless, Redfield ratio kinetics applied to the surveys indicate Narragansett Bay, on average, is regionally partitioned into an upper, Si-sensitive region that extends up bay from Nayatt Point (Station 4) into the Providence River estuary, and a strongly-N-sensitive mid- and lower-bay region that extends down bay from the Narragansett Bay Sanctuary region located near Prudence Island (Fig. 15.1).

The excess of N relative to Si in upper Narragansett Bay is expected to favor flagellate blooms over blooms of Si-requiring diatoms. Flagellates have a greater general capacity to tolerate and utilize high levels of inorganic and organic nitrogen than diatoms, which utilize inorganic nitrogen fractions primarily, and then only to the point of Si exhaustion. The latter physiological limitation results in the residual concentrations of nitrogen and phosphorus becoming more available to flagellates and other non-siliceous species then freed from significant competition for N and P by Si-limited diatoms. Field studies on the phytoplankton composition and abundance in the Si-sensitive region of Narragansett Bay generally support this expectation of a Si-regulated functional group selection (Mitchell-Innes, 1973; Culver-Rymsza, 1988;

Kullberg, 1992). At Station 4, located off Nyatt Point, and surveyed from July 1987 to May 1988, the <10-μm phytoplankton size fraction ($=$ nanophytoplankton) dominated the summer and winter–spring bloom communities, both as chlorophyll and in primary production (Kullberg, 1992). Mitchell-Innes (1973) described the summer phytoplankton flora in the Providence River at a station intermediate between Stations 2 and 3 as "characteristic of polluted estuaries." The flora was dominated by an unidentified chlorophyte, *Nannochloris* sp., whose population level reached 100 million cells L^{-1}, accompanied by a diverse and abundant micro-flagellate population. Intense transient euglenid blooms (10 million cells L^{-1}) were also notable, and similar to those that developed in the region extending from Station 2 to Station 5 during the 1985–1986 surveys (Culver-Rymsza, 1988; Smayda, unpublished). In contrast, Mitchell-Innes (1973) found diatom blooms and succession dominated the phytoplankton community in lower Narragansett Bay, as reported also by Pratt (1959; 1965).

The distribution of the Si:P ratio and along the salinity gradient differed from the N:P and N:Si ratios, and between surveys (Fig. 15.3c; Tables 15.4 and 15.5). The Si:P ratio increased with the down bay increase in mean salinity, unlike the N:Si and N:P ratios. The stations sampled during the 1985–1986 surveys clustered into two subgroups: Stations 1, 2, and 3, and Stations 4, 5, 6, and 7. Each subgroup was correlated with mean salinity, but *directly* rather than *inversely* as in the case for N:Si and N:P. The high correlations ($r = 0.95$ and 0.89, respectively) decreased to $r = 0.58$ pooling all stations (Table 15.5). For the 1986–1987 surveys, mean Si:P ratios (10.9:1–14.5:1) substantially exceeded those during 1985–1986, and were more or less invariant along the salinity gradient, but still positively correlated (Fig. 15.3c; Table 15.5). The elevated Si:P ratio during the 1986–1987 survey reflects the significantly greater loading of $Si(OH)_4$ compared with the 1985–1986 survey (Table 15.3). The lower correlation ($r = 0.67$) grouping all stations (Table 15.5) reflects diminution of the strong correlation ($r = 0.92$) for Stations 1, 4, 5, 6, and 7, which are clustered when grouped with Providence River estuary Stations 2 and 3, which fall outside this cluster (Fig. 15.3c).

The gradient patterns in the three nutrient ratios indicate that upper Narragansett Bay, on average, is more sensitive to P and Si relative to N, whereas mid to lower Narragansett Bay is more sensitive to N limitation relative to P and Si, and more responsive to NO_3-rich incursions of bottom water into Narragansett Bay.

15.3.9 Phytoplankton Biomass

15.3.9.1 Chlorophyll–Nutrient–Salinity Gradient

Mean surface chlorophyll concentrations exhibited a well-defined hyperbolic pattern in their distribution along the salinity gradient during both survey periods (Fig. 15.4). Mean concentrations at all stations were higher

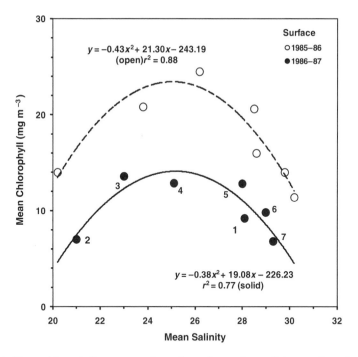

Fig. 15.4 Mean chlorophyll concentrations along the surface salinity gradient during the surveys.

($4.0–12.0$ mg m^{-3}) during the 1985–1986 survey (Table 15.6), reflecting the intense brown tide bloom that occurred during that period. In this hyperbolic distribution, chlorophyll concentrations were conspicuously lower within the Providence River estuary (Station 2), increased to a maximum in the region between Gaspee Point (Station 3) and Providence Point on Prudence Island (Station 5), then progressively decreased down bay (Fig. 15.4). The station alignment in mean chlorophyll levels along the gradient was: $4 > 3 > 5 > 1 > 6 = 2 > 7$ during 1985–1986 and $3 > 4 = 5 > 6 > 1 > 2 = 7$ during 1986–1987.

Table 15.6 Mean chlorophyll concentrations (mg m^{-3}) at surface during the 1985–1986 and 1986–1987 surveys.

	1985–1986	1986–1987	
Station	Chl	Chl	Δ
3	21.0	13.6	7.4
4	24.6	12.9	-11.7
5	20.5	12.5	-8.0
1	16.2	9.3	-6.9
6	13.9	9.8	-4.1
2	13.9	7.2	-6.7
7	11.2	7.1	-4.1

Chlorophyll concentrations and their distribution along the salinity gradient are not osmotically regulated since the phytoplankton assemblage in Narragansett Bay is euryhaline. Rather, the concentrations and their gradient pattern are related to the nutrient and light gradients, among other variables, that are embedded within the salinity gradient (Fig. 15.2; Tables 15.1 and 15.2). Mean annual surface chlorophyll levels plotted against mean $NH_4 + NO_3$, PO_4, and $Si(OH)_4$ concentrations along the spatial salinity gradient exhibit a bell-curve pattern of high statistical significance (Fig. 15.5). This hyperbolic relationship persisted during both survey periods despite differences in mean chlorophyll.

Several distinct features characterize the chlorophyll gradient pattern. In the region extending up bay from lower Narragansett Bay to mid bay at Stations 1, 4, 5, 6, and 7, chlorophyll increases with nutrient concentration, i.e., there is a yield–dose relationship. Maximal chlorophyll concentrations occurred at an intermediate, not the highest nutrient concentration along the gradient, for each chlorophyll–nutrient regression (Fig. 15.5). This maximum, which we term the "saturation concentration," usually occurred at Station 4 (Fig. 15.5). In contrast, the yield–dose relationship in the region north of Prudence Island in the Providence River estuary (Stations 2 and 3) is the reverse of that in the lower and mid bay segments (Fig. 15.1). Nutrient concentrations at Stations 2 and 3 exceeded saturation concentrations, and chlorophyll decreased. This suggests that either the assimilatory capacity of the Providence River estuary is exceeded, or nutrient uptake is impeded. This local, inverse response to nutrient concentration transforms the general yield–dose relationship along the chlorophyll–nutrient–salinity gradient from a linear to hyperbolic pattern (Fig. 15.5). The saturation concentrations that separate the region of positive yield–dose (lower and mid bay regions) from the negative relationship in the Providence River estuary region, suggested by the regressions are: 25–30 μM $NH_4 + NO_3$ and 3.0–3.5 μM PO_4 (Fig. 15.5a,b). For $Si(OH)_4$, the nutrient required by diatoms, the yield–dose relationship is also strongly hyperbolic, but unlike N and P diverged between survey years. The chlorophyll yield–Si dose response was more circumscribed and elevated during the 1985–1986 brown tide year (Fig. 15.5c). This suggests that the Si saturation concentration is more influenced by species composition and abundance than are the N and P saturation concentrations.

15.3.10 Impediments to Biomass Accumulation in the Providence River Estuary

The progressive increase in mean chlorophyll concentrations proceeding up bay from Fox Island (Station 7) to Conimicut Point (Station 4) (Fig. 15.1) is probably a nutrient-stimulated response to the down bay advection of nutrient-enriched upper Narragansett Bay water. The statistically significant correlations

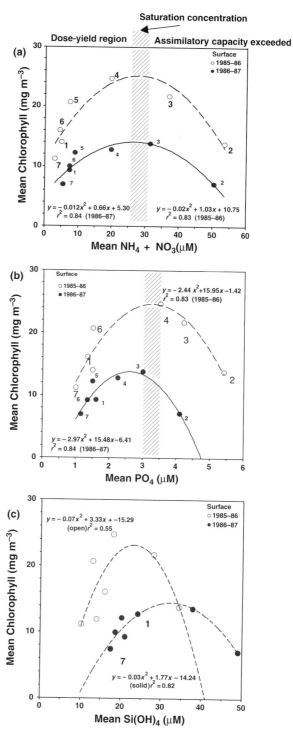

Fig. 15.5 (a) Mean chlorophyll concentrations vs. mean $NH_4 + NO_3$ concentrations along the surface salinity gradient during the surveys, with three features recognized: Dose–yield region where biomass increases with nutrient concentration; the Saturation Concentration (hatched area): the $NH_4 + NO_3$ concentration at which mean chlorophyll concentrations are maximal; and the range in $NH_4 + NO_3$ concentrations over which mean chlorophyll levels decrease, i.e., the Assimilatory Capacity of the ecosystem for $NH_4 + NO_3$ is exceeded. See text for further discussion. (b) Mean chlorophyll concentrations vs. mean PO_4 concentrations along the surface salinity gradient; hatched area corresponds to the Saturation Concentration. Equation given applies to 1985–1986 survey. (c) Mean chlorophyll concentration vs. mean $Si(OH)_4$ concentration along the surface salinity gradient during the surveys.

found for the chlorophyll–nutrient–salinity gradient relationship support this conclusion. This nutrient stimulation of growth contrasts with the apparent suppression that accompanies the higher nutrient loadings into the Providence River estuary between Conimicut Point and Field's Point. The assimilatory capacity of this highly enriched, innermost bay region appears to be exceeded. The cause(s) of this apparent suppression, which results in reduced chlorophyll levels at the highest nutrient concentrations found in Narragansett Bay, are unknown. Several potential processes, each amenable to future field and experimental evaluation, may be responsible. Phytoplankton growth is possibly inhibited by the high nutrient concentrations and/or chemical growth-inhibitors accreted into the eutrophic Providence River estuary. Alternatively, washout may be reducing the residence time of the phytoplankton and preventing greater utilization of nutrients. Washout may prevent the increase in chlorophyll standing stock more in line with that expected from the yield–dose kinetics found in the mid and lower bay regions. The potential influence of washout on chlorophyll accumulation is considered later in this chapter.

Less likely, but not ruled out, the lower salinity and degraded water quality of this region might influence the composition and abundance of the planktonic grazer community, leading to increased grazing pressure and a reduction in chlorophyll levels. Undoubtedly, several interactive processes are involved given that the measured chlorophyll concentrations are residual, i.e., the concentrations at the time of sampling and remaining after *in situ* grazing, sinking, and advective losses of prior cellular growth that occurred in response to the nutrient supply. A consequence of the reduced nutrient uptake in the Providence River estuary is that higher concentrations of nutrients become available for down bay export—the Providence River estuary functions as a nutrient pump—a major feature of Narragansett Bay.

15.3.11 Spatial and Seasonal Patterns in Chlorophyll Dynamics

An intense, novel brown tide bloom of *Aureococcus anophagefferens* dominated the annual bloom cycle from May to October during the 1985–1986 survey (Smayda and Villareal, 1989a,b; Smayda and Fofonoff, 1989). The brown tide was notable for its persistence, numerical abundance (population densities exceeded 10^9 cells L^{-1}), and harmful effects (Smayda and Villareal, 1989a,b; Smayda and Fofonoff, 1989; Tracey *et al.*, 1988; Durbin and Durbin, 1989). Blooms of *Aureococcus anophagefferens,* a picoplanktonic species (diameter 2–5µm), are classifiable as high density low biomass (HDLB) events. Because of its small cell size, biomass (chlorophyll) levels are not commensurate with the high numerical abundance that *Aureococcus* achieves during its blooms (Smayda, 1997). During the 1985 brown tide event, diatoms, dinoflagellates, and other flagellates also proliferated at much lower population densities, but because of their larger cell size disproportionately augmented the chlorophyll levels relative to *Aureococcus* (Smayda and Villareal, 1989b).

Despite the 5-month *Aureococcus* bloom during the 1985–1986 survey period, which did not recur during the 1986–1987 survey, major, persistent similarities in chlorophyll dynamics and bloom behavior of ecological importance characterize the 62 transects comprising the surveys. The interannual persistence of the salinity–nutrient–chlorophyll gradients and the hyperbolic chlorophyll yield–nutrient dose relationship has already been described (Figs 15.2, 15.4, and 15.5). There are also significant annual and regional patterns in phytoplankton growth and biomass embedded within the hyperbolic chlorophyll distribution that are lost in averaging the data. There is an annual chlorophyll (biomass) cycle; an annual winter–spring bloom; a regional synchrony in bloom formation and progress; intense and prolonged summer blooms occur, and chlorophyll levels of great magnitude occur at all stations irrespective of their location along the nutrient–salinity gradients (Figs 15.4 and 15.5).

15.3.12 Summer Blooms

The July 1986–June 1987 surveys illustrate the intense, bay-wide summer blooms that develop (Table 15.7). Summer chlorophyll maxima exceeded winter–spring bloom levels, excluding Greenwich Bay (Station 1) and the Fox Island station (Station 7) near the entrance into Narragansett Bay. The blooms were intense and persistent; they lasted from July–October at Station 4 (approximately 20–30 mg m^{-3} chlorophyll) and July–September (43–56 mg m^{-3}) at Station 3. The maximum biomass recorded during the 62 surveys, 173 mg m^{-3} chlorophyll, was a summer bloom event recorded in August at Station 5. Maximum summer chlorophyll concentrations during the 1985 brown tide (Table 15.8), as during the 1986–1987 survey, exceeded winter–spring bloom concentrations, which ranged from 23.6 (Station 2) to 55.9 mg m^{-3} (Station 4). Blooms of the indigenous summer flora were equally intense during 1985 and 1986.

Summer blooms in Narragansett Bay, such as the 1985 brown tide, are not anomalous, although the bloom species vary and may even be unusual or novel, such as *Aureococcus anophagefferens* responsible for the 1985 brown tide. Another similarity of ecological importance common to the surveys is the large, summer blooms that develop at all stations and reach chlorophyll levels that considerably exceed the annual average at those sites (Table 15.8). It is worth noting that the 1985–1986 annual means are used as they exceed the 1986–1987 means. This reveals the bay-wide capacity of Narragansett Bay to support blooms of unusually high magnitude irrespective of the average bloom behavior in those regions, or expected from the nutrient–salinity gradient (Fig. 15.5). The greater magnitude of summer blooms in the upper bay regions relative to the lower bay (Table 15.8) partly links this capacity to nutrient

Table 15.7 Maximum monthly surface chlorophyll (mg m^{-3}) concentrations during the 1986–1987 survey, with the maximum concentration recorded for each station in bold.

Station	1	2	3	4	5	6	7
1986							
Jul	**27.0**	**30.7**	**55.6**	**29.4**	51.5	**25.3**	**18.1**
Aug	8.8	3.9	47.0	25.8	**172.6**	3.7	5.3
Sep	20.3	18.4	43.4	21.7	21.2	13.2	12.2
Oct	7.3	5.5	12.3	23.8	11.3	7.9	3.6
Nov	4.0	1.1	0.8	1.1	1.4	1.4	1.2
Dec	2.1	1.0	0.8	1.1	1.5	1.1	2.1
1987							
Jan	6.4	2.8	2.4	2.4	2.8	3.3	2.9
Feb	15.4	15.5	19.3	17.7	**21.0**	20.4	9.8
Mar	19.3	26.2	28.9	23.5	15.9	15.9	18.6
Apr	**25.3**	16.7	17.4	19.3	20.8	**22.6**	**22.6**
May	9.6	9.8	9.8	8.5	12.0	13.3	8.1
June	18.6	**40.9**	**105.7**	**39.2**	14.5	17.1	11.7

Table 15.8 Maximum summer chlorophyll (mg m^{-3}) concentrations recorded during the surveys progressing down bay from Station 2 to Station 7 compared with the higher of the two annual survey means for each station, and the maximum chlorophyll concentrations (mg m^{-3}) during the 1985–1986 brown tide at the stations.

Station	2	3	4	5	1	6	7
Summer Maximum	67.9	137.3	84.5	172.3	37.6	34.8	34.4
Maximum Annual Survey Mean	13.9	21.0	24.6	20.5	16.2	13.9	11.2
Brown tide chlorophyll maximum	67.9	137.3	84.5	46.2	37.6	34.8	34.4

availability. A surge in nutrients is needed to stimulate summer blooms in the nutrient-poorer lower Narragansett Bay, even with a relaxation in grazing pressure during those events. Unusually extensive down bay excursions of the upper bay nutrient plume at those times, particularly nutrient surges driven by storm-runoff as discussed earlier, is one probable source (Fig. 15.2).

15.3.13 Winter–Spring Bloom

The anomalous, 1985 *Aureococcus* brown tide did not have an hysteresis effect disrupting the subsequent winter–spring bloom (Pratt, 1959; 1965; Smayda, 1973a). During the 1985–1986 survey, bay-wide chlorophyll levels were lowest (<1.2 mg m^{-3}) in December. The winter–spring bloom began simultaneously at all stations in early January. Bloom intensity, initially, was greatest in lower Narragansett Bay and Greenwich Bay, then progressed up bay the following 2 weeks. Subsequent dynamics differed between the 1985–1986 and 1986–1987 winter–spring blooms. The station rank order of the 1985–1986 chlorophyll maxima was $4 > 1 = 5 > 6 > 3 > 7 > 2$, with the bloom maximum most intense in Greenwich Bay during January. Termination of the bloom varied regionally, in early March (<10 mg m^{-3} chlorophyll) in lower Narragansett Bay (Stations 1, 6, and 7) and 6 weeks later, in mid-April, in the upper bay (Stations 2, 3, 4, and 5). Within 2 weeks of the collapse of the winter–spring diatom bloom in the upper bay, the brown tide species *Aureococcus anophagefferens,* ichthyotoxic *Heterosigma akashiwo* and a diverse dinoflagellate community developed in mid-May.

In 1986, the seasonal, bay-wide decrease in summer-autumn abundance began in October and reached its nadir (1.0–2.0 mg m^{-3}) in November and December (Table 15.7). The winter–spring bloom then began in January, 1987, reached a March maximum at upper bay Stations 2, 3, and 4, and an April maximum in the mid- and lower bay regions (Table 15.7). Unlike the summer chlorophyll maximum, the magnitude of the winter–spring maximum was similar at all stations; it ranged from about 24–29 mg m^{-3} in upper Narragansett Bay, and 22–25 mg m^{-3} elsewhere. The bloom terminated bay-wide in May, although chlorophyll levels remained above pre- winter–spring bloom levels, ranging from about 8.0–13.0 mg m^{-3}. Following the collapse of the 1987 winter–spring bloom, there was a strong resurgence in growth in June, with a major bloom developing once again at Station 5 (106 mg m^{-3} chlorophyll), the third highest bloom event recorded during the 62 transect surveys (Table 15.7).

15.3.14 Bloom Synchrony

A major bloom feature in Narragansett Bay is the bay-wide synchrony that occurs in the initiation and regional progression of the major seasonal bloom events. The 1986–1987 winter–spring bloom that began in February (1987), and the precedent, pre-bloom demise that began in October–November and then reached a December (1986) nadir, were synchronous bay-wide events. During the 1985–1986 survey, the winter–spring bloom commenced simultaneously at all stations (early January). Bloom intensity, initially greatest in lower Narragansett Bay and Greenwich Bay, then progressively increased up bay the following 2 weeks. The winter–spring bloom also progressed up bay during the 1986–1987 surveys. The bloom began in lower Narragansett Bay (Stations 6 and 7) and reached its maximum 5 weeks later (30 March) in upper Narragansett Bay (Stations 2, 3, and 4) and 8 weeks later (27 April) in Greenwich Bay.

The synchronous bay-wide initiation and up bay progression of the winter spring bloom (as chlorophyll) contrast with Pratt's (1959) study which used cell number as the measure of phytoplankton abundance. He reported the winter–spring bloom began up bay, then spread down bay. This different behavior from our study may reflect the different measures of abundance used—chlorophyll vs. cell numbers. However, in agreement with present findings, Pratt found that the winter–spring bloom during the 1950s was more or less uniform bay-wide. Three additional examples of uniform bloom behavior in Narragansett Bay are: the bay-wide similarity in summer bloom dynamics during the 1986–1987 surveys, notwithstanding variations in abundance along the nutrient–salinity gradient; the prodigious bay-wide brown tide in 1985 that discolored bay waters and extended into Rhode Island Sound (Smayda and Villareal, 1989a,b), and the periodic bay-wide blooms of ichthyotoxic *Heterosigma akashiwo* (Tomas, 1980).

In summary, the evidence collectively suggests that blooms in Narragansett Bay tend to respond uniformly, i.e., as whole ecosystem responses, driven by factors (i.e., nutrients, grazing, etc.) that are operative bay-wide, and do not primarily reflect a mosaic of markedly divergent, independent local ecosystems (i.e., Greenwich Bay, Providence River estuary, etc.) where the blooms are driven by novel habitat growth conditions. The unusual fish and shellfish mortalities observed during 2003 in Greenwich Bay partially challenge this view.

15.3.15 Bloom Intensity

The nutrient-linked fertility of Narragansett Bay and bay-wide bloom potential are evident in the frequency of chlorophyll measurements that exceed 10 and 20 mg m^{-3}, respectively, thresholds useful as indicators of bloom intensity (Table 15.9). Blooms >10 mg m^{-3} chlorophyll were common during the surveys; this threshold (excluding Station 7) was exceeded 47–77% of the sampling year during 1985–1986; 33–61% during 1986–1987. Blooms >10 mg m^{-3} chlorophyll were considerably less frequent during 1986–1987. Blooms >20 mg m^{-3} chlorophyll during the 1985–1986 surveys occurred 40% of the year in Greenwich Bay and mid-bay Station 5; 60% of the year near Conimicut Point (Station 4), and 25–33% at all other stations. The high fertility of Greenwich Bay and Stations 4 and 5 is evident. During 1986–1987, blooms >20 mg m^{-3} chlorophyll were considerably less frequent at all stations, ranging from 4% to 29%. Blooms of this magnitude were most prominent at upper and mid-bay Stations 3, 4, and 5, recorded in 25–30% of the surveys.

15.3.16 Nutrients, Temperature, Stress, and Resilience

The spatial and temporal ecological similarities during the surveys, which indicate the inherent and interactive physical, chemical, and trophic drivers of plankton dynamics in Narragansett Bay, can accommodate anomalous stresses and bloom behavior without ecosystem disequilibration. The question is how long, and at what intensity can stresses such as anthropogenic nutrification and climate change be accommodated without ecosystem disruption? The resilient, intrinsic capacity of the Narragansett Bay ecosystem to restore baseline

Table 15.9 Frequency distribution (%) of chlorophyll concentrations >10 mg m^{-3} and >20 mg m^{-3} at the monitoring stations using all data.

	Station						
	1	2	3	4	5	6	7
>10 mg m^{-3}							
1985–1986	48	55	68	77	62	47	42
1986–1987	43	33	50	57	61	50	25
>20 mg m^{-3}							
1985–1986	39	23	29	61	41	33	23
1986–1987	11	11	25	25	29	14	4

plankton dynamics after exposure to anthropogenic stress is, most likely, sensitive to whether the disruptive stress is acute or chronic. Smayda (1998) has analyzed plankton variability in Narragansett Bay and distinguished between induced and reflected variability, and between "point" and "threshold" events. "Threshold" events that occur as ecological "turning points" were also recognized. Ecological "turning points" may be either transitory or prolonged, and can have a temporary or permanent influence on subsequent plankton dynamics and variability. The 1985–1986 brown tide event was a "point" event, having met the following criteria: it has not recurred, the causative species has not become incorporated into the annual successional cycle, its bloom was both unpredictable and transitory, and its trophic consequences were temporary.

Despite the resilience and capacity of the Narragansett Bay ecosystem to accommodate "point events" without loss of equilibrium, a "threshold" event seemingly leading to more permanent change is the long-term increase in winter water temperatures in progress, and which is currently 3°C above the long-term average (Cook *et al.*, 1998; Keller *et al.*, 1999). This temperature increase has led to a seasonal shift and reduction in the magnitude of the annual winter–spring bloom of the diatom *Skeletonema costatum sensu lato* (Borkman and Smayda, in review), the major bloom species in Narragansett Bay (Borkman, 2002). Altered and persistent changes in winter–spring bloom community structure, dynamics and trophic process, including zooplankton dynamics, have also occurred, which Smayda *et al.* (2004) have linked to long-term changes in temperature and the North Atlantic Oscillation Index. A temperature-induced increase in the abundance and altered seasonal dynamics of the carnivorous ctenophore *Mnemiopsis leidyi* may also be occurring (Sullivan *et al.*, 2001; Sullivan *et al.*, Chapter 16).

The capacity of Narragansett Bay to accommodate "point" events, such as the prolonged and anomalous brown tide, and its vulnerability to "threshold"-driven ecosystem changes are relevant to the issue of what ecosystem changes might result from either sustained nutrient enrichment or oligonutrification of Narragansett Bay. Assessment of this cannot overlook the changing character of plankton dynamics in Narragansett Bay currently being driven by a non-nutritional factor—temperature. The effect of climate change must be considered in the selection of which management strategy options to apply in seeking to alter the nutrient–phytoplankton biomass relationship in Narragansett Bay. This assessment should not be restricted to a yield–dose (biomass) evaluation, since grazing processes are also fundamental to ecosystem dynamics and nutrient cycling, particularly in systems such as Narragansett Bay with strong benthic–pelagic coupling (Officer *et al.*, 1982), and which must be factored into an ecosystem management strategy. The prey suitability of the species that bloom in response to, and are sensitive to nutrient conditions varies. The impact and grazer-value of the bloom species selected for will vary along the spectrum, from highly nutritious-to-unpalatable-to-ichthyotoxic, or having some other inimical effect. Eighteen species are found in Narragansett Bay which have produced harmful blooms locally or elsewhere within their distributional range (Smayda, unpublished).

15.3.17 Primary Production

Mean annual primary production rates were similar for the 1985–1986 and 1986–1987 survey periods (Table 15.10), a striking result given the considerably lower biomass levels during 1986–1987 (Table 15.6). The major differences between the surveys were the 33% higher production rate at Station 5, and 23% lower annual rate at Station 2 during 1986–1987. For other stations, the differences between surveys were $<\pm15\%$. There was a gradient in the mean annual primary production rates similar to that for salinity, nutrients, and chlorophyll (Fig. 15.6a–c). The maximal annual production rates (approximately 300 g C m^{-3} yr^{-1}), similar during both the survey periods, occurred at inner bay Station 3. The minimal annual rates (approximately 75 g C m^{-3} yr^{-1}) in lower Narragansett Bay (Station 7) were fourfold lower. The highest production rates along the gradient occurred in the region between Gaspee Point (Station 3) and Providence Point on Prudence Island (Station 5), similar to that found by Oviatt *et al.* (2002). Annual production rates during the present and Oviatt *et al.* surveys relative to the maximum measured (Station 3), were about 55–60% lower up bay of Station 5 in the Providence River (Station 2), in Greenwich Bay (Station 1) and mid bay (Station 6), and about 75% lower at lower bay Station 7 (Fig. 15.1). The 3.5- to 4.5-fold regional difference in primary production rates between lower and upper Narragansett Bay exceeded the

Table 15.10 Annual primary productions rates (g C m^{-3} yr^{-1}) during the surveys.

Station	1	2	3	4	5	6	7
1986–1987	116.7	123.0	284.0	207.7	201.6	127.2	64.5
1985–1986	120.5	159.0	261.3	231.1	151.5	112.7	76.5
Δ	−3.8	−36.0	+22.7	−23.4	+50.1	+14.5	−12.0
%Δ	−3%	−23%	+9%	−10%	+33%	+13%	−16%

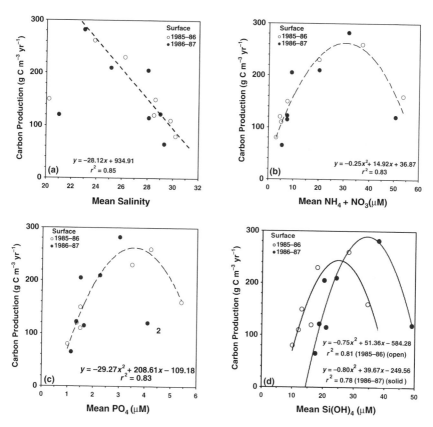

Fig. 15.6 (a) Mean annual carbon production rates along the surface salinity gradient during the surveys. (Equation excludes Station 2). (b) Mean annual carbon production rates vs. mean $NH_4 + NO_3$, (c) mean PO_4, and (d) mean $Si(OH)_4$ concentrations along the surface salinity gradient during the surveys.

twofold difference between the chlorophyll maxima and minima during the surveys (Table 15.7).

15.3.17.1 Assimilation Number and Carbon Growth Rates

The mean daily Assimilation Number (ANo)—the carbon production rate per unit chlorophyll per unit time as mg C mg^{-1} chl d^{-1}—is a biomass-based productivity index that provides insight into the growth vigor and efficiency of the phytoplankton in response to habitat conditions, particularly the nutrient field. It is sensitive to the phylogenetic character of the bloom species. Despite the similar primary production rates during both the surveys (Table 15.10), the mean ANos at the stations during 1986–1987 were 33–118% higher than that in 1985–1986 (Table 15.11). The highest mean ANo (57.2 mg C mg^{-1} chl d^{-1}),

Table 15.11 Mean assimilation number (mg C fixed mg^{-1} chl d^{-1}) at the survey stations, and the difference (%Δ) between the 1985–1986 versus 1986–1987 survey periods.

Station	1	2	3	4	5	6	7
1985–1986	20.4	31.3	34.1	25.7	20.3	22.2	18.7
1986–1987	34.3	46.8	57.2	44.1	44.2	35.6	24.9
%Δ	+68%	+50%	+68%	+72%	+118%	+60%	+33%

recorded at Station 3, was similar to the mean (66 mg C mg^{-1} chl d^{-1}) calculated for that Station from the data (assuming a 12L:12D daily light:dark photoperiod in hours) presented in Table 1 of Oviatt *et al.* (2002), who carried out primary production measurements a decade later, from April 1997–April 1998.

The ANo tended to increase with the magnitude of the annual production rate, and yielded strong, positive linear correlations ($r = 0.68$–0.71) when plotted against the mean NH$_4$ + NO$_3$, PO$_4$, and Si(OH)$_4$ concentration gradients during the 1986–1987 surveys. With Station 2 included in the comparisons against NH$_4$ + NO$_3$ and Si(OH)$_4$, a hyperbolic relationship resulted; exclusion of Station 2 from the comparison yielded a linear relationship and a higher correlation coefficient ($r = 0.89$). For the 1985–1986 regressions, with Station 2 included the correlation ranged from $r = 0.96$–0.98; hyperbolic trends were not evident when regressed against NH$_4$ + NO$_3$ and Si(OH)$_4$. The reasons for the greater carbon production per unit chlorophyll during the 1986–1987 surveys are obscure. The higher ANos are notable given the prolonged and intense brown tide bloom during May–October in 1985.

15.3.17.2 Carbon Growth Rates Based on Primary Production

Mean carbon growth rates were calculated and related to nutrient levels. A carbon:chlorophyll ratio of 50:1, the ratio routinely used by phytoplankton ecologists, was used to transform the chlorophyll data into carbon equivalents, and the carbon growth rate (k) calculated from:

$$k = \ln \frac{C_p + C_b}{C_b} \left[\frac{t}{1 \ \ln 2} \right] \tag{15.3}$$

where C_p is the mean daily C production rate (mg C m^{-3} d^{-1}); C_b the mean daily phytoplankton C standing stock (mg C m^{-3}); and t is time in days. The C-based growth rates ($k = 0.58$–1.10 d^{-1}) at the 1986–1987 stations exceeded 1985–1986 rates ($k = 0.46$–0.75 d^{-1}). The fastest growth rates during both surveys occurred at upper bay Stations 2, 3, 4, and 5 (Table 15.12). Carbon doubling rates in that region during 1986–1987, on average, reached 1.0 d^{-1} ($\mu = 0.97$ d^{-1}). In lower Narragansett Bay (Stations 1, 6, and 7), carbon doubled every 34 h ($\mu = 0.70$ d^{-1}). Rank ordering the stations by growth rate was identical during both survey years.

Table 15.12 Mean daily carbon-based growth rates (k d^{-1}) during the surveys.

Station	1	2	3	4	5	6	7
1985–1986	0.49	0.70	0.75	0.60	0.49	0.53	0.46
1986–1987	0.75	0.95	1.10	0.91	0.91	0.78	0.58

A hyperbolic relationship occurred between the mean daily carbon growth rate and mean $NH_4 + NO_3$, PO_4, and $Si(OH)_4$ concentrations during both surveys (not shown). At <12 mg-at m^{-3} $NH_4 + NO_3$ (Stations 1, 5, 6, and 7), growth rates for both the survey years fell along the same regression line ($r = 0.81$). At >12 mg-at m^{-3} $NH_4 + NO_3$ (Stations 2, 3, and 4), growth rates were more or less invariant up to about 55 mg-at m^{-3} $NH_4 + NO_3$. Mean daily carbon growth rates increased along the nutrient–salinity gradient. The average generation time (e.g., doubling time), calculated from ($1/k$), decreased from 1.7 to 0.9 days between lower and upper bay Stations 7 and 2, respectively, during 1986–1987, and from 2.2 to 1.3 days during 1985–1986. This links the progressive increase in primary production up bay along the salinity gradient to increasing nutrient availability. The second highest growth rates and ANo during both surveys were measured at Station 2 in the Providence River estuary. This rapid growth suggests that the low phytoplankton biomass (chlorophyll) recorded at that station, lower than expected from the relationship between standing stock and nutrients, as discussed previously, was not solely a water quality effect. Washout in this rapidly flushed region (Asselin and Spaulding, 1993), and perhaps secondarily, benthic grazing, may have prevented the accumulation of chlorophyll to a level commensurate with its production. Thus, two mechanisms (excluding accretion) that may be contributing to the high nutrient levels characteristic of Station 2 are repression of phytoplankton growth because of degraded water quality (e.g., the nutrient concentrations, their ratios, and toxicants) and washout.

The data suggest, therefore, that nutrients accreted into upper Narragansett Bay via the Providence River estuary and sewage treatment facility discharge, regulate phytoplankton growth in Narragansett Bay. Along the downstream gradient, the effect of this fertilization is greatest in upper Narragansett Bay. Dilution and phytoplankton assimilation in mid- and lower Narragansett Bay regionally diminish this nutrient flux, the consequence of which is to increase the down bay importance of *in situ* remineralization and offshore nutrient inputs. Hence, two primary nutrient pumps are operative in Narragansett Bay and regulate phytoplankton growth—nutrient accretion into upper Narragansett Bay and *in situ* remineralization and advection in mid- and lower Narragansett Bay. The accompanying gradient in primary grazers (benthos vs. zooplankton) and their collective grazing intensity influences the residual amounts (i.e., after grazing) of chlorophyll that develops in response to these nutrient inputs.

15.3.17.3 Primary Production–Nutrient Relationships

The relationships between primary production and salinity and nutrient gradients are shown in Fig. 15.6a–c. There is a strong, inverse correlation ($r = -0.93$) between mean annual production rate and mean salinity along the gradient, combining data for both surveys at Stations 1, 3, 4, 5, 6, and 7, and which fall along the same regression line (Fig. 15.6a). A progressive up bay increase in primary production occurred along the gradient of decreasing salinity that was abruptly reversed in the Providence River estuary where mean salinity was <23‰. Primary production at Station 2, where the mean surface salinity was 20.3‰ and 21.2‰ during the surveys, respectively, was depressed by 40–60 % (Table 15.10) relative to nearby Station 3. Over the approximately 1.5 km distance separating Stations 2 and 3, mean surface salinity and mean annual surface production during 1985–1986 increased by 3.4 % and 102 g C m^{-3} and by 1.9 % and 161 g C m^{-3} during the 1986–1987 period, respectively. For this reason, divergent Station 2 was excluded from the statistical analyses. The annual production at Providence River Stations 2 and 3 was also reduced (-15 to -25 % y^{-1}) relative to Station 6 (619 g C m^{-2} y^{-1}) during a 1-year survey from April 1997–April 1998 (see Fig. 6 in Oviatt *et al.,* 2002); a gradient in primary production was also found. Annual primary production rates in upper Narragansett Bay were about fourfold greater than at lower bay Station 7 (160 g C m^{-2} y^{-1}). The gradient in primary production is consistent with measurements by Oviatt *et al.* (1981) during 1971–1973, and estimations for a 9-month period in 1976 based on a light–chlorophyll–photosynthesis model (Durbin and Durbin, 1981).

Annual carbon production rates along the gradient during both the surveys are strongly correlated with mean NH$_4$ + NO$_3$, PO$_4$, and Si(OH)$_4$ concentrations (Fig. 15.6b–d). The relationship with each nutrient was hyperbolic, similar to chlorophyll–nutrient regressions (Fig. 15.5). However, unlike the latter, plotting the primary production rates against the NH$_4$ + NO$_3$ and PO$_4$ concentrations clusters the stations from both the surveys into a single regression line, rather than two distinct regressions (Fig. 15.6b,c). When production is plotted against Si(OH)$_4$ concentrations, however, two hyperbolic relationships emerge (Fig. 15.6d), similar to that for chlorophyll against Si(OH)$_4$ (Fig. 15.5c). Each relationship was highly significant, with $r^2 = 0.83$ for the NH$_4$ + NO$_3$ and PO$_4$ regressions and equally strong for the two regressions against Si(OH)$_4$ (Fig. 15.6d). The production rate is depressed at the high NH$_4$ + NO$_3$, PO$_4$, and Si(OH)$_4$ concentrations at Station 2. This inhibition is similar to the depression in mean chlorophyll concentrations observed when plotted against Station 2 nutrient concentrations (Fig. 15.5). The ANos suggest that the reduction in chlorophyll levels in the inner Providence River estuary is not due to "water quality" impairment of photosynthesis (Table 15.11). The productivity index at Station 2, on average, was the second highest among stations during the surveys, falling just below the maximum ANo at nearby Station 3 (Table 15.11). The physiological vigor that is evident suggests washout, rather than growth suppression, was a more important determinant of the reduced chlorophyll concentrations found at Station 2 relative to other stations.

15.3.17.4 Nitrogen Supply Vs. Primary Production Nitrogen Requirements

The mean daily N and P production rates accompanying the primary production (C) rates were estimated applying the Redfield Ratio stoichiometry between C production and N and P utilization. The concentrations of N and P required to support the directly measured C production rates were then summed with the inorganic N and P concentrations measured *in situ*, and the proportion (percentage) of the summed N and P nutrient pools that would be assimilated to support the estimated N and P production rates plotted against the residual N and P concentrations. There was a highly significant, inverse curvilinear correlation between N availability in the $NH_4 + NO_3$ pool along the gradient and its percent utilization in support of the measured carbon production rates. This is shown in Fig. 15.7 for the 1985–1986 surveys. On average, there was a 4- to 7.5-day supply of inorganic N available in upper

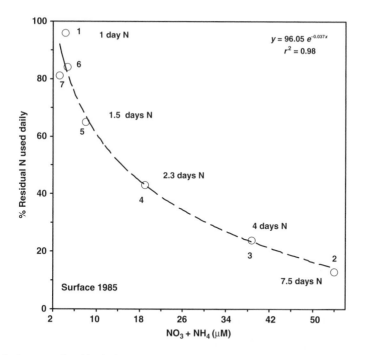

Fig. 15.7 Percent of residual nitrogen ($NH_4 + NO_3$) concentration assimilated by phytoplankton during mean daily carbon production rates vs. mean $NH_4 + NO_3$ concentrations along the surface salinity gradient during the 1985–1986 survey. The corresponding number of days' supply of residual $NH_4 + NO_3$ available, on average, to support measured carbon production rates at those concentrations is entered. See text for further detail.

Narragansett Bay (Stations 2 and 3) to support the measured carbon production rates. The N surplus progressively decreased down bay to a 1.5- to 2.5-day N supply in mid bay (Stations 4 and 5), and a 1- to 1.5-day reserve in Greenwich Bay (Station 1) and lower bay Stations 6 and 7. The PO_4 concentrations along the nutrient–salinity gradient required to satisfy the mean daily carbon production rate ranged from 6% to 28% of the available PO_4 concentration during 1986–1987, and from 6% to 22% during 1985–1986. This is equivalent to a 3.5- to 16-day P supply.

The response (Fig. 15.7) suggests that carbon production in lower Narragansett Bay is much more dependent on recycled nutrients and nutrients advected from Rhode Island Sound than in the down bay flux of the upper Narragansett Bay nutrient enrichment plume originating in the Providence River estuary. A shift in the dominant mode of nutrient flux, from accretion to remineralization, probably occurs in the region of the Narragansett Bay waters near Station 5. The results also suggest that the N resupply rate is more critical than that for P.

15.3.18 Zooplankton

15.3.18.1 Zooplankton Composition, and Influence of a Brown Tide

The zooplankton community in Narragansett Bay is dominated by two congeneric copepods: *Acartia hudsonica*, the winter–spring dominant, and *Acartia tonsa*, the summer dominant (Durbin and Durbin, 1981; Martin, 1965). Benthic larvae also make a large contribution to the zooplankton community, both numerically and in percentage representation. Mean benthic larval abundance during the surveys ranged from about 1,440 to 3,205 larvae m^{-3}; this corresponded to 13–56% and 12–33% of the mean copepod abundance at the seven stations during the 1985–1986 and 1986–1987 surveys, respectively. During 1985–1986, benthic larval abundance generally exceeded *A. tonsa* and *A. hudsonica* abundance in the Providence River estuary (Stations 2 and 3), unlike in 1986–1987.

An annual successional shift from *A. hudsonica* to *A. tonsa* occurs during summer. In 1985, this coincided with the initiation and persistence of the brown tide bloom, which negatively influenced *A. tonsa* abundance. Mean *A. tonsa* numerical abundance m^{-2} plotted against the mean number of brown tide cells m^{-2} was inversely related. The stations clustered into two subgroups of equal correlation ($r = -0.91$): Stations 2, 3, and 4 in upper Narragansett Bay formed one group, and Stations 1 (Greenwich Bay), 5, 6, and 7 the other group. *Acartia tonsa* was much more abundant at the upper bay stations, although the slopes of the regressions for the two subgroups did not differ statistically. The negative impact of the brown tide on *A. tonsa* abundance is consistent with experimental results that the bloom species, *Aureococcus anophagefferens*, is a poor food source for *A. tonsa* (Durbin and Durbin, 1989). During the 1986–1987 surveys, when the brown tide did not occur, mean abundance of *A. tonsa* exceeded 1985–1986 levels bay-wide, by 150% (Station 4) to 348% (Station 1). In contrast, mean *A. hudsonica* abundance was 20–30% lower in Greenwich Bay and at Stations 2, 3, and 6 during 1986–1987 than in 1985–1986, and 47% lower at station 7, but more abundant at Stations 4 and 5: 17% and 36%, respectively.

The most notable zooplankton disruption that accompanied the brown tide was failure of the cladoceran community to develop (Fofonoff, 1994). *Evadne nordmanni* and *Podon* sp. normally exceed 10,000 animals m^{-3} during June–August, but failed to appear in 1985. The mean cladoceran abundance (80 animals m^{-3}) during May–August 1985 at Station 7 was 10- to 75-fold lower than the 6-year mean for an earlier time series available for that station (Smayda and Fofonoff, 1989). During 1986, cladocerans appeared in the spring, and by mid-June reached abundances at the transect stations that ranged from 2,000 to 29,000 animals m^{-3}. Cladocerans remained abundant through late-August, disappeared, then reappeared in April 1987 and became moderately abundant (572–15,500 m^{-3}) by late-June when the 1986–1987 survey terminated.

Benthic larval abundance was also adversely affected by the brown tide. Mean larval abundance m^{-2} at the seven stations was inversely correlated ($r = -0.58$) with brown tide cell numbers (Smayda and Fofonoff, 1989). Benthic invertebrate larvae were least abundant (857 m^{-3}) in 1985 based on the 6-year time series dataset for Station 7 (mentioned earlier), after attaining their greatest recorded abundance (2,385 m^{-3}) in 1984 (Smayda, unpublished). During the 1986–1987 survey, the mean annual benthic larval abundance at Station 3 was 23% lower, but increased by 40–60% in Greenwich Bay and at Stations 2, 5, and 6, and by 10% at Stations 4 and 7. The greater benthic larval recruitment that generally characterized the 1986–1987 survey may partly reflect release from the adverse effects of the 1985 brown tide on the benthic community fecundity (Tracey *et al.*, 1988).

15.3.18.2 Zooplankton Behavior along the Salinity Gradient

The multiple impacts of the brown tide on the copepod, cladoceran, and benthic larval populations during the 1985–1986 surveys, followed by the 1986–1987 recovery period, complicate a detailed analysis of zooplankton behavior during the surveys. However, there were strong inverse correlations between salinity and the abundance of both *A. hudsonica* and benthic larvae. Mean abundance of *A. hudsonica* progressively decreased up bay along the salinity gradient, being strongly correlated with mean salinity during both the survey periods: $r = 0.98$ (1985–1986) and $r = 0.75$ (1986–1987). In contrast, *A. tonsa* abundance did not correlate with salinity. In addition, unlike *A. hudsonica*, mean benthic larval abundance progressively increased up bay during 1985–1986, with a strong inverse correlation found with salinity ($r = -0.67$, but $r = -0.97$ excluding

Station 2). A down bay decrease was also found during 1986–1987, but the correlations were weaker: $r = -0.66$ and -0.40, respectively. Benthic larvae tended to be most numerous in the summer.

The divergent abundance and population behavior of the copepod, cladoceran, and benthic larval communities, including their responses to the brown tide bloom and their distribution along the salinity gradient, are conflated and lost when included within the zooplankton biomass measurements. Zooplankton biomass (dry weight m^{-2}) was correlated with salinity along the gradient, but the relationship is more complex than when numerical abundance is used. Mean biomass levels plotted against mean salinity were strongly and positively correlated at upper bay Stations 2, 3, and 4 ($r = 0.98$), and strongly, but *negatively* correlated along the down bay gradient in the region incorporating Stations 4, 5, 6, and 7, and Greenwich Bay ($r = -0.85$). The data for both the survey periods at the mid- and lower bay stations fall along the same regression line. Thus, mean zooplankton biomass increased with salinity along the gradient from Field's Point in the Providence River estuary to Conimicut Point (Station 4), then progressively decreased with increasing salinity down bay. This highly significant ($r^2 = 0.70-0.73$) zooplankton biomass–salinity relationship is hyperbolic (Fig. 15.8a), similar to the chlorophyll–salinity relationship (Fig. 15.4), chlorophyll and nutrients (Fig. s 15.5a,c), primary production and salinity (Fig. 15.6a), and primary production and nutrients (Figure 15.6b–d). Mean zooplankton biomass during the 1986–1987 surveys exceeded 1985–1986 levels by 58% at Station 2, by 20–40% at Stations 1, 3, 4, 5, and 6, and by 8% at Station 7.

The relationship between zooplankton biomass and salinity (Fig. 15.8a) is probably not osmotically regulated, but a response to the phytoplankton standing stock whose abundance runs in parallel with salinity (Fig. 15.4). Since the dominant copepod taxa sampled are primarily herbivorous, the relationship between mean zooplankton biomass m^{-2} and mean surface chlorophyll was examined (Fig. 15.8b). The slopes of the two annual regressions differed. Zooplankton biomass strongly correlated with chlorophyll during 1985–1986 ($r^2 = 0.92$), and to a lesser extent during 1986–1987 ($r^2 = 0.56$). Mean zooplankton biomass also correlated with mean annual primary production during both the surveys, with $r^2 = 0.72$ and 0.59, respectively, but was depressed at Station 2 during both the surveys relative to the levels expected from the high annual production rates at that location (Fig. 15.6b). Zooplankton biomass did not correlate with brown tide cell abundance. The 1986–1987 surveys indicate that the zooplankton community structure and dynamics returned to "normal" patterns following the anomalous 1985–1986 annual cycle; this may account for the elevated zooplankton biomass at upper bay Stations 2, 3, 4, and 5 during 1986–1987.

Despite the ecologically disruptive 1985 brown tide, the strength of the salinity gradient and associated properties overrode that disruption, and zooplankton abundance remained under regulation of the trophic properties associated with the salinity gradient. The greater abundance of benthic larvae found

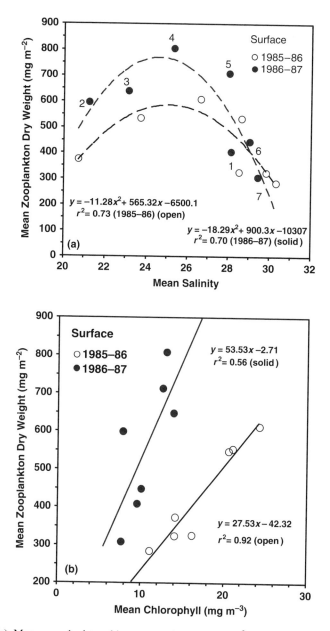

Fig. 15.8 (a) Mean zooplankton biomass as dry weight m^{-2} along the surface salinity gradient during the surveys, and (b) mean zooplankton biomass vs. mean chlorophyll concentrations along the surface salinity gradient.

in upper Narragansett Bay and the larger stocks of benthic, filter-feeding bivalves in that region suggest that there is a two-compartment subdivision of Narragansett Bay with regard to herbivorous grazer regulation. Benthic grazing (including by larvae) would appear to be more important than zoo-plankton grazing in upper Narragansett Bay, whereas in lower Narragansett Bay zooplankton grazing probably exceeds benthic grazing. Yet, the role of benthic grazing in regulating phytoplankton dynamics and nutrient recycling in Narragansett Bay, and the amount of phytoplankton available for predation by zooplankton have been virtually ignored. It has been argued that zooplankton populations in Narragansett Bay are food-limited (Durbin *et al.*, 1983). It is essential that the benthic trophic compartment be included in the development of ecosystem management strategies for Narragansett Bay.

15.3.19 Ctenophores

When the ctenophore *Mnemiopsis leidyi*, a voracious grazer of copepods, is abundant in Narragansett Bay, there is a coincident decrease in zooplankton abundance and increase in phytoplankton (Deason and Smayda, 1982a,b). Higher phytoplankton abundance occurs during ctenophore presence because copepod grazing pressure is reduced through the carnivorous decimation of the copepod population by *Mnemiopsis*. That is, ctenophores can regulate phyto-plankton abundance in Narragansett Bay through top–down grazing on herbi-vorous zooplankton, with the strength of this indirect regulation a function of ctenophore abundance. For this reason, ctenophores were sampled during the surveys to examine the interrelationships between the standing stocks of phy-toplankton, zooplankton, and ctenophores.

Mean numerical abundance per square meter of *Mnemiopsis* (Table 15.13) during the 1986–1987 survey varied threefold, from 25 animals m^{-2} (Station 1) to 74 animals m^{-2} (Station 6), and by sixfold during the 1985–1986 survey, from 15 animals (Station 7) to 95 animals m^{-2} (Station 3). Mean abundance was approximately 50% lower in Greenwich Bay and at upper bay Stations 3 and 4 in 1986–1987 than in 1985–1986, and two- to threefold greater at lower bay Stations 6 and 7. The annual ctenophore population maximum during 1985–1986 occurred in October following disappearance of the brown tide, and ranged from 61 to 806 animals m^{-2} along the salinity gradient. The population then declined, remained at very low levels, and disappeared in March. *Mne-miopsis* reappeared bay-wide in May (1986), persisted in very low abundance through June, usually <6 animals m^{-2}, then surged in abundance in late August and early September (1986) when the population ranged from 42 to 425 animals

Table 15.13 Mean numerical abundance (m^{-2}) of the ctenophore *Mnemiopsis leidyi* during the surveys.

Station	1	2	3	4	5	6	7
1985–1986	50	60	95	66	67	33	15
1986–1987	25	50	46	33	45	74	34

m^{-2} at the individual stations. Thereafter, *Mnemiopsis* declined, resurged in November, reached its annual maximum in December–January, then disappeared in February (1987).

15.3.19.1 Ctenophore Behavior along the Salinity Gradient

Mean ctenophore abundance during 1985–1986 plotted against salinity progressively increased up bay from Station 7 to Station 3, then decreased at Station 2, where it was about 40% less abundant (60 m^{-2}) than at Station 3 (Fig. 15.9; Table 15.13). The hyperbolic relationship between *Mnemiopsis* population density and salinity was highly significant ($r^2 = 0.89$). Excluding Station 2 from the comparison resulted in a strong, inverse correlation between ctenophore numbers and salinity ($r = -0.92$). In contrast, the 1986–1987 population density plotted against mean salinity was neither hyperbolic, nor well-correlated. The trend lines (Fig. 15.9) exhibit two distinct and coherent station groupings. Stations 1, 2, 3, and 4 in Greenwich Bay and upper Narragansett Bay formed one group, and exhibited a strong linear decrease in abundance with increasing mean salinity. This response is similar to the inverse relationship found during the 1985–1986 survey grouping all stations exclusive of Station 2 (Fig. 15.9). Ctenophore abundance in mid- and lower Narragansett Bay (Stations 5 and 6) was not related to salinity.

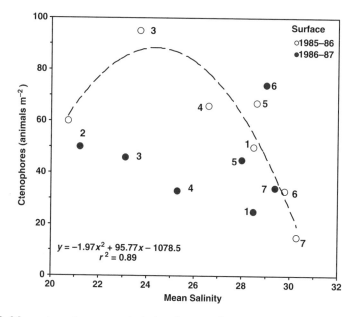

Fig. 15.9 Mean ctenophore numerical abundance m^{-2} along the surface salinity gradient during the surveys. Equation applies to 1985–1986 survey.

Two uncertainties cloud this apparent response. The use of numerical abundance combines adult and juvenile *Mnemiopsis* animals, whereas biomass volume is probably a better indicator of ctenophore abundance since *Mnemiopsis* body size varied considerably in the populations sampled (Smayda and Fofonoff, unpublished). It is also uncertain whether Greenwich Bay (Station 1) is appropriately grouped with Stations 2, 3, and 4 in the 1986–1987 survey, which exhibits a 50% decrease in mean ctenophore abundance over an approximate 7.0‰ decrease in salinity along the gradient. The Greenwich Bay station may be more appropriately clustered with Stations 5, 6, and 7, which collectively exhibit a twofold range in abundance over a narrow salinity range of about 2‰ (Fig. 15.9).

Whatever factors regulate ctenophore abundance along the salinity gradient, the abundance of zooplankton as prey is expected to be important. Comparison of mean *Mnemiopsis* abundance with mean annual zooplankton biomass as dry weight (both m^{-2}) confirms this expectation during the 1985–1986 survey, but not during 1986–1987 (Fig. 15.10). While hyperbolic relationships were found for both the surveys, the correlation for 1985–1986 was very significant ($r^2 = 0.87$), unlike 1986–1987 ($r^2 = 0.29$; $p > 0.05$). *Mnemiopsis* 1985–1986 abundance plotted against copepod numerical abundance yields a strong, highly significant inverse relationship ($r = -0.99$), excluding aberrant Station 2. This inverse relationship is consistent with previous field

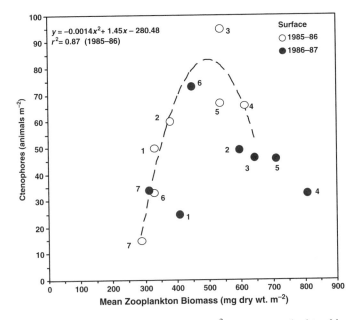

Fig. 15.10 Mean ctenophore numerical abundance m^{-2} vs. mean zooplankton biomass along the surface salinity gradient during the surveys.

observations (Deason and Smayda, 1982a,b) and the well-established grazing of copepods by ctenophores (Kremer, 1994). In contrast, *Mnemiopsis* abundance did not correlate with copepod numerical abundance during the 1986–1987 period.

The different ctenophore–zooplankton relationships for the two survey periods, if not bonafide ecological expressions of the intrinsic variability of Narragansett Bay, may reflect both a sampling artifact and the effect of data averaging. The zooplankton biomass measurements include components not grazed by ctenophores. The measurements also incorporate significant zooplankton biomass levels that develop during the winter–spring bloom, when ctenophores are absent or unimportant—ctenophores are most abundant during summer–early autumn. This seasonal and partly asynchronous annual cycling behavior may partly explain the divergent correlations between ctenophore numerical abundance and zooplankton numerical abundance and biomass. This relationship requires a more detailed study because sudden blooms of harmful algal species in Narragansett Bay appear to be linked to ctenophore–zooplankton dynamics (Pratt, 1966; Deason and Smayda, 1982b; Smayda, 1993).

Phytoplankton standing stock (excluding aberrant Station 2) as carbon ($r = 0.81$) and chlorophyll ($r = 0.84$) was positively correlated with ctenophore abundance during 1985–1986. These correlations reflect the indirect regulation of phytoplankton abundance by the grazing of ctenophores on zooplankton. Correlations were not found using the 1986–1987 survey data.

15.4 Gradients, Ecotones, and the "Nutrient Depuration Zone"

The transect data reveal Narragansett Bay is characterized by six persistent and distinct physical, chemical, and plankton gradients, proceeding down bay from the Providence River estuary: (1) a progressive increase in surface salinity; (2) a decrease in water column turbidity; (3) a decrease in macro-nutrients; (4) a decrease in chlorophyll (Fig. 15.11). The fifth gradient—the transition from a highly stratified water column in the upper bay to a vertically well-mixed lower bay—results from the salinity gradient; a sixth gradient, the nutrient pump gradient, is discussed later.

This interactive combination of physical, chemical, and biological gradients in Narragansett Bay produces a plankton habitat that is regionally transitional and variable, rather than fixed in its physical, chemical, and biological characteristics. That is, Narragansett Bay is an environmental and ecological mosaic; it is not a habitat that can be adequately characterized by the value of a given parameter, such as its bay-wide annual primary production rate, its N:P ratio, or some mass balance feature. Such reductionist characterizations may be suitable for cross-ecosystem comparisons, i.e., Narragansett Bay vs. Delaware

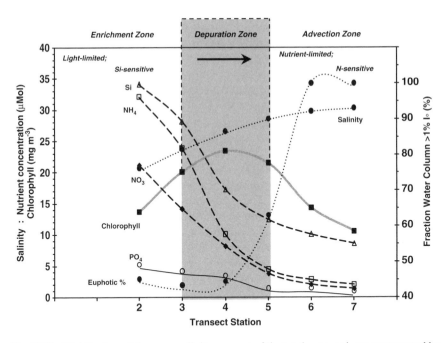

Fig. 15.11 Distribution of the mean salinity, percent of the total water column represented by the euphotic zone, nutrient concentrations and chlorophyll abundance at transect Stations 2 to 7 during the 1985–1986 surveys. (The 1986–1987 data are congruent with the patterns shown). Station 1 in Greenwich Bay excluded from depiction because it lies off the main axis from Stations 2 to 7. The regional transitioning of Narragansett Bay along the salinity gradient from a light-limited, nutrient-enriched, Si-sensitive upper bay to a nutrient limited, N-sensitive lower bay region, and the hypothesized partitioning of Narragansett Bay into three major ecotones—the Enrichment, Depuration and Advection Zones, and their approximate locations along the Narragansett Bay axis depicted. See text for further explanation.

Bay, but are of limited value in the design of a Narragansett Bay ecosystem-based management strategy, which must be whole-system based rather than biased regionally or by parameter.

The interactive physical, chemical, and biological gradients in Narragansett Bay produce an integrated habitat gradient of major ecological importance to phytoplankton growth. Narragansett Bay regionally transitions from a light-limited, nutrient-enriched, Si-sensitive upper bay to a nutrient-limited, N-sensitive lower bay region (Fig. 15.11). These regional patterns and gradients are embedded along a down bay continuum of three distinct, nutrient-defined ecotones—the Enrichment, Depuration, and Advection Zones—that are regionally restricted and partition Narragansett Bay ecologically. These nutrient ecotones and their approximate locations along the gradient are shown in Fig. 15.11.

The Enrichment Zone is located in upper Narragansett Bay, principally in the Providence River estuary, and is characterized by high levels of anthropogenic nutrients accreted through runoff and sewage treatment facility discharge. The Enrichment Zone develops and persists because the nutrient influx, in combination with the flushing characteristics, exceeds the capacity of the phytoplankton to fully utilize the nutrient supply. This results in a surplus of unused nutrient that is advected downstream.

The Depuration Zone, located in the general region of Stations 3, 4, and 5, is nutrient-enriched, and a region of intense phytoplankton growth whose blooms are stimulated by the flux of nutrients being advected down bay from the Enrichment Zone (Fig. 15.11). The chlorophyll abundance pattern along the gradient depicted in Figure 15.11 reflects this locally elevated, nutrient-stimulated growth. Chlorophyll concentrations increase in the Depuration Zone because the phytoplankton growth rate (μ) is elevated relative to flushing and grazing loss rates. A major consequence of the nutrient-stimulated growth in the Depuration Zone is to decrease markedly the ambient nutrient concentrations fluxing through this area. This reduces the magnitude of the nutrient flux down bay and into the Advection Zone during the sequential passage of the nutrient enrichment plume through the Enrichment and Depuration Zones.

The Depuration Zone, consequently, functions as a nutrient biofilter. That is, in its regional photosynthetic conversion of dissolved inorganic (and organic) nutrients to particulate nutrients (i.e., phytoplankton biomass), it regulates the encroachment of nutrient hyper-enrichment and its ecological consequences down bay. The Depuration Zone is an ecosystem property whose dependence goes beyond simple reliance on the phytoplankton as the active biofilterers. Formation of the Depuration Zone requires, and has, a spatial and temporal structure based on a "critical balance" of the rates of advection, nutrient influx and concentrations, phytoplankton growth rate, and grazing rates—interactive rate processes associated with the physical, chemical, and biological gradients found. The Depuration Zone is expected to expand and to contract spatially in response to variations in this "critical balance," with the breadth and duration of these spatial variations dependent on the interactive rates of the co-varying physical, chemical, and biological gradients.

The Advection Zone is located immediately down bay from Station 5 in the Depuration Zone, comprising the mid- and lower bay regions, including both East and West Passages. It is nutrient-poorer and considerably less turbid than the Enrichment and Depuration Zones. The euphotic zone, unlike in the latter zones, extends to the bottom sediments in the Advection Zone (Fig. 15.11). The increased transparency reduces the degree to which phytoplankton growth is light-limited (Hitchcock and Smayda, 1977), and the transmission of light to the bottom sediments allows photosynthesis and growth of the epibenthic microflora (Smayda, 1973b), macroalgae, and *Zostera marina*. In contrast, the euphotic zone of the Enrichment and Depuration Zones comprises <50% of the total water column depth. The down bay reduction in light-limited phytoplankton growth and the development of autotrophic benthic communities in the lower bay segments of the Advection Zone further reduce the progressively diminishing nutrient levels as they flux down bay from their origin in the Enrichment Zone and flow through the Depuration Zone.

The nutrients required to support autotrophic phytoplankton and benthic growth in the Advection Zone are supplied primarily by biological recycling (Hale, 1975a,b; Vargo, 1976; Verity, 1985) and advection from Rhode Island Sound, with NO_3 influx being particularly important (Nixon et al., 1995. Nutrients that escape biofiltration in the Depuration Zone flux down bay and supplement the nutrient supply. The different sources of nutrient along the bay-axis reflect the sixth ecologically important gradient in Narragansett Bay—the Nutrient Pump Gradient. This consists of two different nutrient pumps: one is anthropic and the other natural. Anthropogenic "new" nitrogen in the form of NH_4 and NO_3 is delivered into the Enrichment Zone in upper Narragansett Bay, and "new" nitrogen, principally in the form of NO_3, is advected in bottom water flowing into lower Narragansett Bay from offshore.

15.4.1 Mitigation of Nutrient Enrichment Impacts by the "Depuration Zone"

We hypothesize the Depuration Zone plays a major role in regulating the bay-wide (down bay) ecological impacts of the nutrient enrichment plume originating in the upper Narragansett Bay Enrichment Zone. The habitat features symptomatic of this proposed regulatory function of the Depuration Zone are schematized in Fig. 15.11. The gradients depicted are derived from the salinity–turbidity–nutrient–chlorophyll distributions along the station transects during the surveys (Figs 15.2–15.5). The salient process-feature of the Depuration Zone, and manifestation of its proposed regulatory role, is the coupled increase in chlorophyll (i.e., phytoplankton growth) and the coincident decreases in N, P, and Si concentrations, all very vivid. Although various factors influence this inverse relationship, it conforms to the classical biomass yield–nutrient dose relationship, with the measured consequence of the assimilation of nutrients by the phytoplankton being the decrease in water mass nutrient levels. The Depuration Zone was regionally restricted, located on average in the regions of Stations 3, 4, and 5, and positioned intermediately between the Enrichment and Advection Zones (Fig. 15.11). The stimulation of phytoplankton growth in the Depuration Zone by elevated nutrients following yield–dose kinetics (Figure 15.5a,b) can be likened to a process of biofiltration that leads to nutrient-purification. That is, the Depuration Zone buffers the down bay regions of Narragansett Bay from over-enrichment and its adverse ecological consequences, excluding those segments lying off the axis-main stem, such as Greenwich Bay.

The areal extent of the Depuration Zone was relatively narrow during the surveys, only 10–12 km, but it is probably not fixed in position. It is expected to expand and contract seasonally and interannually, this spatial mobility induced by changes in the interactive rates of advection, nutrient influx, irradiance flux, phytoplankton growth rate, and grazing rates. A highly transient illustration of this chemostat-type behavior is the temporary exceedance of the assimilatory capacity of the Depuration Zone that occurs during excessive rain and runoff events. Plumes of high nutrient levels are then detected in lower Narragansett Bay within the Advection Zone. The magnitudes of the draw-down in nutrients and increase in biomass are also expected to vary. Spatial widening of the "hump" in the hyperbolic distribution of chlorophyll along the gradient (Fig. 15.4), and specifically in the Depuration Zone (Fig. 15.11), is expected to be symptomatic of greater nutrient encroachment downstream from upper bay sources, while a tighter, narrower spatial zone of very high chlorophyll abundance provides a greater barrier to downstream nutrient encroachment.

We suggest that the Depuration Zone is an ecosystem property which must be considered in developing a management strategy to reduce anthropogenic nutrient loading into Narragansett Bay to mitigate undesirable eutrophication-induced ecological disruptions. A key element of this strategy, obviously, should be the reduction of nutrient loading into the Enrichment Zone. Beyond this, habitat manipulations to utilize the spatial features and processes of the Depuration Zone to enhance its natural assimilatory capacity should be considered. The specific objective of this eco-engineering should be to increase the efficiency of the biofiltration capacity of the Depuration Zone to decrease the strength of the down bay flux of anthropogenic nutrient enrichment plume, and its potential negative ecological impacts. Spatial movement of the Depuration Zone down bay toward Station 6 or into the East Passage, both lying with the Advection Zone, is expected to result from, and to reflect, a greater flux of nutrients down bay. This encroachment could develop into the "creeping eutrophication" of Advection Zone waters if the expansion is progressive, chronic, or excessive nutrient concentrations are involved. The corollary of this is—spatial down-sizing of the Depuration Zone and/or its up bay displacement toward the Enrichment Zone would accompany, and reflect reduced nutrient loading. Since the Depuration Zone is enriched by down bay flow from the Enrichment Zone, its spatial retraction during reductions in enrichment plume nutrient levels would probably be in a direction up bay from Station 5 (Fig. 15.11). This zonal repositioning and/or reduction might more effectively filter the nutrient flux, protecting down bay (i.e., Advection Zone) waters from undesirable nutrient-induced impacts and blooms. Hence, this spatial dynamic is expected to reflect a reduced flux of nutrients down bay.

Two habitat modifications that would alter the spatial extent of the Depuration Zone and increase its biofiltration capacity are suggested by the gradients and ecotones depicted and described here: a reduction in turbidity, which would reduce the degree of light limitation, and a reduction in nutrient loading, which would decrease the yield–dose response and resultant biomass accumulation. The increase in irradiance would facilitate increased nutrient assimilation because photosynthesis would be less light-limited. Reductions in nutrients and turbidity are achievable as paired-manipulations of sewage treatment

plant effluent, since a reduction in total suspended solids would be accompanied by a reduction in nutrient delivery, the effect of which would be to reduce
water mass turbidity (Borkman and Smayda, in preparation). In kinetic terms,
the decrease in substrate (nutrients) and increase in irradiance would increase
the amount and fraction of the available nutrient pool taken up, and reduce the
down bay flux of unutilized nutrient. This mitigation strategy might be com-
promised by phytoplankton self-shading during growth, which would increase
the turbidity and suppress the yield–dose response. This prospect and the
proposed consequences of reducing (or increasing) the spatial extent of the
Depuration Zone could be examined in a simulation model, perhaps that
considered by Chen *et al.* in Chapter 9.

Reduction of the Depuration Zone through manipulated reductions in
nutrients and turbidity should curb the current tendency for bottom water
hypoxia to develop in upper Narragansett Bay (Altieri and Witman, 2006).
The mechanism behind this potential and consequential diminution of hypoxic
stress is as follows. The decease in phytoplankton biomass anticipated to
accompany a reduction in nutrient loading should increase the effectiveness
of benthic grazing (G) in controlling the biomass accumulation produced by
nutrient-stimulated growth. The filter-feeding capacity of benthic bivalves is a
recognized "eutrophication control" mechanism (Officer *et al.,* 1982). Hypoxia
amelioration is anticipated because a greater proportion of the phytoplankton
biomass produced would be grazed by the benthic filter feeders (Altieri and
Witman, 2006). This would reduce the mass of ungrazed cells sinking to the
bottom sediments, where the oxygen demands of their decomposition currently
often lead to hypoxia. That is, the current situation in upper Narragansett Bay,
where $\mu > > > G$, would reduce to $\mu \cong G$, with the ungrazed portion of the
biomass then less threatening to hypoxia development during biochemical
decomposition.

Benthic suspension feeders are highly resistant to noxious phytoplankton
bloom species, unlike zooplankton which are more vulnerable. The Depuration Zone is the primary location of intense flagellate blooms in Narragansett
Bay (Tables 15.7–15.9), where their stimulation partially reflects the elevated
N:Si ratio found (Figs 15.3, 15.4, and 15.11), and which selects against the
more nutritious diatoms, as discussed earlier. Nutrient-reduction manipulations designed to decrease the N:Si ratio to promote blooms of the nutri-
tionally more desirable diatoms over flagellate species should also be
considered. The objective of this manipulation would be to increase benthic
grazing control and diminish the tendency toward hypoxia that can result
from the decomposition of ungrazed cells. The efficacy of establishing protected beds, i.e., gardens of benthic suspension feeders within the Depuration
Zone, such as mussel or hard clam which would auto-regulate to high
population densities because of the high phytoplankton biomass available,
should be evaluated. The objective of that ecological manipulation would be
to increase water clarity to reduce light limitation of photosynthesis, thereby
increasing nutrient uptake, and to increase benthic grazing in response to the

resultant increase in phytoplankton biomass. The specific ecological consequences of the increase in benthic filter-feeding and fecundity being sought by this manipulation is the mitigation of hypoxia outbreaks and enhancement of nutrient biofiltration to depurate the enrichment plume flowing down bay from the Depuration Zone.

Whatever ecosystem-based management strategy is adopted, it should be designed to utilize and enhance the suggested purgative benefits of the Depuration Zone, a major and previously unrecognized ecosystem property of Narragansett Bay. The potential economic benefits, and costs that would accompany incorporating the Depuration Zone processes into the ecosystem management plan to be developed should be evaluated.

Acknowledgements This research was funded by the EPA's Science to Achieve Results (STAR) Program, supported by EPA Grant Nos R82-9368-010 and RD83244301 awarded to Dr. Smayda. STAR is managed by the EPA's Office of Research and Development (ORD), National Center for Environmental Research, and Quality Assurance (NCERQA). STAR research supports the Agency's mission to safeguard human health and the environment. This study was also carried out as a component of the Brown Tide Research Initiative (BTRI) of the Coastal Ocean Program of the National Oceanic and Atmospheric Administration, award #NA66RGO368, coordinated by New York Sea Grant, and with the support of U.S. National Science Foundation Grant No. OCE 95-20300. We thank Dr. Paul Fofonoff and Dr. Tracy Villareal for their assistance in the field sampling program during the transect cruises.

References

Altieri, A., and Witman, J. 2006. Local extinction of a foundation species in a hypoxic estuary: integrating individuals to ecosystem. *Ecology* 87(3):717–730.

Asselin, S., and Spaulding, M.L. 1993. Flushing times for the Providence River based on tracer experiments. *Estuaries* 16(4):830–839.

Bergondo, D.L., Kester, D.R., Stoffel, H.E., and Woods, W. 2005. Time-series observations during the low sub-surface oxygen events in Narragansett Bay during summer 2001. *Marine Chemistry* 97:90–103.

Borkman, D.G. 2002. Analysis and simulation of *Skeletonema costatum* (Grev.) Cleve annual abundance patterns in lower Narragansett Bay 1959 to 1996. Ph.D. Thesis, Graduate School of Oceanography, University of Rhode Island, Narragansett, RI, 395 pp.

Borkman, D., and Smayda T.J. [in review] Multidecadal Variations in bloom behavior of *Skeletonema costatum* (Grev.) Cleve *sensu lato* in Narragansett Bay. 1. Reduced annual abundance and shift in annual bloom cycle. *Marine Ecology Progress Series*.

Cook, T., Folli, M., Klinck, J., Ford, S., and Miller, J. 1998. The relationship between increasing sea-surface temperature and the northward spread of *Perkinsus marinus* (Dermo) disease epizootics on Oysters. *Estuarine and Coastal Shelf Science* 46:587–597.

Cosper, E.M., Carpenter, E.J., and Bricelj, V.M. (eds.). 1989. Novel Phytoplankton Blooms: Causes and Impacts of Recurrent Brown Tides and Other Unusual Blooms. Coastal and Estuarine Studies No. 35. Berlin: Springer-Verlag.

Culver-Rymsza, K. 1988. Occurrence of nitrate reductase along a transect of Narragansett Bay. MS Thesis, Oceanography, University of Rhode Island, Narragansett, RI, 135 pp.

Deason, E.E., and Smayda, T.J. 1982a. Ctenophore-zooplankton-phytoplankton interactions in Narragansett Bay, Rhode Island, USA, during 1972–1977. *Journal of Plankton Research* 4:203–217.

Deason, E.E., and Smayda, T.J. 1982b. Experimental evaluation of herbivory in the ctenophore *Mnemiopsis leidyi* relevant to ctenophore-zooplankton interactions in Narragansett Bay, Rhode Island, USA. *Journal of Plankton Research* 4:219–236.

Durbin, A.G., and Durbin, E.G. 1981. Standing stock and estimated production rates of phytoplankton and zooplankton in Narragansett Bay, RI. *Estuaries* 4(10):24–41.

Durbin, A.G., and Durbin, E.G. 1989. Effect of the brown tide on feeding, size and egg laying rate of adult female *Acartia tonsa*. *In* Novel Phytoplankton Blooms: Causes and Impacts of Recurrent Brown Tides and Other Unusual Blooms, pp. 625–646. Cosper, E.M., Carpenter, E.J., and Bricelj, V.M. (eds) *Coastal and Estuarine Studies No. 35*. Berlin: Springer-Verlag.

Durbin, E.G., Durbin, A.G., Smayda, T.J., and Verity, P.G. 1983. Food limitation of production by adult *Acartia tonsa* in Narragansett Bay, RI. *Limnology and Oceanography* 28:1199–1213.

Fofonoff, P.W. 1994. Marine cladocerans in Narragansett Bay. Ph.D. Dissertation, Graduate School of Oceanography, University of Rhode Island. 317 pp.

Friederich, G.O., and Whitledge, T.E. 1972. Autoanalyzer procedures for nutrients. *In* Phytoplankton Growth Dynamics, Technical Series 1, Chemostat Methodology and Chemical Analyses, pp. 38–55. Pavlou, S. (ed.) Seattle: University of Washington, Department of Oceanography Special Report No. 52.

Furnas, M.J., Hitchcock G.L., and Smayda, T.J. 1976. Nutrient-phytoplankton relationships in Narragansett Bay during the 1974 summer bloom. *In* Estuarine Processes, Uses, Stresses and Adaptation to the Estuary, Vol. 1. pp. 118–133. Martin L. Wiley (ed.) New York: Academic Press.

Granger, S., Brush, M., Buckley, B., Traber, M., Richardson, M., and Nixon, S.W. 2000. An assessment of eutrophication in Greenwich Bay. Paper No. 1, Restoring water quality in Greenwich Bay: a whitepaper series. Rhode Island Sea Grant, Narragansett, RI. 20 pp.

Grasshof, K. 1966. Automatic determination of fluoride, phosphate and silicate in seawater. *In* Automation in Analytical Chemistry, 1965 Technicon Symposium, pp. 304–307. Skiggs, L.T., Jr (ed.) New York: Mediad, Inc.

Hale, S.S. 1975a. The role of benthic communities in the nitrogen and phosphorus cycles of an estuary. *Recent Advances in Estuarine Research* 1:291–308.

Hale, S.S. 1975b. The role of benthic communities in the nitrogen and phosphorus cycles of an estuary. Proceedings of a Symposium-Mineral Cycling in Southeastern Ecosystems, 1975. (CONF-740513), EPA T900140-04, 291–308.

Hicks, S.D. 1959. The physical oceanography of Narragansett Bay. *Limnology and Oceanography* 4:316–327.

Hitchcock, G.L. 1978. Labeling patterns of carbon-14 in netplankton during a winter–spring bloom. *Journal of Experimental Marine Biology and Ecology* 31:141–153.

Hitchcock, G.L., and Smayda, T.J. 1977. The importance of light in the initiation of the 1972–1973 winter–spring diatom bloom in Narragansett Bay. *Limnology and Oceanography* 22:126–131.

Holmes, R.W. 1970. The Secchi disk in turbid coastal waters. *Limnology and Oceanography* 15:688–694.

Jeffries, H.P., and Terceiro, M. 1985. Cycles of changing abundances in the fishes of the Narragansett Bay area. *Marine Ecology Progress Series* 25:239–244.

Keller, A.A., Oviatt, C.A., Walker, H.A., and Hawk, J.D. 1999. Predicted impacts of elevated temperature on the magnitude of the winter–spring phytoplankton bloom in temperate coastal waters: a mesocosm study. *Limnology and Oceanography* 44:344–356.

Kennedy, W., and Lee, V. 2003. Greenwich Bay: An Ecological History. Narragansett, RI: Rhode Island Sea Grant. 32 pp.

Kincaid, C., and Pockalny, R.A. 2003. Spatial and temporal variability in flow at the mouth of Narragansett Bay. *Journal of Geophysical Research* 108:3218–3235.

Kremer, P. 1994. Patterns of abundance for *Mnemiopsis* in US coastal waters: a comparative overview. *ICES Journal of Marine Science* 51:347–354.

Kullberg, P.G. 1992. Primary productivity, biomass, and carbon turnover in size-fractionated phytoplankton in upper Narragansett Bay, Rhode Island. MS Thesis, Graduate School of Oceanograhpy, University of Rhode Island, Narragansett, RI, 220 pp.

Li, Y., and Smayda, T.J. 2000. *Heterosigma akashiwo* (Raphidophyceae): On prediction of the week of bloom initiation and maximum during the initial pulse of its bimodal bloom cycle in Narragansett Bay. *Plankton Biology and Ecology* 47:80–84.

Lorenzen, C.J. 1966. A method for the continuous measurement of in vivo chlorophyll concentration. *Deep-Sea Research* 13:223–227.

Martin, J.H. 1965. Phytoplankton-zooplankton relationships in Narragansett Bay. *Limnology and Oceanography* 10:185–191.

Mitchell-Innes, B.A. 1973. Ecology of the phytoplankton of Narragansett Bay and the uptake of silica by natural populations and the diatoms *Skeletonema costatum* and *Detonula confervacea*. Ph.D. Dissertation, Graduate School of Oceanography, University of Rhode Island, Narragansett, RI, 212 pp.

Nixon, S.W., Granger, S.L., and Nowicki, B.L. 1995. An assessment of the annual mass balance of carbon, nitrogen, and phosphorus in Narragansett Bay. *Biogeochemistry* 31:15–61.

Officer, C.B., and Ryther, J.H. 1980. The possible importance of silicon in marine eutrophication. *Marine Ecology Progress Series* 3:83–91.

Officer, C.B., Smayda, T.J., and Mann, R. 1982. Benthic filter feeding: a natural eutrophication control. *Marine Ecology Progress Series* 9:203–210.

Oviatt, C.A., Buckley, B., and Nixon, S. 1981. Annual phytoplankton metabolism in Narragansett Bay calculated from survey field measurements and microcosm observations. *Estuaries* 4:167–175.

Oviatt, C.A., Keller, A., and Reed, L. 2002. Annual primary production in Narragansett Bay with no bay-wide winter-spring bloom. *Estuarine, Coastal, and Shelf Science* 54:1013–1026.

Oviatt, C., Olsen, S., Andrews, M., Collie, J., Lynch, T., and Raposa, K. 2003. A century of fishing and fish fluctuations in Narragansett Bay. *Reviews in Fisheries Science* 11:221–242.

Pratt, D.M. 1959. The phytoplankton of Narragansett Bay. *Limnology and Oceanography* 4:425–440.

Pratt, D.M. 1965. The winter–spring diatom flowering in Narragansett Bay. *Limnology and Oceanography* 10:173–184.

Pratt, D.M. 1966. Competition between *Skeletonema costatum* and *Olisthodiscus luteus* in Narragansett Bay and in culture. *Limnology and Oceanography* 11:447–455.

Rhode Island Department of Environmental Management (RIDEM). 2003. The Greenwich Bay fish kill—August 2003 Causes, Impacts and Responses. Providence, RI. (*www.dem.ri.gov/pubs/fishkill.pdf*).

Sakshaug, E. 1977. Limiting nutrients and maximum growth rates for diatoms in Narragansett Bay. *Journal of Experimental Marine Biology and Ecology* 28:109–123.

Sieburth, J.M., Johnson, P.W., and Hargraves, P.E. 1988. Ultrastructure and ecology of *Aureococcus anophagefferens* gen. et sp. nov. (Chrysophyceae): the dominant picoplankter during a bloom in Narragansett Bay, Rhode Island, summer 1985. *Journal of Phycology* 24:416–425.

Smayda, T.J. 1973a. A survey of phytoplankton dynamics in the coastal waters from Cape Hatteras to Nantucket. *In* Coastal and Offshore Environmental Inventory, Cape Hatteras to Nantucket, Saul B. Saila (ed.) Chapter 3. University of Rhode Island Marine Publication Series No. 2. 100 pp.

Smayda, T.J. 1973b. The growth of *Skeletonema costatum* during a winter–spring bloom in Narragansett Bay, Rhode Island. *Norwegian Journal of Botany* 20:219–247.

Smayda, T.J. 1974. Bioassay of the growth potential of the surface waters of lower Narragansett Bay over an annual cycle using the diatom *Thalassiosira pseudonana* (13-1). *Limnology and Oceanography* 19:889–901.

Smayda, T.J. 1990. Novel and nuisance phytoplankton blooms in the sea: evidence for a global epidemic. *In* Toxic Marine Phytoplankton, pp. 29–40. Granéli, E., Sundström, B., Edler, L., and Anderson, D.M. (eds.) New York: Elsevier.

Smayda, T.J. 1993. Experimental Manipulations of Phytoplankton + Zooplankton + Ctenophore Communities and Foodweb Roles of the Ctenophore *Mnemiopsis leidyi*. ICES Biological Oceanography Committee, CM 1993/L:68. International Council for the Exploration of the Sea, Copenhagen, Denmark, 13 pp.

Smayda, T.J. 1997. Harmful algal blooms: their ecophysiology and general relevance to phytoplankton blooms in the sea. *Limnology and Oceanography* 42:1137–1153.v

Smayda, T.J. 1998. Patterns of variability characterizing marine phytoplankton, with examples from Narragansett Bay. *ICES Journal of Marine Science* 55:562–573.

Smayda, T.J. 2004. Eutrophication and phytoplankton. *In* Drainage Basin Nutrient Inputs and Eutrophication: an Integrated Approach, pp. 89–98. Wassmann, P., and Olli, K. (eds) e-book available at: (*www.ut.ee/~olli/eutr/*).

Smayda, T.J., and Fofonoff, P. 1989. An extraordinary, noxious "brown tide" in Narragansett Bay. II. Inimical Effects. *In* Red Tides: Biology, Environmental Science and Toxicology, pp. 131–134. Okaichi, T., Anderson, D.M., and Nemoto, T. (eds) New York: Elsevier.

Smayda, T.J., and Villareal, T.A. 1989a. An extraordinary, noxious "brown tide" in Narragansett Bay. I. The organism and its dynamics. *In* Red Tides: Biology, Environmental Science and Toxicology, pp. 127–130. Okaichi, T., Anderson, D.M., and Nemoto, T. (eds) New York: Elsevier.

Smayda, T.J., and Villareal, T. 1989b. The 1985 "brown tide" and the open phytoplankton niche in Narragansett Bay during summer, *In* Novel Phytoplankton Blooms: Causes and Impacts of Recurrent Brown Tides and Other Unusual Blooms, pp. 159–187. Cosper, E.M., Carpenter, E.J., and Bricelj, V.M. (eds.) *Coastal and Estuarine Studies No. 35*. Berlin: Springer-Verlag.

Smayda, T.J., Borkman, D.G., Beaugrand, G., and Belgrano, A.G. 2004. Ecological effects of climate variation in the North Atlantic: phytoplankton. *In* Ecological Effects of Climatic Variations in the North Atlantic, pp. 49–58. Stenseth, N.C., Ottersen, G., Hurrell, J., and Belgrano, A. (eds) Oxford University Press, Oxford, England.

Steemann Nielsen, E. 1952. The use of radioactive carbon (^{14}C) for measuring organic production in the sea. *Journal of the Counsel of Exploration of the Sea* 18:117–140.

Strickland, J.H., and Parsons, T.R. 1972. A Practical Handbook of Seawater Analysis. Bulletin 167 (2nd edition) Fisheries Research Board of Canada, Ottawa, Canada, 310 pp.

Sullivan, B.K., Van Keuren, D., and Clancy, M. 2001. Timing and size of blooms of the ctenophore *Mnemiopsis leidyi* in relation to temperature in Narragansett Bay, RI. *Hydrobiologia* 451:113–120.

Tomas, C.R. 1980. *Olisthodiscus luteus* (Chrysophyceae) V. Its occurrence, abundance and dynamics in Narragansett Bay. *Journal of Phycology* 16:57–166.

Tracey, G.A., Johnson, P.A., Steele, R., Hargraves, P.E., and Sieburth, J.M. 1988. A shift in photosynthetic picoplankton composition and its effect on bivalve mollusc nutrition: the 1985 brown tide in Narragansett Bay, Rhode Island. *Journal of Shellfish Research* 7:671–675.

Vargo, G.A. 1976. The influence of grazing and nutrient excretion by zooplankton in the growth and reproduction of the marine diatom *Skeletonema costatum* Greville, Cleve, in Narragansett Bay. Ph.D. Dissertation, Graduate School of Oceanography, University of Rhode Island, Narragansett, RI, 162 pp.

Verity, P.G. 1985. Grazing, respiration, excretion and growth rates of tintinnids. *Limnology and Oceanography* 30:1268–1282.

Yentsch, C.S., and Menzel, D. 1963. A method for the determination of phytoplankton chlorophyll and phaeophytin by fluorescence. *Deep-Sea Research* 10:221–231.

Chapter 16
Narragansett Bay Ctenophore-Zooplankton-Phytoplankton Dynamics in a Changing Climate

Barbara K. Sullivan, Dian J. Gifford, John H. Costello and Jason R. Graff

16.1 Introduction

The 1950s to the 1980s were a heyday for studies of plankton dynamics in Narragansett Bay, during which multiple investigators produced a nearly continuous record of seasonal and inter-annual patterns of phytoplankton, zooplankton, and ctenophores (e.g., Smayda, 1957; 1973; Pratt, 1959; 1965; Martin, 1965; Kremer, 1979). A good understanding of nutrient–plankton interactions and trophic dynamics was developed, especially for the central and lower regions of the bay. The copepods *Acartia hudsonica* and *Acartia tonsa* were clearly the numerical dominants in the zooplankton (Frolander, 1955; Hulsizer, 1976), but the summer species *A. tonsa* was the dominant secondary producer because higher temperatures allowed higher rates of production than for the winter–spring species *A. hudsonica* (Durbin and Durbin, 1981). The summertime production occurred during a predator-free window prior to ctenophore population pulses in late summer. Deason and Smayda (1982) hypothesized that late summer phytoplankton blooms resulted from the reduction in grazing as ctenophore predation reduced copepod abundances. Martin (1965) and Li and Smayda (1998) have also suggested that zooplankton grazing limits phytoplankton production in Narragansett Bay in both summer and winter.

There is an unfortunate hiatus in the record of ctenophores and zooplankton from 1990 to 1999. However, in 1999, Sullivan *et al.* (2001) noted the unusual appearance of the ctenophore *Mnemiopsis leidyi* in late May, much earlier than had been previously recorded for the mid-bay region. This led to renewed sampling, and subsequently Sullivan *et al.* (2001) reviewed the available published and unpublished records, which revealed a significant change in ctenophore seasonality and abundance as compared to earlier periods, probably beginning some time in the 1990s. They reported a statistically significant relationship between spring temperatures and timing of the

Barbara K. Sullivan
University of Rhode Island, Graduate School of Oceanography, 11 Aquarium Road, Narragansett, RI 02882
bsullivan@gso.uri.edu

A. Desbonnet, B. A. Costa-Pierce (eds.), *Science for Ecosystem-based Management.*
© Springer 2008

ctenophore bloom, the change being correlated with warmer winters and springs that have resulted in an increase in the average temperature of Narragansett Bay (Hawk, 1998; Nixon *et al.,* 2004; see also Chapter 2 in this volume). Narragansett Bay is located at the northern edge of the geographic range of the ctenophore *M. leidyi*; thus, it is not unexpected that warming should favor this species.

This chapter explores the consequences of increases in ctenophore populations to copepod abundance and phytoplankton species composition and biomass in mid-Narragansett Bay, for which a historic record is available. We also explore a highly eutrophic embayment, Greenwich Cove, which has not previously been sampled for ctenophores. Greenwich Cove and other regions of Narragansett Bay with high loads of anthropogenic nutrients already experience chronic hypoxia in summer months (see Chapters 11 and 12 of this volume). Hypoxia is related to increased phytoplankton production associated with intermittent periods of stratification caused by increased river input, thermal heating, and reduced tidal range (e.g., neap tides) in Narragansett Bay (Bergondo *et al.,* 2005). If changes in ctenophore seasonality and biomass, through top-down effects on the phytoplankton, result in increased incidence or size of summer phytoplankton blooms, there will be increased respiratory demand, which could increase the risk of anoxia (Bergondo, 2004).

16.2 Methods

Zooplankton, ctenophores, chlorophyll *a*, and hydrographic data were collected weekly at three stations (Fig. 16.1). The Greenwich Cove station (1) was sampled from a marina dock where it enters Greenwich Bay (lat. 41°40.0′ N, long. 71°26.6′ W) and has a high nutrient load due to its location down-stream from a sewage treatment facility. There are no historical records of zooplankton or ctenophores from the Greenwich Cove region. The second station (2), off Fox Island (lat. 41°34.2′ N, long. 71°23.4′ W), is located in mid-Narragansett Bay's West Passage, and is the site of long-term historical plankton time series data (e.g., Hulsizer, 1976; Kremer and Nixon, 1978; Karentz and Smayda, 1984; 1998; Li and Smayda, 1998). It is close to the sites sampled by Deason and Smayda (1982) and Durbin and Durbin (1981). A third station (3), Dutch Island (lat. 41°30.6′ N, long. 71°24.0′ W), was also sampled weekly, which showed virtually the same trends and patterns as found at the Fox Island station (Costello *et al.* 2006b), and so the data are not presented here. This latter station was in the same position as sampled by Frolander (1955) and Jeffries (1962), and just north of a station sampled by Martin (1965). All data were available for 2002 and 2003 for all stations. In 2004, phytoplankton, zooplankton, and ctenophore data were available for Fox Island Station 2, but not for other stations.

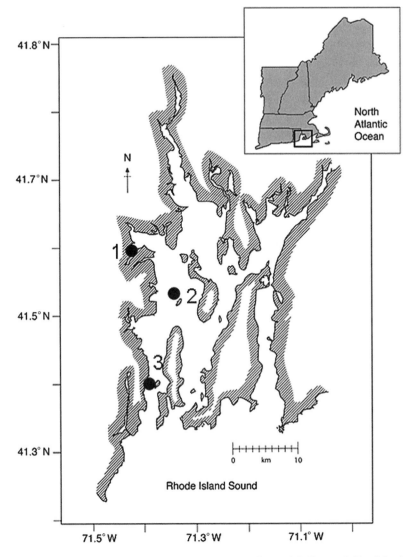

Fig. 16.1 Map of stations sampled in Narragansett Bay. Greenwich Cove = 1; Fox Island = 2; Dutch Island = 3.

Vertical profiles of temperature, salinity, and oxygen were collected with a YSI sonde model 600XLM-M in 2002 and 2003. The samples were collected at approximately 10:00 AM at Fox Island and 11:30 AM at Greenwich Cove. In all the years, zooplankton were enumerated from 64 μm mesh net tows with a 0.25-m diameter mouth-opening and equipped with a flow meter. One tow per station sampled the entire water column with the volume filtered varying between 0.4 and 1 m^3 depending on station depth. The samples were preserved in 5% formalin buffered with sodium borate ($Na_2B_4O_7$). Two hundred individual subsamples were processed under a dissecting microscope at 50× magnification. *M. leidyi* were collected with two replicate vertical or oblique tows per stations using a 0.25-m diameter, 1 mm mesh net with flow meter, and were counted live within 3 h of collection. The volume sampled per tow varied from 0.9 to 79 m^3 depending on the abundance of ctenophores— when ctenophores were very numerous, the volume sampled was kept low to prevent net clogging. The samples for chlorophyll were collected from mid-depth and near-bottom using a Niskin bottle, and from the surface with a clean plastic bucket. Fifty milliliters of aliquots were collected onto GF/F filters, extracted in 90% spectranalyzed acetone, and analyzed by fluorometry (Knap *et al.,* 1996). Data for nutrients were obtained from the GSO plankton survey or the literature as indicated.

Data from the GSO plankton time series at Fox Island were used for the analysis of phytoplankton species compositions from weekly surface, bottom, and net tow samples collected within two days of zooplankton and chlorophyll samples. Equal volumes of surface and bottom samples are combined for counting. Phytoplankton cells were enumerated under a compound microscope using a Sedgewick-Rafter counting chamber. The samples were counted live and unconcentrated. A 20-μm net tow sample was also examined. The species observed in the net sample were recorded as present but not counted. Species identifications were based primarily on their appearance in the Sedgewick-Rafter chamber, supplemented with permanent mounts examined with phase contrast and interference contrast optics.

16.3 Results

16.3.1 Salinity, Temperature, Nutrients, and Chlorophyll

Greenwich Cove clearly differs in the physical and nutrient regimes from Fox Island (Table 16.1). Fox Island was on average slightly saltier and cooler (in summer) than the more inshore and enclosed Greenwich Cove station, which experienced somewhat lower salinities and greater extremes of temperature. In 2003, minimum temperatures were <0°C in Greenwich Cove, but in 2002, the temperatures did not fall below 1°C. Overall, 2003 was the coldest year in both winter and summer, in comparison to 2002 and 2004 (NOAA, 2006). The average nitrate-nitrite concentrations at Greenwich Cove were 12 times higher than at Fox Island (Table 16.1). The higher nutrient levels appear to have resulted in significantly higher values of chlorophyll *a* in Greenwich Cove than at Fox Island (Table 16.2; Fig. 16.2).

Table 16.1 Temperature (°C), salinity (‰) and nutrients (μg L^{-1}) at Fox Island and Greenwich Cove Stations. Unless noted, averages are for November 2001–2003.

Water quality	Fox Island	Greenwich Cove
Average surface salinity	29.28	27.10
Average Bottom salinity	29.72	28.20
Minimum surface salinity	25.5 (May 2003)	22.6 (April 2003)
Avg. surface temp	12.5	13.4
Avg. bottom temp	11.7	12.8
Surface Max. temp 2002	24.64 (Summer)	28.1 (Summer)
Surface Max. temp 2003	24.46 (Summer)	26.4 (Summer)
Avg. Jan–Feb temp 2002	3.90 (3.13 min)	2.48 (1.16 min)
Avg. Jan–Feb temp 2003	1.49 (0.21 min)	1.04 (−1.38 min)
Avg. $NO_2 + NO_3$ (range)	0.75 (0.04–2.36)*	12 (6–14)**

* Data from P. Hargraves, GSO Plankton Survey, June–September 2003.
** Data from ASA (2001) for August–September 2000.

Table 16.2 Maximum and average abundances (number m^{-3}) of *Acartia tonsa* stages C1-adult in July in the present study and in published records.

Station	Average	Maximum	Source
1951	5,081*	15,000*	Frolander (1955)
1973	6,350	15,000	Hulsizer (1976)
1976	25,400	30,000	Durbin and Durbin (1981)
2002 Fox Island	274	1,018	This study
2002 GC	24	54	This study
2003 Fox Island	1,199	4,620	This study
2003 GC	259	923	

*Data for 1951 included C4-adult stages only. Data from published records approximated from graphs except Frolander (1955).
GC = Greenwich Cove.

16.3.2 Ctenophores, Copepods, and Phytoplankton

Ctenophores were first found in large numbers during spring at the Fox Island station (Fig. 16.3). The timing of this pulse varied from early to mid-June (2002 and 2004) to early July in 2003, the colder year. Ctenophores were found throughout the winter of 2002 in Greenwich Cove (average abundance January–May was 18.5 m^{-3}), with population pulses evident from May–August (Fig. 16.4). In 2003, when winter temperatures fell below 0°C, ctenophores were rare in winter (average abundance January–May 1.7 m^{-3}), and not abundant until July 2 (Fig. 16.4).

Copepods, primarily the winter-spring species *Acartia hudsonica,* were abundant in all years, at all stations, shortly before the arrival of ctenophores

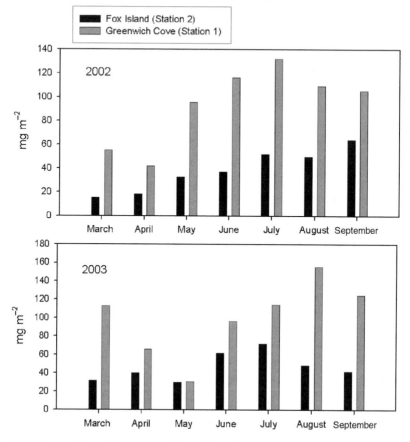

Fig. 16.2 Comparison of mean monthly values of chlorophyll *a* at Fox Island versus Greenwich Cove. Depth averaged water column values were calculated from surface and bottom measurements.

(Figs. 16.3 and 16.4). The decline in the numbers of *A. hudsonica* began just prior to or coincident with large increases in ctenophore numbers in all three years. These decreases were likely initiated by the resting eggs produced when temperatures reach 15°C in spring (Sullivan and MacManus, 1986), rather than by predation of ctenophores. However, the impact of ctenophores is seen in the lack of summer copepod population pulse in all years at all stations (Figs. 16.3 and 16.4). At Fox Island, summer-time abundance of *Acartia tonsa*, the dominant summer copepod, remained extremely low as compared to historic abundances, reaching only 5–25% of the historic levels reported for July (Table 16.2). There are no historic records for Greenwich Cove. Copepods

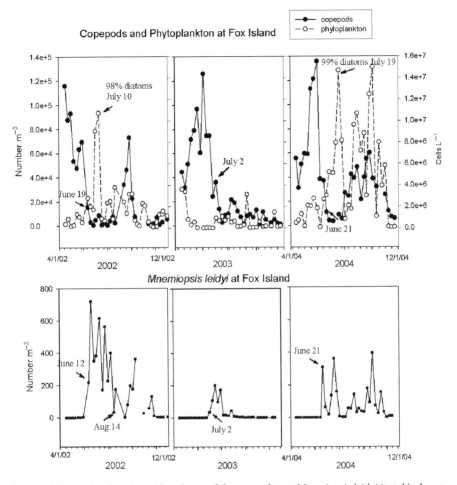

Fig. 16.3 Fox Island station. Abundance of the ctenophore, *Mnemiopsis leidyi* (total in 1 mm mesh net), total copepods (all stages and species), and phytoplankton cell counts during 2002, 2003, and 2004.

became abundant in the fall at Fox Island until after ctenophore numbers declined in August (Fig. 16.3).

Diatom blooms occurred at Fox Island during periods when copepods were low and ctenophores were numerous in 2002 and 2004, but not in 2003 (Fig. 16.3). There was, however, a peak in chlorophyll *a* during the expected period (Fig. 16.5). Ctenophores were not as abundant in July 2003 as in the other years, and copepods were somewhat in abundance. This may explain the lack of a diatom bloom in 2003. At Greenwich Cove, the same general pattern of an inverse relationship between copepods and phytoplankton (chlorophyll *a*) is clear (Fig. 16.4).

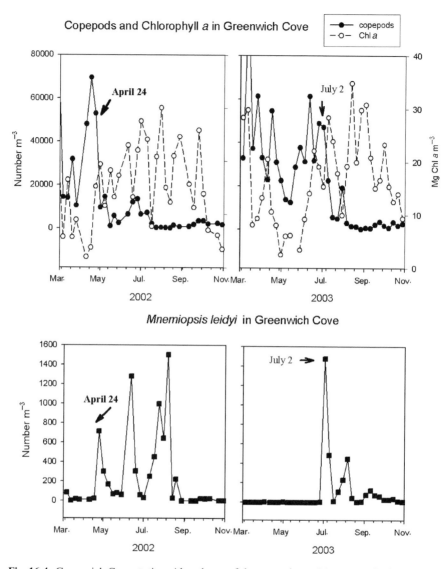

Fig. 16.4 Greenwich Cove station. Abundance of the ctenophore, *Mnemiopsis leidyi* (total in 1 mm mesh net), total copepods (all stages and species), and phytoplankton cell counts during 2002 and 2003.

16.3.3 Dissolved Oxygen

Oxygen stress in the form of hypoxia was clearly evident in Greenwich Cove at nearly all times during May–August of both years, although conditions were worse in 2003 when dissolved oxygen was below 3 mg L^{-1} for nearly

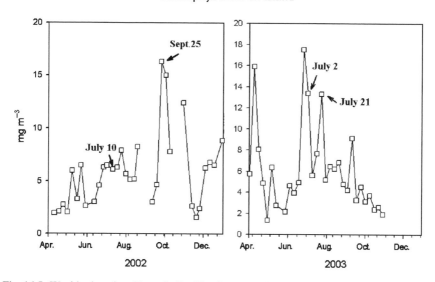

Fig. 16.5 Weekly data for chlorophyll *a* (depth averaged) at Fox Island station.

the entire summer (Fig. 16.6). In that summer there was not only stress from hypoxia, but an anoxic event that covered much of western Greenwich Bay in August (RIDEM, 2003). It is noteworthy that, at Fox Island, there were also periods of oxygen stress (values below 3 mg L^{-1}) near the bottom in both years, although it is possible that low oxygen resulted from transport of water originating upstream, rather than from *in situ* metabolic processes.

16.4 Discussion

Since at least the year 2000, when monitoring of zooplankton in Narragansett Bay was resumed after a decade of neglect, the abundance of summer zoo-plankton, primarily the copepod *A. tonsa,* has undergone dramatic reductions in numbers as compared to values reported in 1950s to 1980s (Costello *et al.,* 2006a). This change has occurred in association with earlier seasonal appearance and increasing numbers of ctenophore predators, and is evident in both eutrophic and lower nutrient regions of Narragansett Bay. However, where nutrient levels are highest, the affect of ctenophores may be even greater. The higher levels of nutrients in Greenwich Cove appear to fuel greater primary and secondary production than the Fox or Dutch Island stations, as evidenced by higher levels of chlorophyll *a* and ctenophores (Table 16.2). In contrast, copepod abundance is significantly lower at Greenwich Cove than in

Daytime Oxygen Values at Depth

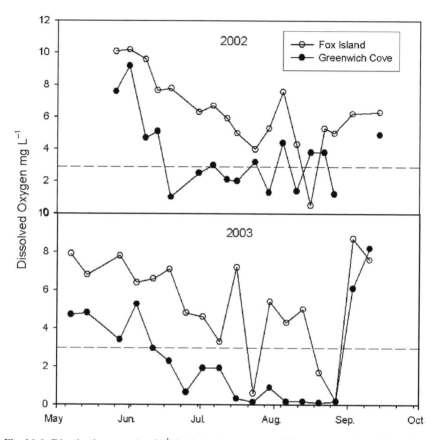

Fig. 16.6 Dissolved oxygen (mg L^{-1}) in the bottom waters of Narragansett Bay at Fox Island and Greenwich Cove Stations. Data were collected at approximately 10:00 AM at Fox Island and 11:30 AM at Greenwich Cove. Dotted line indicates stressful levels of oxygen (3 mg L^{-1}) based on Miller *et al.* (2002).

mid-Narragansett Bay stations. The higher intensity ctenophore predation due to their increased abundance in Greenwich Cove is likely a cause for reduced copepod numbers. Thus, copepod grazing pressure on phytoplankton is even lower than at the mid-bay stations. The increased productivity of this eutrophic station apparently increases metabolic demand, as is evidenced by summer-long hypoxia at Greenwich Cove. Indeed, the hypoxic conditions typical of summer in eutrophic regions may act to maintain a balance favoring ctenophores, because *M. leidyi* is more tolerant of hypoxia than its prey or predators, for example, in increasing its feeding rates on *A. tonsa* at 1–3 mg L^{-1} dissolved oxygen (Decker *et al.,* 2004).

Blooms of summer phytoplankton (including diatoms) occurred at both Greenwich Cove and the less-nutrient-enriched stations in the lower bay. In both cases, these blooms appear inversely related to copepod abundance. When copepods were absent during summer, phytoplankton biomass was high. Thus, while bottom-up effects of nutrients are evident in comparing abundance of plant and animal populations across the more- and less-nutrient-enriched stations, it appears that the top-down predator has a similar effect on copepod populations in both regimes. We argue that trophic dynamics involving ctenophores appear to operate similarly bay-wide to reduce the grazing pressure on phytoplankton. Moreover, the dynamics have changed in a similar direction at both nutrient-enriched and mid-bay stations over time, with earlier seasonal appearance of ctenophores and reduced summer copepod populations during the period of increased risk for hypoxia.

While a two- to three-year record is by no means conclusive evidence of changes in phytoplankton bloom patterns, and preclude a statistical comparison with published patterns of past data (e.g., Li and Smayda, 1998), our data suggest increases in the number and size of summer phytoplankton blooms as a result of expansion of the seasonal range of *M. leidyi*. The potential for such a trophic cascade is already noted in the literature (Deason and Smayda, 1982; Turner et al., 1983; Oguz et al., 2001; Graneli and Turner, 2002; Purcell and Decker, 2005). While there may be other factors influencing the timing and size of summer phytoplankton blooms, such as increasing nutrient load over time or reduced water column mixing during extended warm periods, these influences do not preclude the additional effect of reduced impact from grazers. At the very least, the shifts in ctenophore and copepod abundances described here are significant (Table 16.3), and must be associated with significant changes

Table 16.3 Results of statistical tests comparing weekly station data for two years combined.

Parameter	Median	Stations	Significant
Chlorophyll *a* (mg m^{-3})			
GC	7.3	GC vs F, D	Y, Y
Fox	4.4	F vs D	Y
Dutch	3.0		
Ctenophores (# m^{-3})			
GC	23.6	GC vs F, D	Y, Y
Fox	2.1	F vs D	N
Dutch	2.7		
Copepods (# m^{-3})			
GC	617.5	GC vs F, D	Y, Y
Fox	1,629.3	F vs D	N
Dutch	2,060.1		

GC = Greenwich Cove; F = Fox Island; D = Dutch Island.
Values were compared using a repeated measures on ranks ANOVA. Chi-square values for the ANOVAs were significant at $p < 0.001$. Dunn's Test for pairwise multiple comparisons was used to isolate differences among the stations and was significant at <0.05.

elsewhere in the ecosystem, for example increased competition by ctenophores with larval fish for food resources (Sullivan *et al.,* 2001) as well as altered response by phytoplankton.

Changes in the trophic dynamics of plankton will make it more difficult to predict the response of the ecosystem to the current or future levels of nutrient input. While there was a long experience with high levels of summer copepod production and grazing during the heavily monitored period of 1950–1990, we have little experience with the community composition and functioning of this new system. The uncertainties about new biotic and hydrographic conditions that may result from climate change compound the problem of prediction. There is strong evidence that the changes in plankton reported here are associated with climatic warming of the bay waters (Sullivan *et al.,* 2001). Consideration must also be given to other changes related to climatic variation that might also exacerbate the effects of nutrients and tip the already hypoxic systems toward anoxia, such as increased periods of stratification due to more intense summer warming or increased pulses of nutrient inputs due to increase in rainfall (Harley *et al.,* 2006).

We urge that evidence of changes to planktonic community structure, such as those described here, should be taken into consideration when modeling the response of Narragansett Bay to nutrient loads, in planning for monitoring of biota, and in making decisions about reductions of nutrient loads. Our current knowledge is the best source we have for making future decisions. To quote from an article by Mastrandrea and Schneider (2004) on how to evaluate the need for regulatory action in the face of uncertainties of climate change: "Despite great uncertainty...prudent actions can substantially reduce the likelihood and thus the risks of dangerous anthropogenic interference with natural systems." Along with serious consideration of how ctenophore-zooplankton-phytoplankton dynamics may alter ecosystem–nutrient responses, we urge greater attention to adequate monitoring of zooplankton communities in the bay than has been typical of the period 1990–2006.

References

Applied Science Associates (ASA). 2001. Project Report for Greenwich Bay Water Quality Summer 2000. ASA Project No. 00-029 for RIDEM.

Bergondo, D.L. 2004. Examining the processes controlling water column variability in Narragansett Bay: time-series data and numerical modeling. Ph.D. Dissertation, University of Rhode Island, Narragansett, RI. 187 pp.

Bergondo, D.L., Kester, D.R., Stoffel, H.E., and Woods, W. 2005. Time-series observations during the low sub-surface oxygen events in Narragansett Bay during summer 2001. *Marine Chemistry* 97:90–103.

Costello, J.H., Sullivan, B.K., and Gifford, D.J. 2006a. A physical-biological interaction underlying variable phonological responses to climate change by coastal zooplankton. *Journal of Plankton Research* 128:1099–1105.

Costello, J.H., Sullivan, B.K., Gifford, D.J., Van Keuren, D., and Sullivan, L. 2006b. Seasonal refugia, shoreward thermal amplification and metapopulation dynamics of the ctenophore *Mnemiopsis leidyi* in Narragansett Bay, RI, USA. *Limnology and Oceanography* 51:1819–1831.

Deason, E.E., and Smayda, T.J. 1982. Ctenophore-zooplankton-phytoplankton interactions in Narragansett Bay, Rhode Island, USA, during 1972–1977. *Journal of Plankton Research* 4:203–217.

Decker, M.B., Breitburg, D., and Purcell, J. 2004. Effects of low dissolved oxygen on zooplankton predation by the ctenophore *Mnemiopsis leidyi*. *Marine Ecology Progress Series* 280:163–172.

Durbin, A.G., and Durbin, E.G. 1981. Standing stock and estimated production rates of phytoplankton and Zooplankton in Narragansett Bay, RI. *Estuaries* 4(10):24–41.

Frolander, H. 1955. The biology of the zooplankton of the Narragansett Bay area. Ph.D. Thesis, Brown University, RI. 94 pp.

Graneli, E., and Turner, J.T. 2002. Top-down regulation in ctenophore-copepod-ciliate-diatom-phytoflagellate communities in coastal waters: a mesocosm study. *Marine Ecology Progress Series* 239:57–68.

Harley, C.D.G., Randall Hughes, A., Hultgren, K.M., Miner, B.G., Sorte, C.J.B., Thornber, C.S., Rodriguez, L.F., Tomanek, L., and Williams, S.L. 2006. The impacts of climate change in coastal marine systems. *Ecology Letters* 9:228–241.

Hawk, J.D. 1998. The role of the North Atlantic Oscillation in winter climate variability as it relates to the winter-spring bloom in Narragansett Bay. MS Thesis in Oceanography, University of Rhode Island, Narragansett, RI. 148 pp.

Hulsizer, E.E. 1976. Zooplankton of lower Narragansett Bay, 1972–1973. *Chesapeake Science* 17:260–270.

Jeffries, H.P. 1962. Succession of two *Acartia* species in estuaries. *Limnology and Oceanography* 7:354–364.

Karentz, D., and Smayda, T.J. 1984. Temperature and seasonal occurrence patterns of 30 dominant phytoplankton species in Narragansett Bay over a 22-year period (1959-1980). *Marine Ecology Progress Series* 18:277-293.

Karentz, D., and Smayda, T.J. 1998. Temporal patterns and variations in phytoplankton community organization and abundance in Narragansett Bay during 1959–1980. *Journal of Plankton Research* 20:145–168.

Knap, A., Michaels, A., Close, A. Ducklow, H., and Dickson, A. (eds). 1996. Protocols for the Joint Global Ocean Flux Study (JGOFS) Core Measurements. JGOFS Report Nr. 19, vi + 170 pp. Reprint of the IOC Manuals and Guides No. 29, UNESCO 1994.

Kremer, P. 1979. Predation by the ctenophore *Mnemiopsis leidyi* in Narragansett Bay, Rhode Island. *Estuaries* 2:97–105.

Kremer, J.N., and Nixon, S.W. 1978. A Coastal Marine Ecosystem: Simulation and Analysis. New York: Springer Verlag. 217 pp.

Li, Y., and Smayda, T.J. 1998. Temporal variability of chlorophyll in Narragansett Bay, 1973–1990. *ICES Journal of Marine Science* 55:661–667.

Martin, J.H. 1965. Phytoplankton-zooplankton relationships in Narragansett Bay. *Limnology and Oceanography* 10:185–191.

Mastrandria, M.D., and Schneider, S.H. 2004. Probabalistic integrated assessment of "Dangerous Climate Change." *Science* 304:571–575.

Miller, D.C., Poucher, S.L., and Cairo, L. 2002. Determination of lethal dissolved oxygen levels for selected marine and estuarine fishes, crustaceans and bivalves. *Marine Biology* 140:287–296.

Nixon, S.W., Granger, S., Buckley, B.A., Lamont, M., and Rowell, B. 2004. A one hundred and seventeen year coastal water temperature record from Woods Hole, Massachusetts. *Estuaries* 27:397–404.

NOAA. 2006. *(http://www.ncdc.noaa.gov/oa/climate/research/anomalies/anomalies.html)*.

Oguz, T., Ducklow, H.W., Purcell, J.E., and Malanotte-Rizzoli, P. 2001. Modeling the response of top-down control exerted by gelatinous carnivores on the Black Sea pelagic food web. *Journal of Geophysical Research* (C: Oceans) 106:4543–4564.

Pratt, D.M. 1959. The phytoplankton of Narragansett Bay. *Limnology and Oceanography* 4:425–440.

Pratt, D.M. 1965. The winter-spring diatom flowering in Narragansett Bay. *Limnology and Oceanography* 10:173–184.

Purcell, J.E., and Decker, M.B. 2005. Effects of climate on relative predation by scyphomedusae and ctenophores on copepods in Chesapeake Bay during 1987–2000. *Limnology and Oceanography* 50:376–387.

Rhode Island Department of Environmental Management (RIDEM). 2003. The Greenwich Bay Fish Kill—August 2003 Causes, Impacts and Responses. Providence, RI. *(www.dem.ri.gov/pubs/fishkill.pdf)*.

Smayda, T.J. 1957. Phytoplankton studies in lower Narragansett Bay. *Limnology and Oceanography* 2:342–359.

Smayda, T.J. 1973. The growth of *Skeletonema costatum* during a winter-spring bloom in Narragansett Bay, Rhode Island. *Norwegian Journal of Botany* 20:219–247.

Sullivan, B.K., and MacManus, L.T., 1986. Factors controlling seasonal succession of the copepods *Acartia hudsonica* and *A. tonsa* in Narragansett Bay, RI: temperature and resting egg production. *Marine Ecology Progress Series* 28:121–128.

Sullivan, B.K., Van Keuren, D., and Clancy, M. 2001. Timing and size of blooms of the ctenophore *Mnemiopsis leidyi* in relation to temperature in Narragansett Bay, RI. *Hydrobiologia* 451:113–120.

Turner, J.T., Bruno, S.F., Larson, R.J., Staker, R.D., and Sharma, G.M. 1983. Seasonality of plankton assemblages in a temperate estuary. *P.S.Z.N. I: Marine Ecology* 4:81–89.

Chapter 17
Coastal Salt Marsh Community Change in Narragansett Bay in Response to Cultural Eutrophication

Cathleen Wigand

> *Along the Eastern Coast of North America, from the north where ice packs grate upon the shore to the tropical mangrove swamps tenaciously holding the land together with a tangle of roots, lies a green ribbon of soft, salty, wet, low-lying land, the salt marshes.*
>
> John and Mildred Teal, 1969

17.1 Introduction

In "*Life and Death of the Salt Marsh,*" the authors describe how the activities of people, beginning with the first settlers from Europe, are a major force in causing either loss or preservation of both the quantity and quality of salt marshes (Teal and Teal, 1969). Human activities such as draining, filling, diking, and ditching had fragmented and destroyed over 50% of the wetlands in the conterminous US between the 1780s and 1980s (Dahl, 1990; Mitsch and Gosselink, 2000). A recent report that assessed New England's salt marsh losses for the period between the 1700s and the 1900s describes an average loss of 37% for the region, with the highest loss in Rhode Island at 53% (Bromberg and Bertness, 2005). In addition, cultural eutrophication—especially the over-enrichment of nitrogen (N)—is identified as a major stressor of macrophyte-based ecosystems (Short and Burdick, 1996; Valiela *et al.*, 1997, 2000; Carpenter *et al.*, 1998; Morris and Bradley, 1999; Deegan, 2002). Over-enrichment of nitrogen is shown to alter salt marsh structure (e.g., dominant plant species, plant species richness, and shoot density and height) and function (e.g., plant productivity, trophic interactions, denitrification, and soil respiration) (Sullivan and Daiber, 1974; Gallagher, 1975; Valiela *et al.*, 1975, 1976; Mendelssohn, 1979; Levine *et al.*, 1998; Morris and Bradley, 1999; Emery *et al.*, 2001; Minchinton and Bertness, 2003; Wigand *et al.*, 2003; 2004a,b).

Cathleen Wigand
US Environmental Protection Agency, Office of Research and Development, National Health and Environmental Effects Research Laboratory, Atlantic Ecology Division 27 Tarzwell Drive, Narragansett, RI 02882
wigand.cathleen@epa.gov

A. Desbonnet, B. A. Costa-Pierce (eds.), *Science for Ecosystem-based Management.*
© Springer 2008

These changes will alter the ability of the marsh to provide key ecosystem services such as water quality maintenance, erosion and flood control, and habitat and food to fish and wildlife (Teal and Howes, 2000; Wigand et al., 2001; Deegan, 2002). Additional stressors such as global warming, sea level rise, watershed development, and the introduction of non-native species may amplify the effects of cultural eutrophication on marsh structure and function.

The northeast is the most densely populated coastal region in the US, and wastewater effluent loading is often one of the dominant N sources to northeastern estuaries. In contrast, phosphorus (P) inputs to northeastern estuaries have decreased in recent decades due to improved treatment facility operations and declines in the levels of P used in detergents (Nixon et al., 2005; see also Chapter 5 of this volume). The Narragansett Bay estuary is a 381-km^2 coastal embayment (including Mount Hope Bay) that dominates the Rhode Island landscape. For the period of 1988–1994, wastewater N accounted for 73% of the total N loading into Narragansett Bay (Roman et al., 2000). In spite of the tremendous growth in population in the Narragansett Bay watershed over the past century, a review of the past and present data on N levels in direct sewage and rivers discharging into the bay, surprisingly, shows little change in the N inputs since mid-1980s (Nixon et al., 2005; see also Chapters 5 of this volume). It is unclear why there are no detectable increases in N inputs during the 25-year period, but it is likely that the remaining natural sinks such as inland wetlands, forests, and coastal salt marshes act as efficient buffers and transformers of land-derived N. Other explanations that have been suggested are that atmospheric N in the watershed may be declining or that improved fertilizer application and wastewater treatment has reduced N inputs (Nixon et al., 2005; see also Chapters 3, 5, and 6 of this volume).

Based on published reports, the current status and areal extent of coastal salt marshes in Narragansett Bay will be described. Then, the changes in plant and animal community structure and system-level functions, due partly to N overenrichment of salt marshes, will be reported. Finally, some successful approaches used to protect, restore, and manage coastal wetlands are discussed.

17.2 Areal Extent of Coastal Salt Marshes in Narragansett Bay

When compared to the landscape signature of salt marshes of the middle and south Atlantic coasts, northeast salt marshes, including those of Narragansett Bay, are small in spatial extent and often exist as narrow fringing systems (Roman et al., 2000). An estimate of the mid-1990s status of Narragansett Bay coastal wetlands (including estuarine marsh, oligohaline marshes, and scrub-shrub wetlands) was 1,463 ha (Tiner et al., 2004). Over 689 ha of these remaining coastal wetlands have been altered by ditching and/or impoundment. Of the mid-1990s coastal wetland total, *Phragmites australis* accounted for about 200 ha or 14%. Saltonstall (2002) reports that most of the native haplotypes of *P. australis* in New England have been displaced by an introduced haplotype (Saltonstall, 2002). Recent studies report only one native population of *P. australis* in Block Island in the state of Rhode Island (Lambert, 2005; Lambert and Casagrande, 2006).

Tiner *et al.* (2004) further report that between the 1950s and 1990s Narragansett Bay showed a net loss of about 124 ha of estuarine marshes (excluding oligohaline marshes), which was nearly 9% of the coastal wetland total. Over 50% of the recent loss of coastal wetlands was due to filling that created upland, and an additional 40% due to conversion into open-water, palustrine wetland or tidal flats. During the same period, there was an increase of about 93 ha of *P. australis,* which is about 47% of coverage of *P. australis* in mid-1990s (Tiner *et al.*, 2004). In the northeast, increases in the extent of *P. australis* and other changes in plant community structure are attributed to the introduction of non-native species, increase in land-derived N loads, hydrologic alterations, filling, climate change, and upland buffer disturbances (Roman *et al.*, 1984; Bart and Harman, 2000; Minchinton, 2002; Saltonstall, 2002; Silliman and Bertness, 2004). Less understood than the loss in acreage is the loss of quality in the remaining coastal wetlands of New England—changes in the quality of coastal wetlands are evident by changes in community structure and ecosystem functions of these systems.

17.3   Changes in Community Structure Due to Nitrogen Over-enrichment

17.3.1   Changes in Plant Community Structure

The human population in the Rhode Island portion of the Narragansett Bay watershed has been increasing linearly since the 1850s (Nixon *et al.*, 2005). In a recent field study, 10 sub-estuaries in Narragansett Bay were selected to span a range of anthropogenic stress defined by watershed land development and N loadings (Wigand *et al.*, 2003; see also Table 17.1). Vegetation changes on the marsh landscape were found in salt marshes subject to increasing N loads in Narragansett Bay (Wigand *et al.*, 2003). At sites that received greater N loads, there was significantly lower areal extent and shoot density of high marsh dominant species *Spartina patens*, lower plant species richness, and significantly greater areal extent, density, and height of the tall form of *S. alterniflora* (Wigand *et al.*, 2003; Fig. 17.1a–e). An earlier field survey in Narragansett Bay during the 1970s showed increased height of *S. alterniflora* with increasing watershed development (Nixon and Oviatt, 1973). The natural effects of flooding regime and the marsh slope and width can contribute to vegetation changes on the marsh landscape in Narragansett Bay salt marshes (Wigand *et al.*, 2003). However, the likelihood of N enrichment causing, at least in part, changes on

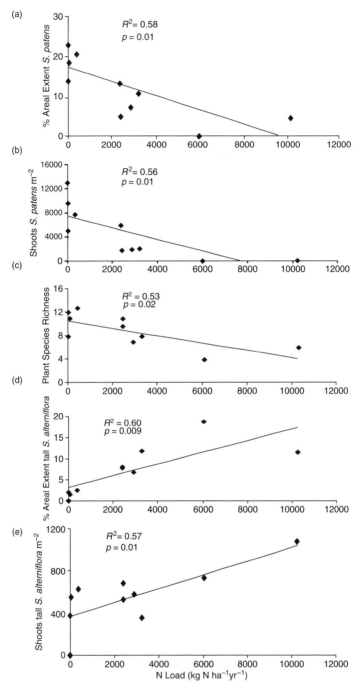

Fig. 17.1 Relationships of watershed nitrogen loads in Narragansett Bay with marsh plant structure: (a) areal extent of high marsh dominant, *S. patens*, (b) shoot density of *S. patens*, (c) plant species richness, (d) areal extent of tall *S. alterniflora*, and (e) shoot density of tall *S. alterniflora*. Figures based on data previously reported in Wigand *et al.*, (2003).

the marsh landscape is supported by separate findings in manipulative, fertilization experiments in Narragansett Bay salt marshes (Levine *et al.*, 1998; Emery *et al.*, 2001). In these manipulative studies, researchers demonstrated that *S. patens*, which is a superior competitor in the high marsh when N is limiting, is out-competed and displaced by *S. alterniflora*, the low marsh dominant plant under N-enriched conditions (Levine *et al.*, 1998; Emery *et al.*, 2001). These fertilization studies suggest that under nutrient-enriched conditions, where *S. alterniflora* is the best competitor for N and light and most tolerant of physical stresses, an entire salt marsh could become dominated by one species (Levine *et al.*, 1998; Emery *et al.*, 2001). A description of the expected plant structure and natural marsh landscape in comparison with the stressed condition (at least in part due to N over-enrichment) of a coastal salt marsh is summarized in Table 17.1.

An additional stressor on coastal salt marshes in New England is the rise of sea level. The effect of sea level rise on salt marsh ecosystems will depend upon marsh elevation, sediment supply, nutrient inputs, plant productivity, and tidal range (Morris *et al.*, 2002). The inland migration of salt marshes is often blocked by human development (e.g., sea walls and fill) along the coast of the northeastern US. In some salt marshes in Connecticut, sea level rise is thought to contribute to changes in plant communities due to increased frequency and duration of flooding (Warren and Niering, 1993). In Narragansett Bay, it is thought that accretion rates for most coastal salt marshes are keeping up with sea level rise (Bricker-Urso *et al.*, 1989); however, a recent report suggests that some Narragansett Bay marshes may not be keeping pace with sea level rise (Donnelly and Bertness, 2001). Nevertheless, the results of manipulative, fertilization studies and field surveys in Narragansett Bay support the hypothesis that cultural eutrophication, at least in part, is contributing to plant community shifts on the marsh landscape. These results are further supported by increasing $\delta^{15}N$ in Narragansett Bay salt marsh biota [*Geukensia demissa* (ribbed mussels), *S. alterniflora*, and *Fundulus* spp. (mummichogs)] collected from sites with increasing wastewater N inputs (Fig. 17.2; McKinney *et al.*, 2001; Wigand *et al.*, 2001; Wigand, unpublished data). Nitrogen derived from wastewater is relatively enriched in $\delta^{15}N$, which results in an enriched stable nitrogen isotopic ratio in the biotic components that process the water (Kreitler *et al.*, 1978; Gormly and Spalding, 1979; Arevena *et al.*, 1993). These results suggest that at least some wastewater N from watersheds with high residential development in Rhode Island is entering coastal salt marshes and is being assimilated by plants and consumers.

Over the past century, non-native *P. australis* has been aggressively invading the coastal marshes of North America (see reviews in Chambers *et al.*, 1999; Meyerson *et al.*, 2000). Shoreline development with concomitant nutrient inputs at the upland borders of salt marshes in New England are correlated with increasing cover of *P. australis* (Bertness *et al.*, 2002; Silliman and Bertness, 2004). At sites where *P. australis* has invaded, there is significantly decreased plant species richness on the marsh landscape (Warren *et al.*, 2001; Silliman and Bertness, 2004). In the manipulative studies on Narragansett Bay, Minchinton

Table 17.1 Description of the marsh landscape and community structure of natural and stressed southern New England salt marshes. The symptoms of the stressed condition could at least in part be due to N over-enrichment, but might also be caused by other anthropogenic stressors.

Natural condition	Stressed condition	References
High marsh: Patchy mosaic of *S. patens, D. spicata*, short *S. alterniflora*, and forbs.	Marsh vegetation changes— *Scenario 1*: Dense, extensive, homogenous stands of non native *P. australis* in the high marsh; *Scenario 2*: Remnant, narrow belts of *S. patens* with extensive monocultures of *S. alterniflora* and/or non native *P. australis* in the high marsh; *Scenario 3*: Freshwater marsh stands and non native *P. australis* stands in the high marsh.	*Natural refs:* Miller and Egler, 1950; Redfield, 1972; Niering and Warren, 1980; Nixon, 1982; Bertness and Ellison, 1987; Bertness, 1991, 1992; Roman *et al.*, 2000; *Stressed refs*: Levine *et al.*, 1998; Chambers *et al.*, 1999; Bertness and Pennings, 2000; Emery *et al.*, 2001; Wigand *et al.*, 2003; Silliman and Bertness, 2004.
Low marsh: *S. alterniflora* (short and tall forms).	Only tall form *S. alterniflora* in the low marsh.	Valiela *et al.*, 1975; Howes *et al.*, 1986; Mendelssohn and Morris, 2000; Wigand *et al.*, 2003.
Natural vegetated upland buffers.	Loss of some or all of vegetated upland buffer.	Bertness *et al.*, 2002; Minchinton and Bertness, 2003; Silliman and Bertness, 2004
Diverse assemblage of infaunal species on the marsh and in the adjacent mudflats; high biomass and density of ribbed mussels; high abundance of mummichogs and shrimp.	Increase in detritivores on the salt marsh and in the adjacent mudflats; increase in taxa richness and number of total infaunal species in adjacent mudflats; increase in growth rates of soft-shelled clams and the biomass and density of ribbed mussels; for most tidal marshes, no change or an increase in abundance of mummichogs and shrimp; for *Phragmites*-dominated marshes a decrease in larval and young-of-the year mummichogs.	Oviatt *et al.*, 1977; Chalfoun *et al.*, 1994; Fritz *et al.*, 1996; Tober *et al.*, 1996, 2000; Able *et al.*, 2003; Chintala *et al.*, 2005; Chintala and Wigand, unpublished data.

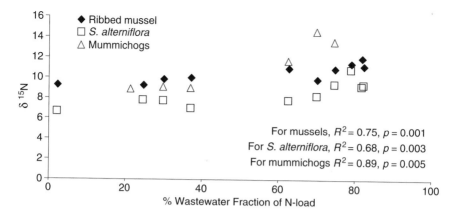

Fig. 17.2 Relationships of watershed nitrogen loads in Narragansett Bay with the $\delta^{15}N$ in ribbed mussels, *S. alterniflora*, and mummichogs. Figure based on previously data reported in McKinney *et al.* (2001), Wigand *et al.* (2001), and Wigand, unpublished data.

and Bertness (2003) demonstrate that the removal of border vegetation (i.e., marsh grasses and rushes) adjacent to *P. australis* promotes the spread of this invasive plant apparently because there may be no competitors for nutrients. Associated with the invasion of *P. australis* in brackish tidal marshes in New Jersey are lower soil salinities at the soil surface, lower water levels, less pronounced microtopographic relief, and higher redox potentials as compared with neighboring shortgrass communities (Windham and Lathrop, 1999).

Nitrogen fertilization of salt marsh plants has shown an apparent shift toward more allocation of carbon to aboveground than to belowground structure (Valiela *et al.*, 1976; Morris, 1982; Minchinton and Bertness, 2003). In addition, the belowground structure varies with species and, therefore, with plant community shifts. For example, *S. patens* has a tightly woven root matrix in the surface soils, and *S. alterniflora* has coarser roots that dwell deeper in the soil. Changes to the belowground structure when one species displaces another may affect the cycling and storage of nitrogen and carbon and the marsh's ability to provide erosion control and keep pace with sea level rise. Less organic matter belowground will reduce peat accumulation and adversely affect the marsh's ability to maintain elevation in equilibrium with sea level (Roman and Daiber, 1984; Morris *et al.*, 2002).

17.3.2 Changes in Animal Community Structure

Changes in the structure of fish communities due to nutrient enrichment are difficult to assess because other potential stressors, such as freshwater inputs, other pollutants and toxicants, global temperatures, and fishing pressure, may also cause adverse effects (Nixon and Buckley, 2002). Nevertheless, increased N loading is important to the structure and function of coastal salt marshes because low N often limits the production of phytoplankton and other plants, which are ultimately the food sources for fish, shellfish, and wildlife. In Waquoit Bay (Massachusetts), fish species richness and abundance of certain benthic-feeding fish and shrimp increased with increasing N loadings (Fritz et al., 1996; Tober et al., 1996; 2000). In the recent field surveys of Narragansett Bay salt marshes, a significantly higher percentage of deposit-feeders in marsh soils at sites with highest N loads was found (Fig. 17.3a), but there was no detectable relationship of N loading with the total number of infaunal individuals, species, or taxa richness among the marshes (M. Chintala and C. Wigand, unpublished data). However, significant positive relationships of N loads with the number of total infaunal species and taxa richness in the mud flats adjacent (within 1 m) to the marsh bank were found (Fig. 17.3b,c). Some deposit-feeders in the samples included *Capitella capitata, Heteromastus filiformis, Leitoscoloplos fragilis, Mediomastus ambiseta*, and other oligochaetes. Moreover, the density and biomass of ribbed mussels (*Geukensia demissa*), significantly increased with increasing N loads among the salt marshes (Chintala et al., 2006; Fig. 17.4a,b). In the MERL (Marine Ecosystems Research Laboratories) mesocosms, researchers have demonstrated that most polychaetes increase in abundance in response to food supply, both from nutrient and organic matter additions (Grassle et al., 1985; Levin, 1986). A description of some likely changes in animal community structure in response, at least in part, to cultural eutrophication is given in Table 17.1.

Cultural eutrophication is likely to contribute to the occurrence of hypoxic and anoxic events in shallow and coastal estuarine areas, and these events can affect the condition and survival of fish, crustaceans, and the macrobenthic community (D'Avanzo and Kremer, 1994; Diaz and Rosenberg, 1995; D'Avanzo et al., 1996; Lerberg et al., 2000; Wannamaker and Rice, 2000; Deegan, 2002). Eutrophication acts as an accelerant or enhancing factor to hypoxia and anoxia, and when coupled with adverse meteorological and hydrodynamic events, low oxygen events increase in frequency and severity (Diaz and Rosenberg, 1995; see also Chapters 11 and 12 of this volume). Low dissolved oxygen near the bottom prevents some fish from feeding on benthic fauna (Nestlerode and Diaz, 1998).

Under nutrient-enriched conditions in salt marshes, a possible food web shift from one dominated by emergent plants to one dominated by macroalgae might occur and alter fish species composition and abundance (Deegan, 2002). It is also proposed that changes in the density of plant shoots, due to either within-species responses to N enrichment or the displacement of a dominant plant species, could affect predator–prey interactions and alter the refugia for juvenile fish and other macrobenthic organisms (Deegan, 2002).

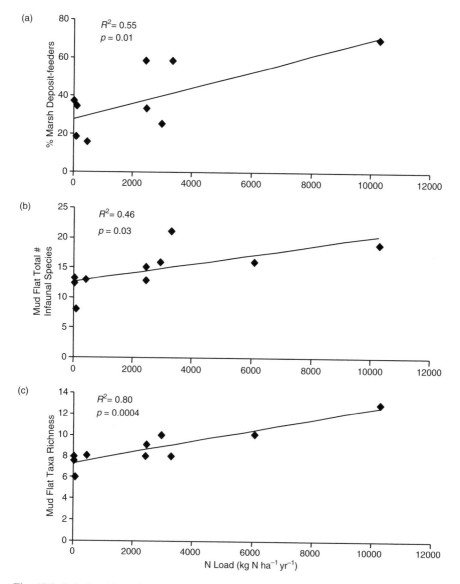

Fig. 17.3 Relationships of watershed nitrogen loads in Narragansett Bay with metrics of infauna: (a) percentage of deposit feeders in the marsh sediments, (b) number of total infaunal species in the adjacent mudflats, and (c) taxa richness in the adjacent mudflats. Figure is based on unpublished data of M. Chintala and C. Wigand. Triplicate cores (diameter = 7 cm; height = 15 cm) were sampled in the low marsh and in the adjacent mudflat (1 m from the marsh bank) during the summer 1998.

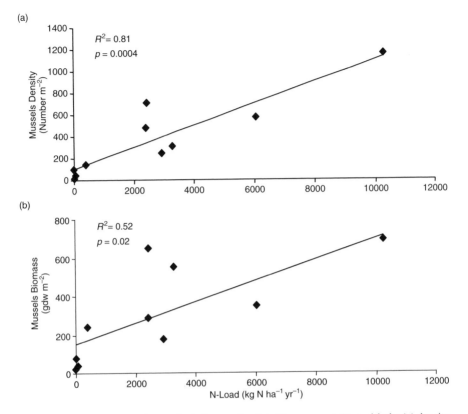

Fig. 17.4. Relationships of watershed nitrogen loads in Narragansett Bay with the (a) density and (b) dry weight biomass of ribbed mussels. Figure based on previously reported data in Chintala *et al.* (2006).

As discussed in the Changes in Plant Community Structure section of this chapter, *Phragmites*-dominated marshes are often prevalent in hydrologically altered systems and under nutrient-enriched conditions. In some cases, restoring tidal action to New England salt marshes has resulted in increased abundance of *Spartina* species, and has restored nekton densities and species richness (Roman *et al.*, 2002). However, not all tidally restricted salt marshes dominated by *P. australis* show lower nekton densities and species richness (Raposa and Roman, 2001). In some tidally restricted as well as other non-restricted tidal marshes, the nekton species composition associated with *P. australis* does not appear to differ relative to other vegetation types (Fell *et al.*, 1998; Meyer *et al.*, 2001; Raposa and Roman, 2001; Warren *et al.*, 2001). In addition, there are reports that *P. australis* might be contributing to some food chains and webs of marine transient and resident finfish (Fell *et al.*, 1998; Wainright *et al.*, 2000; Weinstein *et al.*, 2000). Nevertheless, it is also reported that *P. australis* has deleterious effects on larval and juvenile fish use of some marsh surfaces (Able and Hagan, 2000; 2003; Able *et al.*, 2003; Raichel *et al.*, 2003).

These somewhat contradictory reports on fish use of *P. australis* tidal marshes are explained by Able *et al.* (2003) as responses of fish to various stages in the chronology of *P. australis* invasion, particularly in marshes surrounding Delaware Bay and elsewhere in southern New Jersey. They propose that, as *P. australis* initially begins to replace *Spartina* and other plant species, there are few changes in the marsh topography and elevation (Able *et al.*, 2003). Later, as *P. australis* turns into a monoculture on the marsh landscape, it is proposed that an extensive root and rhizome mat is developed, which raises the elevation of the marsh and reduces marsh flooding (Windham and Lathrop, 1999; Able *et al.*, 2003). Furthermore, litter would accumulate on the surface clogging depressions that would usually be prime habitat for fish. Therefore, Able *et al.* hypothesize that observations of fish composition and use of *P. australis* marshes early in an invasion will show few or no differences as compared with reference *Spartina* marshes, but in the later stages of *P. australis* invasion, there would be a reduction in the feeding and reproduction of important forage fish (e.g., mummichogs) and an elimination of the nursery function for fish (Able *et al.*, 2003).

Sea level rise has been documented as a serious problem for many coastal salt marshes (Stevenson *et al.*, 1988; Day and Templet, 1989; Warren and Niering, 1993; Morris *et al.*, 2002). Although ironically a problematic invasive species, *P. australis* is proposed to provide marsh accreting processes to combat the rising sea level at some sites where marshes are severely eroding or where sea level is rising and there are hardened shorelines (Rooth *et al.*, 2003). However, it is still unclear how resistant will *P. australis* be to erosion under rising sea level. In addition, *P. australis* does have adverse effects on habitat quality for some fish and wildlife (e.g., Benoit and Askins, 1999; Able *et al.*, 2003). The role of *P. australis* in providing marsh maintenance clearly needs more research.

17.4 Changes in System-level Functions Due to Nitrogen Over-enrichment

17.4.1 Primary and Secondary Production

Unlike the shifts in marsh plant community structure observed in Narragansett Bay in recent salt marsh surveys and attributed at least in part to N enrichment (Wigand *et al.*, 2003), surveys in the 1970s and in late 1990s reported variable *S. alterniflora* production (peak summer biomass for late 1990s, means $\pm$ SE: 1025 ± 202 g DW m^{-2} aboveground, 7835 ± 1402 g DW m^{-2} belowground) among sites throughout the bay (Oviatt *et al.*, 1977; Table 17.2). There were no relationships of plant biomass with watershed land development or N loadings in either survey, and in the late 1990s the biomass of *S. alterniflora* ranged

Table 17.2 Watershed description, calculated nitrogen loadings, and *Spartina* production for ten Narragansett Bay coastal fringe marshes. Watershed data and calculated nitrogen loads were previously reported in Wigand et al. (2003), and plant biomass data are unpublished (C. Wigand). Triplicate plant samples (aboveground: 0.0625 m²; belowground: 0.002 m²) for each *Spartina* species were collected, dried, and weighed at the ten salt marsh sites in late August/September, 1998.

*Site	Watershed			Calculated nitrogen loads			1998 *Spartina* biomass			
	Watershed area, ha	Percent residential	Marsh area, ha	Total N load (kg N yr⁻¹)	Percent wastewater	Marsh N-load (kg N ha⁻¹ yr⁻¹)	*S. alterniflora* aboveground g m⁻²	*S. alterniflora* belowground g m⁻²	*S. patens* aboveground g m⁻²	*S. patens* belowground g m⁻²
JEN	41	4.0	12	29	37.4	2	786	6,680	367	11,665
FOX	62	0.3	10	103	2.3	10	277	3,965	998	8,198
FOG	30	14.9	4	280	25.0	63	1,104	6,679	510	8,723
DON	2,975	10.0	29	11,593	30.2	400	821	6,177	1,056	8,287
PAS	314	65.4	4	9,917	82.2	2,418	2,403	13,162	675	6,626
BRU	781	61.8	9	22,344	81.9	2,440	591	3,480	197	13,709
BIS	2,296	22.1	4	11,235	62.6	2,922	501	9,938	148	1,029
OLD	1,505	44.5	10	31,587	79.1	3,282	1,310	6,397	699	8,634
WAT	402	56.0	2	11,920	70.0	6,037	726	17,395	275	5,823
APP	1,738	43.3	3	32,472	74.5	10,253	1,726	4,474	350	6,898

*Sites are listed from lowest to highest marsh N load (JEN = Jenny Pond, FOX = Fox Hill Salt Marsh, FOG = Fogland Marsh, DON = Mary Donovan Marsh, PAS = Passeonkquis Cove, BRU = Brush Neck Cove, BIS = Bissel Cove, OLD = Old Mill Creek, WAT = Watchemoket Cove, APP = Apponaug Cove).

from 277–2,403 g DW m^{-2} aboveground and 3,480–17,395 g DW m^{-2} below-ground (Table 17.2). In one marsh survey in Narragansett Bay, a trend of increasing standing crop of *S. alterniflora* with increasing development and eutrophication was reported (Nixon and Oviatt, 1973). Furthermore, in a number of fertilization experiments of *S. alterniflora*, N enrichment signifi-cantly increased aboveground production in experimental plots (Valiela *et al.*, 1973; 1975; 1976; Sullivan and Daiber, 1974; Valiela and Teal, 1974; Gallagher, 1975; Mendelssohn, 1979; Wigand *et al.*, 2004a) and sediment trapping ability (Morris *et al.* 2002) as compared to control plots. Furthermore, when *Spartina* is subjected to N enrichment, reports suggest that the above/belowground biomass ratio of *Spartina* species increases because of more allocation of energy and resources to aboveground production (Valiela and Teal, 1974; Valiela *et al.*, 1976; Morris, 1982).

In a review of the effect of N enrichment on fisheries in coastal systems (i.e., Scottish Sea Lochs; Baltic Sea; Kattegat), Nixon and Buckley (2002) concluded that N enrichment will ultimately increase the production of pelagic fish species. In addition, there is a significant, direct relationship of phytoplankton produc-tion with fishery yields in coastal systems (Nixon *et al.*, 1986; Nixon, 1988). Using the MERL mesocosms, Keller *et al.* (1990) demonstrated that the growth and size of juvenile Atlantic menhaden were significantly and positively corre-lated with food availability, which was in turn positively correlated with nutri-ent loading. Some field studies in southern New England salt marshes reported increases in the growth of soft-shelled clams and ribbed mussels, the abundance of some fish and shrimp, and the biomass of ribbed mussels with increasing N loadings (Chalfoun *et al.*, 1994; Fritz *et al.*, 1996; Tober *et al.*, 1996; 2000; Evgenidou and Valiela, 2002; Chintala *et al.*, 2006). These field studies support the hypothesis that increases in N supply lead to increased phytoplankton production and particulate organic matter, and therefore an increased food supply for some fish and shellfish (Nixon, 1988; Tober *et al.*, 2000; Evgenidou and Valiela, 2002; Chintala *et al.*, 2006). However, top-down controls (e.g., predation) and hypoxia may also influence secondary production rates (D'Avanzo and Kremer, 1994; Diaz and Rosenberg, 1995; Tober *et al.*, 2000). In one field study of Narragansett Bay marshes with varying watershed devel-opment, no significant differences in the production of fish, shrimp, crab, or insects were found (Oviatt *et al.*, 1977).

17.4.2   *Microbial Processes and the Transformations of Nitrogen and Carbon*

Nitrogen over-enrichment can have direct effects on microbial processes in marsh soils. Denitrification is a bacterially mediated process common in estuarine soils, in which nitrate is removed from porewater and converted to gaseous nitrogen (N_2O, N_2), rendering it unavailable to most primary producers and transferring it from the coastal system to the atmosphere. Denitrifying bacteria are facultative anaerobes requiring both an organic carbon and a nitrate source. In general, denitrification rates (-2.9 to 8.0 mmol N $m^{-2} d^{-1}$) in Rhode Island coastal salt marshes are within a similar range as those reported in other tidal marshes (Valiela *et al.*, 2000; Davis *et al.*, 2004). In a coastal salt marsh at Brush Neck Cove in Narragansett Bay, Addy *et al.* (2005) detected groundwater denitrification capacity down to depths of 300 cm. The upwelling groundwater flow paths in the high and low marshes support the contention that fringing salt marshes are an important consideration in year-round watershed scale nitrogen budgets for the bay (Addy *et al.* 2005). Nowicki and Gold, in Chapter 4 of this volume, provide greater insight into N behavior along the land–water interface, concluding that groundwater can be of great importance at local scales. In a study of 10 salt marshes in Narragansett Bay, researchers found a significant positive relationship between fall denitrification enzyme activity in the high marsh zone and watershed N loadings (Wigand *et al.*, 2004b). The denitrification enzyme activity in the low marsh zone did not significantly correlate with watershed N loadings, and those rates may more closely relate to nutrient inputs from the estuary through tidal exchange (Wigand *et al.*, 2004b).

In contrast, Davis *et al.* (2004) found a significant inverse relationship between the net flux of N gas from high marsh soils and watershed N loadings, but a positive relationship between the net N flux and soil organic matter. It appears that there may be a threshold level of nitrogen inputs at which coastal wetlands can no longer denitrify and transform elevated nitrate inputs. Davis *et al.* (2004) propose that N fixation rates may increase at more impacted sites, possibly because of the occurrence of more bare spots, and/or actual denitrification rates may decrease because labile organic matter is limiting (Table 17.3). Algal N fixation rates are high on the marsh landscape, but are usually not important in an overall marsh budget because they are associated with pannes and bare spots on the marsh that are usually only a small part of the total marsh area (Valiela and Teal, 1979). However, in an impaired marsh with more bare spots, N fixation may be more prevalent, and N fixation rates might offset the benefit of N losses due to denitrification rates normally associated with salt marshes (Davis *et al.*, 2004).

In a long-term (12 years) fertilization experiment in an oligotrophic South Carolina salt marsh, Morris and Bradley (1999) demonstrated that, with increasing nitrogen and phosphorus additions, soil respiration rates increased and soil macro-organic matter in the top 5 cm of the soil was significantly reduced. In Narragansett Bay marshes, increased *in situ* soil respiration rates (as CO_2 production in domes, about 500 ml) were measured at sites with more developed watersheds and higher N loadings (Fig. 17.5; P. Brennan and C. Wigand, unpublished data). The increased soil respiration rates at the Narragansett Bay marshes subject to increased N loads were likely stimulated by N-enriched, labile organic matter. Nutrient loading to coastal salt marshes may change the amount of carbon stored because of the effects on the

Table 17.3 Description of some ecosystem-level processes of natural and stressed southern New England salt marshes. The symptoms of the stressed condition could at least in part be due to N over-enrichment, but might also be caused by other anthropogenic stressors.

Natural condition	Stressed condition	References
Productive, co-dominant *Spartina alterniflora* and *S. patens* on the marsh landscape.	Increased aboveground productivity of *S. alterniflora* and *P. australis*; no change or an increase in the above/belowground biomass ratio of some dominant plants; associated with exotic *P. australis* monocultures, lower salinities at the soil surface, lower water levels, less pronounced microtopographic relief, and higher redox potentials compared to natural communities.	Nixon and Oviatt, 1973; Valiela *et al.*, 1973; 1975; 1976; Sullivan and Daiber, 1974; Valiela and Teal, 1974; Gallagher, 1975; Mendelsohn, 1979; Morris, 1982; Windham and Lathrop, 1999; Minchinton and Bertness, 2003; Wigand *et al.*, 2004a.
Denitrification is driven primarily by coupled nitrification-denitrification; usually only a small portion of the nitrogen annually cycled through the marsh is denitrified, the majority is recycled into new plant biomass.	Denitrification enzyme activity increases; increased denitrification in creek bottom sediments.	Howes *et al.*, 1996; Nowicki *et al.*, 1999; Teal and Howes, 2000; Wigand *et al.*, 2004b.
Nitrogen fixation confined to sparse bare patches on marsh landscape.	Nitrogen fixation possibly more common because of increased number of bare patches.	*Natural ref:* Valiela and Teal, 1979; *Stressed ref:* Davis *et al.*, 2004.
Marsh soil respiration rates are moderate.	Marsh soil respiration rates are elevated.	Morris and Bradley, 1999; Brennan and Wigand, this report.
Accretion rates greater or equal to decomposition rates and apparent sea level rise; peat growing or building.	Accretion rates less than decomposition rates and apparent sea level rise; peat not building; loss of macro-organic matter in the surface soils; subsidence, ponding, and fragmentation.	Bricker-Urso *et al.*, 1989; Warren and Niering, 1993; Morris and Bradley, 1999; Roman *et al.*, 2000; Donnelly and Bertness, 2001; Deegan, 2002; Morris *et al.*, 2002.

decomposition rate of labile (N-rich) organic carbon, the stability of the existing refractory carbon pool, or the allocation of carbon between above- and below-ground plant production (Morris and Bradley, 1999). The changes in the soil respiration rates may have important implications for marsh elevation and

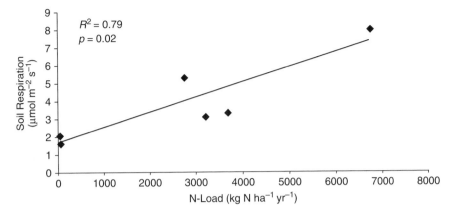

Fig. 17.5 Relationships of watershed nitrogen loads in Narragansett Bay with high marsh soil respiration rates. Figure is based on unpublished data of P. Brennan and C. Wigand. *In situ* soil respiration was measured in triplicate at six marshes during the summer 2004.

maintenance of coastal wetlands threatened by rising sea level. Some ecosystem-level changes in stressed salt marshes are summarized in Table 17.2.

Changes in the large refractory pool of carbon in wetland soils might have broader significance than contributing to coastal marsh loss. Wetlands on a global basis, including peatlands, have been reported to contain a large inactive pool of soil carbon that is roughly equivalent to the carbon content of atmosphere (Morris and Bradley, 1999). Any change in the mobility of this carbon or the carbon balance of peat soils could have important implications on global climate.

17.5 Actions to Protect and Restore Coastal Wetlands

Ecosystem services provided by coastal wetlands (e.g., water quality maintenance, flood and erosion control, and habitat and food for fish and wildlife) are difficult to value economically, because they are often outside the market. However, in a review of the world's ecosystem services and natural capital, Costanza *et al.* (1997) conservatively estimated that wetlands throughout the world contribute an average of US $5 trillion worth of ecosystem services annually, or US $15,000 worth of services per hectare per year. The authors caution that these monetary values are probably underestimates, and that as these ecosystems become scarcer the value of their services will become much greater (Costanza *et al.*, 1997). Although ecosystem valuation is difficult and fraught with uncertainties, some sense of monetary value of these non-market goods and services of wetlands will help in weighing social, economic, and environmental costs and benefits in various societal decisions (e.g., demonstration of restoration success, prioritization of restoration efforts, developing budgets for conservation and management efforts).

Given the range of human activities and the propagation of their effects, it is most likely that ecosystems are exposed to multiple stressors of anthropogenic origin, rather than to a single discrete one (Breitburg et al., 1998). Cultural eutrophication apparently has some adverse effects on the structure and function of coastal wetlands, and reductions of watershed nutrient inputs should help protect and facilitate recovery of some of these systems. Protection, restoration, and creation of vegetated buffers and inland sinks (e.g., forested lands, inland wetlands) in the watershed that intercept diffuse N sources, which might otherwise reach the coastal salt marshes, may be an important part of a coastal wetlands management plan. In a review of the estimated percent removal of various pollutants by vegetated buffers in coastal areas, Desbonnet et al. (1995) reported that over 70% of the non-point source N inputs were removed by vegetated buffers of 25 m or greater, and 50% of N was removed by buffers of only 3.5 m width. However, for many impaired salt marshes, restoration efforts and interventions (e.g., removing dikes and/or sea walls, restoring hydrologic connection with the sea, etc.) will be necessary to begin recovery (e.g., Roman et al., 2002).

In addition, if sediment accretion does not keep up with sea level rise, noticeable changes in the plant structure and marsh landscape will occur in response to the increased duration and frequency of flooding (Stevenson et al., 1988; Day and Templet, 1989; Warren and Niering, 1993). Accelerated rates of sea level rise may have serious consequences in both the distribution and possible overall loss of coastal wetlands in the northeastern US (Warren and Niering, 1993). The deleterious effects of sea level rise on coastal salt marshes may negate the positive effects of reducing non-point source N pollution. Therefore, a strategy to address multiple stressors, especially non-point source nutrient pollution and sea level rise, is needed to protect coastal wetlands in Narragansett Bay. Interagency partnerships and community involvement are necessary for successful protection, restoration, and management of coastal resources, and some standardized methods to monitor coastal wetlands and assess restoration success are now available to assist in developing monitoring and assessment strategies, and to allow for year-to-year and among-site comparisons (Short et al., 2000; Roman et al., 2001; Neckles et al., 2002). With an informed community making decisions to sustain and restore coastal wetlands using an adaptive management approach, future generations will be able to reap the economic and environmental benefits of coastal wetlands, and also enjoy the natural beauty of the "green ribbon of soft, salty, wet, low-lying land along the coast."

Acknowledgments I would like to thank NHEERL-EPA-AED colleagues, academic partners, and student interns (Rick McKinney, Glen Thursby, Jenny Davis, Barb Nowicki, Mark Stolt, Earl Davey, Rex Tien, Deborah Fillis, Kenny Raposa, Steve Ryba, Charley Roman, Peter Groffman, Heather Smith, Brandon Keith, and Cassius Spears) who have cooperated, assisted, and collaborated with me on various research projects in Narragansett Bay. Special thanks to Marty Chintala and Patricia Brennan for allowing me to report on some unpublished results. Also, I am grateful to Barry Costa-Pierce for the invitation to participate in the nutrients workshop and contribute to this book. Charley Roman, Dan Campbell, Roxanne Johnson, and Giancarlo Cicchetti provided insightful comments to an early draft of this manuscript. Mention of trade names or commercial products does not constitute endorsement or recommendation for use by the US Environmental Protection Agency. This report, contribution number AED-05-082, has been technically reviewed by the US EPA's Office of Research and Development, National Health and Environmental Effects Research Laboratory, Atlantic Ecology Division, Narragansett, RI, and approved for publication. Approval does not signify that the contents necessarily reflect the views and policies of the Agency.

References

Able, K.W., and Hagan, S.M. 2000. Effects of common reed (*Phragmites australis*) invasion on marsh surface macrofauna: responses of fishes and decapod crustaceans. *Estuaries* 23:633–646.

Able, K.W., and Hagan, S.M. 2003. Impact of common reed, *Phragmites australis*, on essential fish habitat: influence on reproduction, embryological development, and larval abundance of mummichog (*Fundulus heteroclitus*). *Estuaries* 26:40–50.

Able, K.W., Hagan, S.M., and Brown, S.A. 2003. Mechanisms of marsh habitat alteration due to *Phragmites*: response of young-of-the-year Mummichog (*Fundulus heteroclitus*) to treatment for *Phragmites* removal. *Estuaries* 26:484–494.

Addy, K., Gold, A.J., Nowicki, B., McKenna, J., Stolt, M., and Groffman, P. 2005. Denitrification capacity in a subterranean estuary below a Rhode Island fringing salt marsh. *Estuaries* 29:896–908.

Aravena, R., Evans L., and Cherry J.A. 1993. Stable isotopes of oxygen and nitrogen in source identification of nitrate from septic systems. *Ground Water* 31:180–186.

Bart, D., and Harman, J.M. 2000. Environmental determinants of *Phragmites australis* in a New Jersey salt marsh: an experimental approach. *OIKOS* 89:59–69.

Benoit, L.K., and Askins, R.A. 1999. Impact of spread of *Phragmites* on the distribution of birds in Connecticut tidal marshes. *Wetlands* 19:194–208.

Bertness, M.D. 1991. Zonation of *Spartina patens* and *Spartina alterniflora* in a New England salt marsh. *Ecology* 72:138–148.

Bertness, M.D. 1992. The ecology of a New England salt marsh. *American Scientist* 80:260–268.

Bertness, M.D., and Ellison, A.M. 1987. Determinants of pattern in a New England salt marsh plant community. *Ecological Monographs* 57:129–147.

Bertness, M.D., and Pennings, S.C. 2000. Spatial variation in process and pattern in salt marsh plant communities in eastern North America. *In* Concepts and Controversies in Tidal Marsh Ecology, pp. 39–58. Weinstein, M.P., and Kreeger, D.A., (eds.), The Netherlands: Kluwer Publishers.

Bertness, M.D., Ewanchuk, P., and Silliman, B.R. 2002. Anthropogenic modification of New England salt marsh landscapes. *Proceedings of the National Academy of Sciences of the United States of America* 99:1395–1398.

Breitburg, D.L., Baxter, J.W., Hatfield, C., Howarth, R.W., Jones, C.G., Lovett, G.M., and Wigand, C. 1998. Understanding effects of multiple stressors: ideas and challenges. *In* Successes, Limitations and Frontiers in Ecosystem Science, Pp. 416–431. Pace, M.L., and Groffman, P.M. (eds), New York: Springer-Verlag. 499 p.

Bricker-Urso, S., Nixon, S.W., Cochran, J.K., Hirschberg, D.J., and Hunt, C. 1989. Accretionrates and sediment accumulation in Rhode Island salt marshes. *Estuaries* 12:300–317.

Bromberg, K.D., and Bertness, M.D. 2005. Reconstructing New England salt marsh losses using historical maps. *Estuaries* 28:823–832.

Carpenter, S., Caraco, N.F., Correll, D.L., Howarth, R.W., Sharpley, S.N., and Smith, V.H. 1998. Nonpoint pollution of surface waters with phosphorus and nitrogen. *Issues in Ecology* 3:1–12.

Chalfoun, A., McClelland, J., and Valiela, I. 1994. The effect of nutrient loading on the growth rate of two species of bivalves, *Mercenaria mercenaria* and *Mya arenaria*, in estuaries of Waquoit Bay, Massachusetts. *Biological Bulletin* 187:281.

Chambers, R.M., Meyerson, L.A., and Saltonstall, K. 1999. Expansion of *Phragmites australis* into tidal wetlands of North America. *Aquatic Botany* 64:261–273.

Chintala, M.M., Wigand, C., and Thursby, G. 2006. Comparison of *Geukensia demissa* (Dillwyn) populations in Rhode Island fringe salt marshes with varying nitrogen loads. *Marine Ecology Progress Series* 320:101–108.

Costanza, R., d'Arge, R., de Groot, R., Farber, S., Grasso, M., Hannon, B., Limburg, K., Naeem, S., O'Neill, R.V., Raskin, R.G., Sutton, P., and van den Belt, M. 1997. The value of the world's ecosystem services and natural capital. *Nature* 387:253–260.

Dahl, T.E. 1990. Wetland Losses in the United States, 1780s to 1990s. Washington, DC: US Department of the Interior, Fish and Wildlife Service. 21 pp.

D'Avanzo, C., and Kremer, J. 1994. Diel oxygen dynamics and anoxic events in an eutrophic estuary of Waquoit Bay, Massachusetts. *Estuaries* 17:131–139.

D'Avanzo, C., Kremer, J., and Wainwright, S.C. 1996. Ecosystem production and respiration in response to eutrophication in shallow temperate estuaries. *Marine Ecology Progress Series* 141:263–274.

Davis, J.L., Nowicki, B., and Wigand, C. 2004. Denitrification in fringing salt marshes of Narragansett Bay, Rhode Island, USA. *Wetlands* 24(4):870–878.

Day, J.W., and Templet, P.H. 1989. Consequences of sea level rise: implications from the Mississippi Delta. *Coastal Management* 17:241–257.

Deegan, L.A. 2002. Lessons learned: the effects of nutrient enrichment on the support of nekton by seagrass and salt marsh ecosystems. *Estuaries* 25:727–742.

Desbonnet, A., Lee, V., Pogue, P., Reis, S., Boyd, J., Willis, J., and Imperial, M.1995. Development of coastal vegetated buffer programs. *Coastal Management* 23:91–109.

Diaz, R.J., and Rosenberg, R. 1995. Marine benthic hypoxia: a review of its ecological effects and the behavioral responses of benthic macrofauna. *Oceanography and Marine Biology Annual Review* 33:245–303.

Donnelly, J.P., and Bertness, M.D. 2001. Rapid shoreward encroachment of salt marsh cordgrass in response to accelerated sea-level rise. *Proceedings National Academy of Science* 98:14218–14223.

Emery, N.C., Ewanchuk, P.J., and Bertness, M.D. 2001. Competition and salt-marsh plantzonation: stress tolerators may be dominant competitors. *Ecology* 82:2471–2485.

Evgenidou, A., and Valiela, I. 2002. Response of growth and density of a population of *Geukensia demissa* to land-derived nitrogen loading, in Waquoit Bay, Massachusetts. *Estuarine, Coastal and Shelf Science* 55:125–138.

Fell, P., Weissbach, S.P., Jones, D.A., Fallon, M.A., Zeppieri, J.A., Faison, E.K., Lennon, K.A., Newberry, K.J., and Reddington, L.K. 1998. Does invasion of oligohaline tidal marshes by reed grass, *Phragmites australis* (Cav.) Trin. ex Steud., affect the availability of prey sources for the mummichog, *Fundulus heteroclitus* L. *Journal of Experimental Marine Biology and Ecology* 222:59–77.

Fritz, C., Labrecque, E., Tober, J., Behr, P.J., and Valiela, I. 1996. Shrimp in Waquoit Bay: effects of nitrogen loading on size and abundance. *The Biological Bulletin* 191:326–327.

Gallagher, J. 1975. Effect of an ammonium nitrate pulse on growth and elemental composition of natural stands of *Spartina alterniflora* and *Juncus roemerianus*. *American Journal of Botany* 62:644–648.

Gormly J.R., and Spalding, R.F. 1979. Sources and concentrations of nitrate-nitrogen in groundwater of the central Platte region, Nebraska. *Ground Water* 17:291–301.

Grassle, J.F., Grassle, J.P., Brown-Leger, L.S., Petrecca, R.F., and Copley, N.J. 1985. Subtidal macrobenthos of Narragansett Bay. Field and mesocosm studies of the effects of eutrophication and organic input on benthic populations. *In* Marine Biology of Polar Regions and Effects of Stress on Marine Organisms, pp. 421–434. Gray, J.S., and Christiansen, M.E. (eds.), New York: Wiley.

Howes, B.L., Dacey, W.H., and Goehringer, D.D. 1986. Factors controlling the growth form of *Spartina alterniflora*: feedbacks between above–ground production, sediment oxidation, nitrogen and salinity. *The Journal of Ecology* 74:881–898.

Howes, B.L., Weiskel, P.K., Goehringer, D.D., and Teal, J.M. 1996. Interception of freshwater and nitrogen transport from uplands to coastal waters: the role of salt marshes. *In* Estuarine Shores: Evolution, Environments and Human Alterations, pp. 287–310. Nordstrom, K. F., and Roman, C.T. (eds.), John Wiley and Sons Ltd., New York, NY, USA.

Keller, A.A., Doering, P.H., Kelly, S.P., and Sullivan, B.K. 1990. Growth of juvenile Atlantic menhaden, *Brevoortia tyrannus* (Pices: Clupeidae) in MERL mesocosms: effects of eutrophication. *Limnology and Oceanography* 35(1):109–122.

Kreitler, C.W., Ragone, S., and Katz, B.G. 1978. $^{15}N/^{14}N$ ratios of ground water nitrate, Long Island, NY. *Ground Water* 16:404–409.

Lambert, A.M. 2005. Native and Exotic *Phragmites australis* in Rhode Island: Distribution and Differential Resistance to Insect Herbivores. PhD Dissertation, University of Rhode Island, RI, USA.

Lambert, A.M., and Casagrande, R.A. 2006. Distribution of native and exotic *Phragmites australis* in Rhode Island. *Northeastern Naturalist* 13:551–560.

Lerberg, S.B., Holland, A.F., and Sanger, D.M. 2000. Responses of tidal creek macrobenthic communities to the effects of watershed development. *Estuaries* 23:838–853.

Levin, L.A. 1986. Effects of enrichment on reproduction in the opportunistic polychaete *Streblospio benedicti* (Webster): a mesocosm study. *Biological Bulletin* 171:143–160.

Levine, J., Brewer, S.J., and Bertness, M.D. 1998. Nutrient availability and the zonation of marsh plant communities. *The Journal of Ecology* 86:285–292.

McKinney, R.A., Nelson, W.G., Charpentier, M.A., and Wigand, C. 2001. Ribbed mussel nitrogen isotope signatures reflect nitrogen sources in coastal salt marshes. *Ecological Applications* 11:203–214.

Mendelssohn, I.A. 1979. The influence of nitrogen level, form and application method on the growth response of *Spartina alterniflora* in North Carolina. *Estuaries* 2:106–112.

Mendelssohn, I.A., and Morris, J.T. 2000. Ecophysiological controls on the growth of *Spartina alterniflora*. *In* Concepts and Controversies in Tidal Marsh Ecology, pp. 59–80. Weinstein, M.P., and Kreeger, D.A. (eds.), The Netherlands: Kluwer Publishers.

Meyer, D.L., Johnson, J.M., and Gill, J.W. 2001. Comparison of nekton use of *Phragmites australis* and *Spartina alterniflora* marshes in Chesapeake Bay, USA. *Marine Ecology Progress Series* 209:71–84.

Meyerson, L.A., Saltonstall, K., Windham, L., Kiviat, E., and Findlay, S. 2000. A comparison of *Phragmites australis* in freshwater and brackish marsh environments in North America. *Wetlands Ecology and Management* 8:89–103.

Miller, W.B., and Egler, F.E. 1950. Vegetation of the Wequetequock-Pawcatuck tidal marshes, Connecticut. *Ecological Monographs* 20:143–172.

Minchinton, T.E. 2002. Precipitation during El Niño correlates with increasing spread of *Phragmites australis* in New England, USA coastal marshes. *Marine Ecology Progress Series* 242:305–309.

Minchinton, T.E., and Bertness, M.D. 2003. Disturbance-mediated competition and the spread of *Phragmites australis* in a coastal marsh. *Ecological Applications* 13:1400–1416.

Mitsch, W.J., and Gosselink, J.G. 2000. Wetlands, 3rd edition. New York: John Wiley and Sons, Inc.

Morris, J.T. 1982. A model of growth responses by *Spartina alterniflora* to nitrogen limitation. *The Journal of Ecology* 70:25–42.

Morris, J.T., and Bradley, P.M. 1999. Effects of nutrient loading on the carbon balance of coastal wetland sediments. *Limnology and Oceanography* 44:699–702.

Morris, J.T., Sundareshwar, P.V., Nietch, P.T., Kjerfve, B., and Cahoon, D.R. 2002. Responses of coastal wetlands to rising sea level. *Ecology* 83:2869:2877.

Neckles, H.A., Dionne, M., Burdick, D.M., Roman, C., Buchsbaum, T., and Hutchins, E. 2002. A monitoring protocol to assess tidal restoration of salt marshes on local and regional scales. *Restoration Ecology* 10:556–563.

Nestlerode, J.A., and Diaz, R.J. 1998. Effect of periodic environmental hypoxia on predation of a tethered polychaete, *Glycera americana*: implications for trophic dynamics. *Marine Ecology Progress Series* 172:185–195.

Niering, W.A., and Warren, R.S. 1980. Vegetation patterns and processes in New England saltmarshes. *Bioscience* 30:301–307.

Nixon, S.W. 1982. The ecology of New England high salt marshes: a community profile. FFWS/OBS-81/55. Washington, DC, USA: United States Fish and Wildlife Service.

Nixon, S.W. 1988. Physical energy inputs and the comparative ecology of lake and marine ecosystem. *Limnology and Oceanography* 33(4):1005–1025.

Nixon, S.W., and Buckley, B. 2002. "A strikingly rich zone"-Nutrient enrichment and secondary production in coastal marine ecosystems. *Estuaries* 25(4b):782–796.

Nixon, S.W., and Oviatt, C.A. 1973. Analysis of local variation in the standing crop of *Spartina alterniflora. Botanica Marina* 16:103–109.

Nixon, S.W., Oviatt, C.A., Frithsen, J., and Sullivan, B. 1986. Nutrients and the productivity of estuarine and coastal marine ecosystems. *Journal of the Limnology Society of South Africa* 12(1/2):43–71.

Nixon, S.W., Buckley, B., Granger, S., Harris, L., Oczkowski, A., Cole, L., and Fulweiler, R. 2005. Anthropogenic nutrient inputs to Narragansett Bay: a twenty five year perspective. A Report to the Narragansett Bay Commission and Rhode Island Sea Grant. Rhode Island Sea Grant, Narragansett, RI. (*www.seagrant.gso.uri.edu/research/bay_commission_report.pdf*.)

Nowicki, B.L., Requintina, E., Van Keuren, D., and Portnoy, J. 1999. The role of sediment denitrification in reducing groundwater-derived nitrate inputs to Nauset Marsh Estuary, Cape Cod, Massachusetts. *Estuaries* 22(2):245–259.

Oviatt, C.A., Nixon, S.W., and Garber, J. 1977. Variation and evaluation of coastal salt marshes. *Environmental Management* 1:201–211.

Raichel, D.L., Able, K.W., and Hartman, J.M. 2003. The influence of *Phragmites* (common reed) on the distribution, abundance, and potential prey of a marsh resident fish in the Hackensack Meadowlands, New Jersey. *Estuaries* 26:511–521.

Raposa, K.B., and Roman, C.T. 2001. Seasonal habitat-use patterns of nekton in a tide-restricted and unrestricted New England salt marsh. *Wetlands* 21:451–461.

Redfield, A.C. 1972. Development of a New England salt marsh. *Ecological Monographs* 42:201–237.

Roman, C.T., and Daiber, F.C. 1984. Aboveground and belowground primary production dynamics of two Delaware Bay tidal marshes. *Bulletin of the Torrey Botanical Club* 3:34–41.

Roman, C.T., Niering, W.A., and Warren, R.S. 1984. Salt marsh vegetation change in response to tidal restriction. *Environmental Management* 8:141–150.

Roman, C.T., Jaworski, N., Short, F.T., Findlay, S., and Warren, R.S. 2000. Estuaries of the northeastern United States: habitat and land use signatures. *Estuaries* 23:743–764.

Roman, C.T., James-Pirri, M.J., and Heltshe, J.F. 2001. Monitoring salt marsh vegetation. Technical Report, Long-term Coastal Ecosystem Monitoring Program, Cape Cod National Seashore, Wellfleet, MA.

Roman, C.T., Raposa, K.B., Adamowicz, S.C., James-Pirri, M., and Catena, J.G. 2002. Quantifying vegetation and nekton response to tidal restoration of a New England salt marsh. *Restoration Ecology* 10:450–460.

Rooth, J.E., Stevenson, J.C., and Cornwell, J.C. 2003. Increased sediment accretion rates following invasion by *Phragmites australis*: the role of litter. *Estuaries* 26:475–483.

Saltonstall, K. 2002. Cryptic invasion by a non-native genotype of common reed, Phragmites australis, into North America. *Proceedings of the National Academy of Sciences USA* 99:2445–2449.

Short, F.T., and Burdick, D.M. 1996. Quantifying eelgrass habitat loss in relation to housing development and nitrogen loading in Waquoit Bay, Massachusetts. *Estuaries* 19:730–739.

Short, F.T., Burdick, D.M., Short, C.A., Davis, R.C., and Morgan, P.A. 2000. Developing success criteria for restored eelgrass, salt marsh and mud flat habitats. *Ecological Engineering* 15:239–252.

Silliman, B.R., and Bertness, M.D. 2004. Shoreline development drives invasion of *Phragmites australis* and the loss of plant diversity on New England salt marshes. *Conservation Biology* 18:1424–1434.

Stevenson, J.C., Ward, L.G., and Kearney, M.S. 1988. Sediment transport and trapping in marsh systems: implications of tidal flux studies. *Marine Geology* 80:37–59.

Sullivan, M.J., and Daiber, F.C. 1974. Response in production of cord grass, *Spartina alterniflora*, to inorganic nitrogen and phosphorus fertilizer. *Chesapeake Science* 15(2):121–123.

Teal, J.M., and Howes, B.L. 2000. Salt marsh values: retrospection from the end of the century. *In* Concepts and Controversies in Tidal Marsh Ecology, pp. 9–22. Weinstein, M.P., and Kreeger, D.A. (eds.), The Netherlands: Kluwer Publishers.

Teal, J., and Teal, M. 1969. Life and Death of the Salt Marsh. New York: Ballantine Books. 274 pp.

Tiner, R.W., Huber, I.J., Nuerminger, T., and Mandeville, A.L. 2004. Coastal Wetland Trends in the Narragansett Bay Estuary During the 20th Century. US Fish and Wildlife Service, Northeast Region, Hadley, MA. In cooperation with the University of Massachusetts–Amherst and the University of Rhode Island. National Wetlands Inventory Cooperative Interagency Report. 37 pp. plus appendices.

Tober, J., Fritz, C., Labrecque, E., Behr, P.J., and Valiela, I. 1996. Abundance, biomass, and species richness of fish communities in relation to nitrogen-loading rates of Waquoit Bay estuaries. *Biological Bulletin* 191:321–322.

Tober, J.D., Griffen, M., and Valiela, I. 2000. Growth and abundance of *Fundulus heteroclitus* and *Menidia menidia* in estuaries of Waquoit Bay, Massachusetts exposed to different rates of nitrogen loading. *Aquatic Ecology* 34:299–306.

Valiela, I., and Teal, J.M. 1974. Nutrient limitation in salt marsh vegetation. *In* Ecology of Halophytes, pp. 563-574. Reimold, R.J., and Queen, W.H. (eds.), New York: Academic Press.

Valiela, I., and Teal, J.M. 1979. The nitrogen budget of a salt marsh ecosystem. *Nature* 280:652–656.

Valiela, I., Teal, J.M., and Sass, W.J. 1973. Nutrient retention in salt marsh plots experimentally fertilized with sewage sludge. *Estuarine and Coastal Marine Science* 1:261–269.

Valiela, I., Teal, J.M., and Sass, W.J. 1975. Production and dynamics of salt marsh vegetation and the effects of experimental treatment with sewage sludge. *Journal of Applied Ecology* 12:973–981.

Valiela, I., Teal, J.M., and Persson, N.Y. 1976. Production and dynamics of experimentally enriched salt marsh vegetation: belowground biomass. *Limnology and Oceanography* 21:245–252.

Valiela, I., Collins, G., Kremer, J., Lajtha, K., Geist, M., Seely, B., Brawley, J., and Sham, C.H. 1997. Nitrogen loading from coastal watersheds to receiving estuaries: new method and application. *Ecological Applications* 7:358–380.

Valiela, I., Cole, M.L., McClelland, J., Hauxwell, J., Cebrian, J., Joye, S.B. 2000. Role of salt marshes as part of coastal landscapes. *In* Concepts and Controversies in Tidal Marsh

Ecology, pp. 23–38. Weinstein, M.P., and Kreeger, D.A. (eds.), The Netherlands: Kluwer Publishers.

Wainright, S.C., Weinstein, M.P., Able, K.W., and Currin, C.A. 2000. Relative importance of benthic microalgae, phytoplankton and the detritus of smooth cordgrass Spartina alterniflora and the common reed Phragmites australis to brackish-marsh food webs. *Marine Ecology Progress Series* 200:77–91.

Wannamaker, C.M., and Rice, J.A. 2000. Effects of hypoxia on movements and behaviour of selected estuarine organisms from the southeastern United States. *Journal of Experimental Marine Biology and Ecology* 249:145–163.

Warren, R.S., and Niering, W.A. 1993. Vegetation change on a northeast tidal marsh: interaction of sea –level rise and marsh accretion. *Ecology* 74:96–103.

Warren, R.S., Fell, P.E., Grimsby, J.L., Buck, E.L., Rilling, G.C., and Fertik, R.A. 2001. Rates, patterns, and impacts of Phragmites australis expansion and effects of experimental Phragmites control on vegetation, macroinvertebrates, and fish within tidelands of the lower Connecticut River. *Estuaries* 24:90–107.

Weinstein, M.P., Litvin, S.Y., Bosley, K.L., Fuller, C.M., and Wainright, S.C. 2000. The role of tidal salt marsh as an energy source for marine transient and resident finfishes: a stable isotope approach. *Transactions of the American Fisheries Society* 129:797–810.

Wigand, C., Comeleo, R., McKinney, R., Thursby, G., Chintala, M., and Charpentier, M. 2001. Outline of a new approach to evaluate ecological integrity of salt marshes. *Human and Ecological Risk Assessment* 7:1541–1554.

Wigand, C., McKinney, R.A., Charpentier, M.A., Chintala, M.M., and Thursby, G.B. 2003. Relationships of nitrogen loadings, residential development, and physical characteristics with plant structure in New England salt marshes. *Estuaries* 26(6):1494–1504.

Wigand, C., Thursby, G., McKinney, R.A., and Santos, A. 2004a. Response of *Spartina patens* to Dissolved Inorganic Nutrient Additions in the Field. *Journal of Coastal Research* Special Issue, 45:134–149.

Wigand, C., McKinney, R.A., Chintala, M.M., Charpentier, M.A., and Groffman, P.M. 2004b. Denitrification enzyme activity of fringe salt marshes in New England (USA). *Journal of Environmental Quality* 33:1144–1151.

Windham, L., and Lathrop, R.G., Jr. 1999. Effects of *Phragmites australis* (Common Reed) invasion on aboveground biomass and soil properties in brackish tidal marsh of the Mullica River, New Jersey. *Estuaries* 22:927–935.

Chapter 18
Impacts of Nutrients on Narragansett Bay Productivity: A Gradient Approach

Candace A. Oviatt

18.1 Introduction

In recent years, summer hypoxia and anoxic events have received increasing attention in Narragansett Bay. Starting in 1999, bay-wide surveys and fixed site monitoring of dissolved oxygen and other parameters were initiated. Bergondo *et al.* (2005) found that summer monthly low oxygen concentrations in the upper bay coincided with the low mixing neap tide period. The Rhode Island Department of Environmental Management (RIDEM) initiated volunteer surveys in the upper bay during these high risk periods to evaluate the extent and intensity of monthly low oxygen events (Deacutis *et al.,* 2005). As part of the NOAA Bay Window program, a bay-wide spatial survey conducted by the National Marine Fisheries Service complemented other surveys in revealing the horizontal and vertical patterns of oxygen concentration (Melrose, 2005). The events of low oxygen concentration did not occur in every summer month in the survey years from 1999 to the present. Severe events continued to be correlated with neap tide, but also with the intensity of stratification, stratification-induced phytoplankton blooms, temperature, recent rainfall, water residence time, wind direction, and strength (Bergondo, 2004).

Severe events in summer 2003, which included masses of rotting macroalgae and fish kills, aroused public concern and accelerated the ongoing plans by RIDEM to initiate nutrient reduction measures at Rhode Island sewage treatment facilities. Mesocosm studies have indicated that nitrogen limits primary production in the bay (Oviatt *et al.,* 1995). In 2004, denitrification processors were installed in one of the largest plants, and regulations to decrease nitrogen output to 5 mg L^{-1} (357 μM) in the effluent have been established in all Rhode Island facilities (RIDEM, 2004; A. Liberti, personal communication). RIDEM has also been working with regulators in Massachusetts to reduce nitrogen input from plants upstream of Narragansett Bay. While RIDEM

Candace A. Oviatt
University of Rhode Island, Graduate School of Oceanography, 11 Aquarium Road, Narragansett, RI 02882
coviatt@gso.uri.edu

A. Desbonnet, B. A. Costa-Pierce (eds.), *Science for Ecosystem-based Management.* 523
© Springer 2008

managers estimated the reduction in TN concentrations to about 50% that of 2004 levels, other authorities, including the Chair of the Nutrient Monitoring Committee of The Rhode Island Governor's Narragansett Bay and Watershed Planning Commission, suggested reduction on the order of 20% (Pryor, 2004). The aggressive regulations have stirred controversy in the scientific community with respect to how reduced nutrients might reduce the productivity of resource species in the bay, as well as excited hopes in conservation groups that eelgrass meadows might return to areas now long dominated by abundant macroalgae.

Increasing amounts of nutrients entered Narragansett Bay as an unintended consequence of the introduction of a public water supply to the City of Providence in 1871 (Nixon et al., 2005). According to this study, the number of people served by the evolving sewer systems that discharge directly to Narragansett Bay rose steadily from 1870 to about 1950, and has remained more or less constant up to the present. Estimations of the pre-industrial nitrogen inputs as compared to that of the late 20th century suggested that nitrogen input had increased on the order of 15 times the pre-industrial input (Nixon, 1997). The TN input from sewage treatment plants and rivers to Narragansett Bay has been estimated for 2003 with each source accounting for 50% of the TN (Nixon et al., 2005). TN from sewage treatment plants was estimated at 174 million moles per year when small sewage plants were included; TN from rivers was estimated at 173. If totals of TN from sewage treatment plants and rivers were taken per day and by area and volume for the bay (Kremer and Nixon, 1978), the resulting concentration increase in the bay on average would be 3.6 mm TN m^{-2} d^{-1} and 0.37 μm TN L^{-1} d^{-1}. In Narragansett Bay, many rivers and most of the effluents from sewage treatment plants enter the northern urbanized portion of the bay. These nutrients decline exponentially with dilution and distance toward the mouth of the bay, resulting in persistent nutrient gradient along the north–south axis of the bay (Kremer and Nixon, 1978; Oviatt et al., 2002).

As in Narragansett Bay, nutrient increases in aquatic, estuarine, and coastal systems have occurred since the Industrial Revolution; in some locations, the situation has been reversed over the past 35 years. In systems where nutrient supply has been reduced, productivity levels dropped and recovery was evident in the natural systems (Edmondson, 1970; Smith et al., 1981; Carstensen et al., 2006), experimental systems (Oviatt et al., 1984), and mathematical simulation models (Jensen et al., 2006). Over several decades, numerous studies have been conducted along the Narragansett Bay nutrient gradient, eventually defining the responses of many dominant flora and fauna to the gradient of nutrients. At one time or another, the surveys have been, or are being, conducted on zooplankton, benthos, shellfish, ichthyoplankton, and fish. If nutrients were reduced along this gradient, primary productivity will decrease according to the studies cited above. A goal of the analysis presented here was to use the data collected earlier as a basis for predicting the impact of the impending nutrient reduction. Primary productivity and chlorophyll, but perhaps also higher trophic levels, might respond to the nutrient gradient with decreases in secondary productivity. In addition to surveys within the bay, an experimental study on nutrient impacts had been conducted in large estuarine mesocosms, providing additional verification of the relationship between nutrient concentration and impacts on productivity (Nixon *et al.*, 1986; Oviatt *et al.*, 1986).

18.2 Methods

18.2.1 Nutrients

An independent variable that was used to anchor ecological variables was TN concentration (Oviatt *et al.*, 1984, 2002). While many studies have been conducted on nutrients, only one has measured TN at 16 stations on a biweekly basis over an annual cycle (Table 18.1). Total nitrogen was measured using the persulphate digestion technique of Valderrama (1981), and concentrations were determined colorimetrically on a Technicon Autoanalyzer with a precision of ± 0.1 μM N at 10 μM level of N and ± 0.3 μM at 60 μM level of N (Oviatt *et al.*, 2002). Annual means from this study were tabulated along a north–south axis for Narragansett Bay to provide TN concentration as a function of distance from Fox Point in the Providence River to the mouth of the bay and as a

Table 18.1 Sources of data for nutrient gradient impacts in Narragansett Bay.

Parameter	Funding Agency	Study	Reference
Nutrients	RI Sea Grant	1997–1998 Bay Survey	Oviatt *et al.*, 2002
[14]C Prod.	RI Sea Grant	1997–1998 Bay Survey	Oviatt *et al.*, 2002
Chlorophyll	RI Sea Grant	1997–1998 Bay Survey	Oviatt *et al.*, 2002
Zooplankton	EPA	1979–1980 Bay Survey	Oviatt *et al.*, 1984
	RI Sea Grant	1971–1973 Bay Survey	Kremer and Nixon, 1978
Benthos	NOAA	Bay Window Program	Ellis, 2002
	State of RI		Calebretta, unpub.
Quahog	RIDEM	Providence R. Survey	Saila *et al.*, 1967
	USF&W		Stickney and Stringer, 1957
	EPA	Bays Program	Frithsen *et al.*, 1989
	URI-AES, RI Sea Grant		Rice *et al.*, 1989
	RIDEM		Russell, 1972
Oyster	EPA	Intertidal Survey	Oviatt, 2004
Fish larvae	RIDEM	1989-1990 & ongoing surveys	Keller *et al.*, 1999
Fish	RIDEM	Spring & Fall Surveys	Lynch, 2000; Oviatt *et al.*, 2003
Several	EPA	Nutrient Grad. Exp.	Frithsen, 1989; Oviatt *et al.*, 1986; Nixon *et al.*, 1986; Kelly *et al.*, 1985

function of the down-bay dilution gradient. In the mesocosm nutrient gradient experiment, TN was also measured (Kelly *et al.*, 1985). Thus, the same annual mean independent variable of TN was available from the field and from the experiment.

18.2.2 Ecological Variables

All ecosystem variables, compiled from several studies, according to station locations along the north–south axis of Narragansett Bay, were tabulated and plotted as functions of distance down bay. Distance south in the bay was used in exponential regressions as a proxy for TN for all ecological variables along the gradient. In a spreadsheet, a column for distance south was followed by a column for TN concentration and columns for ecological variables. All variables were regressed using an exponential regression as a function of distance south and estimated at 5 km intervals. Ecological variables at 5 km intervals were then regressed as a function of estimated TN at 5 km intervals. Once the regression for TN versus ecological variable was derived, new values were estimated at 80 and 50% of the regression for 100% TN. In figures, the ecosystem variables, as functions of usually two levels of nitrogen, were regressed with distance south in the bay. The ecological data were also plotted to show the goodness of fit (R^2) to distance of the proxy for 100% TN. The curve for the variable at 100% TN was a function of distance. The regressions at 80 and 50% were functions of TN, and the shapes of the curves were slightly different.

Several assumptions were made in conducting these analyses. The sources of ecosystem variables were derived from studies conducted over the past 50 years (Table 18.1). Total nitrogen was measured as of 1997–1998, and a question arises as to how constant nutrient concentrations have been over time. According to a recent analysis, the nitrogen levels in the bay have not varied much between the mid-20th century and 2004 prior to nutrient reductions (Nixon *et al.*, 2005). Another assumption was that factors like salinity, temperature, mixing depth, water residence time, currents, sediment type, turbidity, and oxygen were uniform along the gradient to have no effect on the ecological variables. In the Providence River, the salinity differences were large, but below the Providence River in Narragansett Bay, salinity gradient was slight at four to five parts per thousand between the northern bay and the southern bay (Kremer and Nixon, 1978). Thus, different types of biota do not exist in the northern bay as compared to the southern bay due to salinity differences. For other variables, this assumption may not be true. However, at least in the experimental system used to verify the field results, this assumption was true. An underlying assumption of this procedure was that location along the down-bay gradient was a surrogate for time; that is, future mid-Narragansett Bay productivity values might be equivalent to the 1997–1998 productivity values in southern Narragansett Bay.

The variables included ^{14}C primary production, phytoplankton biomass as chlorophyll, zooplankton, benthic infauna, shellfish, and fish (Table 18.1). The primary production estimates were based on chlorophyll and light extinction data from 16 stations sampled biweekly. The locations of the 16 stations are shown in Fig. 18.1, and the methods were presented in Oviatt et al (2002). Briefly, photosynthesis versus irradiance (P vs I) curves were estimated from 2-h incubations with water from station locations representing the Providence River, the mid West Passage, and the mid East Passage. Incubation was conducted on water samples from three locations between approximately noon and 2:00 PM at 17 depths in the water column to precisely define production response to a natural light gradient with P vs I curves. Light data were obtained for incubation days from the Epply Laboratory, Newport, RI. A fitted P vs I relationship was used to estimate primary production for the light data at 15-min intervals, which was integrated over depth and the day time period; these values were then integrated over the annual cycle (Oviatt et al., 2002). Phytoplankton biomass as chlorophyll was determined from acetone extractions using Lorenzen's equations (1966).

Zooplankton biomass data from vertical tows were available from two surveys (Table 18.1). During the 1972–1973 survey, dry weight biomass was determined by oven-drying to constant weight; during the 1979–1980 survey, dry weight was determined by freeze-drying, and the data from the two surveys seem to be offset, perhaps as a consequence of the difference in methods. No attempt was made to reconcile the data sets, and they were separately plotted along the north–south axis of the bay. The freeze dry weight data of the latter survey was used for comparison with the freeze dry weight data from the mesocosm experiment.

Over the past five years, four stations in the upper bay have been sampled in June with diver-collected cores for benthic infauna using 300 and 500 μm mesh sizes to a depth of 10 cm (for methods, see Ellis, 2002). The June samples provided a maximum estimate of abundance because of the inclusion of the annual recruitment of juveniles at this time of the year.

Quahog abundance, especially adults of harvest size, was determined from several surveys conducted over the past 50 years (Table 18.1). Only Rice et al. (1989) determined the abundance of juvenile sizes.

A few years after a bay-wide set, the abundance of inter-tidal oysters was determined; the oysters failed to recruit, and have since died of diseases and fishing pressure (Oviatt, 2004). This ephemeral event might be expected to reflect the nutrient gradient within the bay.

Fish larvae abundance and adult fish abundance and biomass distributions were examined. Fish larvae surveys with net tows have documented abundance and species distribution during a 1989–1990 survey (Keller et al., 1999). Raw data, from this larval fish survey, were made available to this study by Grace Klein-MacPhee (Keller et al., 1999). RIDEM has conducted spring and

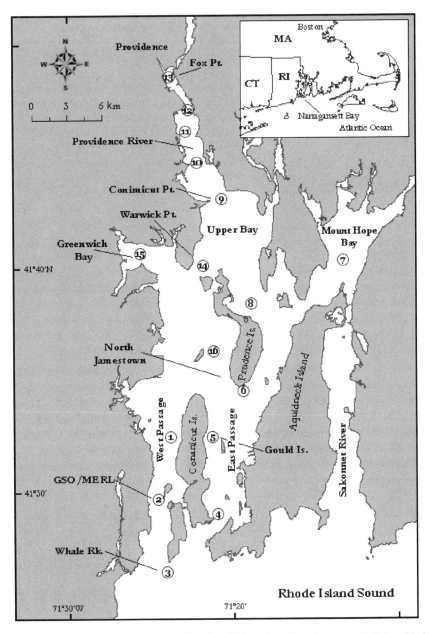

Fig. 18.1 A map of Narragansett Bay showing the location of stations sampled biweekly in 1997–1998 for TN used to create the north-south axis of nitrogen concentration. Key to station locations and north–south distances (km): Fox Point, Sta 13, 0.2 km; Conimicut, Sta 10, 9.0 km; Mount Hope, Sta 7, 13.9 km; Warwick Point, Sta 14, 18.7 km; East Greenwich, Sta 15, 19.0 km; North Jamestown, Sta 16, 24.0 km; Gould Island, Sta 5, 33.0 km; GSO, Sta 2, 36.8 km; Whale Rock, Sta 3, 41.8 km.

fall fish trawl surveys (13 bay stations; 20-min tows at 2.5 kts; cod mesh size 0.95 cm) for fish abundance since 1979, and the data were available electronically or in a report (Lynch, 2000).

18.2.3 Mesocosm Experiment

From June 1981 to October 1983, a nutrient addition experiment was conducted in mesocosms (13 m^3) containing bottom sediment communities (2.5 m^2) (Table 18.1) (Oviatt et al., 1986). Six 5 m-deep mesocosms received inorganic nutrients daily in a logarithmic series representing the average sewage effluent-nutrient input to Narragansett Bay on an area basis ($1X = 2.88$ mM N, 0.23 mM P, 0.19 mM Si $\text{m}^{-2} \text{ d}^{-1}$). Three other mesocosms acted as controls with no nutrient additions. All mesocosms received sediments collected from mid bay using a 0.25 m^2 box corer, which retrieved correctly oriented and undisturbed sediment (Oviatt et al., 1986).

Wherever possible, comparisons were made between the ecosystem parameters from the field (above) and from the experiment. These comparisons were linear regressions of TN versus the ecosystem parameters from the field and from the mesocosms. The linear regression was used because, in the experimental system, the nutrient additions were made in a logarithmic series of 1X, 2X, 4X, 8X, 16X, and 32X. Because of the difference in average depth (9 m bay vs. 5 m mesocosm), comparisons were made on an area basis (mM N m^{-2}). Dependant variables included net production, chlorophyll, zooplankton, and benthos. The production was measured weekly as the difference between dawn and dusk oxygen concentrations and converted to carbon with a photosynthetic quotient (PQ) of 1.2 to be comparable to Bay ^{14}C values (Oviatt et al., 1986). Chlorophyll (acetone extraction), zooplankton (vertical tows; freeze dry weights), and benthos (diver collected cores) were determined as above for field data. For primary production, chlorophyll, and zooplankton, data from the annual cycle in 1982 were used as integrated total (primary production) or annual means (chlorophyll, zooplankton). For the benthic community data, abundances were maximum values over the experimental period, as mesocosms were not necessarily sampled in June, the time of field samples.

18.3 Results

18.3.1 The Bay Nutrient Gradient

Exponential regression equations for distance along the north–south axis of the bay versus nutrient and ecological variables had percent variation ranging from

40% for quahogs to 94% for TN (Table 18.2). The distance regressions were used to estimate TN and dependent variables at 5 km intervals. A new set of exponential regressions with TN versus dependent variables was developed from these values (Table 18.2). Since these data have been smoothed by using distance regression predictions, the percent variation was artificially high and over 94% in all cases (not shown). The percent variation explained and the 95% confidence intervals of the unsmoothed data were shown only for distance regressions (Table 18.2).

Total nitrogen declined exponentially down bay with dilution and distance (Fig. 18.2). The value at Fox Point represented the nutrient loading from the Blackstone and Seekonk Rivers and the Bucklin Point Wastewater Treatment Facility (WWTF). This concentration of about 65 µM increased to about 75 µM south of the Field's Point WWTF (Station 12; Fig. 18.1). From this high point, the values for TN decreased to about 10 µM at Whale Rock at the mouth of the West Passage. While the regression did not fit the higher points in the Providence River, the fit was close for stations south of Conimicut Point, including Mount Hope Bay, which received Taunton River nutrient inputs, and Gould Island in the lower East Passage (Fig. 18.2).

Primary production values followed the TN gradient, showing more scatter than TN, but increasing also south of the Field's Point WWTF, and decreasing with dilution and distance down bay (Fig. 18.3; Table 18.2). While the regression line for a 20% reduction intersected about five data points, the line for a 50% reduction in TN lay beneath all values for production, suggesting that a 20% decrease in TN would be difficult to detect. A 50% decrease might be detected in the northern bay but not at the 95% confidence level (Fig. 18.3).

Phytoplankton biomass as chlorophyll annual means displayed a flat response to TN in the Providence River before decreasing down bay, making the regression relationships invalid for the Providence River (Fig. 18.4). For the northern and southern bay, the regression lines for 20% and 50% nutrient reductions lay beneath all data points. Some points were also beneath the 95% confidence interval, suggesting that, with nutrient reduction, declines in phytoplankton biomass would be detected at 50% nutrient reduction, particularly in mid to upper bay regions (Fig. 18.4).

Zooplankton biomass as per the 1972–1973 and 1979–1980 surveys decreased from the northern bay to southern bay with values from the latter survey much lower than the former, perhaps due to differing methods in determining dry weight as described in the methods section (Fig. 18.5). The zooplankton annual means from oven dry weight in the 1972–1973 survey reached maximum values at 10 km at the mouth of the Providence River and decreased south in the bay (Fig. 18.5a). Zooplankton annual means from freeze dry weight in 1979–1980 displayed a maximum value below the Providence River and decreased south in the bay (Fig. 18.5b). The regressions for both data sets were based only on data south of Conimicut Point. The regression lines for 20% reduction in TN for both surveys indicated overlap.

Table 18.2 Regression equations.

Exponential regressions as a function of distance (Km) South in Narragansett Bay				
Total nitrogen	1997–1998	$y = 75e^{-0.0485x}$	$R^2 = 0.94$	$F = 224.1, P < 0.001$
^{14}C production	1997–1998	$y = 622e^{-0.31x}$	$R^2 = 0.70$	$F = 32.2, P < 0.001$
Chlorophyll	1997–1998	$y = 30e^{-0.0594x}$	$R^2 = 0.92$	$F = 99.7, P<0.001$
Zooplankton	1972–1973	$y = 185e^{-0.0216x}$	$R^2 = 0.54$	$F = 12, P = 0.007$
Zooplankton	1979–1980	$y = 141e^{-0.0380x}$	$R^2 = 0.73$	$F = 8.23, P = 0.064$
Macrofauna	2002	$y = 225e^{-0.0311x}$	$R^2 = 0.83$	$F = 9.67, P = 0.090$
Quahog		$y = 73e^{-0.1050x}$	$R^2 = 0.40$	$F = 4.74, P = 0.066$
Oyster	1995	$y = 2086e^{-0.1034x}$	$R^2 = 0.49$	$F = 7.59, P = 0.025$
Derived exponential regressions as a function of TN, μM				
^{14}C production		$y = 123e^{0.0305TN}$		
Chloropyll		$y = 2.25e^{0.041TN}$		
Zooplankton		$y = 15e^{0.0514TN}$		
Macrofauna		$y = 65e^{0.0207TN}$		
Quahog		$y = .0257e^{0.0811TN}$		
Oyster		$y = 13.6e^{0.0805TN}$		
Comparison of field and experimental regressions as a function the standing stock of TN, mM m^{-2}				
Net production	Field	$y = 0.77x + 125$	$R^2 = 0.68$	$F = 30, P < 0.001$
	Experimental	$y = 0.79x + 23$	$R^2 = 0.98$	$F = 79, P = 0.012$
Chlorophyll	Field	$y = 0.053x - 3.03$	$R^2 = 0.95$	$F = 171, P < 0.001$
	Experimental	$y = 0.09x - 9.7$	$R^2 = 0.94$	$F = 80, P < 0.001$
Zooplankton	Field	$y = 0.10x + 24$	$R^2 = 0.43$	$F = 5, P = 0.077$
	Experimental	$y = 0.30x + 15$	$R^2 = 0.69$	$F = 11, P = 0.021$
Benthos	Field	$y = 0.31x + 40$	$R^2 = 0.81$	$F = 9, P = 0.095$
	Experimental	$y = 0.36x + 34$	$R^2 = 0.56$	$F = 5, P = 0.087$

Bay Survey 1997–98

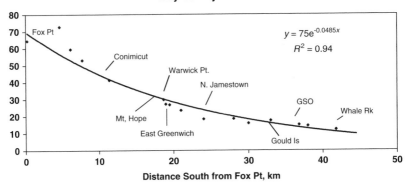

Fig. 18.2 Annual mean TN concentrations (Total Nitrogen, μM) plotted on the north–south axis of Narragansett Bay and regressed as a function of distance from Fox Point at the head of the Providence River to Whale Rock at the mouth of the West Passage (F = 436, P = <0.001) (see Fig. 18.1).

The lines for 50% reduction in TN for both data sets indicated separation but not always at the 95% confidence level (Fig. 18.5).

Benthic infauna demonstrated a relationship to the TN gradient as it was represented by distance south in the bay (Fig. 18.6). In this case, only data from the northern portion of the bay was available. The regression lines for 20% and 50% reductions in TN were separated from the data for the main regression,

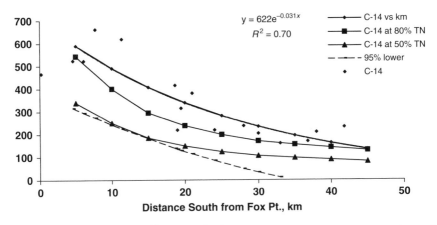

Fig. 18.3 Primary production (^{14}C), $gCm^{-2}y^{-1}$ as a function of distance along the north–south axis of Narragansett Bay: the data, from stations shown in Fig. 18.1, a regression line for the data as a function of distance (F = 32, P < 0.001), regression lines for 20% and 50% reductions in TN on primary production as functions of distance and the lower 95% confidence interval (95% lower).

Bay Survey 1997-98

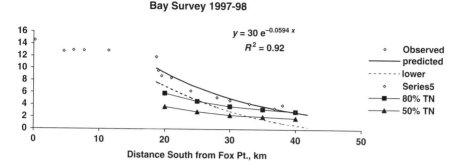

Fig. 18.4 Annual mean chlorophyll, mg m^{-3} concentration as a function of distance along the north–south axis of Narragansett Bay: the data, from stations shown in Figure 18.1, a regression line for the data as a function of distance (F = 336, P < 0.001), regression lines for 20% and 50% reductions in TN on chlorophyll as function of distance and the lower 95% confidence interval (lower).

suggesting that benthic macrofauna abundance would decrease with 50% reduced nutrient inputs although not at the 95% confidence level.

While shellfish resources tend to be higher in the northern bay, they tend to decrease down bay abruptly rather than in a regular manner. Since the demise of the oyster industry in early 1900s, quahogs have been the main shellfish resource species. Numerous surveys of their abundance have been conducted over the past 50 years or so (Table 18.1). Probably due to harvest pressures, two population levels appear in the bay. The quahogs have always been high at the mouth of the Providence River, where fishing is prevented, and low in the rest of the bay (Fig. 18.7), where harvest pressures have been intense. While a slight down bay gradient has existed, survey data overlap with the 50% nutrient reduction regression, suggesting that no change in population would be detectable with nitrogen reduction. All surveys except one were of adult shellfish; a survey of juveniles, which have not been subject to harvest, might show more detectable changes in abundance (Fig. 18.7). Oysters had a range of abundance in the northern bay, which would obscure a response to nutrient reduction. Low abundances were found in the northern Providence River and in the southern bay, while highest abundances were found at the north end of Prudence Island (Fig. 18.8).

Larval fish, trawl fish abundance, and biomass showed no relationship to the north–south distance or gradient in TN in Narragansett Bay (fish species shown in Table 18.3; Fig. 18.9). The total larval fish in the 1989–1990 survey, dominated by anchovy larvae, had variable abundances along the north–south axis of the bay (Fig. 18.9a). The highest numbers were found in the mid to northern bay. In trawl surveys, fish abundance had an irregular tendency to decrease down bay; whereas biomass had no trend throughout the bay (Fig. 18.9b,c). Fish showed no correlation with the north–south axis of distance or the TN gradient.

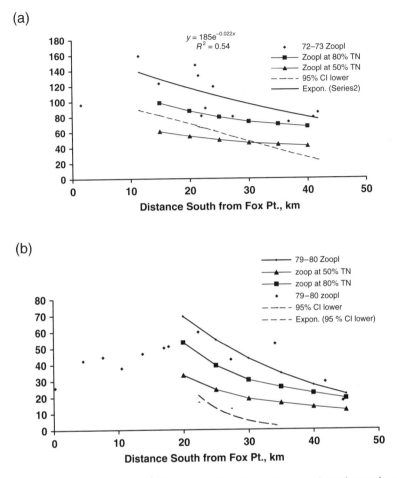

Fig. 18.5 Zooplankton, mg dw m^{-3} biomass as a function of distance along the north–south axis of Narragansett Bay from surveys in (a) oven dry weight 1972–1973, and (b) freeze dry weight 1979–1980 (Table 18.1): the data, a regression line as a function of distance, regression lines for 20% and 50% reductions in TN on zooplankton biomass as a function of distance and the lower 95% confidence interval.

18.3.2 A Comparison of Field and Experimental Values

The field and experimental data exhibited similar relationships to TN. The values estimated from nitrogen regression above, and values from a nutrient addition experiment in mesocosms seeking to mimic the bay, were compared for primary production, chlorophyll, zooplankton biomass, and abundance of benthic infauna as a function of total nitrogen (Fig. 18.10). Regression lines for field and experimental values of primary production were parallel with

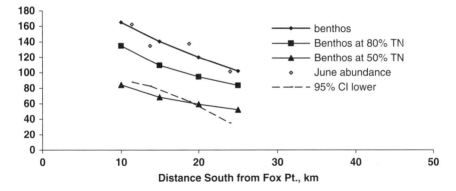

Fig. 18.6 Benthic infauna, m^{-2} × 1000 abundance as a function of distance along the north-south axis of Narragansett Bay from a survey in 2002 (Table 18.1): the data, a regression line as a function of distance, regression lines for 20% and 50% reductions in TN on animal abundance as a function of distance and the lower 95% confidence interval.

nearly identical slopes, but with the experimental values lower than the field values (Fig. 18.10a). The experimental values derived from dawn-dusk changes in oxygen represent net daytime production, whereas the ^{14}C values used for

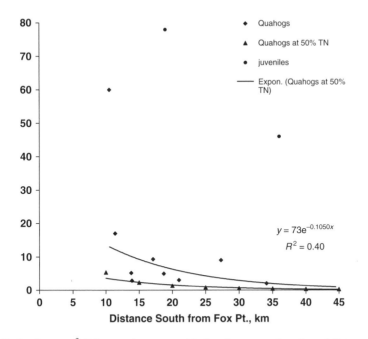

Fig. 18.7 Quahog, m^{-2} (*Mercenaria mercenaria*) abundance as a function of distance along the north-south axis of Narragansett Bay from surveys listed in Table 18.1: the data, a regression line as a function of distance and regression lines for 20 and 50% reductions in TN on shellfish abundance as a function of distance.

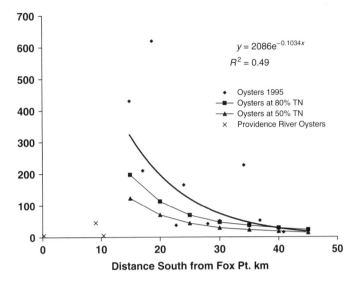

Fig. 18.8 Oyster, m^{-2} (*Crassostrea virginica*) abundance as a function of distance along the north-south axis of Narragansett Bay from a 1995 survey (Table 18.1): the data, a regression line as a function of distance and regression lines for 20 and 50% reduction in TN on shellfish abundance as a function of distance.

field estimations, based on short incubations, were intermediate between net and gross values. Thus, the experiment values of net production were lower than the field values of gross production. Both data sets fall within the 95% confidence intervals of the field data.

In the phytoplankton biomass comparison, the annual mean data points tend to overlap over the range of TN appropriate to both data sources, indicating an agreement for field and experimental data at the 95% confidence level

Table 18.3 A list of fish species from spring and fall trawl surveys in Narragansett Bay conducted by RIDEM for which abundance and biomass data was compiled.

Alewife	*Alosa pseudoharengus*
Atlantic Herring	*Clupea harengus*
Menhaden	*Brevoortia tyranus*
Bay Anchovy	*Anchoa mitchilli*
Butterfish	*Peprilus tricanthus*
Scup	*Stenotomus chrysops*
Little Skate	*Leucoraja erinacea*
Summer Flounder	*Paralichthys dentatus*
Tautog	*Tautoga onitis*
Weakfish	*Cynoscion regalis*
Windowpane Flounder	*Scopthalmus aquosus*
Winter Flounder	*Pseudopleuronectes americanus*

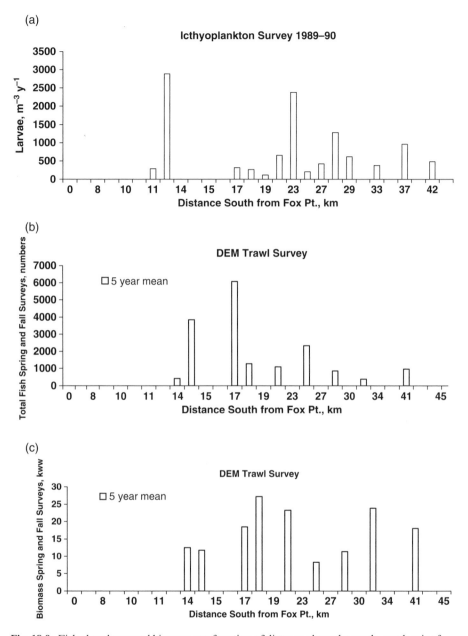

Fig. 18.9 Fish abundance and biomass as a function of distance along the north–south axis of Narragansett Bay (Table 18.1): (a) total fish larvae during 1989–1990, (b) 1996–2000 mean annual number of fish trawled at 13 bay stations, (c) mean fish biomass for the same period and stations.

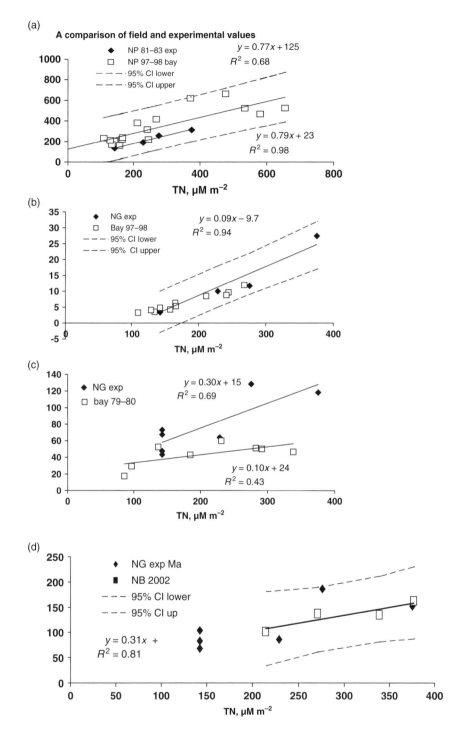

(Fig. 18.10b). The bay values appear to show a slightly lower slope than experimental values, with higher values at lower TN and lower values at higher TN. In contrast to primary production and chlorophyll comparisons, zooplankton annual means showed little overlap between the two sources of data, with 1979–1980 field data values generally lesser than experimental values for the same concentration of TN (Fig. 18.10c). In addition, both data sets were more variable than the other two parameters with an R^2 of 43% for the bay data and of 69% for the experimental data. As with the chlorophyll comparison, the field response of zooplankton to TN appears to have a lower slope than the experimental data. The agreement between field and experimental data was exhibited by abundance of benthic infauna showing an overlap of data points within the 95% confidence interval of the field data (Fig. 18.10d). Generally, the data comparison was in agreement for primary production, chlorophyll, and benthic fauna, suggesting that, for these parameters, a decrease in TN would result in a decrease in the ecological variables.

18.4 Discussion

The Providence River portion of the bay was omitted from several regressions because of its low salinity and difference in its physics and ecology. In the Providence River, several of the measured variables appear to be not related to TN. In this portion of the bay, light limitation was high (Oviatt, 2002). Chlorophyll showed no response over this portion of the gradient (Fig. 18.3). The Providence River was also fresher and more stratified than the southern portions of the bay (Kremer and Nixon, 1978). Salinity or factors associated with fresh water inputs may account for decreased zooplankton in this portion of the bay (Fig. 18.5). Oyster abundance was another variable that was low in the Providence River, perhaps due to lack of hard substrate or factors associated with fresh water inputs (Fig. 18.8). In these cases, TN regressions were limited to the areas south of the Providence River.

Zooplankton annual means, although variable between survey years, were correlated with the concentrations of TN (Fig. 18.5). The steeper slope and higher R^2 of experimental data from the mesocosm experiment as compared to the 1979–1980 bay survey suggested that some characteristic of bay waters, not present under experimental conditions or during the 1972–1973 survey, was

Fig. 18.10 A comparison of Narragansett Bay field and mesocosm experimental values for (a) net primary production, $g C m^{-2} y^{-1}$ (TN vs field C14: F = 30, P < 0.001), (b) chlorophyll, $\mu g l^{-1}$ concentration (TN vs exp chl: F = 80, P < 0.001), (c) zooplankton, mg dw m^{-3} biomass, (d) benthos, $m^{-2} \times 1000$ abundance as a function of TN mean annual standing stock per square meter (TN vs field macrofauna: F = 9, P < 0.09).

decreasing the response of zooplankton to TN in the bay (Figs 18.10c and 18.5b). Enhanced predation in the field due to ctenophores might explain the decrease in zooplankton response (Sullivan *et al.,* 2007, Chapter 16, this volume). Ctenophores, which have shifted its dominance from fall to summer, have been cited as responsible, through their grazing activities, for decreased summer zooplankton in recent years. The relationship of zooplankton to TN was sufficient to suggest a 50% decrease in TN although not with 95% statistical confidence (Fig. 18.5).

Shellfish abundance showed no clear relationships to TN although quahogs and oysters had greater abundances at higher TN in the northern bay. In order to detect the impact of reduced TN on shellfish, surveys of juvenile stages would likely be more successful than surveys of harvest-size organisms (Rice *et al.,* 1989; Fig. 18.7). In the case of oysters, which depend on a rocky intertidal habitat, habitat availability may always cause wide variation. Thus, low oyster values in the Providence River may be a function of habitat (Fig. 18.8). Alternatively, some pollution factor may be decreasing oysters in this region of the bay.

Habitat preference has probably obscured any direct relationship between TN and fish in the bay. Ichthyoplankton, which were dominated by bay anchovy, showed no relationship to the TN gradient (Fig. 18.9a). Total fish abundance tended to be higher in the northern bay, but biomass was evenly distributed throughout the bay (18.9b,c). Examination of the abundance and biomass data of other ichthyoplankton and fish species also did not reveal any relationship to the nutrient gradient. If abundance and biomass of the fish community is related to nutrient level, a cross system comparison, rather than a within-system gradient comparison, may be required to reveal the relationship. Any relationship may be obscured by exploitation and harvest of fish populations and by trends in abundance driven by climate changes. The mean late 1990s biomass of 2 gww m^{-2} can be compared to historical biomass values five to six times higher; the current biomass levels may have little relationship to nutrient levels in the system (Oviatt, 2004).

Decreases of ecological variables with 50% decrease in TN were usually not detectable with a 95% level of statistical confidence. Since the survey data sets were not gathered for the purpose of revealing gradients in ecological variables, this should not be surprising. However, the agreement between the experimental gradient and the field gradient helps to confirm that nitrogen was controlling the levels of productivity of several ecological variables at different trophic levels. The experimental verification indicated that, when nitrogen was decreased, the values for these variables would also decrease. The regression equations provided some guidelines on the degree of the decrease (Table 18.2). For areas south of Conimicut Point, a 50% reduction in TN may result in an average decrease of 50% in primary production, chlorophyll, zooplankton, and benthic animals, or 150 g C m^{-2} y^{-1} in ^{14}C primary production, 4.1 μg l^{-1} in chlorophyll, 60 mg dw m^{-3} of zooplankton biomass (oven dry method), and a decrease of 71,000 benthic animals per square meter (Figs 18.3, 18.4, 18.5b, and 18.6).

A simple calculation is presented to estimate the impact of reduced primary production on oxygen in the water column. An assumption has been made that sea grasses or macroalgae would not replace phytoplankton production. Assuming an oxygen-to-carbon ratio (O:C) of 2, the decrease of 150 g C m^{-2} y^{-1} would be equivalent to saving or not respiring 800 g O$_2$ m^{-2} y^{-1}. In eutrophied systems, benthic respiration has accounted for less than 40% of water column respiration (Oviatt et al., 1986). Both benthic and water column respiration would be included in the calculation of respiration of organic matter. If it was assumed that much of the respiration takes place in the six warmer months in the average depth of the bay of 8 m, savings can be estimated at 0.55 mg O$_2$ L^{-1} d^{-1} at a saturation oxygen value of about 7 mg L^{-1}. Over the stratification period of 5 days associated with a neap tide, the unrespired oxygen in bottom water would be equal to 2.8 mg O$_2$ L^{-1} d^{-1} per 5 days, or an amount sufficient to keep the water column from becoming hypoxic, as compared to the previous demand of 5.5 mg O$_2$ L^{-1} per 5 days. (For this calculation, oxygen in demand and the oxygen saved would be equal, assuming a new average primary production rate of 150 g C m^{-2} y^{-1}). Of course, the savings might be greater in the northern bay and in shallower regions of the bay.

The gradient approach has been applied in Narragansett Bay with its persistent north–south distribution of unchanging nutrients over the period of data accumulation. This approach will work for any estuary with a similar well-defined gradient of unchanging nutrients and lacking gradients of other chemical and physical driving variables. However, many estuaries are still experiencing large changes in nutrients with ongoing cultural eutrophication (Seitzinger et al., 2002; Fisher et al., 2006). An alternative promising approach with steady accumulation of information on more and more estuaries would be cross-system comparisons. Studies of some 92 coastal zone ecosystems have shown that average concentrations of chlorophyll were strongly dependent on the average concentrations of total nitrogen and total phosphorus in the water column (Smith, 2006). In our Narragansett Bay study, the correlation between nitrogen and resource species was low, but in other studies of cross-system comparisons, these correlations have been convincingly strong (Nixon and Buckley, 2002).

Acknowledgments This paper was originally presented at the Rhode Island Sea Grant sponsored symposium on Block Island in November 24, 2004. Electronic data sets of RI DEM fish trawl surveys up to 2000 from Timothy Lynch and Kenneth Raposa and the ichthyoplankton survey of 1989–1990 from Grace Klein-MacPhee are gratefully acknowledged. Christopher Calebretta provided benthic infauna data from his incomplete dissertation research supported by the Environmental Protection Agency Cooperative Agreement R-83060801 and NOAA Bay Window grant number NAO 7FE0425. The mesocosm experiment was supported by grants from the Environmental Protection Agency. Barbara Sullivan and five anonymous reviewers provided comments that greatly improved the manuscript.

References

Bergondo, D. 2004. Examining the processes controlling water column variability in Narragansett By: time series data and numerical modeling PhD Dissertation in Oceanography, University of Rhode Island, Narragansett, RI 02882. 187 p.

Bergondo, D., Kester, D., Stoffel, H., and Woods, W. 2005. Time-series observations during the low sub-surface oxygen events in Narragansett Bay during summer 2001. *Marine Chemistry* 97:90–103.

Carstensen, J., Conley, D., Andersen, J., and Aertebjerg, G. 2006. Coastal eutrophication and trend reversal: a Danish case study. *Limnology Oceanography*, 51(1, part 2):398–408.

Deacutis, C.F., Murray, D., Prell, W., Saarman, E., and Korhun, L. 2006. Hypoxia in the upper half of Narragansett Bay, RI during August 2001 and 2002. *Northeastern Naturalist*, Special Issue on Mount Hope Bay, 13 (Special Issue 4): 173–198.

Edmondson, W. 1970. Phosphorus, nitrogen and algae in Lake Washington after diversion of sewage. *Science* 169:690–691.

Ellis, G. 2002. An examination of the benthic macrofauna of Narragansett Bay and the possible implications of the winter-spring bloom intensity on population size. Masters Thesis in Oceanography, University of Rhode Island, Narragansett, RI 02882. 181 p.

Fisher, T., Hagy, J. III, Boynton, W., and Williams, M. 2006. Cultural eutrophication in the Choptank and Patuxent estuaries of Chesapeake Bay. *Limnology and Oceanography* 51(1, part 2):435–447.

Frithsen, J.B. 1989. The benthic communities within Narragansett Bay. An assessment completed for the Narragansett Bay Project, RIDEM, Providence, RI 92 p.

Jensen, J., Pedersen, A., Jeppesen, E., and Sondergaard, M. 2006. An empirical model describing the seasonal dynamics of phosphorus in 16 eutrophic lakes after external loading reduction. *Limnology and Oceanography* 51:791–800.

Keller, A.A., Klein-MacPhee, G., Burns, J.S.O. 1999. Abundance and distribution of Ichthyoplankton in Narragansett Bay, Rhode Island, 1989–1990. *Estuaries* 22:149–163.

Kelly, J.R., Berounsky, V.M., Nixon, S.W., and Oviatt, C.A. 1985. Benthic-pelagic coupling and nutrient cycling across an experimental eutrophication gradient. *Marine Ecology Progress Series* 26:207–219.

Kremer, J.N., and Nixon, S.W. 1978. A Coastal Marine Ecosystem: Simulation and Analysis. Springer-Verlag, NY, 271p.

Lorenzen, C.J. 1966. A method for the continuous measurement of in vivo chlorophyll concentration. *Deep Sea Research* 13:223–227.

Lynch, T. 2000. Assessment of recreationally important finfish stocks in Rhode Island waters. Coastal Fishery Resource Assessment Trawl Survey. RIDEM Division of Fish and Wildlife, Government Center, Wakefield, RI.

Melrose, D.C. 2005. Underway profiling of photosynthesis and dissolved oxygen in Narragansett Bay, RI PhD Dissertation in Oceanography. University of Rhode Island, Narragansett, RI. 179 p.

Nixon, S. 1997. Prehistoric nutrient inputs and productivity in Narragansett Bay. *Estuaries* 20:253–261.

Nixon, S. and Buckley, B. 2002. "A strikingly rich zone"—nutrient enrichment and secondary production in coastal marine ecostystems. *Estuaries* 25(4b):782–796.

Nixon, S.W., Oviatt, C.A., Frithsen, J., and Sullivan, B. 1986. Nutrients and the productivity of estuarine and Coasal marine ecosystems. *Journal of the Limnology Society of South Africa* 12(1/2):43–71.

Nixon, S., Buckley, B. Granger, S. Harris, L. Oczkowski, A. Cole, L. and Fulweiler, R. 2005. Anthropogenic nutrient inputs to Narragansett Bay—A twenty-five year perspective. A report to the Narragansett Bay Commission and Rhode Island Sea Grant. 29pp.

Oviatt, C.A. 2004. The changing ecology of temperate coastal waters during a warming trend. *Estuaries* 27:895–904.

Oviatt, C.A., Pilson, M.E.Q., Nixon, S.W., Frithsen, J.B., Rudnick, D.T., Kelly, J.B., Grassle, J.F., and Grassle, J.P. 1984. Recovery of a polluted estuarine ecosystem: a mesocosm experiment. *Marine Ecology Progress Series* 16:203–217.

Oviatt, C.A., Keller, A.A., Sampou, P.A., and Beatty, L.L. 1986. Patterns of productivity during eutrophication: a mesocosm experiment. *Marine Ecology Progress Series* 28:69–80.

Oviatt, C., Doering, P. Nowicki, B. Reed, L. Cole, J. Frithsen, J. 1995. An ecosystem level experiment on nutrient limitation in the temperate coastal marine environment. *Marine Ecology Progress Series* 116:171–179.

Oviatt, C.A., Keller, A., and Reed, L. 2002. Annual Primary production patterns in Narragansett Bay in a year with no bay-wide winter-spring phytoplankton bloom. *Estuarine, Coastal and Shelf Science*. 54:1013–1026.

Oviatt, C.A., Olsen, S., Andrews, M., Collie, J., Lynch, T., and Raposa, K. 2003. A century of fishing and fish fluctuations in Narragansett bay. *Reviews in Fisheries Science* 11:1–22.

Pryor, D. 2004. Nutrient and Bacteria Pollution Panel Report, March 2004. Rhode Island Governor's Narragansett Bay and Watershed Planning Commission. 33p. *http://www.ci.uri.edu/GovComm/Documents/Phase1Rpt/Does/Nutrient-Bacteria.pdf*

Rice, M.A., Hickox, C., and Zehra, I. 1989. Effects of intensive fishing effort on the population structure of quahogs, *Mercenaria mercenaria* (Linnaeus 1758), in Narragansett Bay. *Journal of Shellfish Research* 8:245–354.

Russell, H.H., Jr. 1972. Use of a commercial dredge to estimate a hardshell clam population by stratified random sampling. *Journal of the Fisheries Research Board of Canada* 29:1731–1735.

Saila, S.B., Flowers, J.M., and Canario, M.T. 1967. Factors affecting the relative abundance of *Mercenaria mercenaria* in the Providence River, Rhode Island. *Proceedings of the National Shellfisheries Association* 57:83–89.

Seitzinger, S., Kroeze, C., Bouwman, A., Caraco, N., Dentener, F., and Styles, R. 2002. Global patterns of dissolved inorganic and particulate nitrogen inputs to coastal systems: recent conditions and future projections. *Estuaries* 25(4b):640–655.

Smith, V. 2006. Responses of estuarine and coastal marine phytoplankton to nitrogen and phosphorus enrichment. *Limnology and Oceanography* 51(1. part 2):377–384.

Smith, S., Kimmerer, W., Laws, E., Brock, R., and Walsh, T. 1981. Kaneohe Bay sewage diversion experiment: perspectives on ecosystem responses to nutritional perturbation. *Pacific Science* 35:279–385.

Stickney, A.P., and Stringer, L.D. 1957. A study of the invertebrate bottom fauna of Greenwich Bay, Rhode Island. *Ecology* 38:111–122.

Valderrama, J.C. 1981. The simultaneous analysis of total nitrogen and phosphorous in natural waters. *Marine Chemistry* 10:109–122.

Chapter 19
An "Ecofunctional" Approach to Ecosystem-based Management for Narragansett Bay

Barry A. Costa-Pierce and Alan Desbonnet

Ecosystem-based management is an integrated approach that takes into consideration the entire ecosystem, including humans (McLeod *et al.*, 2005). The goal of ecosystem-based management is to maintain a healthy, productive, and resilient ecosystem that provides services that humans want and need. The information presented in this volume can be considered a baseline of the latest scientific knowledge on Narragansett Bay, and as an inflection point for ecosystem-based management of the bay ecosystem. Despite our attempts to synthesize and integrate ideas across disciplines in this chapter, it is only a beginning. We point to some specific needs for successful ecosystem-based management in Narragansett Bay using the concept of "ecofunctional zoning." The bay's scientific community is much more robust and sophisticated today than it was 25 years ago when "The Narragansett Bay Comprehensive Conservation and Management Plan" (NBP, 1992) was completed. Much good has arisen from the implementation of that plan. Now, on the heels of this volume, we have an unprecedented opportunity to coalesce around this science and make a first, major "adaptive iteration" of an ecosystem-based management plan for Narragansett Bay.

19.1 A Complex Shallow Water Bay

Narragansett Bay is atypical of the coast of southern New England—it is long relative to its width, and is bifurcated into two narrow passages due to the location of several major, mid-bay islands. The story of Narragansett Bay, therefore, begins with its geology.

To tell that story, Boothroyd and August take the reader back 25,000 years to look at Narragansett Bay's geological origins. The complex array of bedrock

Barry A. Costa-Pierce
University of Rhode Island, Rhode Island Sea Grant College Program and the Department of Fisheries Animal & Veterinary Science, Narragansett RI 02882USA
bcp@gso.uri.edu

A. Desbonnet, B. A. Costa-Pierce (eds.), *Science for Ecosystem-based Management.*   545
© Springer 2008

and glacial debris created over a millennia of geological activity has caused a differential weathering of the landscape that created the East and West Passages—the bay's defining characteristics—giving each a unique depth and sediment composition. This geological configuration (details in the chapters by Spaulding and Swanson, Chen *et al.,* and Kincaid *et al.*) has profound influence on the overall circulation characteristics exhibited within Narragansett Bay, between Narragansett Bay and Mount Hope Bay, and with Rhode Island Sound.

Once nutrients, and for that matter any "pollutant", enter bay waters, the unique geological characteristics of Narragansett Bay set the stage for the movement of water in, out, and through the ecosystem. While Narragansett Bay displays classic estuarine circulation—surface flow down and out of the bay, bottom water inflow back into the bay—the unique location of islands at mid-bay in the system creates a dynamic atypical of other east coast estuaries. The mid-bay islands create both clockwise and counter-clockwise circulation patterns that cycle water into the bay, around the islands, and then back out. These patterns leave the upper bay area somewhat isolated, though circulation driven by tides and freshwater inputs to the upper bay moves water into the mid-bay area where it joins the overall bay circulation. Kincaid *et al.* note that Narragansett Bay behaves, in many ways, as several tightly coupled sub-systems.

Spaulding and Swanson conclude that, despite this complexity, the current state of our knowledge on Narragansett Bay circulation is quite good, and that accurate predictions of tidal impulses and flows can be made, given the varying depths of the East and West Passages. The problems in predictive modeling may arise, however, when modeling attempts move into shallower waters that are more influenced by wind, or into constricted regions where rapid changes in bathymetry—where a dredged channel rises quickly to meet a shallow tidal flat, for instance—create a complex structure whose dynamic circulation is difficult to re-create in a model.

Chen *et al.* find that Narragansett Bay and Mount Hope Bay are tightly coupled as a unit, so much so that they function essentially as an integrated sub-system with regard to circulation dynamics. Chen *et al.* find that parcels of water originating in Mount Hope Bay readily make their way into upper Narragansett Bay, and even into, as Smayda and Borkman put it, that "poorly flushed cul de sac", Greenwich Bay. While Mount Hope Bay and Narragansett Bay have traditionally been modeled as separate entities, Chen *et al.* link the two into one model in order to better illustrate and explore the circulation at an ecosystem scale. For this sub-system, 70–80% of the current energy comes from semi-diurnal tides, and it is this energy that forces strong flushing between them through the narrow channel that links them. Once waters move out of the connecting channel, winds and other forces, such as freshwater inputs from the upper bay, incorporate the Mount Hope Bay waters into the complex circulatory patterns inherent in Narragansett Bay.

Chen *et al.* further find that small-scale circulation events, such as eddies, are formed at the ends of the narrow channel connecting Mount Hope and Narragansett Bay. Kincaid (2006) also provides observations of these phenomena. Eddies can persist beyond a given tidal cycle, and they may entrain nutrients, plankton, parcels of warmer or cooler water, or affect larval transport and other ecological dynamics within the ecosystem. These small-scale events are possibly less important at the scale of the bay as an entire ecosystem, but may be dramatically important at local scales.

Spaulding and Swanson, Chen *et al.,* and Kincaid *et al.* find that winds have a major influence on water movement, and can have profound impacts upon the overall circulation within the bay. Several authors have noted instances where short-lived but intense wind events have reversed bay circulation patterns. Such wind events are not the norm, and their impacts due to infrequency and short duration probably have little impact on the flora and fauna of the bay. However, these events may have profound short-term effects, such as those noted by Saarman *et al.* and Deacutis regarding hypoxia. Wind is critical to the alleviation of oxygen-stressed conditions, and, according to Kincaid *et al.,* is a major factor in driving the circulation and exchange of water between Narragansett Bay and Rhode Island Sound.

Interestingly, Pilson finds that wind patterns over Narragansett Bay appear to have changed over time, particularly that wind speed has decreased. How this change may affect circulation has not been well considered at ecosystem levels or in modeling exercises. Winds impact vertical stratification in the upper bay, with a reduction in wind speed decreasing the potential to break down water-column stratification and alleviate oxygen stress.

At the mouth of the bay, Kincaid and Bergondo observed an overall counter-clockwise flow pattern that moves water from Rhode Island Sound up the East Passage of Narragansett Bay, then back down through the West Passage. This flow is strongest during summer months, when it is fueled by persistent south-west winds (sea breezes), and less during winter, when wind speeds are more variable, yet more consistent from the north and west.

These authors have also observed, at least during summer months, a persistent movement of water along the coast outside of Narragansett Bay from the east to west. Summer winds move water traveling from Cape Cod and the Elizabeth Islands into Block Island Sound, up the East Passage and into Narragansett Bay.

Vertical mixing is a critical variable, but it is complex and often simplified for use in models. Due to the complex vertical structure of Narragansett Bay—well stratified in the Providence River, partially stratified to well-mixed in the mid-bay, and well-mixed in the lower bay—models need multiple vertical layers to capture the horizontal and vertical variation in velocities. Vertical mixing due to turbulence remains a challenge to modelers, and the data needed to validate models that incorporate turbulent mixing are largely unavailable for Narragansett Bay. Although much is known and existing models perform well overall, more work still needs to be accomplished in circulation modeling to link it more directly to ecosystem functions, and to develop models that perform better in near-shore areas that, despite a challenging task for modelers, are of great importance to ecosystem-based managers. Circulation models are of importance to ecosystem-based management at larger scales as well because they are used to depict and predict the impacts of anthropogenic changes. The availability of robust circulation models capable of linking to ecological parameters is paramount because Narragansett Bay, for over a century and a half, has been a human-dominated coastal ecosystem.

Boothroyd and August stress that it was only about 10,000 years ago, as salty ocean water spilled into a large glacial lake in the East Passage, that Narragansett Bay could be considered a marine environment. Over the past century and a half, however, the footprint of humankind has left deep imprints on the shoreline of the bay ecosystem. Boothroyd and August point out that, at the onset of the 21st century, more than 30% of the Narragansett Bay shoreline is human engineered in the form of revetments, bulkheads, seawalls, groins, breakwaters, and jetties. "Hardened" shoreline, however, is only one example of human domination along this geologically unique New England estuary. The impacts of human domination are readily found not only in the highly modified and dredged upper bay, but also in the swaths of seemingly undisturbed green fringing the bay, in the air, and contained in the rural lower bays' sparkling waters.

19.2 A Sewage-fueled and Oxygen-starved Bay

Human population density around Narragansett Bay, according to Boothroyd and August, currently ranges from 1,075 (urban coast) to 168 (rural coast) people per square kilometer, with most of the population clustered in the Providence metroplex at the head of the bay and in Fall River-Somerset for Mount Hope Bay. This population distribution is the result of several centuries of historical change in land use and life style in the watershed of the bay ecosystem.

Hamburg *et al.* tell a story of the Narragansett Bay watershed focusing on human and animal population growth and how that changed the input of nitrogen to the system over time and space. The early bay residents who settled in the region spent much time clearing the stony New England soils into pasture, mainly for diary cows and cattle. As agriculture became more widespread and more livestock populated the watershed, more animal nitrogen was deposited on the landscape. Hamburg *et al.* note that the landscape, cleared of trees, shifted from being a nitrogen sink to being a major source of nitrogen to bay waters from agricultural runoff. The impacts of this shift were local in extent, not bay wide, and occurred in the mid- and lower bay areas where agriculture was most intensive. Agricultural expansion continued until the turn of the 20th century, when it began a slow but steady decline throughout the watershed of Narragansett Bay.

Hamburg *et al.* note that with the waning of agriculture came a decline in the number of livestock kept, and hence nitrogen inputs to the bay declined. Human population grew readily in the urban Providence area of the bay, and as farmland was fallowed it became reforested, turning again into a nitrogen sink. This trend continued until the mid-20th century when population growth in Providence declined, and urbanites moved into the surrounding towns and forested suburban areas. This recent trend once again shifted the forested areas from a nitrogen sink to a nitrogen source, mainly due to nitrogen inputs from groundwater as most rural areas were not served by sewer systems. Furthermore, the trend of population growth in the suburbs has resulted in increased impervious surface areas as well as more vehicle miles traveled since much of the suburban population worked in the Providence metroplex. More recent trends, however, see a resurgence of interest in the Providence metro region as the city is revitalized, drawing population back into the urban confines. How this trend might alter nitrogen dynamics is unclear.

Nixon *et al.* bring the reader back in time to trace the ebb and flow of nutrient enrichment in Narragansett Bay with a focus on the upper bay and urbanized region. The authors reach as far back as the European discovery of Narragansett Bay by Verrazzano, and, from that historical baseline, rebuild the flow of nutrients to the bay through the onset of the Industrial Revolution in America, and then into contemporary times. In the authors' historic construction of pre-development, they describe a well-flushed, nutrient-poor ecosystem fueled largely by nitrogen and phosphorus inputs from the offshore waters of Rhode Island Sound. Nixon *et al.* move forward in time, developing some intriguing insights into Narragansett Bay's rich history. For instance, the authors cite an historic nautical chart showing eelgrass thriving in the Providence River in the mid-1860s, a time when degradation of the bay was typically considered to be well underway and quite intense. According to Nixon *et al.*, the reason for this conundrum was that nutrients did not reach the bay waters in a focused manner until the installation of a centralized municipal sewage system around the turn of the century. The municipal sewer system, as opposed to the sheet flow movement of nutrients across the landscape in a diffuse, non-point manner, concentrated and conveyed nutrients to the bay as point sources. This singular event moved Narragansett Bay from having Rhode Island Sound for thousands of years as its major, but limited, "nutrient benefactor" to having a new, seemingly unlimited source of nutrients arriving at its northerly terminus.

For over 100 years, the Narragansett Bay ecosystem continued to be a "sewage fueled" ecosystem, particularly in the urban, upper bay area. The nitrogen dynamics of the bay are driven by the input of wastewater treatment facilities, and not agriculture or livestock sources as is typical of southern estuaries such as the Chesapeake Bay or Pamlico Sound. Narragansett Bay has been "well fed" by this source of nutrients; and the ecosystem, for better and for worse, has shifted over time in response to seemingly unlimited nutrient availability from wastewaters.

Nixon *et al.* continue to develop a detailed analysis of nutrient loadings for Narragansett Bay from pre-colonization through present times, showing that the most intense time of nutrient input increase for Narragansett Bay was between 1865 and 1925, with degradation limited mainly to the upper bay. The areas of the bay outside of the upper bay region were better mixed and better flushed, and did not show symptoms of eutrophication to the degree found in the Providence and Seekonk Rivers. However, Hamburg *et al.* make a case for nutrient enrichment outside of the upper bay confines, particularly along the areas of the bay where agriculture was a predominant land use. In Jamestown and Newport, for instance, Hamburg *et al.* cite historical references to nitrogen-contaminated drinking water supplies.

The nutrient enrichment of upper Narragansett Bay continued into the 1950s when, according to data presented by Nixon *et al.,* the flow of nitrogen to bay waters stabilized. Detailed synopses of nutrient flows from sewage treatment facilities and from rivers feeding into Narragansett Bay are presented, but the bottom line, according to these authors, is that total nitrogen loading has remained relatively constant since the 1980s.

Interestingly, Hamburg *et al.* arrive at a different conclusion based on their modeling scheme and interpretation of data—they suggest that the nitrogen flux to Narragansett Bay has increased over the past few decades. While this may be confusing at first consideration, it must be kept in mind that each author has taken different analytical routes to their end points, making different assumptions and calculations along the way. Despite that, both authors agree that nutrient impacts are currently concentrated in the upper reaches of the Narragansett Bay ecosystem, and that those impacts are mainly sewage-fueled. This is acknowledged by most other authors of this volume, and is challenged by none.

A final, but by no means trivial, finding presented by Nixon *et al.* is that there has been a distinct change in the form of nitrogen entering the ecosystem, shifting from organic nitrogen to dissolved inorganic nitrogen. As a result, dissolved inorganic nitrogen now makes up a much larger percentage of total nitrogen discharge from sewage treatment facilities. This is an important shift in nutrient loading; the nitrogenous "fuel" has been converted to a more bio-readily available form, or metaphorically, into a "higher octane" fuel.

Narragansett Bay, fueled by human-generated sewage nutrients, has a very productive upper bay region. Eelgrass disappeared long ago, water clarity declined as plankton growth and suspended solids increased, low dissolved oxygen concentrations are often found, and macroalgal biomass has increased to the point of being considered a nuisance.

Saarman *et al.* and Deacutis undertook detailed studies of dissolved oxygen concentrations in Narragansett Bay during the critical time of late summer when waters are warmest, and conditions that foster the onset and persistence of water column oxygen stress are most probable. The authors focus their analysis in the Providence River area—the most urbanized and heavily impacted region of Narragansett Bay—and in upper and mid-bay regions where stressful oxygen conditions have been observed previously or are likely to occur.

Saarman *et al*. and Deacutis find that hypoxic events tend to set up over a period of several days around the time of neap tides, with dissolved oxygen concentrations in bottom waters decreasing over that span, and lasting between 4 and 9 days. Saarman *et al*. estimate that nearly 70% of the Providence River becomes oxygen stressed (<4.8 mg L^{-1}) during worst case scenarios, with up to 45% of the area becoming hypoxic (<2.9 mg L^{-1}). The authors found that hypoxic and/or oxygen-stressed conditions persist nearly the entire summer at some stations. The authors, however, report no incidence of hypoxia in the Providence River and upper bay areas outside of the months of July and August.

Saarman *et al*. found scattered incidences of hypoxic conditions south of the Providence River and upper bay area, although all sampled stations showed at least one incidence of oxygen stress (<4.8 mg L^{-1}). Hypoxia appears to be most intense in the urban and upper bay region; however, Saarman *et al*. and Deacutis note hypoxia and oxygen stress, often severe, in several smaller bays and coves (e.g., Greenwich Bay).

Similar to Narragansett Bay, Mount Hope Bay experiences oxygen stress in bottom waters during summer, with oxygen-stressed conditions common in deeper waters, such as dredged channels, and areas with consistent freshwater inputs, such as the mouth of the Taunton River. Krahforst and Carullo relate the incidence of oxygen stress to water column stratification events, with critical times, as for Narragansett Bay proper, being late summer on neap tides when water column mixing is minimum.

Of particular interest in Mount Hope Bay, Spaulding and Swanson (this volume), Swanson *et al*. (2006), Deacutis (this volume), Krahforst and Carullo (this volume), and Deacutis *et al*. (2006) report the occurrences of low-dissolved oxygen concentrations in surface waters, which at times attain oxygen readings less than the bottom waters. Unfortunately, the phenomenon is not well documented, and authors could not, given existing data, point toward a cause other than that the event seems to be correlated to freshwater inputs from the Taunton River. The authors also note that the mixing dynamics between Narragansett Bay and Mount Hope Bay aid in the breakdown of water column stratification, thus making Mount Hope Bay less susceptible to low oxygen stress.

Deacutis, using Rhode Island Department of Environmental Management's (RIDEM) fisheries trawl surveys as a proxy, describes a convincing story for macroalgal increases in the upper bay in recent decades. The author considers this phenomenon, along with low dissolved oxygen findings, evidence of increasing nutrient impacts to the upper bay. In the mid-1980s, trawls in waters less than 6 m deep were halted due to interference by macroalgal biomass, which rapidly filled and choked the trawl. For this reason, in 2004, all trawl survey stations in the upper bay were abandoned. It is clear that macroalgal biomass has increased significantly, and mostly in the past few decades.

In light of the long history of anthropogenic nutrient inputs to Narragansett Bay, particularly in the upper bay region, one would expect a trend of worsening conditions over time. If nutrient inputs have indeed stabilized over time as suggested by Nixon *et al.,* why haven't oxygen stressed conditions (such as hypoxic occurrences) stabilized as well? Wouldn't macroalgal biomass have peaked at some earlier point in time? Or has some nutrient input "tipping point" been reached, or exceeded, with oxygen stress and macroalgae becoming increasingly more widespread and/or severe in recent times? Given the limited time, from an historical perspective, of monitoring focused on dissolved oxygen and/or macroalgae in Narragansett Bay, it is not possible to determine whether or not these events are becoming more common.

Deacutis also delves into the benthos, looking for ecological evidence of oxygen stress by examining benthic infaunal communities and comparing Narragansett Bay to classic Pearson-Rosenberg patterns. He finds, moving in a south to north direction, increasing densities of small, opportunistic infaunal species occurring in the lower reaches of the upper bay and upper reaches of the mid-bay, increasing to its maximum in the Providence River where the bay's major sewage treatment outfall is located. The author suggests, based on findings of other research studies, that the spatial shifts in benthic infaunal species composition occurs in parallel with the spatial patterns of nutrient loadings in the upper bay. Opportunistic infauna are also present, though not dominant, in the mid-bay section of Narragansett Bay. Deacutis suggests that this indicates disturbed and stressed ecological conditions, and that oxygen stress occurs at least regularly enough outside of the Providence River and upper bay region to cause shifts in benthic infaunal community structure.

Deacutis also explores pelagic fish abundance data, noting that there has been a shift from demersal to pelagic species in the past few decades, a shift that is consistent with observations from other estuaries experiencing oxygen stress. This shift also correlates with nutrient loadings and shifts in benthic infaunal species composition. However, Oviatt, based on her modeling studies, suggests that changes in fish populations appear to be due more to commercial and recreational fishing impacts, and to changing climate or habitat loss, than to nutrient trends. Deacutis studied some fish species in his analysis that are outside of commercial or recreational fishing efforts, but his findings for these species are similar to regional trends overall (Roundtree and MacDonald, 2006), and are therefore inconclusive.

There are clear signs that upper Narragansett Bay, and some of its major studied bays, is oxygen stressed at least during intermittent, short-lived intervals. Stresses appear to occur at predictable times—late summer neap tides—lasting up to a week, although typically less. How this stress may affect the bay ecosystem is not well understood. It could be, as Nixon *et al.* point out, that the infaunal opportunistic species may be good food for the fishes of the bay. Focused research needs to occur to look toward how low oxygen conditions, macroalgal mats, and opportunistic benthic species affect the bay's ecosystem, if at all, and if these phenomena are spreading and/or becoming more frequent, and if these are linked in any way to fish abundance trends.

19.3 A Climate Changed Bay

Climate change is a major external forcing function with broad implications for altering the circulation and ecology of the Narragansett Bay ecosystem. There is a consensus among the scientists contributing to this volume that climate change is distinct and very strong in Narragansett Bay, and that it is changing the bay's ecology. Bay water temperatures have increased over the past few decades, and additional climate-change-induced temperature swings could raise Rhode Island's air temperatures by an expected 6°C; and it is projected that Narragansett Bay will thermally approximate the coastal ecosystems currently found in South Carolina (USGCRP, 2006).

Warming bay waters as a result of changing climate, along with sewage nutrients entering the head of the estuary, are combining to exert major new forces of change on the Narragansett Bay ecosystem. Increases in the average annual sea surface temperatures confound changes in the biological structure and ecological function attributed to nutrient loading, and temperature changes may also have synergistic effects with nutrient loads. For instance, nitrogen loading to the upper bay area, as noted by Nixon *et al.* and Deacutis, is far beyond the threshold considered acceptable for eelgrass. In view of warming waters due to climate change, Deacutis concludes that eelgrass restoration efforts in that area would not be viable. Krahforst and Carullo observed the loss of eelgrass for Mount Hope Bay long ago, and concur that the ecosystem can no longer support this species.

Pilson points to the shifts in precipitation being a major element of climate change and provides evidence that precipitation to the bay has increased over time. Freshwater inputs significantly influence the residence time of water in the bay, as well as in smaller coves and arms of the bay. Narragansett Bay experiences a range of residence times of 14 to 49 days for the overall bay depending upon the season, with summer having longer residence times than other seasons due to lower freshwater inputs. Flushing times in the Providence and Seekonk Rivers average 3.6 days, with a range of 1.3–8 days. Pilson shows that the Narragansett Bay watershed has seen as much as a 30% increase in precipitation since 1900, which would tend to increase the rate at which nutrients are moved out of the Providence River and into Narragansett Bay proper via increased flushing rates. The impacts this may have on flushing dynamics, and how this would influence bay functions, are unexplored.

Climate change may be a key to predicting the effects of nutrients on shallow, well-mixed ecosystems such as Narragansett Bay. An increase by a few degrees of surface water temperature is significant in shallow water systems such as Narragansett Bay, exacerbating nutrient impacts to plankton, infauna, macroalgae, eelgrass, and fish populations. How this broad-scale, temperature-mediated change will play out in the larger bay ecosystem is nearly impossible to predict, but changes will certainly occur at various, if not all, ecosystem levels. For example, the $1°C$ warming currently affecting the bay is implicated in a major shift in the magnitude of the annual lower bay winter–spring diatom bloom. It is not entirely clear if this may be a result of temperature change directly, or through increased cloudiness due to increased temperatures. Borkman (2002) found that warmer winters over Narragansett Bay are cloudier winters, which would tend to decrease primary production.

The observed multi-decadal warming trend is also implicated in the suppression of winter blooms during warmer winters, which is vitally important because winter recycling has traditionally fueled summer production in Narragansett Bay. Oviatt sees this as a fundamentally important shift in the overall ecology of the bay ecosystem, and Nixon et al. speculate that, if nutrient levels decline because of reduction from wastewater treatment facilities, then spring–summer, and perhaps fall blooms, may be suppressed further, resulting in greater ecosystem stress during a period when there is already evidence of food limitation in the ecosystem of the lower bay.

Sullivan et al. examine the impacts of warming on the bay by describing the increase in abundance of a ctenophore, Mnemiopsis leidyi, which has extended its range as well as its seasonal cycle of abundance in Narragansett Bay. Due to warming, Mnemiopsis becomes abundant earlier in the season than before, and being voracious predators, they reduce the population densities of bay zooplankton during summer months, allowing summer phytoplankton blooms to occur. As a result, researchers in contemporary times are finding almost a complete lack of copepods during summer months in Narragansett Bay. With this link in the trophic web removed, what might be the impacts upon other components of the ecosystem? Late summer phytoplankton blooms might add fuel to oxygen stress by providing more organic matter to bottom waters for decomposition during that period of summer when the upper bay is most susceptible to oxygen stress. An additional concern, but not well reported at present, is that the warming of Narragansett Bay may allow southern species to expand their range northward, incurring new impacts and further altering the ecology of Narragansett Bay, much as Mnemiopsis has done.

The questions that arise from and the problems posed to coastal ecosystem managers due to climate change are numerous, significant, and immediate. There is every indication that the warming trend will continue, and it appears that New England will continue to get wetter as well (USGCRP, 2006). The impacts of climate change are far reaching—they impact the biological, chemical, and physical attributes and functions of the bay, often having a "ripple effect" where one change may impart further change in other attributes and functions. Ecosystem-based management must capture the changing climate over Narragansett Bay, and incorporate it into predictive scenarios and modeling efforts to better comprehend possible future conditions, particularly with regard to nutrient reduction management and the impacts this may have on an ecosystem experiencing climatic change.

19.4 A Bay on a Diet

The primary objective of mandated nitrogen reduction is to reduce seasonal occurrences of algal blooms and low dissolved oxygen conditions, mainly in upper Narragansett Bay. Save The Bay, a primary environmental advocate for Narragansett Bay, and the Governor's Nutrient Task Force, among others, believe that new regulations to reduce sewage derived nitrogen loading to the bay by 50%, which reduces the overall nitrogen loading to Narragansett Bay by about 20%, will decrease the frequency of oxygen stressed conditions in the bay. Oviatt, based on her gradient model, notes that a 20% reduction in nutrient input to Narragansett Bay will probably be undetectable for many variables (but not all); a 50% reduction in nutrients, however, would be detectible for a suite of ecological variables.

Given a 50% reduction in nutrient input to the upper bay, which is greater than the anticipated reduction of implemented nutrient removal measures, Oviatt suggests that benthic and water column respiration might be reduced enough to alleviate much of the oxygen stress in the upper bay area. This is based on model assumptions that all oxygen demand is biological, and that macroalgae (or seagrasses) do not replace phytoplankton production after nutrient reduction. Nor does this include any estimate of the probable impacts of climate change, such as the effect of increased temperatures and perhaps heightened water column stratification (though perhaps with increased flushing from increased precipitation), or ecological changes in trophic dynamics resulting from nutrient availability or climate change.

Some questions that beg to be asked are: If 20% nitrogen reduction will be mainly undetectable in the ecosystem, why bother? If 20% nitrogen reduction is not detectable, but 50% is, where between 20 and 50% will ecosystem responses become detectable? These questions are not insignificant, and need answers for an ecosystem-based management approach for Narragansett Bay to proceed logically and effectively. Nitrogen reduction is an expensive undertaking; if little ecosystem change can be shown with a 20% nitrogen reduction, and with urban taxpayers footing the bill, might they be unwilling to commit funds for further reduction in the future? Will they feel "cheated" and create some form of backlash that could hamper further ecosystem-based management efforts for Narragansett Bay? What if 25 or 30% reduction would provide ample detectible change in the bay ecosystem? Might this be a better ecosystem-based management strategy than to proceed with 20% nitrogen reduction when the best available science suggests ecosystem change will be noticeable? There are many important policy questions to consider, and more input, agreement, and advice from the scientific community would be useful in defining an equitable ecosystem-based management approach. Ecosystem-based management urges managers and scientists to consider the ecosystem as an integrated unit, and not simply a nitrogen-impacted upper bay.

For instance, if changing climate over Narragansett Bay is driving the ecosystem dynamics in a direction that favors phytoplankton blooms by removing predators, then anticipated improvements in water column oxygen stress may not be realized. Oxygen stress conditions may increase as a result of an ecosystem shift in species abundance, interleafed with other changing ecological variables that promote increased use of oxygen in the water column. Scientific understanding of an expected response of the Narragansett Bay ecosystem to nitrogen reduction is not very good, largely because experimental studies have tended to focus on nitrogen enrichment rather than nitrogen reduction, and current models do not have these capabilities. Such models, however, would allow probabilities of success to be realized and considered. Currently this is not possible, thus leading to much speculation.

Nitrogen limitation in lower Narragansett Bay is most extreme during summer months, which is exactly when the planned nitrogen reductions will be implemented. Without models to provide probable scenarios for consideration, we are left with many questions. What will happen to the ecology of the already nitrogen-limited lower bay in light of the upper bay nitrogen reduction scheme? Will the nitrogen-limited lower bay become a nitrogen-starved lower bay? Nixon *et al.* question more generally if the bay needs to be "fixed" by reducing the nitrogen inputs. "Is it broken?" they query. The authors point out that, despite the occasional fish kill or algal bloom in the upper bay, there is no indication that the bay is sick and/or is dying. With this said, we must consider that blooms of nuisance macroalgae, such as those plaguing Rhode Island Department of Environmental Management fish trawl survey efforts are persistent in the upper bay. Macroalgae blooms are often linked to anthropogenic nitrogen inputs—why would this be not the case for upper Narragansett Bay?

There are more concerns that need to be addressed to carry out ecosystem-based management successfully. For instance, the slowing of the rise of sea level from 2,500 years ago allowed the formation of salt marsh communities along the narrow fringing shoreline of Narragansett Bay. Wigand notes that Rhode Island had lost 53% of its salt marshes since colonization, mostly during the early industrialization era, with little change having occurred over recent decades. Today, salt marshes along Narragansett Bay are limited, most often found as small, isolated fringing marsh habitats.

Wigand finds it difficult to pin a number on salt marsh nitrogen removal in Narragansett Bay, but notes instances when high rates of nitrogen removal have been found. Given the possibilities of robust nitrogen removal, and their relative rarity in the Narragansett Bay ecosystem, salt marsh preservation, restoration, and rehabilitation should be considered an important part of the overall ecosystem-based management approach for Narragansett Bay, as found for other estuarine ecosystems (NRC, 2007).

Atmospheric nitrogen is also poorly understood, but it may be critical to ecosystem-based management efforts. There is very little data available to assess the impacts of atmospheric deposition of nitrogen to Narragansett Bay and its surrounding watershed. As such, Howarth approaches his assessment by looking at atmospheric deposition of nitrogen at a larger scale, and then uses that to focus on atmospheric nitrogen inputs to the Narragansett Bay ecosystem. Howarth explores three possible scenarios to estimate atmospheric deposition of nitrogen to Narragansett Bay, and suggests that atmospheric deposition of nitrogen makes up about 30% of the total nitrogen load to Narragansett Bay. If his estimate is correct, this source of nitrogen is the third most important source to the Narragansett Bay ecosystem, after treated sewage and river inputs. While being a challenge for local resource managers, atmospheric deposition must be considered in ecosystem-based management efforts because it is assessed as a major source of nitrogen, and is expected to gain more importance, particularly as sewage derived nitrogen inputs are curtailed.

Once atmospheric nitrogen is deposited onto the landscape, it moves over the land in surface water flow, while some of it moves into the soils and merges with other nitrogen contained in groundwater. Nowicki and Gold describe how groundwater flows and nutrient dynamics are tightly linked to the basic geology of the bay area. The eastern shores of Narragansett Bay, for instance, restrict groundwater movements due to geologic impermeability of the soils, while the western shores are much more permeable and conducive to groundwater movements.

The complex glacial outwash soils vary considerably in their ability to promote nitrogen movement and removal, making nitrogen control very difficult to apply in any generalized manner. For instance, nitrogen inputs to Greenwich Bay have been considered of such importance that recent ecosystem-based management efforts (RICRMC, 2005) have resulted in widespread sewering of the watershed to alleviate symptoms associated with excess nitrogen inputs. Groundwater input to Greenwich Bay is about 16% of the total freshwater input, and provides about 12% of the dissolved inorganic nitrogen load to Greenwich Bay (Dimilla, 2006). In contrast, in the Pettaquamscutt (Narrow) River, a steep-walled estuarine arm of lower Narragansett Bay, groundwater input, depending upon location along the bay, makes up 33 to 100% of the total freshwater input (ASA, 1995). With the entire watershed being served by septic systems at that time, it is likely that the majority of nitrogen delivered to the estuary was of groundwater origin.

King *et al.* delve deeper into the groundwater story for the Pettaquamscutt River through observation and analysis of sediment cores, which shows heavy anthropogenic eutrophication occurring from human wastes. As noted earlier, since the entire region has historically relied upon septic systems, groundwater flow was the presumed culprit. Ecosystem-based management initiatives for the

Pettaquamscutt mandated sewering as a remedy to excess nitrogen loading (RICRMC, 1999). Interestingly, King *et al.* did not find ecosystem "rebound" resulting from the sewering initiative in their sediment core analysis. They suggest that this may be due to the time lag typical of a "dose–response" reaction in groundwater, and that further time is needed before the response might be noticeable.

For groundwater nitrogen, too little is known at the scales needed for effective management. Further research should be targeted at those subsystems of the larger bay ecosystem that show eutrophication symptoms coupled with the geologic potential for groundwater nitrogen impacts. Further effort should be focused in the Pettaquamscutt to determine if sewering had the anticipated management outcome in this sub-bay ecosystem. Is it "lag time delay" as King *et al.* suggest or are other forces at work? The answer to this question will be critical in considering the incorporation of groundwater dynamics into ecosystem-based management strategies for the western shoreline of Narragansett Bay.

Finally, viral communities are also not often considered in ecosystem management, and are not well studied overall in many estuarine systems. Marston, however, sheds light upon the viral community of the Narragansett Bay ecosystem, noting that marine viruses are extraordinarily abundant, and that the marine viral community of Narragansett Bay, in general terms, exhibits the same sort of seasonality as that exhibited by the bacterial and phytoplankton communities of the bay. Furthermore, the composition of the viral community seems to be similar between Narragansett Bay, Mount Hope Bay, and Rhode Island Sound, suggesting that there is great ecological connectedness throughout the bay ecosystem. Marston notes that viruses can shape the abundance and distribution of host communities in marine systems where they have been studied in detail, and so there would be no reason to believe that the same could not hold true for the Narragansett Bay ecosystem.

19.5 "Ecofunctional" Zones in Narragansett Bay

While much remains to be learned, it is common to hear, and to read comments about "The Bay" as if it is a singular, uniform, well-understood ecosystem. When a million fish, primarily juvenile menhaden died in Greenwich Bay on 20 August 2003, Governor Carcieri declared "Rhode Island is Narragansett Bay" (RIDEM, 2003). But, in the minds of many, Narragansett Bay is divided into two bays: an urbanized upper bay—the Providence metroplex—and a rural, "South County" bay. Most scientists, however, recognize that Narragansett Bay is a complex ecosystem with a number of ecologically distinct functional zones.

There is much discrepancy however, amongst scientists regarding the division of Narragansett Bay into upper, mid, and lower bays. As a case in point, the editors of this volume had to "neuter" most references to the various sections of "The Bay." The differing definitions of upper, mid, and lower Narragansett Bay lead to confusion in the discussion of the processes and impacts, and often lead to wonderment in the eyes of the public when scientists cannot agree upon something so seemingly simple. Since nearly all dialog about "The Bay" refers to one or more of these divisions, there is a need for scientists to come to an accepted definition of divisions for Narragansett Bay to instill some consistency.

We believe that there is merit in considering "ecofunctional zones" for Narragansett Bay. Deacutis, Oviatt, and Smayda and Borkman emphasize that Narragansett Bay shows a biochemical and ecological gradient of conditions along a north–south axis driven in part by the discharge of rivers and sewage treatment facilities into the Providence and Seekonk Rivers at the northern terminus of the bay. These authors describe Narragansett Bay according to various gradients and/or functional divisions, which are iterations of what we will refer to as "ecofunctional zones."

In general terms, Narragansett Bay is characterized by low freshwater flows and high effluent discharges into the head of the estuary, making the Providence River an area of high nutrient availability. Dilution of the nutrient load begins just outside the Providence River, and proceeds south along a gradient that decreases in magnitude down bay. The bay area from the Providence River south to the northern tip of Prudence Island appears to be a transitional zone between upper and lower bay regions, with a unique set of water column biochemistry and benthic conditions. Below this transition zone is an area of much lower nutrient availability, progressively decreasing down bay due to both dilution and biological uptake. This nutrient gradient closely tracks the down-bay salinity gradient. In general terms, Narragansett Bay comprises a "nutrient enriched" upper bay zone and a "nutrient poorer" lower bay zone, linked by a transitional zone.

Smayda and Borkman describe three distinctive "ecofunctional zones" for Narragansett Bay (Fig. 15.11): an Enrichment Zone—a region of high nutrient input levels in the Providence River; a Depuration Zone—a region of intense phytoplankton growth in the area from below the Providence River to the northern tip of Prudence Island acting as a "biofilter" to nutrient inputs; and an Advection Zone—a region of nutrient limitation in the lower reaches of the bay where growth is fueled by remineralization of nitrogen inputs from Rhode Island Sound.

Deacutis describes six nutrient classification divisions (Fig. 12.4), which we collapse into three "ecofunctional zones." In the lower bay, from the mouth northward to the Jamestown–Newport region, the author finds no obvious expressions of eutrophic conditions, corresponding roughly to the Smayda-Borkman Advection Zone. In the mid-bay, Deacutis describes a zone of transition from non-impacted areas to those that experience rare hypoxic events. This area can be considered moderately eutrophic, experiencing high primary productivity, short periods of seasonal oxygen stress, complete loss of submerged aquatic vegetation with a replacement by nuisance macroalgae, and increased abundance of opportunistic benthic infauna. This region corresponds roughly to the Smayda-Borkman Depuration Zone. The upper bay, from the mouth of the Providence River northward into the Seekonk River, is considered by Deacutis as highly eutrophic and severely impacted, with high macroalgal biomass and experiencing seasonal hypoxia that is becoming more frequent and persistent. This area corresponds roughly to the Smayda-Borkman Enrichment Zone.

Defining, understanding, and managing the ecological functional zones for Narragansett Bay may be the most effective way to understand the response of the bay to changes in nutrient dynamics, climate change, and other forcing functions. Attempting to replicate the functions of these zones in models, as suggested by Chen *et al.* as a component to a general circulation model, would be a significant accomplishment. Smayda has a 30-plus-year record of physical and ecological variables for a station located in lower Narragansett Bay, which perhaps could be used in the construction and/or validation of models. Such models could be used to estimate the changes in ecosystem productivity, and perhaps function, with regard to nutrient management regimes. As such, they should be able to assist in determining if nitrogen reduction in upper Narragansett Bay will alter the occurrence, duration, and extent of oxygen stress. Such models will assist in predicting the possible impacts of further nutrient limitation on the ecosystem of lower Narragansett Bay. The current lack of such models hampers the application of ecosystem-based management efforts for Narragansett Bay.

19.6 An Ecofunctional Zoning Plan for Narragansett Bay

When we examine the horizon of knowledge on Narragansett Bay contained in this book, we find that much is still lacking that will allow effective management at ecosystem scales. Seemingly simple, basic questions, such as "Is hypoxia increasing or decreasing in the bay?" and "Do hypoxic events have any significant, lasting impact on the ecology and function of the bay ecosystem?" cannot readily or reliably be answered for "The Bay" with present scientific knowledge. Circulation modeling, although quite good at whole ecosystem scales, is lacking at the scale of application required for ecosystem managers. Circulation models also need to be coupled to ecological modules that integrate the biotic functions of the bay into the physics of the bay.

We suggest the starting point for a discussion on how to better model and manage the Narragansett Bay ecosystem is the "ecofunctional zone" concept arising from the integration of Deacutis's and Smayda and Borkman's interpretations of bay response and functions to nutrient inputs, overlain with the nutrient gradient scheme presented by Oviatt (Fig. 18.1). The concept of "ecofunctional zones" provides a beginning by which to array the existing knowledge base contained in this volume as well as incorporate restoration, rehabilitation, conservation, and engineering practices alongside policy and governance for Narragansett Bay.

If managers embrace the ecofunctional zone concept for Narragansett Bay, ecological engineering should be considered an important element of ecosystem-based management approaches (Mitsch and Jorgensen, 2004). For instance, incorporating expanded shellfish aquaculture leasing and public shellfish aquaculture approaches pioneered by the Rhode Island Aquaculture Initiative (RIAI, 2004) into the area along the southern terminus of the "Enrichment Ecofunctional Zone" could remove and export nitrogen from the bay ecosystem while creating jobs and new business opportunities. Innovative ecological engineering for planting eelgrass (Granger *et al.*, 2002) in the "Advective Ecofunctional Zone" could be considered and deployed to create positive feedback loops that lead toward improved sustainability in the lower bay. Lastly, managing the bay according to its ecofunctional zones and using sophisticated ecological engineering concepts sets the stage for developing and applying a more sophisticated governance system (Olsen, 2003). An ecofunctional zoning approach will provide a robust, stable platform for decision-making and research, while maintaining the flexibility to allow for adaptive management, targeted research, and ecological engineering applications, even in the face of climate change.

Nature knows no stasis. "The Bay" that Verrazzano saw is not "The Bay" we see today, and what we see today will not be "The Bay" that our grandchildren will inherit. The unique qualities of Narragansett Bay are what attracted humankind to settle upon its shores, and subsequently spurred the Industrial Revolution that shaped the future of the bay, the region, and the nation. Narragansett Bay and its unique qualities have been and continue to be a dominant cultural and economic influence upon the human population within its watershed. In turn, human actions have significantly influenced the Narragansett Bay ecosystem. Today, we consider the unique qualities of Narragansett Bay through "sustainability" and "ecosystem-based management" to improve upon how human populations can live in better harmony with the environment of which they are a part.

We propose a dialog be opened regarding the development of an ecofunctional zoning approach to ecosystem-based management for Narragansett Bay. The opportunity to do so is excellent as the Narragansett Bay Comprehensive Conservation and Management Plan comes up for update and revision. Here lies the opportunity, through the application of science for management, to chart out a truly sustainable future for Narragansett Bay, a future that utilizes ecofunctional zoning concepts and ecological engineering to the benefit of both people and place.

References

Applied Science Associates (ASA). 1995. Narrow River Stormwater Management Study, Problem Assessment and Design Feasibility. ASA, Narragansett, RI.

Borkman, D. 2002. Analysis and simulation of *Skeletonema costatum* annual abundance patterns in lower Narragansett Bay 1959–1996. Ph.D. Thesis. University of Rhode Island Graduate School of Oceanography, Narragansett, RI.

Deacutis, C.F., Murry, D., Prell, W., Saarman, E., and Korhun, L. 2006. Hypoxia in the upper half of Narragansett Bay, RI, during August 2001 and 2002. *Northeastern Naturalist* 13(Special Issue 4):173–198.

Dimilla, P.A. 2006. Using stable nitrogen isotopes to characterize and evaluate nitrogen sources to Greenwich Bay, RI and their influence on isotopic signatures in estuarine organisms. MS Thesis, University of Rhode Island Graduate School of Oceanography, Narragansett, RI. 162 pp.

Granger, S., Traber, M., Nixon, S.W., and Keyes, R. 2002. A Practical Guide for the use of Seeds in Eelgrass (*Zostera marina L.*) Restoration. Part 1. Collection, Processing, and Storage. Rhode Island Sea Grant, Narragansett, RI. 20 pp.

Kincaid, C. 2006. The exchange of water through multiple entrances to the Mount Hope Bay estuary. *Northeast Naturalist* 13(Special Issue 4):117–144.

McLeod, K.L., Lubchenco, J., Palumbi, S.R., and Rosenberg, A.A. 2005. Scientific consensus statement on marine ecosystem-based management. Communication Partnership for Science and the Sea. *http://compassonline.org/? = EBM*.

Mitsch, W.J., and Jorgensen, S.E. 2004. Ecological Engineering and Ecosystem Restoration. John Wiley amp; Sons, New York.

Narragansett Bay Project (NBP). 1992. The Narragansett Bay Comprehensive Conservation and Management Plan. US Environmental Protection Agency, Washington, DC.

National Research Council. 2007. Mitigating Shore Erosion along Sheltered Coasts. National Research Council of the National Academies. The National Academies Press, Washington. 174 pp.

Olsen, S.B. 2003. Framework and indicators for assessing progress in integrated coastal management initiatives. *Ocean and Coastal Management* 46:347–361.

Rhode Island Aquaculture Initiative (RIAI). 2004. Rhode Island Aquaculture Initiative: A Shared Vision for the Future. Rhode Island Sea Grant, Narragansett, RI. 16 pp.

Rhode Island Coastal Resources Management Council (RICRMC). 1999. The Narrow River Special Area Management Plan. Stedman Government Center, Wakefield, RI. 212 pp.

Rhode Island Coastal Resources Management Council (RICRMC). 2005. Greenwich Bay Special Area Management Plan. Stedman Government Center, Wakefield, RI. 475 pp.

Rhode Island Department of Environmental Management (RIDEM). 2003. The Greenwich Bay Fish Kill—August 2003 Causes, Impacts and Responses. Providence, RI. (*www.dem.ri.gov/pubs/fishkill.pdf*).

Roundtree, R.A., and MacDonald, D.G. 2006. Introduction to the Special Issue: Natural and anthropogenic influences on the Mount Hope Bay ecosystem. *Northeastern Naturalist* 13(Special Issue 4):1–13.

Swanson, C., Kim, H.S., and Sankaranarayanan, S. 2006. Modeling of temperature distributions in Mount Hope Bay due to thermal discharges from the Brayton Point Station. *Northeastern Naturalist* 13(Special Issue 4):145–172.

US Global Change Research Program (USGCRP). 2006. The New England regional assessment of the potential consequences of climate variability and change. *http://www.necci. sr.unh.edu/assessment.html*.

Index

Acadia National Park, 54
Accelerator mass spectrometry (AMS), 217
Acoustic Doppler Current Profiler (ADCP),
    243, 259, 296, 304, 306, 318
Advection zone, 476–479, 559
Advective ecofunctional zone, 561
Albuminoid, 120–121, 130, 133, 153
Alleghenian orogeny, 1
American Association for the Advancement
    of Science, 119
Ammonia volatilization of, 47, 80, 225
Animal community structure, 500, 505–509
    changes in, 500, 505
Animal waste, 47, 70, 110–111, 128, 179–180,
    185, 190, 194, 196–198
Annual average water temperatures, 355
Anoxia, 70, 113, 136, 214, 350, 363, 369, 370,
    496, 506
Anthropogenic activities, impacts of, 180
Anthropogenic enrichment, 445
Anthropogenic eutrophication, 211–212,
    214–216, 225–230
    indicator of, 212–214
Apponaug Cove, 264–265, 359, 510
Aquidneck Island, 11, 21–23, 89, 240, 250
Aquifer zones, 80
Argon plasma atomic emission
    spectrometer, 219
Assimilation number, 463
Atlantic Ecology Division, 499, 516
Atmospheric loading, 185
Aureococcus anophagefferens, 419, 432,
    456–459, 468
Autonomous water quality
    monitoring, 397
Avalon zone, 1
Axial channels for navigation by commercial
    shipping, 17
Azoic zones, 370

Baltic Sea, 48, 511
Baltimore Long Term Ecological Research
    Project, 68–69
Barnegat Bay, 57, 433
Barrington River, 235
Bathymetry, 25, 233, 271, 285, 290, 304,
    329, 340, 350, 356, 358, 367,
    371, 441, 443, 445, 466, 524,
    526, 527, 540
Bay nutrient gradient, 529, 559
Beach and barrier spit, 19
Benthic
    abundance, 506, 527–529, 539
    animals, 540
    community change, 349, 365, 372
    diversity, changes in, 69, 356
    feeding, 506
    geologic habitats, 16–17
    grazing, 465, 472, 480
    infauna, 166, 527, 532, 534, 539,
        552, 560
    larvae, 435, 468–470
    macroalgae, 356, 440
    organism density, 372
    suspension feeders, 480
Biofiltration capacity, 478–479
Biological gradient, 475–477
Biological lamination thickness, 218–220,
    225, 229
Biological recycling, 443, 554
Biological remineralization, 445, 465, 559
Biomass accumulation, 454
Blackstone River, 12, 60, 110, 115, 119–120,
    120–121, 125, 142, 145, 146, 149,
    156, 161, 185–188, 202, 235, 288
    historical monitoring of the, 117
    phosphate concentrations in, 148
    Valley Sewer Commission, 163
Block Island, 5, 7

Bloom
    intensity, 458
    synchrony, 459
Bonnet Point, 13
Boston Harbor, 60
Boundary-fitted hydrodynamic model, 246
Brayton Point, 246, 265–267, 273, 284, 290,
        293, 298, 384, 390, 393, 397, 402,
        409, 413–414
Bridgewater State College, 407–408
Brown tide, 166, 425, 432, 434, 446,
        456–461, 468
    bloom, 453, 456, 464
    trophic effects of the, 434
    year, 454
Bucklin Point, 102, 116, 134, 137–138, 140,
        160, 189, 213, 216, 222, 228–230,
        236, 261, 350, 530
Buffer zone, 440
Bullock's Reach data, 331–332
Buoy data, 337–339, 342, 345
    application of, 341
Bureau of Transportation Statistics, 24

C/N ratio, 211–212
Carbon growth rates, 464–465
Census Bureau, 25, 131, 199
Chesapeake Bay, 47–50, 57–60, 72–73, 87,
        108, 163, 228, 355, 362, 549
Chicago sewer system, 112
Chipuxet River, 77–78
Chlorophyll, 166, 219, 225, 358–359, 396,
        452–454, 460–461, 466–467, 470,
        488, 530
    concentrations, 141, 165, 452–455, 461,
        466, 471, 477, 541
    levels, 352, 441, 453, 455–457, 459, 466
    monitoring of, 373
Circulation modeling, 291, 547, 560
Citizen's Charitable Foundation, 140
Clean Air Act Amendments, 61
Climate-induced oligotrophication, 165–166
Climate change, 148, 165, 366, 496, 541,
        553–556
Coastal drainage, 138
Coastal geomorphology, 67, 82
Coastal groundwater
    aquifers, 87
    discharge, 82, 92
Coastal marine ecosystems, 48, 56, 59, 135, 159
Coastal plain sediment, 5
Coastal transition zones, 81
    geomorphology of, 81

Coastal water tables, 86
Coastal wetlands, 500–501, 514–516
Coast survey, 108–111
Cold Spring Point, 246
Colored dissolved organic matter, 409
Combined sewer overflows (CSO), 262,
        371, 389
Common Fence Point, 19, 22
Conanicut Island, 13, 19, 21, 23, 182,
        239, 281
Concentration measurements, 124, 138,
        141, 160
Confined aquatic disposal (CAD), 267
Conimicut Point, 13, 102, 244, 250, 254, 257,
        269, 289, 325, 365, 454, 530, 540
Copepods, 166, 435–436, 468–470, 472,
        474–475, 485, 489, 493–496, 554
Copper concentration, 217, 219–222, 226–227
Cormorant Point, 21
Cryptic nutrient-phytoplankton
        relationship, 447
Ctenophores, 432, 435–436, 461, 472–475,
        485–486, 489, 490–491, 493–495,
        540, 554, 491
    zooplankton relationships, 475
Cultural eutrophication, 499
Cyanobacteria, 419–420, 425, 448
Cyanophage, 420–426
    communities, 423
    diversity, 423
    genotypes, 421, 423–425
    genotypic analysis of, 423
    infected chemostats, 424

Darcian flow, measurements of, 70
Darcy's Law, 76
Davisville Channel, 259
Deforestation, 180
Delaware Bay, 57
Deltas, 9–11
Demersal ratio, 352, 367
Denitrification, 80–81, 85, 87, 101, 134, 159,
        160, 161, 195, 213, 225, 227, 229, 408
    enzyme activity, 83, 512–513
    rates, 80, 85–86, 512
    of sewage effluent, 205
Density-induced
    circulation, 254
    flows, 272
Deposit-feeders, 166, 506
Depositional
    environment, 14
    velocities, 50

Deposition monitoring data, 50
Depuration zone, 475–478, 479–481,
        559–560
Direct discharge zones, 89
Discontinuous bedrock, 22
Dissolved Inorganic Nitrogen (DIN),
        71–72, 104–107, 117, 122–125,
        138, 140, 146, 150, 156–157, 179,
        186–189, 198, 202, 319, 350, 361,
        396, 403
    deposition of, 104
Dissolved nutrient concentrations, 405
Dissolved Organic Carbon (DOC), 68, 81,
        84, 86
Dissolved Oxygen, 85, 92, 101, 125, 136, 162,
        167, 255–256, 261, 282, 326–335,
        336–337, 340, 342, 344–345, 363,
        372, 395–396, 399–400, 402, 412,
        492, 506, 523, 550–551
Double-peaked flood, 241, 247–248, 258, 271
Down-bay dilution gradient, 526, 559
Dredged channel, 17, 217, 221–222, 227, 234,
        236, 244, 248, 254, 257, 259,
        271–272
Dredging, 17, 217, 222, 227, 233, 267, 269,
        273, 301, 365
Dry deposition, 49–50, 52–53, 56, 58,
        104, 105

East Passage, 12, 138, 233, 240, 243, 247,
        249–250, 256, 261, 271, 281, 293,
        302, 304–305, 307–311, 315–321
        337, 359, 368, 383, 402, 439, 479,
        527, 547
Ebb, 113, 131, 244, 247, 258, 262, 267, 271
Ecofunctional zones, 545, 558–560
Ecosystem-based management, 1, 35, 47, 67,
        101, 177, 211, 233, 281, 383, 390,
        410, 412, 434, 448, 481, 545, 548,
        555–558, 560–561
    principles of, 414
Ecotones, 439, 475–478, 479
Eddies, 285, 292–293, 298, 547
Eelgrass, 108–109, 111–112, 282, 325,
        352, 354–356, 368, 370, 372,
        410, 440–441, 524, 549–550,
        553–554, 561
    loss of, 352
Ekman transport, 306, 317, 320–321
Emission-based modeling, 52–53
Enrichment zone, 476–477, 559–560
Environmental management, 390
Environmental monitoring, 393

Environmental Protection Agency (EPA),
        108, 134, 188, 193, 195, 327, 331,
        334, 341–342, 354, 389–390, 408,
        499, 541
Environmental regulations, 221, 227
Environmental sensitivity index, 18
Epply Laboratory, 527
Estuarine cove, 16
    groundwater discharge, 85
Euphotic zone, 165, 435, 441, 476–477
European settlement, 179–180
Eutrophication, 62, 68, 118, 135–137, 159,
        211–212, 214, 216, 225–227, 229,
        325, 350, 352, 358, 363, 367,
        392–393, 412, 431, 448, 449, 479,
        506, 511, 515, 550, 557
    control mechanism, 480
Evapotranspiration, 39, 41, 69, 79, 83, 86,
        88, 153
Exotic organisms, 206

Fall river, 116, 134, 138, 161, 183, 237,
        239–240, 243, 289, 350, 363, 386,
        387, 389
Federal Census of Agriculture, 195
Field's Point, 102, 112–114, 130–134, 161,
        180, 189, 213, 216, 228–229, 235,
        246, 254, 267, 350, 437, 446, 470
First World War, 130, 131–132
Flux, 48, 59, 70–74, 151, 156, 190, 198–200,
        202–203, 205, 408
    atmospheric, 59
    freshwater, 70
    measurements, 74, 141
Food web dynamics, 443
Foredune zone, 19
Fossil fuel combustion, 47–48, 50, 51, 62,
        104, 128, 159, 185
Fossil pigment analyses, 218
Fox Island, 10, 165, 359, 437, 454, 457,
        486–491, 493–495
Fox Point, 267, 525, 528, 530
Fresh groundwater input, 88

Gamma counting methods, 217
Gaspee Point, 242, 244, 257, 259, 360,
        453, 462
Gauging stations, 35, 51, 389
GCTM model, 51
General ocean turbulence model, 290
Genotypic composition, 423
Ghyben-Hertzberg model of high-density
        seawater intrusion, 86

Glacial
    delta plains, 13–14
    geology, 74–75
    ice, 5, 7
    outwash, 74
Gould Island, 13, 259, 530
Graphic furnace atomic absorption
    spectrometer, 219
Gray-water discharge, 92
Great Depression, 227
Green Airport T.F., 37–38, 45, 246, 253, 257,
    308–309
Green tide, 356
Greenwich Bay, 8, 10, 12, 15–16, 68, 82,
    90–93, 145, 153, 163, 167, 233, 240,
    247, 264–265, 272, 282, 325, 334,
    337, 340, 344, 353, 355–357, 359,
    362, 364, 368–369, 370, 431–432,
    440, 445–446, 451, 457, 459–460,
    468, 469, 470, 474, 478, 487, 493,
    546, 551, 557
    Special Area Management Plan, 91
Greenwich Cove, 92, 264–265, 486–495
Groton sewage treatment wastewater
    discharge, 372
Groundwater
    aquifers, 68, 77, 80, 87, 93
    bypass flow, 78
    carbon interactions, 87
    denitrification, 80–81, 93, 512
    discharge, 68–70, 74, 82, 85, 93,
        193, 304
    enrichment, 117
    flux, 70, 74, 85
    hydrology, 91
    nutrient transport, 62
    withdrawals, 198
Gulf of Mexico, 163

Hardened shorelines, 22
High Density Low Biomass (HDLB), 456
High Performance Liquid Chromatography
    (HPLC), 218
Hudson River, 57, 60, 105
Hurricane barrier, 267
Hydraulic
    conductivity, 70, 76–77
    gradient, 70, 77, 83
Hydrodynamic model, 246, 253, 256–257,
    264–265, 273, 370
    boundary-fitted, 243, 246, 264
    three-dimensional, 243, 249, 257, 259
    vertically-averaged, 241, 253

Hydrodynamics and thermal structure of
    Mount Hope, 266
Hydrologic modifications, 79
Hydrostatic primitive equations, 290
Hyperbolic distribution of chlorophyll, 479
Hypereutrophic threshold, 358
Hypersaline soils, formation of, 83
Hypoxia, 69, 136, 162, 163, 166, 167, 227,
    284, 325, 326–328, 330, 331–332,
    334, 339, 342, 349, 350, 356, 359,
    361–362, 363–364, 366, 367–369,
    371, 395, 431, 439, 480, 486, 492,
    494–495, 506, 511, 523, 547,
    551, 560

Ichthyoplankton, 524, 540–541
Industrial Revolution, 101, 524
Intergovernmental Panel on Climate Change
    (IPCC), 37, 39, 45
Isostatic adjustment, 11

Jamestown water supply, 183

Kjeldahl digestion, 133

Labile Organic Carbon, 81
Lamination thicknesses of the biological
    layer, 214
Landsat thematic mapper satellite imagery, 23
Larval fish, 525, 527, 533, 537
Laurentide ice, 5, 7–8
Lawrence Experiment Station, 119–120,
    122, 127
Least-squares harmonic analysis
    toolbox, 305
Liverpool University Environmental
    Radioactivity Laboratory, 217
Lorenzen's equations, 527
Low impact development, 93
Lysogenic infections, 426

Macroalgal
    biomass, 349, 352, 357, 410, 432, 440–441,
        477, 506, 541, 550–552, 555
    blooms, 92, 167, 325, 327, 354, 356,
        357, 370, 372, 550, 551, 523,
        556, 560,
Macrobenthic
    biomass, 365
    species, 370, 506
Marine Ecosystems Research Laboratories
    (MERL), 506
Marsh zone, 512

Mass Accumulation Rate (MAR), 211, 219
Massachusetts Office of Coastal Zone
    Management, 395
Merrimack River, 57, 119
Mesocosm, 166, 506, 511, 523, 525, 527, 529,
    534, 539, 541
Meso-scale atmospheric model, 290
Metamorphosed sedimentary rocks, 1, 4, 21
Meta-sedimentary bedrock, 21
Meteorological measurement stations, 289
Microbiological indicators of pollution, 135
Microscopic epibenthic algae, 441
Minimum variance-based estimation
    technique, 396
Mississippi River, 48, 60
Monitoring stations, 188, 304, 359, 393,
    399, 461
Morphosequences, 5
Moshassuck River, 31, 102, 108, 112, 127,
    137, 142–147, 148–151, 235
Mount Hope Bay, 11, 13, 14, 102, 107, 118,
    137–138, 151, 156, 233, 235, 240,
    243, 246, 247, 249, 256, 258, 264,
    265–267, 271, 273, 281, 284–285,
    288, 289–291, 292, 296, 297, 302,
    317, 319, 325, 329, 337, 339, 340,
    345, 354, 363, 383, 386, 388, 390,
    393, 395, 402, 405, 409, 412–413,
    420, 421, 423, 500, 530, 546, 548,
    551, 553, 558
    ecological integrity of, 413

Narragansett Bay
    annual river flow to, 39
    bathymetric data for, 25
    circulation dynamics, 236
    circulation modeling of, 281
    Colt State Park in, 420
    Commission, 102, 131–132, 138, 140–141,
        222, 284, 364, 446
    direct discharges to, 154, 189
    ecofunctional zoning plan for, 560
    ecology of, 162, 554
    Estuary Program, 328
    fertilization of, 160
    fresh groundwater input to, 88
    geological characteristics of, 546
    geologic history of the, 1
    nitrogen inputs to, 108, 177
    nutrient gradient in, 350
    particle size of sediment, 15
    patterns of hypoxia in, 362
    phosphorus inputs to, 101

phytoplankton patterns in, 359
Project, 138
residence time of water in, 44
salt marshes fringing, 83
Sanctuary, 451
sediment, 14
National Atmospheric Deposition Program,
    49, 55
National Climate Data Center, 305
National Land Cover Dataset (NLCD), 23
National map database, 24
National Pollution Discharge Elimination
    System, 384, 390, 408, 414
National Water-Quality Assessment
    Program, 68
New England, geomorphology of, 4
New England Regional Assessment, 37–38
Nitrate depletion, 79
Nitrogen
    availability, 449
    budget, 184
    concentrations, 117, 121–122, 127,
        145–146, 147, 162, 185, 188, 221,
        224, 325
    cycle, 177–179, 180
        drivers of, 179
        models of, 177
    deposition, 48–49, 50–51, 56–57, 104,
        117, 128
    discharges, 160–161, 205, 550
    enrichment, 206, 556
    exports, 54, 57, 104
    fertilization of salt marsh, 505
    flux, 48, 59, 105–106, 129, 143, 156, 161,
        165, 179, 186, 205, 412, 550
    isotopes, 212
    limitation, 556
    over-enrichment, 511
    oxidized, 81, 101, 124, 147, 163, 167, 228,
        301, 328, 330, 342, 357–358, 362,
        363, 399, 412, 523, 551
    pollution, 47, 50–51, 58–59, 117
    sewage-derived, 157
    source of, 47–48, 52, 56, 104, 121, 123,
        128, 156, 549, 557
Non-random sampling design, 396
North Atlantic Oscillation index, 41, 461
Numerical models, 162, 179, 184, 190,
    271, 413
Nutrient
    availability, 141, 361, 367, 426, 458, 555
    based ecosystem management strategy, 432
    biofiltration, 481

Nutrient (*cont.*)
    concentrations, 126, 139–140, 150–152,
        432, 437, 443, 445–446, 451,
        454–455, 465, 476
    cycling, 177, 420, 462
    dynamics, 70, 177, 557, 560
    enrichment impacts, mitigation of, 478
    fluxes, 139, 142, 152–153, 301, 356, 441,
        465, 468, 477, 479
    gradient, 350, 358, 367, 369, 371, 372,
        441, 443, 445, 449, 466, 523, 525,
        529, 541
    impacts of, 367, 523, 550, 554
    pumps, 456, 478
    ratios, 448–449, 451–452
    recycling, 434, 472
    supply, 445, 456, 477–478, 524
    surges, 447

Oceanic forcing, 236
Organic Carbon, 87, 395–396, 512–513
Organic matter, 54, 81–83, 101, 119, 120,
    136, 165–167, 211–214, 228, 512
Organic Nitrogen, 104, 119–120, 121,
    128–129, 133, 140, 153–154, 361,
    451, 550
Outwash areas, 74
Oviatt data, 365

Pawcatuck River, 10, 105, 107
Pawtucket, 102, 109, 116, 118, 123, 133, 138,
    153, 235, 239
Pawtuxet River, 1, 12, 30, 102, 105, 108,
    110, 116, 117, 125, 127, 137, 138,
    143, 145–151, 155, 156, 158, 160,
    179, 187, 188–189, 198, 202, 257,
    285, 288
Pettaconsett pumping station, 111, 127
Pettaquamscutt River, 214–216, 219–220,
    223, 225, 229, 557
Phosphorus fluxes, annual estimate of, 144
Photosynthetically active radiation, 396, 408,
    410, 413
Physical Oceanographic Real Time
    System, 237
Phytoplankton, 69, 425, 436, 441, 482, 485,
    487, 491, 519
    biomass, 54, 80, 84, 165, 356–357, 365, 452,
        470, 477, 480, 486, 495, 527, 533
    blooms, 358
    grazing pressure on, 494
    growth, 447, 456, 476, 479
    productivity, 225

response, 364
species, 488
Piezometers, 73
Planktonic
    larvae of scallops, 370
    organisms, 44
Plant community structure, changes in, 501
Pleistocene
    epoch, 74
    glaciations, 1
    kettle-hole ponds, 81
Postglacial drainage systems, 12
Potomac River, 58, 74
Potter Cove, 214–215, 216, 217, 219, 221,
    226, 227, 229, 289
Power loom, invention of, 101
Prehistoric forests, 104, 106–107
Pre-industrial deposition, 104
Productivity measurements, 435
Providence River, 9, 12, 15, 101, 103,
    108–110, 112, 119, 125, 126, 131,
    136, 141, 148, 157, 163, 167, 180,
    214, 216, 217, 221–222, 224,
    227–228, 230, 233, 234, 240, 244,
    248–250, 254, 256–257, 259, 262,
    267, 269, 272, 282, 284, 301, 325,
    327, 332, 334, 337, 339, 342, 350,
    354, 355, 357, 363–364, 365, 368,
    369, 372, 412, 432, 435, 437, 441,
    443, 447, 452, 456, 460, 465, 466,
    468, 470, 475, 476, 525, 526, 530,
    533, 539, 540, 547, 549, 551, 552,
    559, 560
Pyrite-rich deposits, 80

Quaternary geology, 5
Quonset, 10, 154, 161, 237–238, 240,
    242–243, 247, 259, 261, 289, 350,
    360, 368

Recreational boating, 17, 226–227, 229
Redfield ratio, 405, 448–449, 467
Redox potential discontinuity depth, 366
Regression analysis, 339, 345, 540
Rhode Island, 5, 77, 109, 115, 123, 194–195,
    262, 362, 384–385, 423
    Board of Purification of Waters
        (RIBPW), 123
    Coastal Resources Management Council,
        22, 93, 269, 354, 365, 557
    Department of Environmental
        Management, 93, 123, 130, 284,
        366, 367, 523, 525, 527, 551

Department of Health Beach Monitoring Program, 93
Division of Water Resources, 284
Sea Grant, 90, 137, 140, 141, 164, 525
Sound, 236, 243, 249, 258, 262, 281, 291, 302, 304, 305–310, 315–321, 420, 421, 423, 439, 451, 460, 468, 478, 546, 547, 549, 558, 559
State Board of Health, 124, 186
Ribbed mussels, biomass of, 511
Riparian wetland forests, 80

Sakonnet River, 5, 13–14, 21, 39–40, 137, 233, 236, 239–241, 243, 246, 247, 249, 250, 256, 258, 271, 272, 281, 292, 302, 306, 317, 319
    railroad bridge, 241
Salinity, 242, 257, 326, 398, 433, 436, 439–440, 445–447, 462, 465–466, 469, 484
    gradient, 329, 432, 439–440, 446–448, 451–452, 454, 457, 465, 469–470, 473
    surface water, 403
Salt marshes, 14, 19, 72, 81, 83–85, 87, 90, 387, 499–500, 503–503, 506, 512, 556
    nitrogen removal, 556
Saturation concentration, 329, 361, 454
Secchi disc measurements, 435, 440
Second World War, 103, 117, 135
Sedgewick-Rafter counting chamber, 488
Seekonk River, 10, 12, 123, 125, 140, 141, 189, 217, 218, 220, 221, 236, 239, 254, 262–263, 265, 272, 327, 350, 358, 369, 530, 550, 553, 559, 560
Seepage, 67–8
    direct groundwater, 85
    localized groundwater, 67
Segreganset River, 385, 405–406
Semi-diurnal
    components, 239, 241
    tides, 84, 237, 239, 241, 244, 249, 254, 259, 270, 546
Sewage, 39, 41, 44, 80, 113, 115, 119, 123–124, 128, 134, 136–137, 139, 180
    disposal systems, individual, 67
    effluents, mandated denitrification of, 205
    outfalls, 204
    treatment facilities, 44, 114, 129, 136–137, 139, 154–156, 160, 167, 197, 202, 549
Shallow groundwater, contamination of, 69
Shellfish abundance, 535–536, 540

Shoreline
    geologic habitats, 18
    habitat rankings, 19
    protection structures, 18, 20–22
Short-wave solar radiation, 266
Soil respiration rates, 512–513
SPARROW model, 48, 56–57, 59, 61
Starve Goat Island, 108
Storm drains, 79, 88, 91–92
Stratified bluffs, 21
Stream-aquifer, 75, 77
Submarine groundwater discharge (SGD), 74, 88
Submerged aquatic vegetation (SAV), 352–353, 355, 368, 369, 371, 372, 409, 410, 412, 560
Subterranean Estuary, 84
Summer Blooms, 325, 334, 457
Synoptic scale fluctuations, 288
Synoptic surveys, 332, 334, 337
Synthetic fertilizer production, 103, 106, 117, 159, 184

Tampa Bay, 48, 372
Taunton River, 5, 10–11, 13, 30, 115, 118, 126, 137, 150, 152, 158, 188, 202, 235, 236, 240, 244, 246, 255, 257–258, 265, 267, 271–272, 273, 285, 288, 302, 363, 383–384, 387, 389, 393, 395, 397, 399, 400, 402, 405–407, 408, 412, 530, 551
    Watershed Alliance, 407
Technicon autoanalyzer, 435, 525
Teflon membranes, 329
Ten Mile River, 102, 126, 138, 141, 143–144, 155–157
Tertiary period, chemical weathering during the, 5
Thermal mapping field program, 273
Thermal stratification, 255, 304
Tidal
    circulation, 233, 237, 241, 246, 254, 261, 264–265, 269–270, 274, 306
    cycle, 239, 262, 269–271, 274, 296, 547
    forcing, 87, 236–237, 247, 250–251, 285, 293, 306
    harmonics, 237, 241, 247, 258, 307
        impact of, 246
    oscillations, 265
    residuals, 248
    scale, 330, 338
Total suspended solids, 269, 274, 361, 397
Trace metal analyses, 218

Transition zone, 81–82, 368, 559
Tukey's probability tables, 147, 149

Underwater seepage, 88
University of Massachusetts-Dartmouth, 290, 296, 298
Urban watershed managers, 93
US Army Corps of Engineers (USACE), 327, 365
US Census Bureau, 25, 131, 199
US Census of Agriculture, 195
US Department of Agriculture Natural Resources Con, 23
US Geological Survey (USGS), 23, 36, 68, 105, 385
US Public Health Service, 135, 163
US Weather Bureau, 35

Vadose zone, 81
Vegetated buffers, 93, 183, 515
Verrazzano, 103, 104, 105, 106, 108, 163, 549
Vertical eddy viscosity, 244, 246, 250, 254, 256
Vertical mixing, 547
    characteristics, 439
Vertical salinity gradient, 438

Waquoit Bay, 68, 70–71, 86, 506
War of 1812, 101
Wastewater treatment, 147, 161, 213, 350, 364, 388, 413, 500
Water analysis, standard methods of, 119
Water column
    characteristics of Mount Hope Bay, 383
    denitrification, 225, 227
    density, calculations of, 438
    nutrient species, 395
    oxygen stress, 550, 556
    stratification, 166, 256, 327, 362, 399, 412, 551
    transparency, 440

Water quality, 62, 186, 188, 254, 259, 267, 273, 327, 372, 386, 390, 393, 395, 403, 413
    degraded, 456, 465
    managers, 61–62
    surveys, 395–396
Watershed
    area, 51, 58, 89, 105, 124, 142, 180–181, 510
    development, 500, 511
    environment, 22
Webber's chart, 117
Wetlands, 19, 28, 30, 79, 80, 81, 82, 91, 93, 103, 290, 387, 499–500, 512, 514, 515
Wickford Cove, 240
Wind-driven
    circulation, 244, 248–249, 259
    variability, 312, 319
Winter-spring bloom, 164, 166, 167, 284, 452, 457, 458, 459, 461, 475
Woods Hole Harbor, 422–423
Woods Hole Oceanographic Institution (WHOI), 217, 296, 298
Woonasquatucket River, 137, 144, 146, 149, 235
Woonsocket wastewater treatment facility, 145
Wye River estuary, 87

Zooplankton, 432, 434, 436, 465, 468, 472, 474, 475, 480, 485–486, 530–531, 534, 540
    behavior, detailed analysis of, 469
    biomass, 434, 470–471, 474, 5, 533, 534, 536, 540, 541
    composition, 468
    dynamics, 461
    grazing, 419, 472, 485
    response, 540

Printed in the United States of America.

#18 09-01-2009 5:52PM
Item(s) checked out to p1107189.

TITLE: Crispin : the cross of lead
BARCODE: 31499003952105
DUE DATE: 09-22 09

TITLE: First they killed my father : a d
BARCODE: 31499003953756
DUE DATE: 09-22-09

TITLE: American born Chinese
BARCODE: 31499004101223
DUE DATE: 09-22-09

TITLE: What a difference a bay makes
BARCODE: 31499002179270
DUE DATE: 09-22-09

TITLE: The history and future of Narraga
BARCODE: 36012000126428
DUE DATE: 09-22-09

TITLE: Science of ecosystem-based manage
BARCODE: 32148001446052
DUE DATE: 09-22-09